# NON-SIMPLE LIQUIDS

## ADVANCES IN CHEMICAL PHYSICS

\\

### VOLUME XXXI

# NON-SIMPLE LIQUIDS

EDITED BY

## I. PRIGOGINE

University of Brussels,
Brussels, Belgium
and
University of Texas,
Austin, Texas

AND

## STUART A. RICE

Department of Chemistry
and
The James Franck Institute
The University of Chicago
Chicago, Illinois

## VOLUME XXXI

AN INTERSCIENCE® PUBLICATION

## JOHN WILEY AND SONS

NEW YORK · LONDON · SYDNEY · TORONTO

An Interscience® Publication

Copyright © 1975 by John Wiley & Sons, Inc.

**Library of Congress Cataloging in Publication Data**

Main entry under title:

Non-simple liquids.

  (Advances in chemical physics; v. 31)
  "An Interscience publication."
  Includes bibliographical references and indexes.
  1. Fluids.  2. Liquids.  I. Prigogine, Ilya.
II. Rice, Stuart Alan, 1932–      III. Series.
QD453.A27 vol. 31 [QC145.2]   541'.08s
ISBN 0–471–69933–0       [530.4'2]      75–11642

Printed in the United States of America

10 9 8 7 6 5 4 3 2 1

# INTRODUCTION

In the last decades chemical physics has attracted an ever-increasing amount of interest. The variety of problems, such as those of chemical kinetics, molecular physics, molecular spectroscopy, transport processes, thermodynamics, the study of the state of matter, and the variety of experimental methods used, makes the great development of this field understandable. But the consequence of this breadth of subject matter has been the scattering of the relevant literature in a great number of publications.

Despite this variety and the implicit difficulty of exactly defining the topic of chemical physics, there are a certain number of basic problems that concern the properties of individual molecules and atoms as well as the behavior of statistical ensembles of molecules and atoms. This new series is devoted to this group of problems which are characteristic of modern chemical physics.

As a consequence of the enormous growth in the amount of information to be transmitted, the original papers, as published in the leading scientific journals, have of necessity been made as short as is compatible with a minimum of scientific clarity. They have, therefore, become increasingly difficult to follow for anyone who is not an expert in this specific field. In order to alleviate this situation, numerous publications have recently appeared which are devoted to review articles and which contain a more or less critical survey of the literature in a specific field.

An alternative way to improve the situation, however, is to ask an expert to write a comprehensive article in which he explains his view on a subject freely and without limitation of space. The emphasis in this case would be on the personal ideas of the author. This is the approach that has been attempted in this new series. We hope that as a consequence of this approach, the series may become especially stimulating for new research.

Finally, we hope that the style of this series will develop into something more personal and less academic than what has become the standard scientific style. Such a hope, however, is not likely to be completely realized until a certain degree of maturity has been attained—a process which normally requires a few years.

At present, we intend to publish one volume a year, and occasionally several volumes, but this schedule may be revised in the future.

In order to proceed to a more effective coverage of the different aspects of chemical physics, it has seemed appropriate to form an editorial board. I want to express to them my thanks for their cooperation.

I. PRIGOGINE

# CONTRIBUTORS TO VOLUME XXXI

S. A. ADELMAN, Department of Chemistry, Massachusetts Institute of Technology, Cambridge, Massachusetts

LESLIE E. BALLENTINE, Simon Fraser University, Department of Physics, Burnaby, British Columbia, Canada

ROGER G. BROWN, Department of Chemical Engineering and Chemistry, University of Minnesota, Minneapolis, Minnesota

JOHN S. DAHLER, Department of Chemical Engineering and Chemistry, University of Minnesota, Minneapolis, Minnesota

TED DAVIS, Department of Chemical Engineering and Chemistry, University of Minnesota, Minneapolis, Minnesota

J. M. DEUTCH, Department of Chemistry, Massachusetts Institute of Technology, Cambridge, Massachusetts

DIETER FORSTER, The James Franck Institute and Department of Physics, The University of Chicago, Chicago, Illinois

NORMAN H. NACHTRIEB, Department of Chemistry, University of Chicago, Chicago, Illinois

FRANK H. STILLINGER, Bell Laboratories, Murray Hill, New Jersey

MARY THEODOSOPULU, Faculté des Sciences, Université Libre de Bruxelles, Belgium

# CONTRIBUTORS TO VOLUME XXVI

# CONTENTS

# NON-SIMPLE LIQUIDS

# ADVANCES IN CHEMICAL PHYSICS

## VOLUME XXXI

# THEORY AND MOLECULAR MODELS FOR WATER

## FRANK H. STILLINGER

### Bell Laboratories, Murray Hill, New Jersey

## CONTENTS

1

# I. INTRODUCTION

## A. Importance of Water

Throughout all the natural sciences, no single other substance approaches water in prominence and indispensibility. This fact obviously derives from the large amount of water present at the earth's surface as vapor, liquid, and solid. It has also meant that phenomena deemed worthy of scientific attention in chemistry, physics, geology, and meteorology have had a decidedly aqueous bias.

Informed opinion holds that life can only arise in, and be supported by, a suitable aqueous medium.[1] It is therefore fitting that high importance has been attached to the detection recently of water on Mars[2] and in galactic clouds.[3] Since it is likely that life exists at many scattered planetary sites throughout the known universe,[4] molecular biology may therefore have to enlarge its scope to literally cosmic proportions, without at the same time foregoing its fundamentally aqueous character.

In view of these matters, it is quite natural to lavish particular attention on water in the form of quantitative theories designed primarily to describe this substance alone. With the possible exception of liquid helium, no other substance deserves such concentrated scrutiny. Thus water historically has elicited frequent, wide-ranging, imaginative, and often mutually contradictory models or theories.[5] Viewing these works in detail and in chronological sequence may eventually provide historians of science with valuable insights into the social dynamics of scientific research.

Only within the last few years (since 1960, roughly) has it become technically feasible to produce quantitative and deductive theory for water without large elements of uncertainty. This period coincides with the general availability of rapid digital computers to the scientific community. These computers have provided essential numerical advances in both the quantum mechanical and statistical mechanical aspects of the fields that underlie present understanding. This should not be interpreted to mean that the future of water theory will remain heavily computational, rather than analytical, but the present era seems to be a natural evolutionary stage of development of the subject with no reasonable alternative.

The scope of this article is neither historical nor comprehensive. The goal instead encompasses a small set of approaches to the subject, each member of which bears a clearly definable relation to the fundamental principles of statistical mechanics. It is our belief that extensions of these approaches hold the greatest promise for future progress, and considerable attention has been devoted to the ways in which that progress can be realized.

## B. Characteristic Properties of Water

Water displays a striking set of physical properties, some apparently unique, which serve to define its unusual "personality." Their existence adds extra zest to the task of developing a viable theory of water, which ultimately is charged with connecting these properties to molecular structure and interactions. We now list a few of the more important attributes.

1. Contraction on melting. At 1 atm pressure, the molar volume decreases from 19.66 cm$^3$ for ice at 0°C, to 18.0182 cm$^3$ for liquid water at the same temperature, a loss of 8.3 %. This property is relatively rare among all substances but, even considering just the elements, it is shared by germanium and bismuth.

2. Density maximum in the liquid. Subsequent to melting, the liquid at atmospheric pressure continues to contract on further heating until a density maximum is achieved at 3.98°C. The molar volume change for the liquid over this narrow temperature interval is only 0.013 %.

No other liquid is known with a corresponding density maximum above its normal melting point. However, fused silica ($SiO_2$) can be supercooled below its melting point (1610°C) so as to pass through a density maximum.[6]

By increasing the external pressure, the temperature of maximum water density shifts downward, reaching 0°C at 190 bars.

Isotopic substitution exerts significant effects on the temperature of maximum density. For $D_2O$ (m.p. 3.81°C), the maximum occurs at 11.19°C, while for $T_2O$ (m.p. 4.48°C) it occurs at 13.40°C.

3. Numerous ice polmorphs.[7] Hexagonal ice Ih is the familiar form that results from freezing the liquid at atmospheric pressure. Cubic ice Ic is a closely related modification, with virtually the same density, which forms by vapor-phase condensation at very low temperatures. Under elevated pressure, a series of dense ice polymorphs designated ices II, III, V, VI, VII, VIII, and IX has been observed. In addition, Bridgeman has reported ice IV with $D_2O$, although this has not been subsequently confirmed and may correspond to a special metastable structure.

None of the thermodynamically stable ice polymorphs displays a close-packed arrangement of molecules. It seems likely therefore that further ice polymorphs are yet to be discovered under extremely high pressures (i.e., the megabar range).

The multiplicity of ice crystal structures alone suggests that water molecule interactions must be rather complicated. It is inconceivable that spherical molecules interacting, say, through the Lennard–Jones 12-6 potential could exhibit anywhere near as many crystal structures.

4. High melting, boiling, and critical temperatures. In contrast with other molecular substances having comparable molecular weight (nitrogen, carbon dioxide, methane, ammonia, hydrogen fluoride, hydrogen cyanide, etc.), these characteristic temperatures are anomalously high. Relatively speaking, water must consequently be a strongly binding many-body system.

5. Compressibility minimum. Normally, liquids become more isothermally compressible as the temperature rises. However, under atmospheric pressure, the isothermal compressibility for water declines with increasing temperature from the melting point to 46°C. This phenomenon disappears at high pressure (above 3 kbars).

6. Large dielectric constant. This results from a complex combination of mutual polarization between neighboring molecules in an external field and their tendency mutually to align their permanent moments. In attempting to understand the nature of water as a dielectric medium (particularly as a solvent for electrolytes), it is valuable to remember that other liquids can have similarly high dielectric constants but very different chemical structure. Some examples are formamide, ethylene carbonate, and N-methylacetamide.[8]

7. Negative pressure coefficient of viscosity. Among all liquids for which the relevant measurements are available, water is exceptional. Below 30°C the initial effect of compression is to increase fluidity. Clearly related is the observation that electrolyte conductances increase with pressure in water, unlike the behavior in other solvents.[9]

8. High molar heat capacity.

9. Negative entropies of transfer for many hydrocarbons from nonpolar solvents into water.[10]

10. Curved Arrhenius plots for kinetic properties. For viscosity, dielectric relaxation time, and the self-diffusion constant, all in the liquid, the respective energies of activation increase as temperature declines. The effect is particularly noticeable when results for supercooled water are included in the Arrhenius plots.

### C. Suitable Goals for the Theory

Three rather distinct objectives can be posed for the development of a satisfactory theory of water. They are:

*a.* Reproduce as faithfully as possible all available experiments.

*b.* Predict novel phenomena, and results of measurements within uncharted territories, to motivate future experimentation.

*c.* Generate results that aid human comprehension (of water), but which are attainable through no conceivable laboratory experiment.

Owing to the striking nature of the characteristic properties of water listed in Section B, objective *a* becomes indeed pressing. Certainly a theory of water that fails ultimately to reproduce most of those properties qualitatively must be branded a pragmatic failure. However, to insist at this present early stage of the subject that the sole legitimate end of theory lies in numerical precision must surely be an unbalanced view.

Category *b* includes properties under extreme conditions of temperature and pressure, or behavior induced by extremely high electric fields. Theory can aid in sifting out expensive and arduous hypothetical experiments with little chance of adding fundamental knowledge, from those with high promise.

In the long run item *c* is probably the most significant of the three objectives. It is possible in principle to calculate quantities such as face distributions and volume distributions for Voroᵢ (nearest-neighbor) polyhedra, curvatures and torsions for diffusive paths, three-molecule distribution functions, and the statistical topology of random hydrogen bond networks. Precise data of these kinds have inestimable value in constructing a vivid picture of what water is like at the molecular level. Without a structurally and kinetically detailed description, one can hardly claim to understand water. Yet it is obvious that experiments alone, however precise, will not suffice to bring this elaborate picture into sharp focus. Theory must move to provide that clarification.

Subsequent sections in this exposition reflect our conviction about the absolute importance of developing theory along proper deductive lines. One is first obliged to obtain a well-defined Hamiltonian for the water system, at some level of molecular detail; then one must proceed to apply known principles of statistical mechanics to predict the consequent ensemble behavior. To this end we begin with a description of individual water molecules, pass on to study of their molecular interactions, and then analyze several recent statistical mechanical formalisms based on the knowledge of those interactions.

## II. PROPERTIES OF THE WATER MOLECULE

### A. Equilibrium Structure

The lowest vertical electronic excitation for the isolated water molecule lies quite high, at 7.49 eV. Consequently, it suffices for the purposes of the present exposition to consider only the ground electronic state.

The ground state is nonlinear, with $C_{2v}$ symmetry. The three nuclei therefore inhabit the vertices of an isosceles triangle, with the apex angle at the oxygen equal to 104.48° and OH bond lengths equal to 0.9576 Å.[11] The center of mass of the molecule is displaced from the oxygen along the bisector

of the HOH apex angle. In the event that the oxygen nucleus is $^{16}O$, these
displacement distances have the values:

$$
\begin{array}{lll}
H_2O & 0.06563 \text{ Å} & \\
D_2O & 0.1179 \text{ Å} & \quad (2.1) \\
T_2O & 0.1603 \text{ Å} &
\end{array}
$$

An important feature of the subsequent theoretical development hinges on
the fact that the internal bond angle in the isolated water molecule is only
slightly smaller than the ideal tetrahedral angle $\theta_t$:

$$
\theta_t = \cos^{-1}(-\tfrac{1}{3}) = 109.4712° \qquad (2.2)
$$

This is the angle subtended by two vertices of a regular tetrahedron at that
tetrahedron's center. Equivalently, it is the angle of intersection between any
two principal diagonals of a cube. The relevance of $\theta_t$ for structural chemistry
of course arises from $sp^3$ hybridization of atomic orbitals for elements in the
first row of the periodic system.[12]

The Hellmann–Feynman theorem[13] requires that the total electrostatic
force on each nucleus, due both to other nuclei and to electron distribution,
vanish identically. This places an important condition on the electron
distribution in the molecule. This distribution cannot be a linear superposition of spherically symmetric components for the three atoms, while
maintaining a mechanically stable nonlinear triatomic configuration. Instead,
the charge distribution must build up along the OH bonds and within the
interior of the isosceles triangle.[14]

## B. Electical Properties

The measured value for the water molecule dipole moment is[15]

$$
\mu = 1.855 \times 10^{-18} \text{ esu cm} \qquad (2.3)
$$

Symmetry requires that the vector moment point along the HOH angle
bisector; the oxygen end of the molecule is negative and the hydrogen end is
positive. Given $\mu$ and the nuclear geometry, it is a trivial exercise to compute
the position of the centroid of the electronic charge distribution (total charge
$-10e$). This position lies along the symmetry axis 0.07866 Å from the oxygen
nucleus. The result should be compared to the corresponding distance
0.1173 Å that would have been implied by a linear superposition of atomic
densities and consequently zero net dipole moment for the molecule.

The traceless electrical quadrupole tensor is conventionally defined to
have the elements:

$$
\Theta_{ij} = \tfrac{1}{2}\int (3x_i x_j - r^2)\rho_e(\mathbf{r})\, d\mathbf{r} \qquad (2.4)
$$

wherein $\rho_e(\mathbf{r})$ represents the molecular charge density, including both electronic and nuclear contributions. In the case of water, $\Theta$ is diagonal in a Cartesian coordinate system whose axes $x_1$, $x_2$, and $x_3$ are respectively parallel to the symmetry axis (and along the angle bisector), in the molecular plane but perpendicular to the symmetry axis, and perpendicular to the symmetry plane. Figure 1 illustrates this coordinate system.

Since water possesses a permanent dipole moment, the elements of $\Theta$ depend on the origin chosen for the diagonalizing coordinate system. Suppose these tensor elements are known in one coordinate system, with values denoted by $\Theta_{ij}^{(0)}$. If an alternative parallel system with origin at $x_1'$, $x_2'$, $x_3'$ is of interest, the transformed tensor elements become:

$$\Theta_{ij} = \frac{1}{2} \int [3(x_i - x_i')(x_j - x_j') - r^2(x_1', x_2', x_3')]\rho_e(\mathbf{r}) \, d\mathbf{r}$$

$$= \Theta_{ij}^{(0)} + \mu_1 x_1' + \mu_2 x_2' + \mu_3 x_3' - \tfrac{3}{2}(\mu_i x_j' + \mu_j x_i') \qquad (2.5)$$

In particular $\Theta_{ij}^{(0)}$ might represent values relative to the oxygen nucleus; then, if the origin is moved forward along the $x_1$ direction by distance $l$ (shown in Fig. 1), we have

$$\Theta_{11}(l) = \Theta_{11}^{(0)} - 2\mu l$$
$$\Theta_{22}(l) = \Theta_{22}^{(0)} + \mu l \qquad (2.6)$$
$$\Theta_{33}(l) = \Theta_{33}^{(0)} + \mu l$$

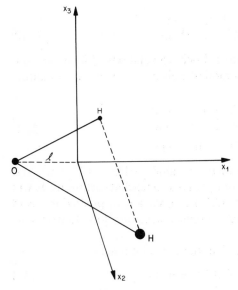

**Fig. 1** Cartesian coordinate system for which the electric quadrupole tensor $\theta$ is diagonal. The molecule lies in the $x_1 x_2$ plane.

The most accurate source of quadrupole moments is molecular beam Zeeman spectroscopy. The moments thus determined refer to an origin at the molecular center of mass. A recent application of the method for $D_2O$ has been carried out by Verhoeven and Dymanus,[16] who conclude

$$\Theta_{11}(0.1179) = -0.321 \times 10^{-26} \text{ esu cm}^2$$
$$\Theta_{22}(0.1179) = \phantom{-}2.724 \times 10^{-26} \text{ esu cm}^2 \qquad (2.7)$$
$$\Theta_{33}(0.1179) = -2.402 \times 10^{-26} \text{ esu cm}^2$$

Equation (2.6) permits us to transfer these to the oxygen origin:

$$\Theta_{11}(0) = \Theta_{11}^{(0)} = 0.116 \times 10^{-26} \text{ esu cm}^2$$
$$\Theta_{22}(0) = \Theta_{22}^{(0)} = 2.505 \times 10^{-26} \text{ esu cm}^2 \qquad (2.8)$$
$$\Theta_{33}(0) = \Theta_{33}^{(0)} = -2.621 \times 10^{-26} \text{ esu cm}^2$$

Note that the axial quadrupole moment $\Theta_{11}(l)$ has changed sign as a result of this origin shift, passing through zero when

$$l = 0.0313 \text{ Å} \qquad (2.9)$$

The water molecule can respond to electric fields at optical frequencies only by polarizing its electron distribution. The resulting linear response may be expressed as a polarizability tensor $\boldsymbol{\alpha}$ which is also diagonal in the coordinate system illustrated in Fig. 1. Separate components of $\boldsymbol{\alpha}$ have not been experimentally established for water, but the mean polarizability $\bar{\alpha}$ can be determined from the Lorenz–Lorentz equation for the refractive index. By this means Moelwyn–Hughes[17] concludes:

$$\bar{\alpha} = \tfrac{1}{3}(\alpha_{11} + \alpha_{22} + \alpha_{33})$$
$$= 1.444 \times 10^{-24} \text{ cm}^3 \qquad (2.10)$$

On the basis of accurate Hartree–Fock calculations, Liebmann and Moskowitz[18] find that $\boldsymbol{\alpha}$ manifests modest anisotropy. These investigators report:

$$\alpha_{11} = 1.452 \times 10^{-24} \text{ cm}^3$$
$$\alpha_{22} = 1.651 \times 10^{-24} \text{ cm}^3 \qquad (2.11)$$
$$\alpha_{33} = 1.226 \times 10^{-24} \text{ cm}^3$$

The agreement between the mean of these values, $1.443 \times 10^{24} \text{ cm}^3$, and result (2.10) is certainly very good and perhaps somewhat fortuitous. It will be interesting ultimately to see how (2.11) compares with predictions from a more elaborate calculation which carefully accounts for electron correlation.

The dipole moment (2.3) is larger than the critical moment:

$$\mu_c = 1.63 \times 10^{-18} \text{ esu cm} \qquad (2.12)$$

above which a molecule automatically binds an extra electron.[19] Thus a stable anion $H_2O^-$ in principle must exist, although doubtless with small ionization potential and a very extended electron cloud. However, there is at present no experimental evidence for this species.

### C. Normal Modes of Vibration

Figure 2 illustrates the directions of nuclear motion in each of the three normal modes. Two of the modes preserve the molecular $C_{2v}$ symmetry, and since they act primarily to change bond angle and bond length, the names symmetric bend and symmetric stretch are appropriate.

The third normal mode rigorously involves hydrogen motions along the OH bond directions, but out of phase. Whereas the oxygen moves parallel to the $x_1$ axis in the two symmetric modes (see Fig. 1), it moves parallel to the $x_2$ axis in the asymmetric mode.

An increase in hydrogen isotope mass causes a decrease in all three vibrational frequencies. The nine specific frequencies for $H_2O$, $D_2O$, and $T_2O$ are shown in Fig. 2 as well, all referring to $^{16}O$. Incomplete isotopic

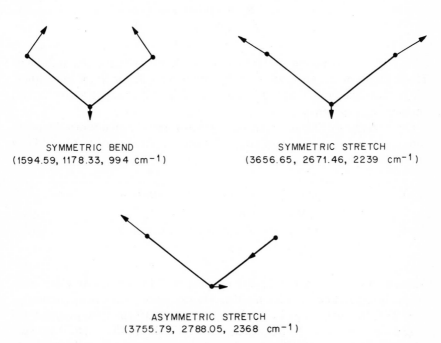

SYMMETRIC BEND
(1594.59, 1178.33, 994 cm$^{-1}$)

SYMMETRIC STRETCH
(3656.65, 2671.46, 2239 cm$^{-1}$)

ASYMMETRIC STRETCH
(3755.79, 2788.05, 2368 cm$^{-1}$)

**Fig. 2** Directions of nuclear motion for vibrational normal modes. The respective frequencies for $H_2O$, $D_2O$, and $T_2O$ (with $^{16}O$) are shown in parentheses.

substitution, to form HDO for instance, destroys the symmetry character of each of the modes.

The intramolecular potential energy surface (in the Born–Oppenheimer approximation) that gives rise to the vibrations can be expressed as a function of the changes in bond angle ($\Delta\theta$) and bond lengths ($\Delta r_1$ and $\Delta r_2$). In the vicinity of the mechanical equilibrium point ($\Delta\theta$, $\Delta r_1$, $\Delta r_2$ all zero), this potential energy $U$ may be developed as a multiple Taylor series in the displacements:

$$\begin{aligned} U(\Delta r_1, \Delta r_2, \Delta\theta) = U(0, 0, 0) &+ \tfrac{1}{2}K_r[(\Delta r_1)^2 + (\Delta r_2)^2] \\ &+ \tfrac{1}{2}K_\theta(r_e\Delta\theta)^2 + K_r'\Delta r_1\Delta r_2 \\ &+ K_{r\theta}(r_e\Delta\theta)(\Delta r_1 + \Delta r_2) + \cdots \end{aligned} \qquad (2.13)$$

where we have used $r_e$ to denote the equilibrium bond length (0.9576 Å). The four harmonic force constants have the values:[20]

$$\begin{aligned} K_r &= 8.454 \\ K_\theta &= 0.761 \\ K_r' &= -0.101 \\ K_{r\theta} &= 0.228 \end{aligned} \qquad (2.14)$$

all in units $10^5$ dynes/cm, under the proviso that $\Delta\theta$ is measured in radians. These numbers demonstrate that, by more than an order of magnitude, water molecules more strongly resist bond length changes than they do angular deformation.

Owing to the small mass of the hydrogen isotopes, the quantum mechanical zero point motion is relatively large. The rms amplitudes for $H_2O$ and $D_2O$ are found to be[21]

$H_2O$:
$$\langle(\Delta r_1)^2\rangle^{1/2} = \langle(\Delta r_2)^2\rangle^{1/2} = 0.0677 \text{ Å}$$
$$\langle(\Delta\theta)^2\rangle^{1/2} = 8.72°$$

$D_2O$:
$$\langle(\Delta r_1)^2\rangle^{1/2} = \langle(\Delta r_2)^2\rangle^{1/2} = 0.0578 \text{ Å} \qquad (2.15)$$
$$\langle(\Delta\theta)^2\rangle^{1/2} = 7.49°$$

Anharmonic terms in the Taylor expansion (2.13) for $U$ therefore need explicitly to be taken into account in precise calculations involving vibrations.

At 25°C thermal energy $k_B T$ is the equivalent of 207 cm$^{-1}$ in spectroscopic usage. Thus reference to frequencies in Fig. 2 shows that water molecule vibrations remain virtually always in their ground states in the ordinary temperature range. By contrast, the free-molecule rotational level spacings are on the order of 20 cm$^{-1}$ for $H_2O$ and 10 cm$^{-1}$ for $D_2O$, hence free rotations are close to the classical limit at room temperature.

In the asymmetric stretch mode, the molecular dipole moment magnitude should remain constant, although it rocks back and forth with the nuclear motion. If the total molecular moment comprises roughly two vector components directed along the OH bonds, the symmetric bend motion will modulate $\mu$, increasing it when the bond angle decreases. Symmetrically stretching the bonds similarly should increase $\mu$. Unfortunately, the magnitudes of these moment modulations are not known, and would be fitting objects for future careful quantum mechanical study.

Formally, an arbitrarily distorted water molecule can be identified uniquely with a fixed and properly oriented molecule without distortion. The centers of mass and the molecular planes of the two will be coincident. The remanent three-dimensional configuration space is spanned by the three normal mode amplitudes; that is, these amplitudes provide an orthogonal system of coordinates within the distortion subspace.

## III. WATER–MOLECULE INTERACTIONS

### A. Potential–Energy Resolution

We now turn our attention to a collection of $N$ water molecules arranged arbitrarily in space. For each a set of nine configurational coordinates is required to locate the precise positions of its three nuclei. These nine coordinates for any molecule $i$ ($1 \leq i \leq N$) are simply denoted by $\mathbf{X}_i$.

If the $N$ molecules are sufficiently well separated from one another to be properly regarded as isolated, the potential energy $V_N$ will consist of a sum of single-molecule energies $V^{(1)}$:

$$V_N(\mathbf{X}_1 \cdots \mathbf{X}_N) \to \sum_{i=1}^{N} V^{(1)}(\mathbf{X}_i) \tag{3.1}$$

Except for a possible difference in choice of zero for energy dictated by convenience, $V^{(1)}$ is essentially the potential energy surface function $U$ discussed above [see (2.13)] in connection with vibrational motion. For the present it is not necessary to fix that energy origin.

When any or all of the $N$ molecules form a compact collection, they interact, causing $V_N$ to deviate from the limit (3.1). We wish to resolve $V_N$ in the most general case uniquely into single-molecule, pair, triplet, quadruplet, ..., $N$-tuplet components:

$$V_N(\mathbf{X}_1 \cdots \mathbf{X}_N) = \sum_{n=1}^{N} \sum_{i_1 < \cdots < i_n = 1}^{N} V^{(n)}(\mathbf{X}_{i_1} \cdots \mathbf{X}_{i_n}) \tag{3.2}$$

The component potentials $V^{(n)}$ may be obtained by successive reversion of identity (3.2) for $N = 1, 2, 3, \ldots$.

$$V^{(1)}(\mathbf{X}_1) \equiv V_1(\mathbf{X}_1)$$
$$V^{(2)}(\mathbf{X}_1, \mathbf{X}_2) = V_2(\mathbf{X}_1, \mathbf{X}_2) - V^{(1)}(\mathbf{X}_1) - V^{(1)}(\mathbf{X}_2)$$
$$V^{(3)}(\mathbf{X}_1, \mathbf{X}_2, \mathbf{X}_3) = V_3(\mathbf{X}_1, \mathbf{X}_2, \mathbf{X}_3) - V^{(1)}(\mathbf{X}_1) - V^{(1)}(\mathbf{X}_2) \qquad (3.3)$$
$$- V^{(1)}(\mathbf{X}_3) - V^{(2)}(\mathbf{X}_1, \mathbf{X}_2) - V^{(2)}(\mathbf{X}_1, \mathbf{X}_3)$$
$$- V^{(2)}(\mathbf{X}_2, \mathbf{X}_3)$$

The quantity $V^{(n)}$ generally may be written as a remainder left after all possible component potentials of lower order have been subtracted from $V_n$:

$$V^{(n)}(\mathbf{X}_1 \cdots \mathbf{X}_n) = V_n(\mathbf{X}_1 \cdots \mathbf{X}_n) - \sum_{j=1}^{n-1} \sum_{i_1 < \cdots < i_j = 1}^{n} V^{(j)}(\mathbf{X}_{i_1} \cdots \mathbf{X}_{i_j}) \quad (3.4)$$

The utility of the potential-energy resolution (3.2) hinges on its rapidity of convergence with respect to $n$. Most of the statistical mechanics of condensed phases has been developed for additive interactions, that is, vanishing $V^{(n)}$ for $n > 2$. This additivity assumption is itself an excellent approximation for nonpolar molecular substances with low polarizability (helium, neon, argon, hydrogen, nitrogen, methane), and for such substances it makes sense to account for nonadditivity via perturbation theory, if at all. Polar substances and ionic materials can generally be expected to possess considerably larger nonadditivity, so special care must be exercised on their behalf. The worst case of all probably is presented by metals (both crystalline and liquid), since the formation of electronic energy bands and a Fermi surface is inevitably a many-nucleus effect. Furthermore, for metals it would not suffice just to consider the ground electronic state potential-energy function $V_N$, since electron excitation above the Fermi surface is an important phenomenon at any positive absolute temperature.

As we shall see, water belongs to an important class of polar molecular substances that engage in hydrogen bonds. For each of these materials, a special type of nonadditivity connected with the partially covalent (chemical) nature of the hydrogen bond requires careful study.

### B. Identification of Molecules and Ions

Thus far we have proceeded under the implicit assumption that $N$ oxygen nuclei and $2N$ hydrogen nuclei are uniquely and completely partitioned into identifiable triads, comprising $H_2O$ molecules. The logical basis for this assumption needs careful scrutiny. In a close encounter between two water molecules, it is quite possible that ambiguity might arise over which two hydrogens "belong" to each of the two oxygens. In addition, we know that water molecules occasionally dissociate into ions:

$$H_2O \rightleftharpoons H^+ + OH^- \qquad (3.5)$$

which drift apart, so that subsequent associations change partners. By drawing on a generalization of the formal theory of ion pairing,[22] we can completely remove the ambiguity.

Consider a specific configuration for the $3N$ nuclei. There are exactly $2N^2$ distinct OH pairs, whose distances will be denoted by $l(i, j)$. By convention $1 \le i \le N$ will refer to the oxygens, and $1 \le j \le 2N$ to the hydrogens. Arrange the $l(i, j)$ in an ascending sequence:

$$l(i_1, j_1) < l(i_2, j_2) < \cdots < l(i_M, j_M)$$
$$M = 2N^2 \qquad (3.6)$$

Since they bear zero measure over the full configuration space, accidental equalities in the ordered list may justifiably be disregarded.

We now proceed to apply the following recursive bonding algorithm.

1. Pair the two nuclei that provide the minimum distance in the list. This OH will henceforth be regarded as bonded.

2. Remove from the list all distances that involve hydrogens previously bonded and/or involve oxygens bonded *twice* previously. This removal leaves a contracted, but still ascending, distance list.

3. Return to step 1 if any distances remain in the list.

The end result of this procedure is that every oxygen is bonded without uncertainty to exactly two hydrogens, and every hydrogen belongs to one and only one oxygen. Furthermore, the procedure is entirely general and can be applied to any initial set of nuclear positions.

If we are to adhere consistently to the bonding algorithm, then it is important to realize that the nine-dimensional vectors $X_i$ used in Section III.B require constraints. These constraints are necessary to avoid the exchange of hydrogens between neighboring water molecules. This remark is not meant to imply that such exchanges cannot occur; physically, it is clear that they can and often do. Instead it means that certain limits on the $X_i$ variables must be obeyed to be consistent with the given bonding scheme.

Figure 3a and b illustrates a pair of water molecules in two relative configurations. In Fig. 3a, they are reasonably well separated, and the choice for associating hydrogens with oxygens is entirely clear. However, the molecules have moved closer together in Fig. 3b to an extent such that the bonding scheme present in Fig. 3a becomes invalid. The bonding algorithm does not permit one of the hydrogens of a neighboring molecule to penetrate

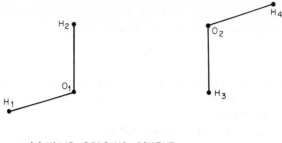

(a) VALID BONDING SCHEME

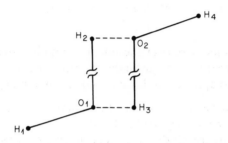

(b) INVALIDATING CLOSE ENCOUNTER

**Fig. 3** Necessity for modified association of nuclei into molecules on close encounter. The bonding algorithm that creates molecules $(H_1O_1H_2)$ and $(H_3O_2H_4)$ in separated configuration (a) gives rise to hydrogen exchange and molecules $(H_1O_1H_3)$ and $(H_2O_2H_4)$ in close encounter (b).

more closely to a given molecule's oxygen than the minimum of the four OH bond lengths in both the molecule and its neighbor. Such penetration would automatically force a redrawing of bonds, as in fact Fig. 3b explicitly shows.

Conversely, it is easy to see that a set of OH bonds which, for each pair of its $H_2O$ molecules, does *not* lead to a nonbonded oxygen–hydrogen separation smaller than the four OH bonds themselves is precisely the bond set that the algorithm would produce. Therefore, the configurational constraint demanded by a given set of molecules consists of the logical conjunction of constraints just for each of the $N(N-1)/2$ pairs of molecules.

There are precisely

$$\Omega = \frac{(2N)!}{2^N} \tag{3.7}$$

distinct ways of bonding $N$ distinguishable oxygens to $2N$ distinguishable hydrogens so as to form $H_2O$ molecules. The full configuration space for the $3N$ nuclei therefore splits naturally into $\Omega$ equivalent, nonoverlapping, and exhaustive regions. It is the interior of such a region to which the collection of nine-vectors $X_1, \ldots, X_N$ must be restricted.

The reader will have noticed that, on its completion, the bonding algorithm will by convention have forced into a long-distance marriage pairs of hydroxyl and hydrogen ions produced by dissociation reaction (3.5). They are then simply interpreted as molecules with a very severely stretched bond. Now there is nothing logically to prevent the theory from proceeding on this basis; especially at low temperature, the dissociation could be neglected anyway. But one of the chemically important aims of the theory ought to be a description of the dissociation reaction itself and of the solvated ions that result, so there should be a provision for rendering ions into distinguishable species.

To this end we now propose a natural modification of the bonding algorithm which is convenient for statistical mechanical study of water dissociation. The modification simply prevents the algorithm from going to completion by putting an upper limit $L$ on the length of bonds. This can be implemented by replacing step 1 of the algorithm above by:

1′. If the minimum distance in the ordered list is less than $L$, pair its two nuclei as a bond. If it is not, stop.

The idea is that any O—H bond stretched to length $L$ by definition breaks at that critical distance. Application of the modified algorithm results in distinct chemical species $H_2O$, $OH^-$, $H^+$, and occasionally $O^{2-}$.

If the theory were to be developed from this stage without subsequent approximation, predictions for all measurable quantities would have to be strictly independent of $L$. But to satisfy tradition, as well as chemical and physical intuition, a preferred range of $L$ values deserves to be identified. We therefore suggest that half the distance between neighboring oxygen nuclei in ice (at $0°K$ and vanishing external pressure)

$$L = 1.375 \text{ Å} \qquad (3.8)$$

is a reasonable choice which agrees with the usual notions of dissociation. Note that this choice entails a bond stretch of 0.417 Å from the equilibrium length before breakage can occur.

## C. Dimer Interaction

Superficially, the pair potential $V^{(2)}(X_1, X_2)$ defined in (3.3) depends on 18 configurational coordinates. But since translational and rotational invariance applies to the complex of six nuclei, $V^{(2)}$ really requires an irreducible minimum of 12 relative configuration coordinates. These 12 coordinates can, for example, be chosen to be

1. The three vibrational normal mode amplitudes for each molecule.
2. Polar coordinates specifying the displacement between fixed points in the two molecules (such as the oxygen nuclei, or the centers of mass).
3. Euler angles describing the rigid-body rotation relating spatial orientation of the undistorted molecules.

In the case of the noble gases, $V^{(2)}$ depends only on the distance between the centers of the structureless, spherical particles. Consequently, it is possible to determine this function of one variable accurately through a combination of experiments on the second virial coefficient, gas-phase transport coefficients, and differential scattering cross sections. An analogous determination of $V^{(2)}$ for water, however, is not feasible; the 12-dimensional character of the water pair potential introduces insurmountable problems of interpretive nonuniqueness. Direct quantum mechanical calculation of $V^{(2)}(\mathbf{X}_1, \mathbf{X}_2)$ is really the only available technique that is comprehensive, reliable, and unambigous. The various experiments, such as those cited, that involve $V^{(2)}$ at best can provide partial checks on the precision of the quantum mechanical results.

At large separation between molecules 1 and 2, $V^{(2)}$ becomes dominated by the electrostatic interaction of the respective vector dipole moments:

$$V^{(2)}(\mathbf{X}_1, \mathbf{X}_2) \sim \boldsymbol{\mu}(\mathbf{X}_1) \cdot \mathbf{T}_{12} \cdot \boldsymbol{\mu}(\mathbf{X}_2)$$
$$\mathbf{T}_{12} = R_{12}^{-3}(1 - 3R_{12}^{-2}\mathbf{R}_{12}\mathbf{R}_{12}) \tag{3.9}$$

$\mathbf{R}_{12}$ is the vector connecting the centers of the molecules and, to the leading asymptotic order required in (3.9), it is irrelevant which location within the molecule serves as center (oxygen nucleus, center of mass, etc.). The moments $\boldsymbol{\mu}(\mathbf{X}_i)$ are those appropriate for isolated molecules, with a mean value for the ground state given earlier by (2.3).

The asymptotic formula (3.9) can formally be extended to higher orders to account for interactions between higher multipoles for the isolated molecules, and changes in those multipoles due to mutual polarization. A necessary and sufficient condition for convergence of multipole expansions is that the respective charge distributions be confined to the interiors of nonoverlapping spheres. However, this cannot be satisfied strictly for molecules, since their electron densities possess tails of infinite extension. Therefore, the infinite series of which (3.9) represents the leading term must either be divergent for all $R_{12} < \infty$, or else it must converge to a function that differs from $V^{(2)}$ more and more as $R_{12}$ decreases. It is for this reason that higher-order terms in (3.9) have limited value.

The present state of development in computational quantum mechanics is such that high-quality Hartree–Fock calculations can be carried out for small sets of $n$ interacting water molecules, certainly for $n = 1, 2, 3,$ and 4. Thus far, extension to $n = 5$ and 6 has required compromises in basis set size with resulting errors which diminish the significance of results.

To date, a large number of Hartree–Fock studies of the water dimer ($n = 2$) has been carried out,[23-29] and as a result a consensus has emerged about the principal characteristics of $V^{(2)}$. Figure 4 displays the structure implied by these studies for the lowest-energy dimer configuration. Two major points become clear, namely, that the individual water molecules suffer little distortion as a result of the interaction, and that an essentially linear hydrogen bond is involved.

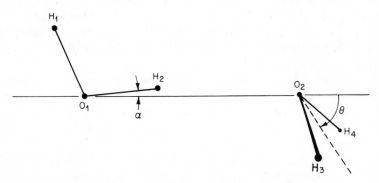

**Fig. 4** Lowest-energy dimer structure. The plane containing the proton donor molecule ($H_1O_1H_2$) is a plane of reflection symmetry for the dimer, and as such contains $O_2$.

Quantitative details vary somewhat from one Hartree–Fock calculation to another, but in all cases the lowest-energy dimer exhibits a phase of symmetry. This plane contains a proton donor molecule ($H_1O_1H_2$ in Fig. 4) which points one of its OH bonds toward the oxygen atom of the proton acceptor molecule ($H_3O_2H_4$ in Fig. 4). The proton acceptor lies in a plane perpendicular to the dimer symmetry plane, with an orientation angle $\theta > 0$ to minimize repulsions between pendant hydrogens. In all calculations the distance $l(O_2H_2)$ is sufficiently large that the bond reconstruction illustrated previously in Fig. 3 is not required—the molecules maintain their individuality.

The specific dimer arrangement shown in Fig. 4 is only one of eight equivalent arrangements leading to an energy minimum. The other seven can be obtained by 180° rotation of the proton acceptor molecule about its symmetry axis and/or use of one of the other three protons in the linear hydrogen bond.

Table I lists the important parameters for the most stable dimer structure as they have been predicted by the available *ab initio* (all-electron) Hartree–Fock calculations. The entries are arranged roughly in increasing order of precision. The first rows historically were earliest, and employed rather restrictive basis sets; those near the end are more recent, and utilized more nearly complete basis function sets. Proceeding down the list, the trend for the most part is toward smaller binding energy, and greater bond length ($R_{OO}$ is the distance between oxygens). It seems reasonable to suppose by extrapolation that the exact Hartree–Fock limit would produce a binding energy of about 4.50 kcal/mole for a linear hydrogen bond with length $R_{OO} = 3.00$ Å.

Motions of the dimer that tend to change $R_{OO}$ from its optimal value, or to move the donated proton ($H_2$ in Fig. 4) substantially off the oxygen–oxygen axis, are costly in terms of stabilizing energy. However, rotation of the proton donor molecule about the linear $O_1H_2O_2$ axis is less costly, as is a change in acceptor rotation angle $\theta$.

The covalent chemical bonds holding the individual water molecules together are much stronger and stiffer than the hydrogen bond. Thus the molecules are only slightly perturbed in internal geometry when forming the optimized dimer, in comparison with isolated molecules. The primary change seems to be a small lengthening of the covalent OH bond forming the hydrogen bond, by an order of 0.005 Å (last column in Table I). At the same time, the stretching force constant for this bond decreases substantially when it forms a linear hydrogen bond, according to all the Hartree–Fock calculations for which this has been tested. An additional small effect of the hydrogen bond seems to be an opening up of the HOH angle in the acceptor molecule by about 0.5°, thereby moving it closer to $\theta_t$ [see (2.2)].[27]

Popkie, Kistenmacher, and Clementi[29] carried out an extensive set of dimer calculations densely spanning the space of relative configurations, especially in the neighborhood of the stable dimer minimum. The separate water molecules, however, were strictly frozen in their isolated-molecule geometries. These investigators carefully fitted their results to a closed-form expression which constitutes a convenient representation of the Hartree–Fock approximation to $V^{(2)}(X_1, X_2)$ within the undistorted molecule subspace. The expression consists of a sum of Coulombic and exponential terms acting between force centers, four in each molecule. These force centers are the three nuclei (with OH bond length 0.957 Å and HOH bond angle 105°), and a fourth center M located 0.2307 Å ahead of the oxygen, along the symmetry axis. The positions and relative charges associated with these centers are shown in Fig. 5.

TABLE I

Parameters Describing Minimum-Energy Dimers Inferred from Various Quantum-Mechanical Calculations[a]

| Investigator | Reference | Basis set character | Binding energy (kcal/mole) | $R_{OO}$ (Å) | $\theta$ (deg) | $\alpha$ (deg) | $\Delta R_{O_1H_2}$ (Å) |
|---|---|---|---|---|---|---|---|
| Morokuma, Pedersen | 23 | G(5, 3\|3) | 12.6 | 2.68 | 18[b] | — | 0.012 |
| Del Bene, Pople | 24 | G[2, 1\|1] | 6.09 | 2.73 | 58 | 1.0 | — |
| Morokuma, Winick | 25 | S (minimal) | 6.55 | 2.78 | 54 | — | 0.0026 |
| Kollman, Allen | 26 | G[3, 1\|1] | 5.27 | 3.0 | 25 | — | 0.010 |
| Hankins, Moskowitz, and Stillinger | 27 | G[5, 3, 1\|2, 1] | 5.00 | 3.00 | 40 | 0 | <0.01 |
| Diercksen | 28 | G[5, 4, 1\|3, 1] | 4.84 | 3.00 | ≈35 | — | 0.0040 |
| Popkie, Kistenmacher, and Clementi | 29 | G[8, 5, 1, 1\|4, 1, 1] | 4.60 | 3.00 | 30 | — | — |

[a] Angles and distances refer to Fig. 4. G, Gaussian basis; S, Slater basis.
[b] Interpolated.

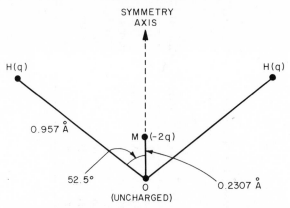

**Fig. 5** Four-force-center model of the water molecule used by Popkie, Kistenmacher, and Clementi to fit the Hartree–Fock approximation to $V^{(2)}$.

The Pokie–Kistenmacher–Clementi (PKC) function may be written:

$$V_{PKC}^{(2)} = a_{OO} \exp\left(-b_{OO} R_{OO}\right) + \sum_{(O, H)} a_{OH} \exp\left(-b_{OH} R_{OH}\right)$$

$$+ \sum_{(H, H)} \left[ a_{HH} \exp\left(-b_{HH} R_{HH}\right) + \frac{q^2}{R_{HH}} \right]$$

$$- 2q^2 \sum_{(H, M)} \left(\frac{1}{R_{HM}}\right) + \frac{4q^2}{R_{MM}} \tag{3.10}$$

$q$, $a$'s and $b$'s are constants, while the $R$'s are the distances between pairs of force centers (one in each molecule). With distances in angstroms, and energy in kilocalories per mole, the parameters are found to be

$$
\begin{aligned}
a_{OO} &= 3.65501 \times 10^5 \text{ kcal/mole} \\
a_{OH} &= 3.43368 \times 10^3 \text{ kcal/mole} \\
a_{HH} &= 90.2576 \text{ kcal/mole} \\
b_{OO} &= 4.76328 \text{ Å}^{-1} \\
b_{OH} &= 3.65973 \text{ Å}^{-1} \\
b_{HH} &= 2.30881 \text{ Å}^{-1} \\
q &= 3.21966 \times 10^{-10} \text{ esu}
\end{aligned}
\tag{3.11}
$$

The variation in $V_{PKC}^{(2)}$ with respect to oxygen–oxygen distance $R_{OO}$ has been plotted in Fig. 6 for the symmetric linear hydrogen bond configuration shown previously in Fig. 4, with $\alpha = 0$ and $\theta = \theta_t/2 = 54.7356°$. Figure 7 presents the $\theta$ variation at $R_{OO} = 3.00$ Å for the same symmetric linear hydrogen bond configuration ($\alpha = 0$).

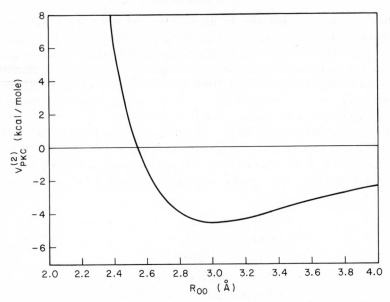

**Fig. 6** Distance variation of $V_{PKC}^{(2)}$ in the symmetric linear hydrogen bond configuration shown in Fig. 4 ($\alpha = 0$, $\theta = \theta_t/2$).

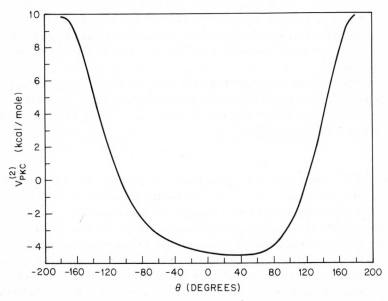

**Fig. 7** Acceptor angle variation for $V_{PKC}^{(2)}$ at $R_{OO} = 3.00$ Å. The configuration is the one shown in Fig. 4, with $\alpha = 0$ (linear hydrogen bond).

21

In the process of forming the linear hydrogen bond, the pair of water molecules experiences a shift in electron density relative to that for isolated molecules. The shift affects the net dipole moment for the interacting dimer, which is no longer the vector sum of the two monomer moments. Hankins, Moskowitz, and Stillinger[27] found that the dimer at its most stable conformation exhibits an 11% enhancement in dipole moment magnitude compared to the magnitude of the isolated-molecule vector sum.

Another measure of the distribution shift is provided by Mullikan atomic population changes relative to isolated monomers. Hankins, Moskowitz, and Stillinger[27] also evaluated these changes for the most stable dimer conformation. In terms of the atomic designations shown in Fig. 4, the changes found are

$$\Delta(H_1) = 0.00963$$
$$\Delta(O_1) = 0.03776$$
$$\Delta(H_2) = -0.03604 \tag{3.12}$$
$$\Delta(O_2) = 0.02212$$
$$\Delta(H_3) = \Delta(H_4) = -0.01674$$

These numbers represent "fractions of an electron," so $\Delta > 0$ implies more negative charge at the given atom on dimerization. Evidently, the pendant hydrogens on the proton acceptor ($H_3$ and $H_4$) have become more positive, while the oxygen of the donor molecule ($O_1$) has become more negative. These atoms can therefore be expected to act as more acidic, and more basic, respectively, in further hydrogen bond formation in larger water molecule aggregates.

One of the errors inherent in the Hartree–Fock approximation applied to water is that the predicted dipole moment for the single molecule is too large. The dipole moment incorporated in $V^{(2)}_{PKC}$, arising from the charge distribution illustrated in Fig. 5, is

$$\mu_{PKC} = 2.27 \times 10^{-18} \text{ esu cm} \tag{3.13}$$

This is 22% larger than the experimental value [see (2.3)]. There is a corresponding modification of $V^{(2)}$ at large separation, in accord with (3.9).

Beside correcting the dipole moment, calculations sufficiently accurate to incorporate the electron correlation neglected by the Hartree–Fock approximation would also begin to account for London dispersion attraction. This attraction arises from correlation between fluctuations in electron distribution for the two molecules, and at large separation it provides for $V^{(2)}$ a negative term proportional to $R_{12}^{-6}$. Since such correlated-electron calculations are not yet available for the water dimer, it is worthwhile to make a rough estimate of the dispersion attraction.

The neon atom is isoelectronic with the water molecule—both have closed-shell, 10-electron singlet ground states. According to Slater and Kirkwood,[30,31] the respective dispersion attraction at a given distance should be proportional to the three-halves powers of the optical polarizabilities. Using the mean value shown in (2.10) for water, and[32]

$$\alpha(\text{Ne}) = 0.39 \times 10^{-24} \text{ cm}^3 \qquad (3.14)$$

is it possible to scale up the accepted neon dispersion interaction[33] to water. The result is

$$\Delta V^{(2)}_{\text{disp}} \cong - \frac{C}{R^6_{12}}$$

$$C = 98.2 \text{ kcal Å}^6/\text{mole} \qquad (3.15)$$

At small separation, the inverse sixth power attractive term is augmented by others which vary as $R^{-8}_{12}$, $R^{-10}_{12}$, and so on. In order to incorporate their effect, we can appeal to the same scaling, but applied to the attractive term of an empirical Lennard–Jones 12-6 potential for neon. In particular we draw on Corner's version.[34] At close range it is thereupon suggested that the constant $C$ in (3.15) be empirically modified to

$$C' = 1023 \text{ kcal Å}^6/\text{mole} \qquad (3.16)$$

The resulting quantity $-C'/R^6_{16}$ can be appended to $V^{(2)}_{\text{PKC}}$ as a crude estimate of the correlation effect in the hydrogen bond distance range (with $R_{12}$ identified as $R_{\text{OO}}$). The net effect (for $\alpha = 0$, $\theta = \theta_t/2$) is that the linear hydrogen bond compresses from 3.007 to 2.855 Å, while its strength increases from $-4.499$ to $-6.123$ kcal/mole.

## D. Nonadditivity

Since the linear hydrogen bond already arises at the water molecule pair interaction level, $V^{(2)}$ suffices to explain the gross aspects of water molecule arrangements in aqueous crystals (and perhaps in the liquid as well). But quantitative understanding of water in its condensed phases clearly requires analysis of potential-energy nonadditivity.

In order to hydrogen-bond as a donor to a neighbor, a water molecule uses only one of its OH groups. The other OH group is free to form its own linear hydrogen bond to the back of a second neighbor. Simultaneously, this doubly donating molecule can accommodate two further neighbors at its back (making four neighbors in all), which themselves donate protons in linear hydrogen bonds. Thus fourfold coordination via linear hydrogen bonds is natural for water molecule aggregates. The most favorable spatial arrangement of these four hydrogen bonds places the five water molecule oxygens at the center and at the vertices of a regular tetrahedron. This pattern is shown in Fig. 8.

The four neighbor molecules in Fig. 8 are available for further hydrogen bonding, so that they too participate in four linear hydrogen bonds emanating in the tetrahedral pattern. Thus a large aggregate of water molecules can form a space-filling network of hydrogen bonds. Since $V^{(2)}$ at least permits relatively free reorientation of the neighbor molecules around the hydrogen bonds shown in Fig. 8, a wide topological variety of three-dimensional networks seems to be possible.

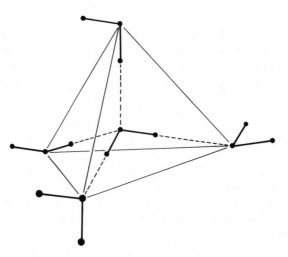

**Fig. 8**  Fourfold tetrahedral coordination of water molecules via linear hydrogen bonds.

The periodic network structure for hexagonal ice Ih is shown schematically in Fig. 9. This crystal utilizes the tetrahedral hydrogen bonding scheme to form puckered hexagons, of both the chair and boat forms. That the hydrogen atoms do in fact reside along the bonds shown (consistent with the fact that molecular HOH angles are nearly $\theta_t$) has been established by neutron scattering from $D_2O$ ice;[35] furthermore, these hydrogens are nearer one end of the bond than the other, since intact molecules are involved.

Closely related to hexagonal ice is the cubic form which consists exclusively of chair form hexagons. This modification also exhibits ubiquitous fourfold coordination, with linear hydrogen bonds in a tetrahedral pattern about

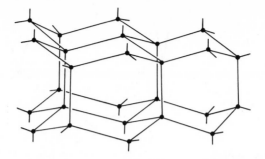

**Fig. 9**  Hydrogen bond network for ice Ih. The vertices (solid circles) are the positions of the oxygen atoms. The hydrogens are located asymmetrically along the bonds, two near each vertex to form the water molecules.

each oxygen vertex. Significantly, this theme persists with minor distortions throughout high-pressure ice polymorphs,[7] and can be seen clearly in the structure of gas hydrates (clathrates).[36]

The three-molecule nonadditivity function $V^{(3)}(\mathbf{X}_1, \mathbf{X}_2, \mathbf{X}_3)$ in the strict sense depends irreducibly on 21 variables, so its full numerical characterization would be a formidable task. The existence of the hydrogen bond networks just mentioned, however, suggests confining attention largely to those trimer configurations that appear in the networks. Furthermore, one would expect the greatest potential nonadditivity to arise when distances are all small, so additionally restricting attention to hydrogen-bonded trimers seems warranted.

Three topologically distinct types of bonded network trimers exist. They are illustrated in Fig. 10. Each one has a central molecule with two hydrogen bonds, and two end molecules with only one hydrogen bond. In an extended hydrogen bond network such as that shown in Fig. 9 for ice Ih, each molecule acts as the center of six bonded trimers. In terms of the trimer categories defined by Fig. 10, the six consist always of one double donor, one double acceptor, and four sequential trimers.

Hankins, Moskowitz, and Stillinger[27] calculated $V^{(3)}$ for the three trimers shown in Fig. 10, using the Hartree–Fock approximation. For their study the individual molecules were maintained at the stable internal structure established by the one-molecule calculations, and the component bonded dimers in each trimer were of the type shown earlier in Fig. 4, with $\alpha = 0$ and $\theta = -\theta_t/2$. The results are plotted in Fig. 11 versus $R_{OO}$, the common hydrogen bond length.

On the basis of quite incomplete information, Hankins, Moskowitz, and Stillinger[27] suggested that, for each of the three types of bonded trimers,

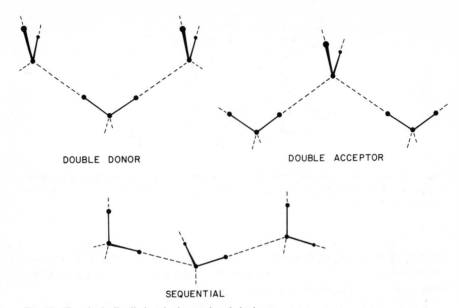

**Fig. 10** Topologically distinct hydrogen-bonded trimers.

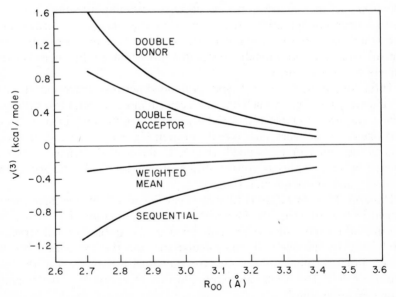

**Fig. 11** Trimer nonadditivities for variable common bond length, $R_{OO}$. The configurations involved are shown in Fig. 10. The weighted mean is $\frac{2}{3}$ (sequential) $+ \frac{1}{6}$ (double donor + double acceptor), appropriate for a four-coordinated network.

26

$V^{(3)}$ is quite insensitive to the rotation of molecules about the linear hydrogen bonds, more insensitive in fact than the separate $V^{(2)}$'s involved. This hypothesis has been confirmed by Lentz and Scheraga.[37] It suggests that the $V^{(3)}$ curves shown in Fig. 11 are representative of *all* bonded trimers in an extended network. Because of the invariant ratios of the three types of trimers, it is therefore relevant to examine the weighted mean

$$\langle V^{(3)} \rangle_{net} = \tfrac{1}{6}[V^{(3)}(\text{double donor}) + V^{(3)}(\text{double acceptor}) + 4V^{(3)}(\text{sequential})] \tag{3.17}$$

The corresponding curve is presented in Fig. 11. It demonstrates that the net effect of $V^{(3)}$, at least for bonded trimers, is to confer extra energetic stability on an extended network. Furthermore, the decrease in $\langle V^{(3)} \rangle_{net}$ with decreasing $R_{OO}$ constitutes a compressive effect which tends to reduce the hydrogen bond length in the network.

The negative values found for $V^{(3)}$ in the sequential trimer may be interpreted as a strengthening of successive hydrogen bonds in a chain after the first is formed. The possibility of this "hydrogen bond cooperativity" was pointed out several years ago by Frank and Wen,[38,39] who observed that the hydrogen atoms of the acceptor molecule in a dimer (see Fig. 4) should become more positive as a result of the hydrogen bond and, consequently, as stronger Lewis acids, should form stronger successive hydrogen bonds. The Mulliken population shifts quoted earlier for the dimer [see (3.12)] document this idea quantitatively.

Very little systematic study of $V^{(4)}$ has thus far been undertaken. Lentz and Scheraga[37] employed the Hartree–Fock approximation for a cyclic tetramer. This tetramer has $S_4$ symmetry; the four pendant OH groups around the hydrogen bond square alternate above and below the plane of oxygens. The three bonded trimers included in the tetramer are all sequential. The results show that about 10% of the tetramer binding energy (at the optimal bond length) is attributable to $V^{(3)}$, but only about 1% to $V^{(4)}$ (it is negative for the cyclic tetramer). The broad implication seems to be that, although three-molecule nonadditivity is important for water, the general potential-energy resolution (3.2) is rapidly convergent with order $n$ for physically relevant configurations of water molecule aggregates.

Del Bene and Pople[24] investigated a series of cyclic water polymers, up to and including the cyclic hexamer. Although they interpreted their results to mean that closure of hydrogen bond polygons entails a special cooperative stabilizing phenomenon, the systematic potential-energy resolution (3.2) was not invoked. It seems likely that their polygon stabilities owe their existence primarily to negative $V^{(3)}$'s associated with sequential trimers, and not to anomalous $V^{(n)}$'s of higher order which come forth under topological closure.

The fractional contribution of potential-energy nonadditivity to the total binding energy of ice Ih is likely typical for all tetrahedrally coordinated networks. The available Hartree–Fock calculations[27,37] imply that about 13 % is attributable to nonadditivity.

A clear need exists for further study of nonadditivity in water clusters, and for incorporation of electron correlation into the quantum mechanical computations.

## IV. SEMICLASSICAL LIMIT

### A. Canonical Partition Function

The connection between the Hamiltonian operator $\mathscr{H}$ for $N$ water molecules (regarded as $2N$ hydrogen nuclei and $N$ oxygen nuclei) and macroscopic thermodynamics can be established through the canonical partition function

$$Q = \mathrm{Tr} \exp\left(-\beta\mathscr{H}\right)$$

$$\beta = \frac{1}{k_B T} \tag{4.1}$$

where the basis over which the trace is computed includes spin and symmetry as appropriate for the isotopes of interest. The potential energy surface appearing in $\mathscr{H}$ is the quantity $V_N(\mathbf{X}_1 \cdots \mathbf{X}_N)$ whose resolution into singlet, pair, triplet, and so on, components has just been considered:

$$\mathscr{H} = -\tfrac{1}{2}\hbar^2 \sum_{j=1}^{3N} m_j^{-1}\nabla_j^{\,2} + V_N \tag{4.2}$$

The partition function $Q$ yields the Helmholtz free energy $F$ directly:

$$Q = \exp\left(-\beta F\right) \tag{4.3}$$

from which other thermodynamic functions may in turn be obtained by standard identities.

For many aspects of the study of water, the full quantum mechanical representation (4.1) for $Q$ is unnecessary. Although the internal normal modes of vibration for the most part remain in their quantum mechanical ground states, it is relevant to examine the rotational and translational degrees of freedom in their classical limits. This point of view is not meant to imply that water (at room temperature, say) achieves this classical limit to high precision. But the required quantum corrections to rotational and translational motion have sufficiently modest magnitudes ($H_2O$ and $D_2O$ differ relatively little) that eventual calculation of leading quantum corrections to the classical limit makes good physical sense. We return to this matter in Section IX within the context of a model for which quantum corrections can be readily generated.

For the present we therefore confine attention to a collection of classical rigid rotor molecules with quantized internal vibrations. The corresponding semiclassical version of the partition function has the appearance:[40]

$$Q = \frac{1}{N!} \left[ \frac{(2\pi k_B T)^3 m^{3/2} (I_1 I_2 I_3)^{1/2} Q_{\text{vib}}}{h^6} \right]^N$$

$$\times \int d\mathbf{x}_1 \cdots \int d\mathbf{x}_N \exp \left[ -\beta \overline{V}_N(\mathbf{x}_1 \cdots \mathbf{x}_N) \right] \tag{4.4}$$

In this expression $m$ stands for the total molecular mass $(m_O + 2m_H)$, $h$ is Planck's constant, and the principal moments of inertia are $I_1, I_2$, and $I_3$; the vibrational partition function is $Q_{\text{vib}}$. The configuration coordinates for each molecule $j$ have now been reduced to six, symbolized by $\mathbf{x}_j$; these are the Cartesian coordinates of a fixed point in the molecule, and three Euler angles $\alpha_j$, $\beta_j$, and $\gamma_j$ describing orientation about that fixed point. Unlike the preceding function $V_N$, the appropriate potential function $\overline{V}_N$ for use in (4.4) employs only these configurational six-vectors $\mathbf{x}_j$ as variables.

Presuming that vibrations are truly harmonic, with angular frequencies $\omega_\alpha$,

$$Q_{\text{vib}} = \prod_{\alpha=1}^{3} \frac{\exp\left(-\frac{1}{2}\beta\hbar\omega_\alpha\right)}{1 - \exp\left(-\beta\hbar\omega_\alpha\right)} \tag{4.5}$$

We noted in Section III.A that molecular vibration force constants are affected by intermolecular interactions, specifically by the formation of hydrogen bonds. We elect to interpret shifts in vibrational energy levels that result from these force-constant changes as a valid part of the interactions $\overline{V}_N$. The frequencies $\omega_\alpha$ in (4.5) thus are strictly isolated-molecule frequencies.

The formal prescription for obtaining the rigid-rotor potential-energy functions $\overline{V}_N$ from the more detailed quantities $V_N$ are now specified. First, split nine-vector $\mathbf{X}_j$ into a direct sum of rigid-body coordinates $\mathbf{x}_j$ and a vibration amplitude part $\mathbf{y}_j$:

$$\mathbf{X}_j = \mathbf{x}_j \oplus \mathbf{y}_j \tag{4.6}$$

Then, for an arbitrary collection of $n$ molecules constrained in their rigid body coordinates to $\mathbf{x}_1, \ldots, \mathbf{x}_n$, the remaining amplitudes $\mathbf{y}_1, \ldots, \mathbf{y}_n$ describe vibrations for the complex, which are now collectively determined normal modes. These $3n$ new normal modes have frequencies shifted from the free-molecule $\omega_\alpha$, but in any event define an $n$-molecule vibrational partition function:

$$Q_{\text{vib}}^{(n)}(\mathbf{x}_1 \cdots \mathbf{x}_n) = \sum_{v=1}^{3n} \frac{\exp\left[-\tfrac{1}{2}\beta\hbar\omega_v(\mathbf{x}_1 \cdots \mathbf{x}_n)\right]}{1 - \exp\left[-\beta\hbar\omega_v(\mathbf{x}_1 \cdots \mathbf{x}_n)\right]} \tag{4.7}$$

The vibrationally averaged interaction $\bar{V}_n$ subsequently is defined:

$$\exp\left[-\beta\bar{V}_n(\mathbf{x}_1 \cdots \mathbf{x}_n)\right] = \left\{\exp\left[-\beta V_n(\mathbf{X}_1 \cdots \mathbf{X}_n)\right]\right\}_{\mathbf{y}_1 \cdots \mathbf{y}_n = 0}$$

$$\times \frac{Q_{\text{vib}}^{(n)}(\mathbf{x}_1 \cdots \mathbf{x}_n)}{(Q_{\text{vib}})^n} \tag{4.8}$$

In principle the resulting function $\bar{V}_n$ should depend on temperature. However, under the regime for which normal modes are thermally unexcited, this temperature dependence vanishes.

If the normal modes $\omega_1$, $\omega_2$, and $\omega_3$ were all to shift substantially downward (beside splitting) when molecules came into interaction, the last factor $Q_{\text{vib}}^{(n)}/(Q_{\text{vib}})^n$ would be greater than unity. Consequently $\bar{V}_n$ would be lower than $V_n$, indicating vibrational stabilization.

A resolution of $\bar{V}_N$ analogous to that shown in (3.2) for $V_N$ is possible. However, the construction definition (4.8) implies that vibrationally averaged pair potentials $\bar{V}^{(2)}$ are the lowest-order components:

$$\bar{V}_N(\mathbf{x}_1 \cdots \mathbf{x}_N) = \sum_{n=2}^{N} \sum_{i_1 < \cdots < i_n = 1}^{N} \bar{V}^{(n)}(\mathbf{x}_{i_1} \cdots \mathbf{x}_{i_n}) \tag{4.9}$$

The inverse relations are

$$\bar{V}^{(2)}(\mathbf{x}_1, \mathbf{x}_2) \equiv \bar{V}_2(\mathbf{x}_1, \mathbf{x}_2)$$
$$\bar{V}^{(3)}(\mathbf{x}_1, \mathbf{x}_2, \mathbf{x}_3) = \bar{V}_3(\mathbf{x}_1, \mathbf{x}_2, \mathbf{x}_3) - \bar{V}^{(2)}(\mathbf{x}_1, \mathbf{x}_2) - \bar{V}^{(2)}(\mathbf{x}_1, \mathbf{x}_3) - \bar{V}^{(2)}(\mathbf{x}_2, \mathbf{x}_3)$$

$$\tag{4.10}$$

with $\bar{V}^{(n)}$ written in general as $\bar{V}_n$ minus all possible component $\bar{V}^{(j)}$ with $j < n$.

We noted earlier that dispersion attraction is one of the important contributions to $V^{(2)}$, and that it arises from correlation of the electron distributions in the quantum mechanically fluctuating dipole moments. Now $\bar{V}^{(2)}$ includes an extra dispersion attraction due to correlation in fluctuations of molecular dipole moments due to normal mode vibrations.

Although the rigid-rotor potentials $\bar{V}_N$ have been developed in the context of the semiclassical partition function $Q$, the detailed classical mechanics with $\bar{V}_N$ is itself a valid object for attention. The coupled Newton–Euler equations of motion provide the time evolution of the system from which linear transport coefficients (viscosity, self-diffusion constant, thermal conductivity, etc.) can be extracted in principle.

## B. Molecular Distribution Functions

The Boltzmann factor integrand $\exp\left(-\beta\overline{V}_N\right)$ for the semiclassical partition function represents the occurrence probability for any arbitrary configuration of all $N$ molecules. It therefore contains all structural information about the system. But normally $N$ is very large, and for mathematical purposes it is permitted to pass to infinity (with fixed system density—the so-called thermodynamical limit). A more compact way of describing equilibrium molecular arrangements is therefore warranted, one that conveys intensive structural information without suffering the divergences of extensive properties in the large-system limit. The generic molecular distribution functions fulfill this role. They also provide an efficient bridge between molecular configurational correlations and macroscopic thermodynamical properties.

Within the canonical ensemble description of our semiclassical water model, the $n$-molecule distribution function $\rho^{(n)}$ is obtained through the definition $(1 \leq n \leq N)$:

$$\rho^{(n)}(\mathbf{x}_1 \cdots \mathbf{x}_n) = \frac{N(N-1)\cdots(N-n+1)\int d\mathbf{x}_{n+1} \cdots \int d\mathbf{x}_N \exp\left(-\beta\overline{V}_N\right)}{\int d\mathbf{x}_1 \cdots \int d\mathbf{x}_N \exp\left(-\beta\overline{V}_N\right)}$$

(4.11)

where integrations cover orientation, and the volume $\mathscr{V}$ made available to each of the molecules by container walls (or mathematical boundary conditions). Considering a set of infinitesimal six-dimensional volume elements $d\mathbf{x}_1, \ldots, d\mathbf{x}_n$, respectively, surrounding configuration points $\mathbf{x}_1, \ldots, \mathbf{x}_n$, the product

$$\rho^{(n)}(\mathbf{x}_1 \cdots \mathbf{x}_n)\, d\mathbf{x}_1 \cdots d\mathbf{x}_n$$

(4.12)

equals the probability that those volume elements simultaneously contain molecules.

If the temperature and density conditions correspond thermodynamically to a single fluid phase, $\rho^{(1)}(\mathbf{x}_1)$ will be independent of $\mathbf{x}_1$:

$$\rho^{(1)}(\mathbf{x}_1) = \frac{N}{8\pi^2\mathscr{V}}$$

(4.13)

except in a thin region near a boundary where forces between water molecules and the wall can induce inhomogeneity and orientational anisotropy. The ordinary number density $N/\mathscr{V}$ comprises molecules of all orientations, and so may be obtained by integrating the homogeneous fluid $\rho^{(1)}$ over Euler angles $\alpha_1, \beta_1,$ and $\gamma_1$:

$$\int_0^{2\pi} d\alpha_1 \int_0^{\pi} \sin\beta_1 \, d\beta_1 \int_0^{2\pi} d\gamma_1 \, \rho^{(1)}(\mathbf{x}_1) = \frac{N}{\mathscr{V}} \qquad (4.14)$$

If the temperature and density correspond instead to the existence of one of the ice polymorphs everywhere throughout $\mathscr{V}$, the boundary conditions *may* serve to produce a single essentially fixed crystal. In that event $\rho^{(1)}(\mathbf{x}_1)$ would possess the periodicity and symmetry of the unit cell involved.

Regardless of which state of aggregation prevails (crystalline, liquid, or vapor), provided that just one phase is present, the distribution functions of order $n > 1$ exhibit an asymptotic factorization property. This occurs when the given set of molecular configuration coordinates $\mathbf{x}_1, \ldots, \mathbf{x}_n$ falls into two widely separated subsets, say, $\mathbf{x}_1, \ldots, \mathbf{x}_m$ and $\mathbf{x}_{m+1}, \ldots, \mathbf{x}_n$. In this circumstance the molecular correlations operate only within the subsets, not between them. The resulting asymptotic behavior of $\rho^{(n)}(\mathbf{x}_1, \ldots, \mathbf{x}_n)$ is found to be

$$\rho^{(n)}(\mathbf{x}_1 \cdots \mathbf{x}_n) \to \left[\frac{(N-m)\cdots(N-n+1)}{N\cdots(N-n+m+1)}\right]\rho^{(m)}(\mathbf{x}_1 \cdots \mathbf{x}_m)\rho^{(n-m)}(\mathbf{x}_{m+1} \cdots \mathbf{x}_n)$$

$$(4.15)$$

The leading numerical factor $[\cdots]$ arises from the normalization inherent in definition (4.11); with fixed $n$ and $m$ it approaches unity as $N$ increases to infinity. By obvious extension, $\rho^{(n)}$ reduces substantially to a product of lower-order functions for any partitioning of $\mathbf{x}_1, \ldots, \mathbf{x}_n$ into any number of widely separated subsets. In any event the rate of approach to factored form is at its slowest with respect to separation distance when the system is at its liquid-vapor critical point.[41]

Apply the six-dimensional gradient operator $\nabla_1$ (corresponding to vector $\mathbf{x}_1$) to both sides of (4.11). After rearrangement, and use of (4.11) itself, the result takes the form

$$\nabla_1 W_n(1, \ldots, n) = \nabla_1 \bar{V}_n(1, \ldots, n)$$

$$+ \sum_{l=1}^n \sum_{m=1}^{N-m} \sum_{i_1 < \cdots < i_l = 1}^n \frac{1}{m!} \int dx_{n+1} \cdots \int dx_{n+m}$$

$$\times [\nabla_1 \bar{V}^{(l+m)}(i_1, \ldots, i_l, n+1, \ldots, n+m)] \frac{\rho^{(n+m)}(1, \ldots, n+m)}{\rho^{(n)}(1, \ldots, n)}$$

$$(4.16)$$

where we have set

$$\rho^{(n)}(1 \cdots n) = \frac{N \cdots (N - n + 1)}{(8\pi^2 \mathscr{V})^n} \exp\left[-\beta W_n(1 \cdots n)\right] \qquad (4.17)$$

Equation (4.16) is a generalized form of the Born–Green–Yvon integro-differential equation that has often been used for the calculation of distribution functions for simple liquids.[42] Its interpretation is straightforward. The first term in the right member of (4.16) is the negative of the generalized force acting on the molecule at $x_1$ due to those held fixed at $x_2, \ldots, x_n$; succeeding terms provide the corresponding effect of the average generalized force at $x_1$ attributable to the material medium surrounding the fixed set of molecules at $x_1, \ldots, x_n$. Thus $-\nabla_1 W_n$ is the total mean force on the molecule at $x_1$, given that $n - 1$ others are present at $x_2, \ldots, x_n$. It is therefore appropriate (and traditional) to call the $W_n$ "potentials of mean force."

The singlet potential of mean force $W_1(x_1)$ vanishes in uniform isotropic fluids, but for crystals it does not vanish if boundary conditions or wall forces serve to clamp the crystal in place. In this case $W_1(x_1)$ expresses the fact that long-range order in the crystal acts through intermolecular forces to confine and orient any molecule to fit properly into the prevailing lattice. In any event a stoichiometric condition applies to $W^{(1)}$, owing to the fact that the integral of $\rho^{(1)}$ over orientations, and inside a unit cell $\tau$, is fixed by structural parameters:

$$\frac{N}{8\pi^2 \mathscr{V}} \int_\tau dx_1 \exp\left[-\beta W^{(1)}(x_1)\right] = n_\tau - n_v + n_i \qquad (4.18)$$

Here $n_\tau$ stands for the number of molecules per unit cell in the perfect crystal, while $n_v$ and $n_i$, respectively, stand for the average number of vacancies and interstitials per unit cell at the ambient temperature and pressure.

In the large-system limit ($N \to \infty$ with $N/\mathscr{V}$ held constant), the asymptotic factorization property (4.15) for $\rho^{(n)}$ is obviously equivalent to an asymptotic subset additivity reduction for the $W$s:

$$W_n(x_1 \cdots x_n) \to W_m(x_1 \cdots x_m) + W_{n-m}(x_{m+1} \cdots x_n) \qquad (4.19)$$

Unlike fluids, crystals are capable of supporting strain fields under the influence of suitable perturbations. For nearly all choices of configurations $x_1, \ldots, x_m$ to be inhabited by molecules, a strain field would be expected to arise which would radiate outward from the location of that subset as a slowly diminishing modulation of the normal crystallographic pattern. This strain field is surely capable of carrying correlation to the other subset, provided that the strain field still has a nonnegligible magnitude at its neighborhood. Thus the rate of attainment of the reduction (4.19) with respect to increasing subset separation is controlled by crystal elasticity.

The implied range of correlation should exceed that applicable in fluids, except at the critical point for the latter.

The pair distribution $\rho^{(2)}(\mathbf{x}_1, \mathbf{x}_2)$ conveys useful information about the modes of packing of molecules in the liquid state. In particular it can be used to calculate the mean number of neighbors $v(R)$ possessed by any given molecule 1, as a function of the radial distance $R$ out to which neighbors are counted. Specifically,

$$v(R) = \frac{8\pi^2 \mathscr{V}}{N} \int_{R_{12} < R} \rho^{(2)}(\mathbf{x}_1, \mathbf{x}_2) \, d\mathbf{x}_2$$

$$= \frac{N}{8\pi^2 \mathscr{V}} \int_{R_{12} < R} \exp\left[-\beta W_2(\mathbf{x}_1, \mathbf{x}_2)\right] d\mathbf{x}_2 \qquad (4.20)$$

where for the second form we have supposed that $N$ is very large. Although this expression is formally correct for any choice of center for all the molecules, it is most informative in our specific application to choose the oxygen nucleus to play this role. Consequently, the running coordination number $v(R)$ counts oxygen nuclei lying within a sphere with radius $R$ drawn about a specific oxygen (although the count excludes the central oxygen).

We know from the generalized Born–Green–Yvon equation [see (4.16)] that an important component of $W_2(\mathbf{x}_1, \mathbf{x}_2)$ is the direct pair interaction $\overline{V}^{(2)}(\mathbf{x}_1, \mathbf{x}_2)$. It is natural to inquire if this is the overwhelmingly predominant contributor to $W_2$ for near neighbors (those close enough to form an unstrained hydrogen bond), or if indirect contributions from the surrounding medium are important.

Information is available to estimate $v(R)$ with $W_2$ simplified to $\overline{V}^{(2)}$ in (4.20).[40] An absurd conclusion follows from this simplification, namely, that in liquid water near its melting point every molecule has about $10^2$ neighbors within 4 Å. Such spontaneous crowding is obviously inconsistent with the strong repulsions operative between molecules closer than 2.5 Å. The problem clearly has its origin in the fact that the pair Boltzmann factor

$$\exp\left[-\beta \overline{V}^{(2)}(\mathbf{x}_1, \mathbf{x}_2)\right] \qquad (4.21)$$

at low temperature takes on enormously large values when $\mathbf{x}_1, \mathbf{x}_2$ lead to a maximally stabilizing hydrogen bond, and the corresponding approximate $v(R)$ picks up a spuriously large contribution.

Evidently, $\overline{V}^{(2)}$ must be largely counterbalanced by indirect medium contributions to $W_2$. The physical necessity for such repulsive contributions to $W_2$ is easy to understand. In order that a molecule at $\mathbf{x}_2$ will be able to approach one held fixed at $\mathbf{x}_1$ so as to form the hydrogen bond described by $\overline{V}^{(2)}(\mathbf{x}_1, \mathbf{x}_2)$, it will almost certainly have to expel another molecule in its way. At a low temperature the other molecule is likely enjoying its own hydrogen bond. Thus one hydrogen bond will have to be broken to form the intended one.

The close balance between direct and indirect interactions in $W_2$ is particularly crucial in the case of low-temperature ice. As temperature $T$ declines toward absolute zero, Boltzmann factor (4.21) diverges to infinity, although the coordination number remains four precisely, that is, $v(R)$ is

$$\lim_{T \to 0} v(R) = 4 \qquad 3 \text{ Å} \leq R \leq 4 \text{ Å} \tag{4.22}$$

It is incumbent on any statistical theory of $\rho^{(2)}$ in water to pay particular attention to the competitive balance between $W_2$ contributions whether the liquid or the solid state is involved.

### C.  Distribution Function Formulas

Within the regime of applicability of the semiclassical approximation, the average energy per particle

$$\frac{E}{N} = \left[ \frac{\partial(\beta F/N)}{\partial \beta} \right]_{\mathscr{V}} \tag{4.23}$$

consists of:

$$\frac{E}{N} = \frac{E_{\text{vib}} + E_{\text{kin}} + \langle \overline{V}_N \rangle}{N} \tag{4.24}$$

[Here we have assumed that $\overline{V}_N$ is temperature-independent, as will surely be the case at about room temperature. However, at high temperatures where vibrational excitations occur, the last term would have to be replaced by $\langle \partial(\beta \overline{V}_N)/\partial \beta \rangle$.] The vibrational energy may be computed from $Q_{\text{vib}}$, [see (4.5)]:

$$\frac{E_{\text{vib}}}{N} = -\frac{\partial \ln Q_{\text{vib}}}{\partial \beta}$$

$$= \sum_{\alpha=1}^{3} \hbar \omega_\alpha \left[ \frac{1}{2} + \frac{\exp(-\beta \hbar \omega_\alpha)}{1 - \exp(-\beta \hbar \omega_\alpha)} \right] \tag{4.25}$$

At about room temperature $\exp(-\beta \hbar \omega_\alpha)$ is negligible for all three normal modes. The kinetic energy $E_{\text{kin}}$ consists of classical rotational and translational parts $k_B T/2$ for each degree of freedom, so that

$$E_{\text{kin}} = 3k_B T \tag{4.26}$$

The average intermolecular potential energy

$$\langle \overline{V}_N \rangle = \frac{\int d\mathbf{x}_1 \cdots \int d\mathbf{x}_N \, \overline{V}_N \exp(-\beta \overline{V}_N)}{\int d\mathbf{x}_1 \cdots \int d\mathbf{x}_N \exp(-\beta \overline{V}_N)} \tag{4.27}$$

may be expressed in terms of molecular distribution functions by inserting relation (4.9) for $\overline{V}_N$ into (4.27), followed by repeated use of the $\rho^{(n)}$ definition (4.11). The final energy formula is found to be

$$\frac{E}{N} = -\frac{\partial \ln Q_{\text{vib}}}{\partial \beta} + 3k_B T + \sum_{n=2}^{N} \frac{1}{n! N} \int \overline{V}^{(n)}(\mathbf{x}_1 \cdots \mathbf{x}_n) \rho^{(n)}(\mathbf{x}_1 \cdots \mathbf{x}_n) \, d\mathbf{x}_1 \cdots d\mathbf{x}_n$$

(4.28)

The thermodynamic pressure $p$ can also be put into molecular distribution function form, starting from the identity

$$p = -\left(\frac{\partial F}{\partial \mathscr{V}}\right)_{\beta}$$

(4.29)

In order to carry out the volume derivative of $\ln Q$ required by this identity, it is convenient to use a position-coordinate scaling originally devised by Green.[43] The volume is taken to have the shape of a cube. Then the positions $\mathbf{R}_j$ of molecular centers are related to reduced positions $\mathbf{s}_j$ by the transformation $(1 \le j \le N)$

$$\mathbf{R}_j = \mathscr{V}^{1/3} \mathbf{s}_j$$

(4.30)

The components of each $\mathbf{s}_j$ have limits 0 and 1, so in terms of these new variables the volume dependence of the configuration integral in $Q$ becomes transferred to the arguments of $\overline{V}_N$. After employing the $\overline{V}_N$ expansion (4.9) once again, and carrying out the $\mathscr{V}$ differentiation under the integrals, the "virial" pressure equation of state is obtained in the form

$$p = \frac{N}{\beta \mathscr{V}} - \frac{1}{3\mathscr{V}} \sum_{n=2}^{N} \frac{1}{n!} \int d\mathbf{x}_1 \cdots \int d\mathbf{x}_n$$

$$\times \left[(\mathbf{R}_1 \cdot \nabla_{R_1} + \cdots + \mathbf{R}_n \cdot \nabla_{R_n}) V^{(n)}(\mathbf{x}_1 \cdots \mathbf{x}_n)\right] \rho^{(n)}(\mathbf{x}_1 \cdots \mathbf{x}_n) \quad (4.31)$$

Next consider a subvolume $\mathscr{V}_0$ contained entirely within $\mathscr{V}$. The average number of molecules $N_0$ to be found with centers in $\mathscr{V}_0$ can be calculated by integrating $\rho^{(1)}$ over all orientations, and over positions in this subvolume:

$$\langle N_0 \rangle = \int_{\mathscr{V}_0} \rho^{(1)}(\mathbf{x}_1) \, d\mathbf{x}_1$$

(4.32)

Analogously, the number of pairs of molecules $\frac{1}{2}N_0(N_0 - 1)$ has an average given by a $\rho^{(2)}$ integral over the same region:

$$\langle \tfrac{1}{2}N_0(N_0 - 1) \rangle = \tfrac{1}{2} \int_{\mathscr{V}_0} d\mathbf{x}_1 \int_{\mathscr{V}_0} d\mathbf{x}_2 \, \rho^{(2)}(\mathbf{x}_1, \mathbf{x}_2)$$

(4.33)

Hence

$$\langle N_0{}^2 \rangle - \langle N_0 \rangle^2 = \langle N_0 \rangle + \int_{\mathscr{V}_0} d\mathbf{x}_1 \int_{\mathscr{V}_0} d\mathbf{x}_2 \left[ \rho^{(2)}(\mathbf{x}_1, \mathbf{x}_2) - \rho^{(1)}(\mathbf{x}_1)\rho^{(1)}(\mathbf{x}_2) \right]$$

(4.34)

Although the last equation is formally exact for any volume $\mathscr{V}$ and subvolume $\mathscr{V}_0$, it is most informative in the $\mathscr{V} \to \infty$ limit, for then the surroundings of $\mathscr{V}_0$ act as an infinite reservoir of particles whose intensive properties are not perturbed by fluctuations in $N_0$. Furthermore, $\mathscr{V}_0$ may subsequently be presumed to have macroscopic dimensions, and the number fluctuation for $\mathscr{V}_0$ shown in (4.34) must then be related in standard fashion to the isothermal compressibility $\kappa_T$:[44]

$$\langle N_0{}^2 \rangle - \langle N_0 \rangle^2 = \frac{\langle N_0 \rangle^2 k_B T \kappa_T}{\mathscr{V}_0}$$

$$\kappa_T = -\left( \frac{\partial \ln \mathscr{V}}{\partial p} \right)_{N,T}$$

(4.35)

Therefore (4.34) may be cast in the following alternative relation:

$$\rho^2 k_B T \kappa_T = \rho + \mathscr{V}_0^{-1} \int_{\mathscr{V}_0} d\mathbf{x}_1 \int_{\mathscr{V}_0} d\mathbf{x}_2 \left[ \rho^{(2)}(\mathbf{x}_1, \mathbf{x}_2) - \rho^{(1)}(\mathbf{x}_1)\rho^{(1)}(\mathbf{x}_2) \right]$$

(4.36)

where we have written $\rho$ for the macroscopic number density $\langle N_0 \rangle / \mathscr{V}_0$. The most striking feature of this result in comparison with the energy and pressure equations (4.28) and (4.31) is that no distribution functions of order greater than two are required, regardless of the nature of $\overline{V}_N$.

In the event that an isotropic liquid or vapor phase inhabits the system, (4.36) may be simplified somewhat:

$$\rho k_B T \kappa_T = 1 + \frac{8\pi^2}{\rho} \int d\mathbf{x}_2 \left[ \rho^{(2)}(\mathbf{x}_1, \mathbf{x}_2) - \left( \frac{\rho}{8\pi^2} \right)^2 \right]$$

(4.37)

Here we have assumed that $\rho^{(2)}$ is strictly the limit function for an infinite system size and, as a consequence, it has been possible to extend the position part of the $\mathbf{x}_2$ integration to infinity.

Finally, we note that standard techniques are available to expand the molecular distribution functions in a density power series[45] valid for description of the vapor. To leading order $\rho^{(n)}$ is of course proportional to $\rho^n$. The specific density series for the infinite system pair distribution function begins with these terms:

$$\rho^{(2)}(\mathbf{x}_1, \mathbf{x}_2) = \left(\frac{\rho}{8\pi^2}\right)^2 \exp\left[-\beta \overline{V}^{(2)}(\mathbf{x}_1, \mathbf{x}_2)\right]$$

$$\times \left\{ 1 + \frac{\rho}{8\pi^2} \int d\mathbf{x}_3 \, f(\mathbf{x}_1, \mathbf{x}_3) f(\mathbf{x}_3, \mathbf{x}_2) \right.$$

$$+ \frac{\rho}{8\pi^2} \int d\mathbf{x}_3 \exp\left[-\beta \overline{V}^{(2)}(\mathbf{x}_1, \mathbf{x}_3) - \beta \overline{V}^{(2)}(\mathbf{x}_3, \mathbf{x}_2)\right]$$

$$\left. \times \left[\exp\left(-\beta \overline{V}^{(3)}(\mathbf{x}_1, \mathbf{x}_2, \mathbf{x}_3)\right) - 1\right]\right\} + O(\rho^4) \qquad (4.38)$$

where the Mayer $f$ function has been introduced following the normal convention:

$$f(\mathbf{x}_i, \mathbf{x}_j) = \exp\left[-\beta \overline{V}^{(2)}(\mathbf{x}_i, \mathbf{x}_j)\right] - 1 \qquad (4.39)$$

By using the density expansion for $\rho^{(2)}$ in the compressibility relationship (4.37), it is possible to generate virial coefficients for the pressure:

$$\frac{\beta p}{\rho} = 1 + B(T)\rho + C(T)\rho^2 + \cdots \qquad (4.40)$$

The second and third virial coefficients have the integral expressions

$$B(T) = -\frac{1}{2(8\pi^2)} \int d\mathbf{x}_2 \, f(\mathbf{x}_1, \mathbf{x}_2) \qquad (4.41)$$

$$C(T) = -\frac{1}{3(8\pi^2)^2} \int d\mathbf{x}_2 \int d\mathbf{x}_3 \, \{f(\mathbf{x}_1, \mathbf{x}_2) f(\mathbf{x}_1, \mathbf{x}_3) f(\mathbf{x}_2, \mathbf{x}_3)$$

$$+ \exp\left[-\beta \overline{V}_3(\mathbf{x}_1, \mathbf{x}_2, \mathbf{x}_3)\right] - \exp\left[-\beta \overline{V}^{(2)}(\mathbf{x}_1, \mathbf{x}_2)\right.$$

$$- \beta \overline{V}^{(2)}(\mathbf{x}_1, \mathbf{x}_3) - \beta \overline{V}^{(2)}(\mathbf{x}_2, \mathbf{x}_3)]\} \qquad (4.42)$$

## V. EFFECTIVE PAIR POTENTIALS

### A. Variational Principle

Both the conceptual and computational aspects of water theory would become significantly streamlined if the total interaction potential were additive, that is, vanishing trimer, tetramer, pentamer, and so on, component potentials. Of course we know this is not strictly the case, and nonadditive components evidently exert considerable influence on at least the thermodynamic properties of water. Even so, it is a legitimate and interesting question to ask if there exists an "effective pair potential," possibly differing substantially from the real molecular pair potential, which alone faithfully reproduces microscopic structure and thermodynamic properties for the true nonadditive water system. We outline a procedure that answers this question affirmatively.[46,47]

The following analysis can actually be carried out at any of several alternative levels. In the most precise version, a full quantum mechanical description applies, for which the partition function $Q$ has previously been displayed as the trace of a density operator. But for present purposes, it suffices in illustrating the general strategy to work at the level of the semiclassical approximation, with $Q$ given by (4.4).

Consider a pair of arbitrary real functions $h_1(x_1 \cdots x_N)$ and $h_2(x_1 \cdots x_N)$ of the $N$ molecular six-vectors. We define their inner product to be

$$\{h_1, h_2\} = (8\pi^2 \mathscr{V})^{-N} \int dx_1 \cdots \int dx_N \, h_1(x_1 \cdots x_N) h_2(x_1 \cdots x_N) \quad (5.1)$$

where the $x_j$ run between the same limits that apply to $Q$, namely, all orientations and positions within volume $\mathscr{V}$. Equation (5.1) is analogous to the ordinary inner product of two vectors, a sum of products of corresponding vector components, in that the integral sums over all differential elements $dx_1 \cdots dx_N$ the product of corresponding function components. In terms of our functional inner product, the semiclassical partition function (4.4) becomes

$$Q(\overline{V}_N) = Q(\overline{V}_N = 0)\left\{\exp\left(\frac{-\beta \overline{V}_N}{2}\right), \exp\left(\frac{-\beta \overline{V}_N}{2}\right)\right\} \quad (5.2)$$

Just as the distance between the end points of two vectors may be calculated in terms of the inner product of their difference with itself, we take the "distance" between the two functions $h_1$ and $h_2$, $D(h_1, h_2)$, to be

$$D(h_1, h_2) = \{h_1 - h_2, h_1 - h_2\}^{1/2} \quad (5.3)$$

The effective pair potential is denoted by $v(x_i, x_j)$. Following the form of (5.2), the effective pair potential approximation to $Q$ can be expressed as

$$Q(\Sigma v) = Q(\overline{V}_N = 0)\left\{\exp\left[-\tfrac{1}{2}\beta \sum_{i<j=1}^{N} v(i,j)\right], \exp\left[-\tfrac{1}{2}\beta \sum_{i<j=1}^{N} v(i,j)\right]\right\} \quad (5.4)$$

We postulate that the optimal choice for the function $v$ is the one for which

$$D\{\exp\left(-\tfrac{1}{2}\beta\overline{V}_N\right), \exp\left[-\tfrac{1}{2}\beta\Sigma v(i,j)\right]\} = \text{minimum} \quad (5.5)$$

In other words, we require that the Boltzmann function's square root $\exp\left(-\tfrac{1}{2}\beta\overline{V}_N\right)$ literally be approached to the minimum possible distance by the corresponding effective pair potential version.

The variational criterion (5.5) forces the best possible fit over the entire configuration space for the $N$ molecules. This is important, for beyond $Q$ itself we want the molecular distribution functions $\rho^{(n)}$ in the effective pair potential approximation:

$$\rho^{(n)}(\mathbf{x}_1 \cdots \mathbf{x}_N | \Sigma v) = \frac{N \cdots (N - n + 1) \int d\mathbf{x}_{n+1} \cdots \int d\mathbf{x}_N \exp\left[-\beta \Sigma v(i, j)\right]}{\int d\mathbf{x}_1 \cdots \int d\mathbf{x}_N \exp\left[-\beta \Sigma v(i, j)\right]},$$

(5.6)

also to represent properly the detailed microscopic structure present in the system.

By setting the first functional derivative of (5.5) with respect to $v$ equal to zero, we derive an integral equation for the determination of $v$:

$$\int d\mathbf{x}_3 \cdots \int d\mathbf{x}_N \exp\left\{-\tfrac{1}{2}\beta\left[\overline{V}_N(\mathbf{x}_1 \cdots \mathbf{x}_N) + \sum_{i<j=1}^{N} v(\mathbf{x}_i, \mathbf{x}_j)\right]\right\}$$

$$= \int d\mathbf{x}_3 \cdots \int d\mathbf{x}_N \exp\left\{-\beta \sum_{i<j=1}^{N} v(\mathbf{x}_i, \mathbf{x}_j)\right\}$$

(5.7)

On account of the nonlinearity of this equation in the unknown function $v$, and because of the high-order multiple integrals, one cannot expect generally to extract the exact solution. However, it should be clear that the effective pair potential can vary somewhat with temperature and density.

Integration of both sides of (5.7) over $\mathbf{x}_1$ and $\mathbf{x}_2$ leads to the relation

$$Q(\tfrac{1}{2}\overline{V}_N + \tfrac{1}{2}\Sigma v) = Q(\Sigma v)$$

(5.8)

or what amounts to the same thing for the Helmholtz free energy:

$$F(\tfrac{1}{2}\overline{V}_N + \tfrac{1}{2}\Sigma v) = F(\Sigma v)$$

(5.9)

Thus the optimally chosen effective pair potential is such that its corresponding free energy is unchanged by the operation of averaging the effective additive interaction with the true nonadditive interaction $\overline{V}_N$. Similar results apply to $\rho^{(1)}$ and $\rho^{(2)}$, which follow directly from (5.7) after dividing by its integral over $\mathbf{x}_1$ and $\mathbf{x}_2$:

$$\rho^{(1)}(\mathbf{x}_1 | \tfrac{1}{2}\overline{V}_N + \tfrac{1}{2}\Sigma v) = \rho^{(1)}(\mathbf{x}_1 | \Sigma v)$$

(5.10)

$$\rho^{(2)}(\mathbf{x}_1, \mathbf{x}_2 | \tfrac{1}{2}\overline{V}_N + \tfrac{1}{2}\Sigma v) = \rho^{(2)}(\mathbf{x}_1, \mathbf{x}_2 | \Sigma v)$$

(5.11)

However, these identities do not extend to higher-order $\rho^{(n)}$ (although they would if the procedure were applied to the determination of $n$-molecule effective interactions $v_n$, with $n > 2$).

The Schwartz inequality[48] implies that

$$Q(\overline{V}_N)Q(\Sigma v) \geq [Q(\tfrac{1}{2}\overline{V}_N + \tfrac{1}{2}\Sigma v)]^2$$

(5.12)

This, in concert with the earlier result (5.8), leads to

$$Q(\overline{V}_N) \geq Q(\Sigma v) \tag{5.13}$$

In terms of free energies, the direction of the inequality is reversed:

$$F(\overline{V}_N) \leq F(\Sigma v) \tag{5.14}$$

Thus the operation of forcing the system of molecules to conform to an effective pair potential never lowers the free energy.

For pedagogical reasons we can write $\overline{V}_N$ as its strictly additive part, plus a perturbation:

$$\overline{V}_N(1 \cdots N) = \sum_{i < j = 1}^{N} \overline{V}^{(2)}(i, j) + \lambda V^{\dagger}(1 \cdots N) \tag{5.15}$$

where $\lambda$ is a formal perturbation parameter destined ultimately to be set equal to unity, and

$$V^{\dagger}(1 \cdots N) = \sum_{n=3}^{N} \sum_{i_1 < \cdots < i_n = 1}^{N} \overline{V}^{(n)}(i_1 \cdots i_n) \tag{5.16}$$

Imagining $\lambda$ to increase continuously from zero to unity, the free energy $F$ as well as $\rho^{(1)}$ and $\rho^{(2)}$ will display first-order changes with $\lambda$. However, (5.9) to (5.11) imply that $v$ changes from $\overline{V}^{(2)}$ during this coupling process in just such a way that $F(\Sigma v)$, $\rho^{(1)}(\mathbf{x}_1|\Sigma v)$, and $\rho^{(2)}(\mathbf{x}_1, \mathbf{x}_2|\Sigma v)$ all manifest exactly the correct leading linear behavior in $\lambda$. Although one can propose other effective pair potentials which selectively fit thermodynamical properties[49] or pair distribution functions,[50] it is only the variationally defined function $v$ satisfying (5.7) that simultaneously eliminates first-order errors in $F$, $\rho^{(1)}$, and $\rho^{(2)}$.

It is possible to develop $v$ in a density power series. The manipulations required are tedious, and are not reproduced here. Instead, we state the result through first order in $\rho$:[47]

$$v(\mathbf{x}_1, \mathbf{x}_2) = \overline{V}^{(2)}(\mathbf{x}_1, \mathbf{x}_2) + \frac{\rho}{8\pi^2 \beta} [\varphi_s(\mathbf{x}_1, \mathbf{x}_2) + \varphi_l] + O(\rho^2) \tag{5.17}$$

$$\varphi_s(\mathbf{x}_1, \mathbf{x}_2) = -2 \int d\mathbf{x}_3 \exp [-\beta \overline{V}^{(2)}(\mathbf{x}_1, \mathbf{x}_3) - \beta \overline{V}^{(2)}(\mathbf{x}_2, \mathbf{x}_3)]$$

$$\times \{\exp [-\tfrac{1}{2}\beta \overline{V}^{(3)}(\mathbf{x}_1, \mathbf{x}_2, \mathbf{x}_3)] - 1\} \tag{5.18}$$

$$\varphi_l = \frac{1}{6\pi^2 \mathscr{V}} \int d\mathbf{x}_3 \int d\mathbf{x}_4 \exp [-\beta \overline{V}^{(2)}(\mathbf{x}_1, \mathbf{x}_3) - \beta \overline{V}^{(2)}(\mathbf{x}_1, \mathbf{x}_4) - \beta \overline{V}^{(2)}(\mathbf{x}_3, \mathbf{x}_4)]$$

$$\times \{\exp [-\tfrac{1}{2}\beta \overline{V}^{(3)}(\mathbf{x}_1, \mathbf{x}_3, \mathbf{x}_4)] - 1\} \tag{5.19}$$

Only clusters of three or fewer molecules can contribute in this density order, so the only nonadditivity that can be involved is that for trimers. Since $\overline{V}^{(2)}$ and $\overline{V}^{(3)}$ drop to zero with increasing separation, so too will $\varphi_s$. However, $\varphi_l$ is independent of separation; although it is inversely proportional to $\mathscr{V}$ and thus very small for a large system, its effect when summed over all $N(N-1)/2$ pairs of molecules is thermodynamically significant.

Using the effective pair potential, the analog of energy expression (4.28) is

$$\frac{E}{N} = -\frac{\partial \ln Q_{\text{vib}}}{\partial \beta} + 3k_B T$$

$$+ \frac{1}{2N} \int \left\{ \frac{\partial [\beta v(\mathbf{x}_1, \mathbf{x}_2)]}{\partial \beta} \right\}_{N,\mathscr{V}} \rho^{(2)}(\mathbf{x}_1, \mathbf{x}_2 | \Sigma v) \, d\mathbf{x}_1 \, d\mathbf{x}_2 \qquad (5.20)$$

while the virial equation of state (4.31) is modified to

$$\frac{\beta p}{\rho} = 1 - \frac{4\pi^2 \beta}{\rho} \int d\mathbf{x}_2 \left\{ \frac{1}{3} \mathbf{R}_{12} \cdot \nabla_{R_{12}} v(\mathbf{x}_1, \mathbf{x}_2) - \rho\left( \frac{\partial v(\mathbf{x}_1, \mathbf{x}_2)}{\partial \rho} \right) \right\}$$

$$\times \rho^{(2)}(\mathbf{x}_1, \mathbf{x}_2 | \Sigma v) \qquad (5.21)$$

In both these expressions it is important to realize that $v$ does not drop quite to zero with increasing separation $R_{12}$, but retains a constant value proportional to $\mathscr{V}^{-1}$. It is the presence of this "tail" on $v$ of indefinite range that invalidates the compressibility theorem (4.36), except as a description of local fluctuations, although a complicated modification can be derived.[46]

## B. Physical Interpretation

In our survey of water molecule interactions (Section III), it was pointed out that the average effect of nonadditive components of the potential $\overline{V}_N$ was to strengthen hydrogen bond networks, while reducing the length of the component hydrogen bonds somewhat. These influences must be felt by the effective pair potential $v$. It seems hard to escape the conclusion that the absolute minimum of $v$ is lower than that of the bare pair potential $\overline{V}^{(2)}$, and that this $v$ minimum occurs at a smaller separation.

Since nonadditivity contributes extra binding energy to an assembly of water molecules, a hypothetical water model in which the molecules interact only via the strict pair interactions $\overline{V}^{(2)}$ would necessarily possess lower melting and boiling points. Evidently, the increased strength of $v$ compared to $\overline{V}^{(2)}$ would tend to bring these phase-transition temperatures back up again.

The melting and boiling transitions are first order, with a discontinuous change in molar volume going from one homogeneous phase to the other. With fixed $N$, $\mathscr{V}$ may be chosen so that $\rho$ lies between the values for coexisting phases (say, liquid and vapor). In this case the system is macroscopically inhomogeneous, for surface tension will cause agglomeration of the phases into large crystals, droplets, or bubbles. The molecular distribution functions $\rho^{(n)}$ will be profoundly affected by this macroscopic inhomogeneity, displaying slow variations with respect to positions over distances comparable to crystal droplet or bubble diameters.

The effective pair potential is permitted full functional freedom within its defining variational principle (5.5). In particular, it can adopt forms with just the proper coupling strength to reproduce the correct phase transitions exhibited by $\overline{V}_N$. Since the distance $D$ involved is always to be minimized, and since it would be expected to be very large if $\overline{V}_N$ caused phase separation and $\Sigma v$ did not (or vice versa), it seems likely that the effective pair potential approximation should reproduce the given first-order phase changes. In fact, we postulate that in a large-system limit $\overline{V}_N$ and $\Sigma v$ will have *exactly* the same phase diagrams in the $T$, $\rho$ plane. Furthermore, the corresponding crystalline phases should have the same symmetry properties.

It is interesting to note that the liquid-vapor critical point for water ($T = 374.15°C$, $p = 221.2$ bars, $\rho = 0.32$ g/cm³) experimentally seems to share with other liquids the conventional set of nonclassical critical exponents.[51] This universality of critical exponents implies insensitivity to most details of the underlying molecular interactions. Consequently, replacement of $\overline{V}_N$ by the proper $\Sigma v$ should involve no change in the critical-point temperature and density, and the required $v$ itself should not entail singular temperature and density variations.

The density expansion (5.17) to (5.19) illustrates the fact that generally $v$ separates into two parts:

$$v(\mathbf{x}_1, \mathbf{x}_2) = v_s(\mathbf{x}_1, \mathbf{x}_2) + \frac{v_l}{\mathscr{V}} \tag{5.22}$$

a short-range part $v_s$ which goes to zero with increasing separation, and a weak constant tail inversely proportional to $\mathscr{V}$ at constant $T$ and $\rho$. Clearly, this tail has no effect on molecular distribution functions, since it has the same value from one configuration to the next. However, the short-range function $v_s$, which is necessary to give optimal $\rho^{(n)}$s, does not by itself give an optimal free energy. The long-range tail provides the necessary correction.

In a dense liquid or crystalline phase, each molecule is surrounded by a rather closely packed multitude of neighbors. From a numerical viewpoint it would probably be feasible to simulate the free-energy effect of $v_l/\mathscr{V}$ by a spatially slowly decaying component of $v_s$ instead. Because the neighbors are closely spaced, they would respond negligibly to this geometric compression of the truly long-range tail of $v$ into $v_s$. Thus, as a practical matter, explicit consideration of $v_l/\mathscr{V}$ may be quite unnecessary for condensed phases. This possibility is especially clear for low-temperature crystals whose structure and vibrational properties would not be changed by constant shifts in $v$ at each of the first few neighbor shell distances, although of course the free energy *would* be correspondingly affected.

An invariant ratio of the different types of bonded trimers was mentioned earlier for all four-coordinated networks, leading to overall stabilization. Temperature rise causes bond breakage in these networks (most notably at the melting point), and at very high temperatures nearly all trimers are disrupted and no enhanced stability for hydrogen bonds applies. But during the course of temperature rise, the ratio of sequential, double-donor, and double-acceptor molecules need not remain fixed. Figure 12 shows how alternative choices for the scission of two hydrogen bonds impinging on a single network molecule can remove different ratios of trimers (although always 11 of all kinds). As a result, we would expect the destabilizing double-donor and double-acceptor trimers to disappear at a greater initial rate than the stabilizing sequential trimers. Just above the melting point, $v$ can in this

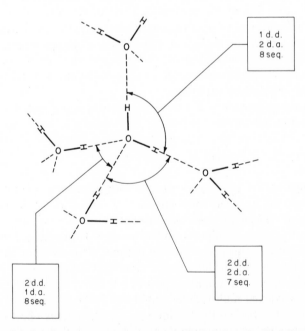

**Fig. 12** Trimers disrupted by breaking two hydrogen bonds emanating from the same network molecule. d.d., Double donor; d.a., double acceptor; seq., sequential trimer. In view of the signs of the trimer nonadditivities, it is energetically least costly to break the two bonds if one and only one incorporates a proton of the central molecule.

fashion maintain its greater hydrogen bond strength and shorter length relative to $\overline{V}^{(2)}$; the temperature variation in $v$ may thus be negligible in this range.

## VI. LATTICE THEORIES

### A. General Formulation

In approximating the definite integral

$$I(a, b) = \int_a^b F(t)\, dt \tag{6.1}$$

by the rectangle rule, the interval $a \leq t \leq b$ is divided into $M$ equal intervals with midpoints

$$t_i = a + \frac{(b - a)(2i - 1)}{2M}, \quad i = 1, 2, \ldots, M \tag{6.2}$$

and then $I$ is represented by a suitably normalized sum of the integrand over these midpoints:

$$I(a, b) \cong \frac{b - a}{M} \sum_{i=1}^{M} F(t_i) \tag{6.3}$$

The lattice gas partition function $Q_l$ bears the same relationship to the semi-classical partition function $Q$ [see (4.4)] that the rectangle rule approximation (6.3) does to the starting integral (6.1). To establish the relationship for the partition functions, the six-dimensional configuration space available to the rigid-rotor molecules must be divided into equivalent cells whose centers are analogs of the midpoints $t_i$.

First let the three-dimensional volume $\mathscr{V}$ be uniformly covered by a lattice of $\mathscr{M}$ equivalent points. We shall always want

$$\mathscr{M} > N \tag{6.4}$$

so that the number of discrete locations exceeds the number of molecules and, as a general matter, the limit of $\mathscr{M}$ tending to infinity is ultimately of interest. For the moment we leave open the choice of specific lattice type (simple cubic, body-centered cubic, diamond, face-centered cubic, etc.), but for every lattice there exists a corresponding division of $\mathscr{V}$ into nearest-neighbor polyhedra.[52] These polyhedra are convex and congruent, contain single lattice points at their inversion centers, and are the loci of all positions closer to this central lattice point than to any other lattice point. On account of the last property, they are bounded by midplanes constructed on line segments connecting lattice points.

The volume of each nearest-neighbor polyhedron is denoted by

$$\tau = \frac{\mathscr{V}}{\mathscr{M}} \tag{6.5}$$

By selecting a very anisotropic lattice, it is possible to produce polyhedra that are very long in one or two directions at the expense of the third. For present purposes we specifically avoid this situation, and suppose that the polyhedra are reasonably compact. This criterion will automatically be fulfilled if the basic lattice has cubic symmetry. Now, if $\mathscr{M}$ is sufficiently large, the distance between two molecules with centers confined to the same polyhedron is necessarily small, and $\overline{V}^{(2)}$ will be positive and large. We proceed under the assumption that $\mathscr{M}$ is in a range in which these molecular repulsions prevent multiple occupancy of the polyhedra.

In the same way the lattice affords $\mathscr{M}$ discrete locations with volume $\mathscr{V}$, we must also divide the orientational space $8\pi^2$ into an integer number $v$ of equal-weight regions. The lattice theory then provides for a coarse-grained description of any arrangement of the $N$ water molecules in space by means of a set of occupation parameters $\xi_i$ $(1 \le i \le \mathscr{M})$, one for each nearest-neighbor polyhedron. In particular,

$$\xi_i = 0 \tag{6.6}$$

indicates that polyhedron $i$ is empty, while

$$\xi_i = 1, 2, \ldots, v \tag{6.7}$$

indicates that it contains the center of a molecule with orientation falling into region 1, 2, ..., or $v$, respectively, of the full $8\pi^2$ orientation space. Among the $\mathscr{M}$ occupation parameters precisely $N$ can be nonzero, but even under this restraint there remain

$$\frac{\mathscr{M}! \, v^N}{N!(\mathscr{M} - N)!} \tag{6.8}$$

distinct sets of values $\{\xi\}$ describing distinguishable coarse-grained system configurations.

The total molecular interaction relevant for the lattice approximation is denoted by $\overline{V}_N\{\xi\}$ for simplicity. It is equal to the vibrationally averaged function $\overline{V}_N(\mathbf{x}_1 \cdots \mathbf{x}_N)$ evaluated for molecules placed at the centers of the polyhedra and given orientations central to the rotational regions specified by the parameter set $\{\xi\}$. Figure 13 illustrates in elementary two-dimensional fashion the centered and discretely oriented water molecules, one to each cell, that might be required by $\{\xi\}$.

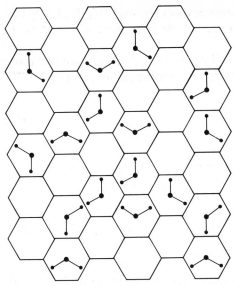

**Fig. 13** Centered molecules, with a discrete set of orientations, used to define the lattice model interaction $\overline{V}_N\{\xi\}$. This is a schematic two-dimensional version of the three-dimensional situation.

After accounting for the $N!$ ways in which the water molecules can be distributed among the occupied lattice sites, the partition function $Q_l$ adopts the form

$$Q_1 = \left(\frac{8\pi^2\tau\Lambda}{v}\right)^N \sum_{\{\xi\}} \exp\left[-\beta\overline{V}_N\{\xi\}\right] \tag{6.9}$$

where

$$\Lambda = \frac{(2\pi k_B T)^3 m^{3/2}(I_1 I_2 I_3)^{1/2} Q_{\text{vib}}}{h^6} \tag{6.10}$$

is a factor carried over without change from the semiclassical partition function (4.4). The sum runs over all distinct sets $\{\xi\}$ consistent with the given number of molecules. As $\mathcal{M}$ and $v$ approach infinity, $Q_1$ approaches $Q$ to be sure, but the advantage of the lattice approximation lies in computational simplifications that obtain for certain finite choices of $\mathcal{M}$ and $v$.

In the coarse-grained lattice representation, the molecular distribution functions $\rho^{(n)}$ are replaced by probability functions $P^{(n)}(i_1 \cdots i_n | \xi_1 \cdots \xi_n)$, giving the chance that simultaneously the lattice sites $i_1, \ldots, i_n$ have the respective occupation parameter values $\xi_1, \ldots, \xi_n$. The defining relations may be written:

$$P^{(n)}(i_1 \cdots i_n | \xi_1 \cdots \xi_n) = \frac{\sum'_{\{\xi\}} \exp\left(-\beta\overline{V}_N\{\xi\}\right)}{\sum'_{\{\xi\}} \exp\left(-\beta\overline{V}_N\{\xi\}\right)} \tag{6.11}$$

The primed numerator summation includes only those sets $\{\xi\}$ displaying the required parameter values at sites $i_1, \ldots, i_n$. By convention we set $P^{(n)} = 0$ if any two $i_\alpha$s happen to be identical.

Both the mean energy expression (4.28) and the compressibility relation (4.37) can be carried over into the lattice approximation, using the $P^{(n)}$ instead of $\rho^{(n)}$:

$$\frac{E}{N} = -\frac{\partial \ln Q_{\text{vib}}}{\partial \beta} + 3k_B T$$

$$+ \sum_{n=2}^{N} \frac{1}{n! N} \sum_{i_1 \cdots i_n = 1}^{\mathcal{M}} \sum_{\xi_1 \cdots \xi_n = 1}^{v} \overline{V}^{(n)} \{\xi_1 \cdots \xi_n\} P^{(n)}(i_1 \cdots i_n | \xi_1 \cdots \xi_n)$$

(6.12)

$$\rho k_B T \kappa_T = 1 + \frac{1}{N} \sum_{i_1, i_2 = 1}^{\mathcal{M}} \sum_{\xi_1, \xi_2 = 1}^{v} [P^{(2)}(i_1, i_2 | \xi_1, \xi_2) - P^{(1)}(i_1 | \xi_1) P^{(1)}(i_2 | \xi_2)]$$

(6.13)

However, there is no way that the virial equation of state (4.31) can be retained in the lattice description, so the pressure must be obtained from another source (such as $\kappa_T$) using suitable thermodynamic identities.

Once again it is necessary to emphasize that the theory has been developed without the necessity for specifying a particular molecular center, and in principle any choice serves as well as any other. Nevertheless, a pragmatic element now intrudes for the first time. We have demanded single occupancy of the space-filling polyhedra. This automatically becomes valid as $\mathcal{M} \to \infty$ ($N$ and $\mathcal{V}$ fixed), but the demand must also be met for finite $\mathcal{M} > N$ as well. If the molecular center were taken eccentrically but by convention to be 10 Å from the oxygen nucleus along the symmetry axis, there would be no energetic prevention of double occupancy for any polyhedron, however small, since the molecular electron clouds and nuclei would be well outside the polyhedron. To avoid this difficulty, one is effectively required to embed the center inside the molecular electron distribution to take full advantage of overlap repulsion in avoiding multiple occupancy. The position of the oxygen nucleus qualifies satisfactorily on this count, as Fig. 13 implicitly acknowledges.

The lattice model was originally advocated to study phase transitions for structureless spherical molecules, or for highly symmetric molecules for which free rotation produces an outward appearance of sphericity.[53] The occupation parameters required in that simpler version needed only two values:

$$\xi_i = 0 \quad \text{empty}$$
$$= 1 \quad \text{singly occupied}$$

(6.14)

It can easily be shown that this two-state lattice model has an inherent symmetry about the half-filled ($N = \mathcal{M}/2$) state that derives from an isomorphism to the Ising model for magnetic-phase transitions. An important consequence of this "particle-hole" symmetry is that the chemical potential along the critical isochore ($N = \mathcal{M}/2$) is an analytical function of temperature at the critical point. Unfortunately, no analogous symmetry can be identified for the lattice models of water, and it remains an open question whether temperature analyticity of the water chemical potential exists at its critical point, along a suitable symmetry line.[54]

## B. Fleming – Gibbs Version

The simplest practical version of the lattice approximation utilizes the body-centered cubic arrangement of sites, with a distance between nearest neighbors corresponding to an unstrained hydrogen bond between two water molecules. Figure 14 shows this lattice structure, for which each site has eight nearest neighbors in a cubical arrangement. The tetrahedral angle between successive hydrogen bonds is realized in a natural way, since four mutually noncontiguous near neighbors (out of the eight) for any given site form vertices of a regular tetrahedron surrounding that site. In fact the fully hydrogen-bonded cubic ice crystal (ice Ic) fits precisely on the

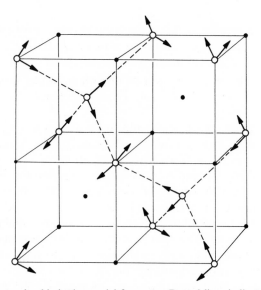

**Fig. 14**  Body-centered cubic lattice model for water. Dotted lines indicate hydrogen bonds between nearest-neighbor molecules. The molecules shown form a portion of the cubic ice Ic crystal.

body-centered cubic lattice, using every other site. The water molecules shown in Fig. 14 constitute a portion of this ice Ic crystal.

The obvious discrete set of orientations to use should be clear from Fig. 14. The two OH bonds of a given molecule should point toward a pair of nearest-neighbor cube vertices, which are located diagonally across a face of the cube from one another. There are 12 such nearest-neighbor pairs, and therefore 24 discrete orientations for a water molecule with distinguishable hydrogens. Melting of the Ic crystal can occur if molecules frequently reorient from the directions shown in Fig. 14 for the crystal, while also moving at random onto the other sublattice of sites which is totally unoccupied by the crystal.

Two interpenetrating cubic networks can coexist on the body-centered cubic lattice, filling it completely and forming the ice VII structure.[7] Each has fourfold hydrogen bond coordination at each site, but no hydrogen bonds connect the two networks.

Fleming and Gibbs have proposed a way of simplifying the lattice-model interaction $\overline{V}_N\{\xi\}$ to allow explicit statistical calculation to be carried out while retaining the essential features of directional bonding that seem to be characteristic of water.[55,56] Only pair interactions between nearest neighbors are permitted, and the value of the nearest-neighbor interaction depends on the relative orientation in such a way that only three distinct magnitudes are involved:

$$\overline{V}^{(2)}\{\xi_1, \xi_2\} = -E_b \quad \text{if } \xi_1, \xi_2 \text{ corresponds to a hydrogen bond}$$
$$= -\varepsilon_1 \quad \text{if rotation of one molecule suffices to create a hydrogen bond}$$
$$= -\varepsilon_2 \quad \text{if rotation of both molecules is necessary to create a hydrogen bond} \quad (6.15)$$

These three energies are treated as adjustable parameters but, clearly, if this simplification is to have meaning, $-E_b$ must be the lowest of the three energies. There are $24^2 = 576$ distinct orientations of a pair of neighboring molecules; of these, 72 lead to interaction $-E_b$, 360 to $-\varepsilon_1$, and 144 to $-\varepsilon_2$. It is important to realize that not all pairs of neighboring molecules, one of which points on OH toward the other, are considered hydrogen-bonded in the Fleming–Gibbs scheme, for it is also necessary that the three pendant OH groups point to other sites of the same ice Ic sublattice (as shown in Fig. 14). In part, pairs with interaction $-\varepsilon_1$ correspond to linear but angularly severely strained hydrogen bonds.

If the lattice is filled with molecules forming ice VII interpenetrating networks, nonbonded nearest neighbors (four of the eight) will always have energy $-\varepsilon_2$. This crystal structure can be thermodynamically destabilized at low pressure by causing $-\varepsilon_2$ to be the most positive of the three interaction energies, thus simulating the strong repulsions that exist between these unbonded neighbors in the real ice VII crystal.[57] In any event use of the body-centered cubic lattice and nearest-neighbor interactions permits at most two distinct crystalline phases in the Fleming–Gibbs formulation, supplemented by liquid and vapor phases.

The Fleming–Gibbs partition function can be written:

$$Q_l = \left(\frac{\pi^2 \tau \Lambda}{3}\right)^N \sum_{\{\xi\}} \exp\left[\beta(N_b\{\xi\}E_b + N_1\{\xi\}\varepsilon_1 + N_2\{\xi\}\varepsilon_2)\right] \qquad (6.16)$$

The numbers of nearest-neighbor pairs of each interaction type have been denoted here by $N_b$, $N_1$, and $N_2$.

Fleming and Gibbs have confined their quantitative studies to the liquid and vapor phases. In order to evaluate the partition function and thermodynamical properties, they have separately used two statistical mechanical approximations, the mean field approximation (MFA),[55] and a more accurate second-order approximation (SOA).[56]

The MFA replaces $N_b$, $N_1$, and $N_2$ in (6.16) by their a priori average values:

$$N_b\{\xi\} \to \tfrac{1}{2}n^2 \mathcal{M}$$
$$N_1\{\xi\} \to \tfrac{5}{2}n^2 \mathcal{M} \qquad (6.17)$$
$$N_2\{\xi\} \to n^2 \mathcal{M}$$

where $n$ is the lattice-filling fraction $N/\mathcal{M}$. With this assumption all terms in the partition function sum become equal, so

$$Q_l(\text{MFA}) = \frac{\mathcal{M}!}{N!(\mathcal{M} - N)!}\left(\frac{\pi^2 \tau \Lambda}{3}\right)^N \exp\left\{\beta n^2 \mathcal{M}\left[\tfrac{1}{2}E_b + \tfrac{5}{2}\varepsilon_1 + \varepsilon_2\right]\right\} \qquad (6.18)$$

Taking logarithms, and differentiating with respect to volume, the MFA equation of state for the lattice model is easily found to be

$$\frac{\beta p \mathcal{V}}{\mathcal{M}} = -\ln(1 - n) - \beta n^2(\tfrac{1}{2}E_b + \tfrac{5}{2}\varepsilon_1 + \varepsilon_2) \qquad (6.19)$$

The liquid-vapor critical point can be located from the horizontal inflection point of the pressure isotherm family (6.19). At this critical point,

$$k_B T_c = \tfrac{1}{4}(E_b + 5\varepsilon_1 + \varepsilon_2)$$
$$n_c = \tfrac{1}{2} \qquad (6.20)$$

Fleming and Gibbs point out that the experimental critical temperature can be reproduced by setting

$$E_b = 4.65 \text{ kcal/mole}$$
$$\varepsilon_1 = \varepsilon_2 = 7.2 \times 10^{-2} \text{ kcal/mole} \qquad (6.21)$$

for the MFA. Certainly, the hydrogen bond energy is not unreasonable, in the light of available quantum mechanical information (Section III).

If the experimental critical density is used to assign the nearest-neighbor distance, the result is 3.93 Å, far too large for a well-formed hydrogen bond. As an alternative, Fleming and Gibbs chose to use the nearest-neighbor separation determined by x-ray scattering measurements on liquid water,[58] which increases slightly (and linearly) with increasing temperature but remains within an acceptable distance range for linear hydrogen bonds. In this manner they derived the following critical constants (experimental values in parentheses):

$$\rho_c = 0.688 \text{ g/cm}^3$$
$$(0.325 \text{ g/cm}^3)$$
$$p_c = 783 \text{ atm} \tag{6.22}$$
$$(218 \text{ atm})$$

That the errors in predicted critical constants are very large no doubt stems primarily from a weakness of the MFA, which assumes each molecule is surrounded by a shell of neighbors having the macroscopic distribution of density and orientation. The SOA explicitly removes this weakness by providing for a distribution of nearest neighbors that is correct to leading order in interactions with the central site. The reader should consult the original reference[56] for details. The SOA equation of state extends the MFA result (6.19) to order $\beta^2$:

$$
\begin{aligned}
\frac{\beta p \mathscr{V}}{\mathscr{M}} =\ & -\ln{(1-n)} - \beta n^2(\tfrac{1}{2}E_b + \tfrac{5}{2}\varepsilon_1 + \varepsilon_2) \\
& - \beta^2 n^2(\tfrac{1}{4}E_b^2 + \tfrac{5}{4}\varepsilon_1^2 + \tfrac{1}{2}\varepsilon_2^2) \\
& + \beta^2 n^3[(\tfrac{1}{2}E_b + \tfrac{3}{2}\varepsilon_1)^2 + (\varepsilon_1 + \varepsilon_2)^2] \\
& - 6\beta^2 n^4(\tfrac{1}{8}E_b + \tfrac{5}{8}\varepsilon_1 + \tfrac{1}{4}\varepsilon_2)^2
\end{aligned}
\tag{6.23}
$$

Using the SOA, it was found that the liquid phase would boil at 100°C under 1 atm pressure if

$$E_b = 4.58 \text{ kcal/mole}$$
$$\varepsilon_1 = 0.915 \text{ kcal/mole} \tag{6.24}$$
$$\varepsilon_2 = -1.19 \text{ kcal/mole}$$

This set of interactions seems somewhat more reasonable than the MFA set (6.21), in that relative configurations leading to $-\varepsilon_2$ are now energetically very unfavorable. By using the same x-ray measurement assignment for the lattice spacing as before, the liquid density at 100°C and 1 atm is found to be

1.323 g/cm$^3$, compared to the experimental value 0.958 g/cm$^3$. Fleming and Gibbs ascribe the discrepancy to remanent correlation error in the SOA, rather than inherent crudeness of the lattice model itself. By implication a more accurate treatment of the statistical mechanical evaluation of $Q_l$ should improve predictions relative to experiment. In any event the SOA improves on the MFA dramatically, for the latter predicts a 30-atm vapor pressure for the liquid at 100°C, using the interaction set (6.24).

The critical parameters obtained for the SOA and energies (6.24) are

$$T_c = 739.1°C$$
$$\rho_c = 0.353 \text{ g/cm}^3 \qquad (6.25)$$
$$p_c = 478.5 \text{ atm}$$

somewhat too large in each case.

The most interesting feature of the SOA calculation is that the liquid, in coexistence with its vapor, exhibits a maximum at 61°C in the lattice-filling fraction $n = N/\mathcal{M}$. The actual mass density requires accounting for the temperature variation of lattice spacing, and as a result the density maximum becomes displaced to roughly $-50°C$. Although one would have liked the mass density to reach a maximum at 4°C to agree with measurements, the significant point to note is that the lattice model evidently possesses the general capacity to produce maxima. No doubt further refinements could achieve better agreement, since it is clear that a delicate balance of opposing influences is present.

Fleming and Gibbs calculated the specific heats $C_V$ and $C_p$ in the SOA, and found the results to be "fairly good." However, the isothermal compressibility $\kappa_T$ is too large by a factor of 2 for the liquid between 0 and 100°C, and increases with temperature instead of showing a minimum as experiment does at 46°C.

Implicit in the SOA is a specification of the mean number of nearest neighbors of each type to a typical molecule. Throughout the normal liquid range, 0 to 100°C, the total mean number of nearest neighbors is close to 6.8. This is significantly larger than the value 4.5 suggested by x-ray scattering work.[58] It would be informative to know if a more precise statistical treatment than SOA would improve this comparison.

## C. Bell's Version

Bell[59] has proposed and analyzed a lattice model for water that is closely related to that of Fleming and Gibbs. It also uses the body-centered cubic array of sites shown in Fig. 14, and each molecule has the same 24 discrete orientations as before. The major difference is that Bell's version incorporates both two *and* three-molecule interactions.

The specific pair interaction used by Bell operates only between nearest neighbors:

$$\overline{V}^{(2)}\{\xi_1, \xi_2\} = -(\varepsilon + w) \qquad \text{if } \xi_1, \xi_2 \text{ corresponds to a hydrogen bond}$$
$$= -\varepsilon \qquad\qquad \text{if no hydrogen bond exists} \qquad (6.26)$$

(This is identical to the Fleming–Gibbs interaction if $E_b = \varepsilon + w$ and $\varepsilon_1 = \varepsilon_2 = \varepsilon$.) The nonadditive three-molecule interaction operates only between compact triangles of molecules:

$$\overline{V}^{(3)}\{\xi_1, \xi_2, \xi_3\} = \tfrac{1}{3}u \qquad (6.27)$$

that is, for three simultaneously occupied sites in a 45°-45°-90° triangle whose legs connect nearest neighbors and whose hypotenuse connects second neighbors. No such compact triangles of molecules exist for the half-filled lattice in an ice Ic configuration, but the completely filled lattice has $12\mathcal{M}$ compact triangles. The incorporation of positive $u$ into the Bell version becomes the mechanism whereby interpenetrating ice VII hydrogen bond networks are thermodynamically relegated to high pressure.

Corresponding to the Fleming–Gibbs expression (6.16), Bell's lattice model partition function has the form

$$Q_l = \left(\frac{\pi^2 \tau \Lambda}{3}\right)^N \sum_{\{\xi\}} \exp\left[\beta(N_b\{\xi\}w + N_p\{\xi\}\varepsilon - \tfrac{1}{3}N_t\{\xi\}u)\right] \qquad (6.28)$$

As before, $N_b$ is the number of hydrogen-bonded pairs; $N_p$ stands for the total number of neighboring pairs of water molecules of all types

$$N_p = N_b + N_1 + N_2 \qquad (6.29)$$

The total number of compact trimers has been denoted by $N_t$.

In order to evaluate $Q_l$ for the fluid phases, Bell invokes a cluster-variation method attributed to Guggenheim and McGlashan.[60] For the present case this method self-consistently calculates the distribution of molecules over compact tetrahedral sets of four sites, one of which is shown in Fig. 15. Two of the edges of the tetrahedron connect second-neighbor sites, while the other four connect first neighbors. The body-centered cubic lattice consists of four face-centered cubic sublattices, and each of the tetrahedral clusters has one vertex belonging to each of the four. Considering the nature of the molecular interactions involved in Bell's version of the lattice model, these clusters should be sufficient to describe the most important aspects of short-range order in the fluid phases.

Bell lists 10 fundamentally distinct ways a cluster of sites can be occupied by molecules or left vacant. Each of these has a degeneracy factor $\delta_i (1 \le i \le 10)$ expressing the number of ways molecules in a given species of

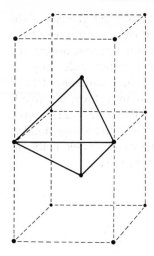

**Fig. 15**  Tetrahedral cluster of sites (outlined with solid lines) used by Bell to evaluate his lattice model partition function [see (6.28)].

cluster can be reoriented and reassigned to sites. The essence of the Guggenheim–McGlashan technique is its approximation for the number of ways $g$ that cluster species present in fractional concentrations $\Psi_i \delta_i (1 \leq i \leq 10)$ can be realized by placing molecules at the lattice sites:

$$\ln g(\Psi_1 \cdots \Psi_{10}) \cong \ln g_0 - \mathcal{M} \sum_{i=1}^{10} \delta_i \Psi_i \ln \Psi_i \qquad (6.30)$$

$g_0$ is a normalization factor which can be assigned from the known result for random molecular distribution over the lattice:

$$\ln g_0 = 3\mathcal{M} \left[ n \ln \left( \frac{n}{12} \right) + (1 - n) \ln (1 - n) \right] \qquad (6.31)$$

Associated with each of the cluster species is a characteristic interaction energy $e_i$, so that the entire system interaction energy is:

$$\mathcal{M} \sum_{i=1}^{10} e_i \delta_i \Psi_i \qquad (6.32)$$

The Helmholtz free energy $F$ then can be obtained by maximizing

$$\ln g(\Psi_1 \cdots \Psi_{10}) - \beta \mathcal{M} \sum_{i=1}^{10} e_i \delta_i \Psi_i = -\beta F - N \ln \left( \frac{\pi^2 \tau \Lambda}{3} \right) \qquad (6.33)$$

with respect to permissible variations in the $\Psi_i$ (namely, those variations consistent with their normalization, and with fixed $n$).

Thermodynamic properties were investigated for several choices of the interaction ratios $\varepsilon/w$ and $u/w$. The most significant conclusion drawn from these studies was that the anomalies of liquid-phase density maximum and compressibility minimum could be produced by a rather wide range of assignments of the interaction ratios. Furthermore, these anomalies tended to be suppressed at elevated pressure, just as they are in real water. Table II presents some results taken from Bell's paper which illustrate these features, along with the corresponding experimental numbers. The average number of nearest neighbors is $2\langle N_p\rangle/N$, and has been included in Table II for the temperature of maximum liquid density at the vapor pressure.

TABLE II

Properties Calculated by Bell[59] for His Lattice Model

| Interaction ratios | | $\dfrac{p_c \mathscr{V}_c}{k_B T_c}$ | $\dfrac{T(\max \rho)}{T_c}$ | $\dfrac{T(\min \kappa_T)}{T_c}$ | $\dfrac{2\langle N_p(\max \rho)\rangle}{N}$ | Ice lattice energy (kcal/mole) |
|---|---|---|---|---|---|---|
| $\varepsilon/w$ | $u/w$ | | | | | |
| 1/2 | 1/2 | 0.197 | 0.525 | 0.455 | 4.07 | 9.0 |
| 1 | 3/4 | 0.208 | 0.477 | 0.447 | 4.42 | 7.2 |
| 2 | 5/4 | 0.218 | 0.364 | 0.381 | 4.82 | 5.7 |
| Experiment: | | 0.243 | 0.427 | 0.494 | 4.4 | 13.4 |

The lattice energy of ice in the present model is $2(\varepsilon + w)$ per molecule. In order to fix the energy scale for each of the energy ratio cases considered in Table II, the calculated critical temperatures were fitted to the measured value. The resulting lattice energies (for cubic ice) are given in the last column of the table. Although they are compared with an experimentally determined value for hexagonal ice, the cubic-versus-hexagonal distinction should have little importance since the two polymorphs have such similar bonding geometry and density. The comparison shows that the lattice models tend to have considerably weaker binding than real ice.

The average coordination numbers look very encouraging, and seem to stay small compared to the Fleming–Gibbs result for a wide range of interaction ratios.

The cluster variation technique employed by Bell is probably more accurate at low temperatures than the Fleming–Gibbs SOA. In order to make decisive comparisons of the two lattice model versions, it would be valuable to have available a set of cluster variation calculations for the Fleming–Gibbs case.

## D. Possible Extensions

The existing calculations demonstrate that the lattice model approach is a valuable theoretical tool in understanding water. It would also be valuable to have research on this class of models widened in scope. We now list several feasible possibilities.

1. In the body-centered cubic lattice models discussed here, the melting transitions and range of crystal stability should be studied for the ice Ic and ice VII structures permitted by the models. Beside comparing the fluid-phase properties to experiment as has already been done, at least the low-pressure melting temperature of cubic ice should be required to agree roughly with that of real ice Ih. This would put an extra constraint on the interaction energy parameters.

Four long-range order parameters would be necessary in an order-disorder treatment of the crystals, corresponding to the four ways of building complete ice Ic networks on the lattice. Their introduction would complicate each of the statistical treatments somewhat, but not to the point of impossibility.

It hardly needs to be mentioned that prediction of a negative melting volume for ice is one of the attractive goals of this extension.

2. In addition to the interactions employed by Fleming and Gibbs, and by Bell, there are other versions that deserve to be tested. Most obvious is the need to rectify the short-range nature of the pair interactions, and for this purpose dipole–dipole interactions could be introduced for second neighbors at least, or perhaps perturbatively for all distances beyond first neighbors using a reasonable value of the molecular dipole moment.

With respect to nearest-neighbor molecule pairs, neither the Fleming–Gibbs nor the Bell version can be reconciled fully with the known behavior of the pair potential-energy function $V^{(2)}(\mathbf{x}_1, \mathbf{x}_2)$ as discussed in Section III.C. A more realistic alternative would involve distinguishing classes of nearest neighbors as: (a) hydrogen-bonded with all relevant angles tetrahedral; (b) single, linear hydrogen bond, but with the acceptor molecule twisted out of one of the tetrahedral arrangements; (c) unlinked by a linear hydrogen bond. The first two of these would have constant interaction energies $-\varepsilon_a$ and $-\varepsilon_b$, the former being more negative. Pairs in category (c) could be treated as interacting via electrostatic dipole moments at their centers.

Bell's three-molecule nonadditive energy [see (6.27)] may indeed be effective in destabilizing interpenetrating networks at low pressure and in holding down the mean coordination number in the liquid. However, it has doubtful validity as a representation of real water molecule interactions. More useful in the long range would be use of trimer nonadditivities only for the three doubly bonded species shown in Fig. 10, with signs and magnitudes agreeing with the available quantum mechanical studies. An inquiry of this kind could help to resolve the long-standing question how hydrogen bond cooperativity (i.e., nonadditivity) influences the microscopic state of aggregation in the liquid.[38,39]

3. We have viewed the lattice models as analogues for the classical configuration integral of the familiar rectangle rule used for numerical integration with one variable. To improve precision for the lattice models, it is desirable to identify specific ways of achieving a finer grid size than the convenient but coarse body-centered cubic lattice. One suggestion is illustrated in Fig. 16. It amounts to augmenting the body-centered cubic array to form a face-centered cubic array with 16 times as many sites per unit volume. The original pairs of sites that were nearest neighbors have become sixth neighbors under the elaboration, and are $6^{1/2}$ times as far apart as the new nearest-neighbor site distance.

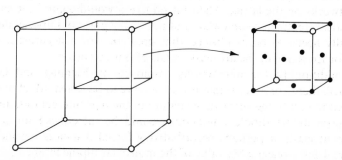

**Fig. 16** Conversion of the body-centered cubic array to a face-centered cubic array by adding new sites (solid circles). The new lattice formed in this way has a site density 16 times that of its precursor.

If the old pairs of neighboring sites in the body-centered lattice are retained as the separation for an unstrained hydrogen bond, it is obvious that several other distances are now available in the augmented lattice for describing stretched and compressed hydrogen bonds. It should also be obvious that the orientation space for a single molecule can now be more finely subdivided according to the new set of ways that OH bonds can point toward nearby sites. The greater number of interaction energies that would have to be specified with the finer grid is a conceptual virtue, but a computational vice.

Hydrogen bond polygons on the body-centered cubic lattice can have only even numbers of sides (6, 8, 10, 12, etc.). The denser face-centered lattice affords sufficient configurational flexibility to allow odd polygons (5, 7, 9, 11, etc.) to exist with relatively little bond strain. This may prove to be an important structural feature for a topologically accurate description of liquid water. In addition, a set of ice polymorphs should become possible larger than that permitted by the coarse body-centered array.

4. The conventional lattice model for simple substances has previously been quantized to provide a description of liquid helium.[61] By introducing creation and annihilation operators for each of the rotational states at a lattice site, with appropriate commutation relations, a quantized lattice model for water could also be constructed. This extension might be a valuable theoretical tool for explaining isotope effects in water and its solutions.

## VII. CELL MODELS

### A. Cell-Cluster Expansion

The principle behind the lattice model approach to evaluation of the water partition function is that a sequence of finer and finer grids enforces convergence toward the desired continuum description of configurations for the full set of molecules. The rate of convergence may be rather slow, however, since hydrogen bonds are not only strong but also markedly concentrated in small regions of relative distance and angle space for a pair of interacting molecules. That the empirically determined bond strengths are low for lattice calculations [see (6.24) and Table II] doubtless stems from the fact that they are averages over finite regions and therefore necessarily include strained configurations.

An alternative route of convergence to the full continuum representation is available, which offers several attractive advantages. This route defines the so-called cell models which attempt in a direct way to account for rapid variation in intermolecular potentials as molecules move away from regularly spaced lattice sites. As we shall see, this approach can be developed as a formally exact method by introducing a cell-cluster expansion. The general character of the cell models is such that the dynamics of the water system can also be investigated, especially in regard to short-time behavior. Of course, the lattice models do not normally possess this capacity.

The cell-cluster general development begins at the same stage as the lattice theory, by using a regular array of $\mathcal{M}$ sites in volume $\mathcal{V}$, on which molecules can adopt a number of $v$ of discrete orientations. In principle, it does not matter what geometric type of array is used, for the subsequent development is always exact. But for practical reasons one will always want to optimize the rate of convergence of the method, and this demands selection of a starting grid that permits the molecules to adopt energetically "natural" configurations relative to each other. The body-centered cubic lattice discussed at length in Section VI fulfills this requirement, and is probably the most sensible starting point for any serious cell model computations.

The basic cell-cluster identity writes the semiclassical partition function $Q$ first in terms of independent-molecule cell factors $Z^{(1)}$, and then incorporates correction factors $Y^{(n)}$ to account for errors attributable to the correlated motion of each subset of $n > 1$ molecules moving in their cells at the same time. Thus we write:

$$Q = \left(\frac{8\pi^2 \tau \Lambda}{v}\right)^N \sum_{\{\xi\}} \left[\prod_{i=1}^{\mathcal{M}} Z_i^{(1)}\{\xi\}\right]\left[\prod_{i<j=1}^{\mathcal{M}} Y_{ij}^{(2)}\{\xi\}\right]$$

$$\times \left[\prod_{i<j<k=1}^{\mathcal{M}} Y_{ijk}^{(3)}\{\xi\}\right] \cdots \left[\prod_{i_1 < \cdots < i_N = 1}^{\mathcal{M}} Y_{i_1 \cdots i_N}^{(N)}\{\xi\}\right] \qquad (7.1)$$

Here we rely on the notation used earlier for the lattice partition function $Q_l$ in (6.9). Subscripts on the $Z$s and $Y$s refer to the lattice sites involved and, by convention, if any one of these sites is unoccupied in state $\{\xi\}$, the corresponding factor should be set equal to unity.

The single-molecule cell factors $Z_i^{(1)}\{\xi\}$ are the integrals:

$$Z_i^{(1)}\{\xi\} = \frac{v}{8\pi^2 \tau} \exp(-\beta \Phi_i^{(0)}\{\xi\}) \int_{(\xi_i)} dx_1 \exp[-\beta \Phi_i^{(1)}(x_1, \{\xi\})] \qquad (7.2)$$

where $\Phi^{(0)}$ is the potential energy of the molecule (numbered 1) when it is located precisely at the lattice site (configuration $x_1^{(0)}\{\xi\}$):

$$\Phi_i^{(0)}\{\xi\} = \sum_{n=2}^{N} \frac{1}{n} \sum_{i_2 < \cdots < i_n = 2}^{N} \bar{V}^{(n)}(x_1^{(0)}\{\xi\}, x_{i_2}^{(0)}\{\xi\} \cdots x_{i_n}^{(0)}\{\xi\}) \qquad (7.3)$$

and where $\Phi^{(1)}$ is the *change* in potential energy experienced by molecule 1 as it wanders off of the lattice site and rotates:

$$\Phi_i^{(1)}(x_1, \{\xi\}) = \bar{V}_N(x_1, x_2^{(0)}\{\xi\}, x_3^{(0)}\{\xi\} \cdots x_N^{(0)}\{\xi\})$$
$$- \bar{V}_N(x_1^{(0)}\{\xi\}, x_2^{(0)}\{\xi\} \cdots x_N^{(0)}\{\xi\}) \qquad (7.4)$$

Although $\Phi_i^{(0)}$ and $\Phi_i^{(1)}$ in the strict sense depend on the status of *all* sites, it is clear that only those close to site $i$ matter substantially. The integration in (7.2) is to be carried out over a six-dimensional cell, denoted by $(\xi_i)$, corresponding to the restriction of the center of molecule 1 to the interior of the nearest-neighbor polyhedron surrounding site $i$, with orientation restricted to the $8\pi^2/v$ region belonging to the standard orientation decreed by $\xi_i$. The $Z_i^{(1)}$ have been defined to reduce to unity in the limit of vanishing interactions.

Aside from normalization the $Z_i^{(1)}\{\xi\}$ are classical configurational partition functions for single molecules moving under the influence of the intermolecular forces of fixed neighbors. In a similar way, for every $2 \le m \le N$

we can define configuration integrals:

$$Z^{(m)}_{i_1, \ldots, i_m}\{\xi\} = \frac{1}{m!}\left(\frac{v}{8\pi^2\tau}\right)^m \exp\left[-\beta(\Phi^{(0)}_{i_1}\{\xi\} + \cdots + \Phi^{(0)}_{i_m}\{\xi\})\right]$$

$$\times \int_{(\xi_{i_1})\cdots(\xi_{i_m})} d\mathbf{x}_1 \cdots d\mathbf{x}_m \exp\left[-\beta\Phi^{(m)}_{i_1, \ldots, i_m}(\mathbf{x}_1\cdots\mathbf{x}_m, \{\xi\})\right] \quad (7.5)$$

for molecules 1, 2, ..., $m$ formally attached to sites $i_1, \ldots, i_m$ by $\{\xi\}$. The appropriate multiple-cell potential energy is

$$\Phi^{(m)}_{i_1\ldots i_m}(\mathbf{x}_1\cdots\mathbf{x}_m, \{\xi\}) = \overline{V}_N(\mathbf{x}_1\cdots\mathbf{x}_m, \mathbf{x}^{(0)}_{m+1}\{\xi\}\cdots\mathbf{x}^{(0)}_N\{\xi\})$$
$$- \overline{V}_N(\mathbf{x}^{(0)}_1\{\xi\}\cdots\mathbf{x}^{(0)}_N\{\xi\}) \quad (7.6)$$

It includes interactions within the movable set of $m$ molecules, as well as their interactions with the surrounding $N - m$ rigidly fixed molecules.

The integration limits in (7.5) are such as to allow the center of each molecule 1, ..., $m$ to inhabit *all* $m$ cells $i_1, \ldots, i_m$. If all surrounding cells are filled, in the noninteracting limit,

$$Z^{(m)}_{i_1\cdots i_m}\{\xi\} \to \frac{m^m}{m!} \quad (7.7)$$

However, if there are vacant sites near $i_1, \ldots, i_m$, we append the restriction that the minimum sum of squares of distances for the $m$ molecules to sites unoccupied by fixed particles be attained *only* with sites $i_1, \ldots, i_m$, and not with the inclusion of local vacant sites.

If sites $i$ and $j$ are widely separated, the motion of molecules in the respective cells will be independent. With cells small enough that multiple occupation is unlikely on energetic grounds, $Z^{(2)}_{ij}$ then factors into $Z^{(1)}$s:

$$Z^{(2)}_{ij}\{\xi\} \sim Z^{(1)}_i\{\xi\}Z^{(1)}_j\{\xi\} \quad (7.8)$$

The cell pair correction term $Y^{(2)}_{ij}$ appearing in the general cell-cluster expansion (7.1) is simply a ratio whose deviation from unity measures the extent to which factorization is inappropriate:

$$Y^{(2)}_{ij}\{\xi\} = \frac{Z^{(2)}_{ij}\{\xi\}}{Z^{(1)}_i\{\xi\}Z^{(1)}_j\{\xi\}} \quad (7.9)$$

Three-cell correction factors are also defined by appropriate ratios of $Z$s:

$$Y^{(3)}_{ijk}\{\xi\} = \frac{Z^{(3)}_{ijk}\{\xi\}Z^{(1)}_i\{\xi\}Z^{(1)}_j\{\xi\}Z^{(1)}_k\{\xi\}}{Z^{(2)}_{ij}\{\xi\}Z^{(2)}_{ik}\{\xi\}Z^{(2)}_{jk}\{\xi\}}$$

$$\equiv \frac{Z^{(3)}_{ijk}\{\xi\}}{Z^{(1)}_i\{\xi\}Z^{(1)}_j\{\xi\}Z^{(1)}_k\{\xi\}Y^{(2)}_{ij}\{\xi\}Y^{(2)}_{ik}\{\xi\}Y^{(2)}_{jk}\{\xi\}} \quad (7.10)$$

These factors differ from unity only when all three sites $i$, $j$, and $k$ are close together, for if one of them (say, $k$) recedes from the other two,

$$Z_{ijk}^{(3)}\{\xi\} \sim Z_{ij}^{(2)}\{\xi\}Z_k^{(1)}\{\xi\} \qquad (7.11)$$

The higher-order correction factors $Y^{(m)}$ have a similar structure. In each case $Z^{(m)}$ is divided by $Y$s of lower order for every proper subset of sites that can be formed out of the full set of $m$ given sites:

$$Y_{i_1\cdots i_m}^{(m)}\{\xi\} = \frac{Z_{i_1\cdots i_m}^{(m)}\{\xi\}}{\prod_{\mu=1}^{m-1}\sum_{\alpha_1\cdots\alpha_\mu\varepsilon i_1\cdots i_m} Y_{\alpha_1\cdots\alpha_\mu}^{(c)}\{\xi\}} \qquad (7.12)$$

where for notational convenience we have set

$$Y_i^{(1)}\{\xi\} \equiv Z_i^{(1)}\{\xi\} \qquad (7.13)$$

The recursive linear relations between logarithms of the $Z$s on the one hand, and the logarithms of the $Y$s on the other hand, are isomorphous to the identities (3.2) to (3.4) relating total potentials $V_n$ to the component potentials $V^{(n)}$. That (7.1) is in fact an identity results from thorough cancelation of factors between the $Y$s for each term of the occupation state sum (over $\{\xi\}$) in (7.1). In fact, the recursive quotient definitions (7.12) for cell correlation factors reduce (7.1) to the elementary form

$$Q = \left(\frac{8\pi^2\tau\Lambda}{v}\right)^N \sum_{\{\xi\}} Z_{i_1\cdots i_N}^{(N)}\{\xi\} \qquad (7.14)$$

which simply resolves $Q$ into separate contributions from distinct modes of occupation of the set of $v\mathcal{M}$ six-dimensional cells by molecules.

Provided the number of lattice sites $\mathcal{M}$ is no less than the number of molecules $N$, the precise value of $\mathcal{M}$ can in no way affect the validity of cell-cluster expansion (7.1); it is rigorously correct independently of the choice of $\mathcal{M}$. But just as an optimal choice exists for the lattice structure itself from the standpoint of convergence rate, there ought to be a "best" value for $\mathcal{M}$. A reasonable choice would be that $\mathcal{M} = \mathcal{M}_1$, which minimizes the free energy (i.e., maximizes $Q$) in the single-cell approximation

$$Q \cong \left(\frac{8\pi^2\tau\Lambda}{v}\right)^N \sum_{\{\xi\}} \prod_{i=1}^{\mathcal{M}} Z_i^{(1)}\{\xi\} \qquad (7.15)$$

On restoring just the pair factors $Y^{(2)}$, a slightly shifted $\mathcal{M} = \mathcal{M}_2$ would minimize the resulting more accurate free energy. Similarly $\mathcal{M}_3$, $\mathcal{M}_4$, and so on, would give minima after restoring triplet, quadruplet, and so on, correction factors, but it must be the case that the free energy near these successive minima becomes a flatter and flatter function of $\mathcal{M}$. This raises an interesting question, yet to be answered, for the general cell-cluster theory of the liquid state: Under what conditions (if any) on the underlying lattice structure does the sequence of formal vacancy concentrations

$$\frac{\mathcal{M}_1 - N}{\mathcal{M}_1}, \frac{\mathcal{M}_2 - N}{\mathcal{M}_2}, \frac{\mathcal{M}_3 - N}{\mathcal{M}_3}, \dots \tag{7.16}$$

converge to a limit with respect to increasing order of included cell correlation? The existence of the limit would define precisely the "holes" in the liquid state.

## B. Weres–Rice Cell Theory

For simple fluids and their mixtures, calculations falling under the general heading "cell theory" have had a long history.[62] By contrast, analogous calculations for water are sparse, and a recent phenomenon. Weissmann and Blum[63] have reported single-molecule partition functions for cells in an expanded (but perfect) ice lattice, and have suggested that the results might be relevant to the liquid state. Nevertheless, the only concerted attack on liquid water using cell theory formalism, and including a realistic assessment of local molecular order, has been published by Weres and Rice.[64] This section outlines their work.

Weres and Rice have based their calculations on an approximate effective pair potential for water molecules that was proposed by Ben-Naim and Stillinger[40] (BNS). This BNS potential $v$ utilizes a spherically symmetric short-range part $v_{LJ}$, plus an angular portion designed to describe formation of linear hydrogen bonds at a small separation:

$$v(\mathbf{x}_1, \mathbf{x}_2) = v_{LJ}(R_{12}) + S(R_{12})v_{el}(\mathbf{x}_1, \mathbf{x}_2) \tag{7.17}$$

The distance $R_{12}$ is measured between the oxygen nuclei. Each water molecule is regarded as a symmetric tetrad of point charges (two are $+q$, and the other two $-q$), as shown in Fig. 17. The positive charges represent

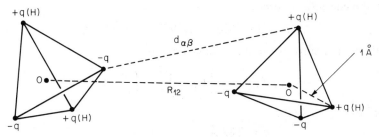

**Fig. 17** Molecular charge tetrahedra used in the BNS effective pair potential. The point charges $\pm q$ are symmetrically located 1 Å from the oxygen nucleus. Only one of the 16 charge pair distances $d_{\alpha\beta}$ used in the potential has been shown explicitly.

partially shielded protons, and the negative charges simulate unshared pairs of electrons at the back side of the molecule. All four charges are precisely 1 Å from the oxygen nucleus. The electrostatic interactions of the 16 pairs of point charges, one in each molecule, make up $v_{el}$:

$$v_{el}(\mathbf{x}_1, \mathbf{x}_2) = q^2 \sum_{\alpha, \beta = 1}^{4} \frac{(-1)^{\alpha+\beta}}{d_{\alpha\beta}(\mathbf{x}_1, \mathbf{x}_2)} \tag{7.18}$$

Here $\alpha$ and $\beta$ index charges (with odd indices belonging to negatives and even indices to positives), and the $d_{\alpha\beta}$ are the appropriate distances. The tendency for tetrahedron vertices with opposite charge signs to align is the mechanism for describing successive linear hydrogen bonds at the tetrahedral angle $\theta_t$. This idea by itself is old, due originally to Bjerrum in his studies of ice,[65] but the BNS application involved for the first time a modulation function $S(R_{12})$ which switches off the electrostatic interactions before the molecules become so close that a charge overlap catastrophe ($d_{\alpha\beta} = 0$) can occur. Specifically, $S$ is a cubic spline function:

$$
\begin{aligned}
S(R_{12}) &= 0 & 0 \leq R_{12} \leq R_L \\
&= \frac{(R_{12} - R_L)^2(3R_U - R_L - 2R_{12})}{(R_U - R_L)^3} & R_L \leq R_{12} < R_U \\
&= 1 & R_U \leq R_{12}
\end{aligned}
\tag{7.19}
$$

with singular points

$$
\begin{aligned}
R_L &= 2.0379 \text{ Å} \\
R_U &= 3.1877 \text{ Å}
\end{aligned}
\tag{7.20}
$$

The original strength parameterization for the BNS potential required a rough simultaneous fit to both ice and water vapor properties. Subsequent detailed studies using molecular dynamics (see Section VIII) have indicated that a renormalization of the original function with multiplier 1.06 throughout materially improves its overall accuracy.[66, 67] After this renormalization the charge $q$ becomes

$$q = 0.19562e = 0.93952 \times 10^{-10} \text{ esu} \tag{7.21}$$

while the Lennard–Jones short-range function

$$v_{LJ}(R_{12}) = 4\varepsilon\left[\left(\frac{\sigma}{R_{12}}\right)^{12} - \left(\frac{\sigma}{R_{12}}\right)^{6}\right] \tag{7.22}$$

requires

$$
\begin{aligned}
\sigma &= 2.82 \text{ Å} \\
\varepsilon &= 5.3106 \times 10^{-15} \text{ erg} \\
&= 7.6472 \times 10^{-2} \text{ kcal/mole}
\end{aligned}
\tag{7.23}
$$

With these values the minimum of the BNS potential is achieved with a linear hydrogen bond ($R_{12} = 2.760$ Å) at energy $-6.887$ kcal/mole.

The Weres–Rice calculation has essentially been carried out with approximation (7.15), that is, with neglect of all cell deviation correlation factors $Y^{(n)}$, $n \geq 2$. The single-cell factors $Z^{(1)}$ were distinguished by the occupation state of the octahedral grouping of 14 first- and second-neighbor sites only; the influence of molecules beyond the second-neighbor shell was taken into consideration by means of a dielectric cavity assumption. The cell-model calculation therefore breaks down into three major parts. The first requires classification of the states of the basic 15-site cluster and construction of an entropy expression for a general set of concentrations for these cluster species. The second part involves evaluation of the distinct $Z^{(1)}$, using the BNS interaction. Finally, the third task requires minimization of the free energy with respect to cluster species concentrations to obtain their equilibrium values.

Just as Fleming and Gibbs had done for the lattice model, Weres and Rice varied the nearest-neighbor spacing linearly with temperature to maintain agreement with x-ray scattering experiments.[58] Specifically, the spacing varies from 2.82 Å at 0°C to 2.88 Å at 100°C.

In accord with Stevenson's spectroscopic evidence that the number of water molecules in the liquid with zero or one hydrogen bond must be small,[68] all $Z^{(1)}$ were required to involve two, three, or four hydrogen bonds to nearest neighbors. Even with this restraint there remained 374 distinct "neighbor environments" to be included in the calculations.

A rather elaborate argument is provided in Ref. 64 for the relevant combinatorial factor $g$, giving the number of ways the cells can be fitted together to form the system. We shall not reproduce that argument here, except to state the result which consists of six contributing factors:

$$g = \prod_{j=1}^{6} g_j \tag{7.24}$$

The first factor gives the number of ways to distribute $N$ molecules over $\mathscr{M}$ sites:

$$g_1 = \frac{\mathscr{M}!}{N!(\mathscr{M} - N)!} \tag{7.25}$$

The second involves the number of hydrogen bonds present in the system, denoted by $NP_h$, and gives the number of ways that they may be distributed over the $4Nn$ nearest-neighbor molecular pairs expected in a random arrangement of molecules ($n = N/\mathscr{M}$):

$$g_2 = \frac{(4Nn)!}{(NP_h)![N(4n - P_h)]!} \tag{7.26}$$

The next factor is present to account for the fact that a random distribution of hydrogen bonds frequently causes molecules to "have too few or too many hydrogen bonds or incorrect angles between the hydrogen bonds";[64] specifically,

$$g_3 = \left(\frac{P_h}{4}\right)^{2NP_h}\left(\frac{1-n}{1-(P_h/4)}\right)^{N(4-2P_h)} \tag{7.27}$$

Since molecules have two, three, or four hydrogen bonds, with respective fractions $P_2$, $P_3$, and $P_4$ (note that $2P_h = 2P_3 + 3P_3 + 4P_4$), it is necessary to account for the multiplicity of ways that these bonding species may be assigned to the $N$ occupied sites:

$$g_4 = \frac{N!}{(NP_2)!(NP_3)!(NP_4)!} \tag{7.28}$$

Taking into consideration the angularly acceptable arrangements of two, three, or four hydrogen bonds about a given central site (numbering 12, 8, and 2, respectively), it is also necessary to include the factor

$$g_5 = 12^{NP_2}8^{NP_3}2^{NP_4} \tag{7.29}$$

Finally, there is the analog, for the random hydrogen bond network, of Pauling's degeneracy factor for ice,[69] giving the number of ways that protons may be distributed along hydrogen bonds asymmetrically so as to leave molecules intact:

$$g_6 = \left(\frac{6}{2^{P_h}}\right)^N \tag{7.30}$$

Several approximations were used to evaluate the cell factors $Z^{(1)}$. We have already mentioned that neighbors beyond the second-neighbor shell were treated as a dielectric continuum. Most importantly, the cell potentials $\Phi^{(1)}$ were replaced by a quadratic form (calculated using the BNS function) about the stable hydrogen-bonding configuration in such a way that translational and librational motions were dynamically independent harmonic oscillators. Furthermore, the relevant potential-energy second derivatives were averaged over the two relative orientations permitted to hydrogen-bonded neighbors by the body-centered cubic lattice of sites. In this locally harmonic approximation, only that subset of the $v$ ($=24$) standard molecular orientations was considered that led to the maximum number of hydrogen bonds possible with the given arrangement of nearest neighbors.

Neighbors in both the first and second coordination shells that do not hydrogen-bond to the central molecule were treated as linear perturbations (with magnitudes provided by the BNS potential) whose effects could be added, as appropriate, to the cell harmonic oscillator partition functions.

Because linear lattice expansion with temperature was used, the calculated potential-energy curvatures and harmonic oscillator frequencies had to be calculated separately at each temperature of interest.

To compensate partially for neglect of the cell distortion correlation factors $Y^{(2)}$, $Y^{(3)}$, and so on, Weres and Rice elected to include a "communal" entropy of vibrational origin. For the diamond lattice a complete phonon calculation[70] yields 0.391 entropy units more than the comparable Einstein approximation for single-particle motion. To scale this result to a random network in which only a fraction $P_h/2$ of the maximum possible number of hydrogen bonds is present, $0.391(P_h/2)$ entropy units were added to the free energy of the independent cells.

The translational and librational frequencies generated in this calculation were sufficiently high that Weres and Rice felt obliged to use quantum mechanical, rather than classical, partition functions for them. This was done in spite of the fact that the BNS interaction was originally devised within the regime of classical statistical mechanics alone.[40]

When the librational frequencies were calculated for the cubic ice structure, they were found to be considerably higher than librational bands measured for ice Ih (the cubic form should be very similar in this respect). Weres and Rice thus concluded that the BNS interaction was too highly curved in the directions of libration about linear hydrogen bonds, and they suggested that this deficiency could be rectified in an ad hoc fashion simply by scaling the BNS librational curvatures downward with a factor 0.458. Consequently, two parallel sets of calculations were carried out, one set using the BNS interaction directly, and the other set using the "curvature-rescaled BNS" with lower librational frequencies.

The effective pair potential is based on the vibrationally averaged component potentials $\overline{V}^{(n)}$. As (4.8) shows, the latter functions (and thus also the former by implication) include effects arising from shifts in vibrational frequencies due to bonding. However, Weres and Rice felt that, because the BNS effective pair potential was originally derived on a purely classical basis, its use for their quantized cell oscillators should be accompanied by a compensating inclusion of extra binding energy due to intramolecular vibration frequency shifts. Using measured vapor- and liquid-phase frequencies, they calculated an extra 0.900 kcal/mole binding energy for the liquid at its melting point, to be added into the final cell approximation results. At the boiling point 0.694 kcal/mole is the corresponding value.

Minimization of the cell model free energy was carried out by a gradient descent technique in the multidimensional space of species fractions for the various permitted cell neighbor types. A constraint was applied to this minimization, to the effect that the various cell species concentrations should display a ratio of 4:3 for the average number of first to second neighbors.

This is the same as the ratio of numbers of sites in the first and second coordination shells (eight and six, respectively). Unfortunately, this constraint seems to inhibit the proper distribution of vacant sites throughout the system. In particular, this ratio is not applicable for the first and second neighbors in the ice Ic structure.

Over the normal liquid range 0 to 100°C, the computed Gibbs free energy, and its component enthalpy and entropy, showed the proper temperature trends and roughly the correct magnitudes. Use of the librationally rescaled BNS interaction tended to reduce the error in comparison with experiment by a factor of 2 at all temperatures. Table III, from Ref. 64, shows the various contributions to the enthalpy and entropy at 0°C and 100°C, and compares their totals to the respective experimental magnitudes. In this table the standard state for enthalpy is infinitely dilute vapor at absolute zero, while for entropy it is that of ice at absolute zero with frozen-in Pauling disorder $Nk_B \ln \frac{3}{2}$ and with nuclear spins disregarded.

Since empty cells were not explicitly considered in this calculation, the liquid density was assumed to be the same as the average density over the

TABLE III

Thermodynamic Properties Calculated by Weres and Rice for Liquid water, using the Uncorrelated-Distortion cell approximation.[a]

| Property | 0°C | 100°C |
|---|---|---|
| Enthalpy (kcal/mole) | | |
|   Lattice ($\Phi^{(0)}$) | −8.296 | −8.088 |
|   Translational | 1.729 | 2.290 |
|   Librational | 2.517 | 2.858 |
|   Nonbonded neighbors | −1.325 | −1.271 |
|   Long-range (diel. approximation) | −1.190 | −1.166 |
|   Intramolecular zero point | −0.900 | −0.694 |
|     Total | −7.465 | −6.071 |
|     Experimental total | −8.594 | −6.791 |
| Entropy (entropy units) | | |
|   Configurational | 4.48 | 4.48 |
|   Orientational | 1.70 | 1.69 |
|   Translational | 7.12 | 9.28 |
|   Librational | 2.00 | 3.44 |
|   Nonbonded neighbors | −1.17 | −0.95 |
|   Vibrational | 0.26 | 0.26 |
|     Total | 14.39 | 18.20 |
|     Experimental total | 15.17 | 20.79 |

[a] The librationally rescaled BNS interaction was employed. (This tabulation was copied from Ref. 64, Table II.)

first and second coordination shells. On a per-site basis, this density remains nearly constant throughout the temperature range of the liquid, with a value of 0.587. Taking into account the lattice expansion, this leads to mass densities which agree with observation to 1 % or better. The greatest error occurs at the lowest temperature, and the cell calculation has a monotonically decreasing density with temperature with no hint of a density maximum.

The average number of hydrogen bonds per molecule remains equal to about 1.35 from 0 to 100°C, and the mean number of nearest neighbors remains 4.7. The latter result is consistent with x-ray scattering measurements.[58] The respective fractions of doubly, triply, and quadruply hydrogen-bonded molecules stay constant at 0.46, 0.38, and 0.16, respectively. Each of these results concerning local structure in the liquid refers specifically to the rescaled BNS interaction. Taken together they imply that very little structural reorganization (beyond lattice expansion) takes place in the liquid between 0 and 100°C. However, this conclusion might perhaps be regarded with some suspicion, since the structurally sensitive heat capacity $C_p$ comes out of the calculations too low (12 cal/mole deg at 0°C instead of the 18 cal/mole deg measured).

Weres and Rice utilized their rescaled BNS cluster concentrations as input for a calculation of the amorphous-medium translational frequency spectrum, according to a method developed by Weres.[71] Strictly speaking, this goes beyond the scope of cell theory itself, but does test the predicted medium structure. At 10°C the frequency spectrum (i.e., the mode density) has two prominent peaks, at 70 cm$^{-1}$ and at 190 cm$^{-1}$, which compare well with the broad bands observed in Raman, infrared, and neutron spectroscopy[72] at 60 cm$^{-1}$ and at 170 cm$^{-1}$.

This cell-model calculation was carried out with a large number of approximations, some of which have uncertain numerical effects. However, the general approach followed seems sound, and at the very least has established that cell-model calculations for water are both feasible and capable of producing nontrivial and interesting results. It would be valuable for the developing field of water theory to exploit cell theory further, using a systematic series of modifications building on the Weres–Rice work. Some possibilities are:

1. The BNS interaction has been superceded by a more accurate effective pair potential (denoted by ST2; see Section VIII). This revised potential has a considerably lower curvature for vibrational motions, and should reduce or eliminate the need for a rescaling operation.

2. The constraint of equal occupation probabilities for first and second coordination shell sites should be removed.

3. A more accurate combinatorial factor $g$ should be sought, perhaps along the lines established by Guggenheim and McGlashan.[60]

4. The equilibrium density should be determined by minimizing free energy with respect to vacancy concentration, rather than by identifying it with the average density over first and second coordination shells. This would be particularly important if the cell model were extended to describe the critical-point region of water, at which local density fluctuations become especially important. Considering the lattice-model results (Section VI), it would not be very surprising if the cell model then produced a liquid-phase density maximum.

5. Since the most frequent number of hydrogen bonds involving the cell molecule turned out to be two, it seems reasonable to expect that singly bonded configurations, and perhaps even unbonded ones, ought to have been permitted at the outset. If Stevenson's[68] ideas are correct, that these species have a low concentration in the liquid, an accurate cell-model calculation should lead to the same conclusion automatically.

6. The assumption of harmonic (and independent) translational and librational motions should be relaxed. Perhaps the requisite cell integrals could be calculated accurately by purely numerical means in the classical limit, and then quantum corrections (as a power series in Planck's constant) explicitly appended.

7. At least a few cell distortion correlation factors $Y^{(2)}$ and $Y^{(3)}$ could be evaluated classically by the Monte Carlo technique[73] to estimate their importance in the full cell cluster development (7.1).

Denley and Rice[74] recently used the configurational results of the Weres–Rice cell theory as a starting point for calculation of the intramolecular frequency spectrum in liquid water. Although they were required to introduce partially untested force-field assumptions, the results seem to account qualitatively for the observed spectra. It will ultimately be rewarding to see if the suggested improvements in the underlying cell theory create major changes in the predicted intramolecular spectra.

## VIII. MOLECULAR DYNAMICS SIMULATION

### A. Techniques of Computer Simulation

The lattice theory, and the cell theory at the dynamically uncorrelated level, have an attractive appeal due to the simple configurational descriptions of local molecular order that they introduce. If they could be followed through to exact solutions, they would provide compelling and remarkably vivid guides to intuition, thus largely satisfying the human urge to "understand" water.

Unfortunately, it is not possible to solve exactly lattice or cell theories, of the type we have encountered, in three dimensions. The results that have been obtained rest on simplifying statistical mechanical approximations, and thus inevitably convey uncertainty. This situation also applies to any other analytical approach to understanding liquid water, such as the integral equation method for predicting molecular distribution functions.[75]

It is therefore fortunate that the complementary technique of direct computer simulation affords a viable and fertile alternative. This is not to imply that computer simulation methods are free of difficulty, for they are restricted to relatively small aggregates of molecules ($N$ in the range $10^2$ to $10^3$), and to classical statistical mechanics at present. However, it is a general rule that the precision and the range of detail available in results of computer simulation for liquids far exceed those of other theoretical methods.

Two distinct simulation techniques have been developed, initially to describe simple liquids. Historically, the Monte Carlo method appeared first.[76] Given the appropriate intermolecular potential, it is designed to generate a large number of system configurations, distributed canonically according to preset values for the temperature and density. The collection of configurations then provides the basis over which arbitrary static properties (energy, virial, fluctuation quantities like $C_V$ and $\kappa_T$, etc.) can be computed as suitable averages. Although some Monte Carlo work relevant to liquid water has been published,[29,77,78] its sum total at present is not very extensive, and therefore is not representative of the inherent capacity of the method to characterize the liquid in depth and detail. Consequently, we concentrate attention instead on the other simulation option.

The molecular dynamics method relies on a powerful digital computer to solve the classical equations of motion for the molecular aggregate, subject to suitable initial and boundary conditions. The temporal evolution of the molecular system is recorded in the course of solving these equations and, unlike the Monte Carlo method, this technique permits the calculation of kinetic properties, such as the self-diffusion constant, rotational relaxation, neutron inelastic scattering, and so on. A single molecular dynamics "run" is representative of a microcanonical ensemble, since total energy is a constant of the motion. But since the canonical (Monte Carlo) and microcanonical (molecular dynamics) equilibrium ensembles are equivalent in the large-system limit, barring first-order phase transitions, both are equally valid sources of structural and thermodynamical information. It is in its capacity to describe molecular motions and irreversible phenomena that the molecular dynamics approach enjoys a major advantage over the Monte Carlo approach.

The first use of molecular dynamics, by Alder and Wainwright,[79] involved spherical structureless particles. In this case the dynamical evolution is prescribed by the Newton equations for each of the $N$ particles with mass $m$

$$\mathbf{F}_j = m\mathbf{a}_j \qquad 1 \leq j \leq N \tag{8.1}$$

which link acceleration $\mathbf{a}_j$ to the total vector force $\mathbf{F}_j$ on each particle $j$ due to all others. In the case of water, the simplest realistic version of molecular dynamics treats each molecule as a rigid asymmetric rotor capable of simultaneous translation and rotation. As a result, it is necessary to supplement (8.1) for the center-of-mass motion by Euler equations[80] for the angular velocities $\boldsymbol{\omega}_j$ in terms of the torques $\mathbf{N}_j$:

$$I_1\dot{\omega}_{jx} - \omega_{jy}\omega_{jz}(I_2 - I_3) = N_{jx}$$
$$I_2\dot{\omega}_{jy} - \omega_{jz}\omega_{jx}(I_3 - I_1) = N_{jy} \qquad (8.2)$$
$$I_3\dot{\omega}_{jz} - \omega_{jx}\omega_{jy}(I_1 - I_2) = N_{jz}$$

Here the Cartesian coordinates are affixed to molecule $j$, diagonalizing the inertial moment tensor $I$ so that $I_{xx} = I_1$, and so on.

Normally, molecular dynamics calculations are carried out with periodic boundary conditions, the unit cell having dimensions fixed by the density of interest. The resulting absence of real boundaries produces an optimal situation for observing bulk water properties. However, this choice need not be the case and, in fact, deliberate insertion of "walls" or of crystal surfaces whose forces and torques appear in the dynamical equations (8.1) and (8.2) would be the means for studying interfacial water.

Temperature is implicitly determined by the amount of total energy given to the system as initial momentum and position data. In the long run the translational and rotational kinetic energies are equipartitioned between molecules, with the well-known mean values

$$\langle \tfrac{1}{2}mv_j{}^2 \rangle = \langle \tfrac{1}{2}\boldsymbol{\omega}_j \cdot I_j \cdot \boldsymbol{\omega}_j \rangle = \tfrac{3}{2}k_B T \qquad (8.3)$$

Molecular dynamics calculations must perforce span a limited time interval. The statistics of fluctuations separately for the translational and rotational terms in (8.3) offers one means of monitoring the quasi-ergodicity of the calculation.

In application to liquid argon, accurate and stable numerical integration of the equations of motion is possible using discrete time increments of $10^{-14}$ sec.[81] This interval is related to the magnitude of the molecular accelerations present in the liquid, and a longer time increment could be used to good practical advantage if the argon atoms interacted more weakly. The situation is quite the opposite in water, however, for the hydrogen-bonding interactions are very strong and highly directional. In conjunction with the small inertial moments possessed by water molecules, this characteristic requires that time increments in the neighborhood of $10^{-16}$ sec be used,[66] making the simulation of water a significantly more arduous task than that of argon. For a modest number of rigid water molecules, the time-dilation factor for a powerful digital computer between the absolute time interval for the molecules on the one hand, and the much slower running time on the computer on the other hand, would be about $10^{16}$. To carry the water sample forward in time by 1 sec would require $3 \times 10^8$ years. Fortunately, most kinetic phenomena of interest in liquid water fall into the picosecond range or shorter, making them fully accessible to the molecular dynamics technique.

Both the Monte Carlo and molecular dynamics simulation methods have frequently been called computer experiments. It is difficult to know if this phrase is offered as a profoundly edifying classification, or as a value judgment. In either event it fails to illuminate. The basic distinction between experiment and theory is that the former manipulates and observes real matter in the laboratory, while the latter constructs algorithms and theorems which may be esthetically pleasing in themselves but which encode numerical operations with varying efficiencies. An exact closed-form solution for the three-dimensional Ising model partition function, for example, would be a remarkable achievement for a variety of reasons, not the least of which would be the rule it gives for high-precision numerical tabulation of the thermodynamical properties for the model at all temperatures. Looked at in this light, it is obvious that the Monte Carlo and molecular dynamics algorithms ought more properly to be classified as computer theory rather than computer experiment.

Only selected portions of the published molecular dynamics work can be covered in this review. The reader may wish to check the cited articles for more detail.

## B. ST2 Interaction

The initial studies of liquid water via molecular dynamics[66,67] used the BNS effective pair potential defined earlier in (7.17) to (7.23). This potential assigns specific positions to the oxygen and hydrogen nuclei (bond length 1 Å and bond angle $\theta_t$), so the inertial moments $I_1$, $I_2$, and $I_3$ are determined completely by the atomic masses. The results of those initial molecular dynamics calculations were very encouraging, but suggested that the BNS potential was too tetrahedral, that is, its hydrogen bonds were too directional. This presumption was supported by the Weres and Rice observation[64] that librational frequencies were much too high for the BNS potential.

Consequently, a "second-generation" effective pair potential was devised[82] to mitigate the difficulty. The new interaction is called ST2, and represents a conservative modification of its predecessor. It has the same generic form as before:

$$v(\mathbf{x}_1, \mathbf{x}_2) = v_{LJ}(R_{12}) + S(R_{12})v_{el}(\mathbf{x}_1, \mathbf{x}_2) \qquad (7.17)$$

with a short-range Lennard–Jones 12-6 part $v_{LJ}$ and a modulated electrostatic portion $Sv_{el}$ again based on charge tetrads in each molecule. The primary geometric change involved in BNS → ST2 is that the two negative point charge ($-q$ in Fig. 17) have been drawn inward from 1 to 0.8 Å, measured along the tetrahedral rays from the oxygen nucleus. The charge tetrahedra thus have lower symmetry than before (the OH bond lengths are still 1 Å, and the angles about the O are still $\theta_t$), but remain consistent with the molecular $C_{2v}$ symmetry.

Changes in the parameters of course were required. For ST2 the Lennard–Jones parameters are

$$\sigma = 3.10 \text{ Å}$$
$$\varepsilon = 5.2605 \times 10^{-15} \text{ erg}$$
$$= 7.5750 \times 10^{-2} \text{ kcal/mole} \qquad (8.4)$$

while the point charges to be used in (7.18) for $v_{el}$ are

$$q = 0.2357e = 1.13194 \times 10^{-10} \text{ esu} \qquad (8.5)$$

The modulation function $S$ has the same cubic spline form (7.19) as before, but now the singular points are

$$R_L = 2.0160 \text{ Å}$$
$$R_U = 3.1287 \text{ Å} \qquad (8.6)$$

The absolute minimum of the ST2 potential is achieved in a mirror-symmetric configuration for the dimer as illustrated in Fig. 4. The displacement angle $\alpha$ of the donating OH (see Fig. 4) is only 1.1°, so the hydrogen bond is virtually linear. The separation $R_{12}$ of the oxygens is 2.852 Å, and at this minimum $v(\text{ST2})$ is $-6.839$ kcal/mole.

It is interesting to trace out the *constrained* minimum for the ST2 potential for fixed oxygen–oxygen separation $R_{12}$. When this distance exceeds 4.964 Å $= R_c$, the molecular symmetry axes are rigorously collinear (the molecular planes are perpendicular as in Fig. 4 for all $R_{12}$). This collinearity reflects the predominant influence of dipole–dipole interactions at these large separations. However, the interactions of higher multipoles succeed in producing an instability at $R_c$, wherein mutual twisting of the molecules at smaller $R_{12}$ leads to configurations more in accord with linear hydrogen bonding. At $R_c$ the dimer can go in either of two ways toward a linear hydrogen bond, depending on which OH bond of the donor molecule begins to rotate toward the acceptor oxygen. Thus $R_c$ is a critical point at which a spontaneous symmetry breaking arises. This point for the constrained energy surface is analogous to the critical point of the free-energy surface (versus magnetization) for a field-free Ising model. In the latter example the spontaneous magnetization also represents a broken symmetry. With respect to water potentials, the existence of a critical separation $R_c$ is not unique with the ST2 interaction; the BNS interaction has qualitatively the same behavior, and so too should the exact water potentials $V^{(2)}(\mathbf{X}_1, \mathbf{X}_2)$, $\overline{V}^{(2)}(\mathbf{x}_1, \mathbf{x}_2)$, and $v(\mathbf{x}_1, \mathbf{x}_2)$.

## C. Nuclear Pair Correlation Functions

Many of the important characteristics of short-range molecular order in liquids are conveniently portrayed in the nuclear pair correlation functions. For water there are three, $g_{OO}(r)$, $g_{OH}(r)$, and $g_{HH}(r)$. They give the probability, relative to random expectation, of the occurrence of distance $r$ as a separation between pairs of nuclei of the subscripted species. The conventional normalization requires ($\mu$, $v$ = O or H)

$$\lim_{r \to \infty} g_{\mu v}(r) = 1 \tag{8.7}$$

in the infinite system limit. These correlation functions obviously represent integral contractions of the pair distribution function $\rho^{(2)}$ introduced earlier; for example,

$$g_{OO}(r) = (32\pi^3 \rho^2 r^2)^{-1} \int d\mathbf{x}_2 \, \delta(R_{12} - r)\rho^{(2)}(\mathbf{x}_1, \mathbf{x}_2) \tag{8.8}$$

where $R_{12}$ is the distance between oxygens.

Figure 8 shows the $g_{OO}(r)$ determined by molecular dynamics for a sample of simulated water[82] (using the ST2 potential) at 10°C and mass density 1 g/cm. The calculation involved $N = 216$ molecules confined to a cube with edge length 18.62 Å, to which periodic boundary conditions applied. For comparison, Fig. 18 also shows a $g_{OO}(r)$ result inferred by Narten[83] from x-ray diffraction intensities for water at 4°C (the small temperature difference is negligible for present purposes). The dynamical simulation spanned 8.1 psec, and for convenience was carried out with neglect of molecular interactions for pairs of molecules having their oxygens more than 8.46 Å apart.

The main features of the two curves agree well. The positions at which the prominant first peak occurs differ by no more than 0.01 Å, and the broad successive maxima exhibit nearly equal positions as well. The experimental curves usually show some short-wavelength ripples, and that in Fig. 18 is no exception. If it is justifiable to consider those ripples artifacts of the experimental data processing, the agreement of the molecular dynamics curve with experiment at the first minimum of $g_{OO}(r)$ (near 3.5 Å) may be somewhat better than Fig. 18 seems to indicate.

The most important structural theme carried by these $g_{OO}(r)$ curves is the persistence of tetrahedral icelike order into the liquid phase. Although the second maxima around 4.5 Å are broad, they occur at about the correct multiple (1.633) of the first peak distance to represent second neighbors

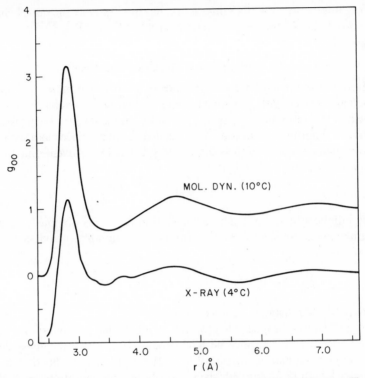

**Fig. 18** Comparison of oxygen nucleus pair correlation functions for liquid water. The molecular dynamics result[82] was based on 216 molecules at mass density 1 $g/cm^3$ interacting through the ST2 effective pair potential. The x-ray diffraction result is due to Narten.[83]

connected along a path of two hydrogen bonds at an angle which on the average is $\theta_t$. In ice Ih or Ic, the second neighbors of course produce a narrow peak; the breadth manifested in the liquid phase indicates frequent and considerable strain.

The molecular dynamics first-neighbor peak is higher and narrower than its experimental counterpart. For the ST2 model the molecular dynamics curve in Fig. 18 is an accurate determination. The experimental curve for real water is probably a less precise determination, considering the interpretive ambiguities one is forced to accept in analyzing the experimental data. Even so, the shape distinction between the respective first peaks is almost certainly real. It probably arises from failure of the molecular dynamics calculations to account for quantum corrections, which tends to delocalize particles somewhat. The same effect on the broader successive $g_{OO}$ maxima is less obvious, and this seems to be the case.

The average number of neighbors computed for the molecular dynamics $g_{OO}$ out to its first minimum is 5.5. The corresponding number for the experimental curve is 5.3, nearly the same. (Note that the average coordination number 4.4 reported by Narten, Danford, and Levy[58] was based on a different definition of "first neighbor.")

ST2 molecular dynamics runs at temperatures both above and below 10°C have also been carried out.[82] The trend observed for $g_{OO}$ agrees with that found by x-ray diffraction,[83] namely, that the amplitude of oscillation of this function about unity diminishes with increasing temperature, while the average coordination number increases somewhat. At least on the basis of $g_{OO}$ evidence, one can conclude that the molecular dynamics approach with the ST2 potential gives a moderately good structural representation of real liquid water.

Some of the characteristic macroscopic anomalies exhibited by water are also qualitatively obtained in the ST2 simulation.[82] It has been found that the liquid density (at the vapour–liquid coexistence line) passes through a

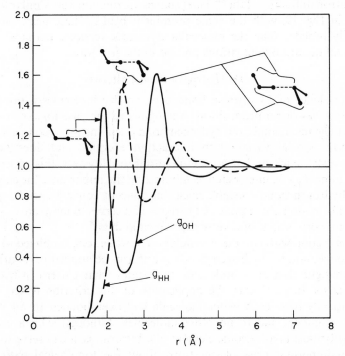

**Fig. 19**  Nuclear pair correlation functions $g_{OH}$ and $g_{HH}$ for ST2 water model at 10°C, 1 g/cm³. The intramolecular pairs are not included. The main contributing structures for the prominent peaks are shown.

maximum at 27°C, at which the mass density reaches 1.0047 g/cm$^3$. Furthermore, the isothermal compressibility passes through a shallow minimum at about 20°C. No doubt these phenomena are related to the remanent tetrahedral order observed in $g_{OO}$, which slowly disappears as the temperature rises.

In principle it should be possible to combine results from x-ray and neutron diffraction experiments (the latter using distinct isotopically substituted waters) to determine all three functions $g_{OO}$, $g_{OH}$, and $g_{HH}$. But in practice this demanding project has not been yet attempted. The molecular dynamics simulations have preceded experiments by calculating $g_{OH}$ and $g_{HH}$ separately and (for the model) precisely. Figure 19 shows these functions determined by the 10°C, 1 g/cm$^3$ run on which the $g_{OO}$ curve in Fig. 18 was based. The prominent peak in both functions at a small distance can be identified as shown in terms of hydrogen bonding between neighbors. As expected, these features diminish in distinctiveness as the temperature rises.

By invoking a plausible assumption about the nature of local order in water, Narten produced tentative $g_{OH}$ and $g_{HH}$ functions from available x-ray and neutron diffraction data.[84] Their shapes are qualitatively similar to those shown in Fig. 19, with the same prominent peaks. These peaks are substantially broader than the molecular dynamics versions, perhaps in part because of quantum fluctuations present in the real water.

### D. Hydrogen Bond Patterns

The existence of a hydrogen bond between two molecules is not fundamentally a yes-or-no proposition. Analogous to the case for conventional covalent chemical bonds, the hydrogen bond phenomenon is connected with continuous spatial variation in interaction energies, and does not discontinuously "click on" at an unique distance. However, this observation should in no way be interpreted as minimizing the importance for chemistry of the hydrogen bond concept, since this concept conveys specific quantitative information about potential surfaces and serves to motivate extremely important correlations of diverse experimental data.[85]

In connection with computer simulation of water, one is obliged to establish a convention for hydrogen bonds, which can be applied to an arbitrary given configuration of $N$ molecules, that states what pattern of hydrogen bonds exists. In particular, the application of this criterion to ice should automatically have each water molecule hydrogen-bonded to all its four nearest neighbors and to no other molecules. Although such a criterion necessarily must involve some element of arbitrariness, it can serve to legitimize an important class of questions about the topological patterns of hydrogen bonds existing in liquid water and aqueous solutions.

Since the primary attribute of hydrogen bonding is the energy of stabilization involved, it is convenient to base the hydrogen bond convention on potential energy alone. For models using an effective pair potential $v$, one decides whether a pair $i$, and $j$ of molecules is bonded or not depending on how $v(i, j)$ compares with a preassigned negative cutoff energy $V_{HB}$:

$$v(\mathbf{x}_i, \mathbf{x}_j) \leq V_{HB} \quad i \text{ and } j \text{ hydrogen-bonded}$$
$$> V_{HB} \quad i \text{ and } j \text{ not hydrogen-bonded} \quad (8.9)$$

The element of arbitariness of course is the magnitude of $V_{HB}$, but once it is assigned the criterion is mathematically unambiguous. A study of the ST2 potential shows that, if $V_{HB}$ lies between $-1.7$ and $-4.5$ kcal/mole, the conventional hydrogen bond pattern in ice will be reproduced.[86]

Having selected $V_{HB}$, it is possible to classify molecules according to the number of hydrogen bonds in which they simultaneously participate. Relative concentrations of nonbonded, single-bonded, double-bonded, and so on, water molecules have been calculated both for BNS[67] and ST2[87] water simulations for a wide variety of $V_{HB}$ choices. For all temperatures and densities investigated to date, the distributions obtained for the liquid have a single maximum. Consequently, one can rule out of serious consideration earlier suggestions[88–91] that liquid water consists of a fully bonded framework heavily invaded by unbonded interstitials for, if that were true, a bimodal distribution would arise for *some* $V_{HB}$ choice. The two maxima of this required bimodal distribution would occur at zero hydrogen bonds (interstitials) and at four hydrogen bonds (framework molecules), with virtually no molecules with one, two, or three hydrogen bonds. Evidently, a more accurate description of liquid water would be "defective, strained, random network," to be consistent with the observed hydrogen bond distributions.

Fixing the concentrations of molecules with different numbers of hydrogen bonds still leaves a wide range of possible topological connections between these bonds. Further specification of the network topology can be achieved by examining the polygons formed by the hydrogen bonds. If it were true, for example, that the liquid networks were equivalent to ice Ih or Ic in which a certain fraction of the nearest-neighbor bonds had been randomly broken, the remaining polygons would have 6, 8, 10, 12, and so on, sides, but no polygons with odd numbers of sides could occur.

Table IV shows polygon counts carried out by computer on the 10°C, 1 g/cm$^3$ sample of water simulated with the ST2 potential.[86] Four alternate values of $V_{HB}$ were employed to illustrate dependence on this arbitrary parameter, from a rather permissive value ($-2.121$ kcal/mole) to a very stringent value ($-4.848$ kcal/mole). Only "non-short-circuited polygons" are included; these are primitive polygons having no hydrogen bond cross-links tending to split them into smaller polygons. For practical reasons it

TABLE IV

Parameters Characterizing the hydrogen bond patterns in Liquid Water at
10°C and Mass Density 1 g/cm$^3$.[a]

| Parameter | I | II | III | IV |
|---|---|---|---|---|
| $V_{HB}$ (kcal/mole) | −2.121 | −3.030 | −3.939 | −4.848 |
| $\langle b \rangle$ | 3.88 | 3.14 | 2.26 | 1.18 |
| $n_0$ | 0 | 0.00331 | 0.0410 | 0.249 |
| $n_1$ | 0.0026 | 0.029 | 0.180 | 0.415 |
| $C_3$ | 0.05952 | 0.002976 | 0 | 0 |
| $C_4$ | 0.1564 | 0.04663 | 0.007606 | 0 |
| $C_5$ | 0.3459 | 0.1362 | 0.03406 | 0.001323 |
| $C_6$ | 0.3548 | 0.1306 | 0.02447 | 0.0006614 |
| $C_7$ | 0.3320 | 0.1280 | 0.01687 | 0 |
| $C_8$ | 0.2715 | 0.1045 | 0.01224 | 0 |
| $C_9$ | 0.1971 | 0.09854 | 0.01224 | 0 |
| $C_{10}$ | 0.1118 | 0.09292 | 0.01422 | 0 |
| $C_{11}$ | 0.07573 | 0.08664 | 0.005952 | 0 |

[a] The mean number of hydrogen bonds terminating at a molecule is $\langle b \rangle$; $n_0$ is the fraction of unbonded molecules, and $n_1$ is the fraction with precisely one bond; $C_j$ stands for the number of non-short-circuited polygons per molecule of the liquid with $j$ sides. (Results from Ref. 86, Table II.)

was necessary to terminate the search-and-count routine after 11-bond polygons.

The entries in Table IV show no preference for either even or odd numbers of sides. Furthermore, by extrapolating the results shown, it is clear that polygons with more than 11 sides exist in nonnegligible concentrations, except when a very strict definition of hydrogen bonds (low $V_{HB}$) is applied. These observations are inconsistent with published opinions[38,92] to the effect that liquid water consists of unconnected, bulky, icelike clusters suspended in a medium of unbonded water molecules. Instead, further support seems to be given to the random, space-filling, hydrogen bond network view, without any large-scale inhomogeneities, that was first developed for the molecular dynamics simulations by examining stereoscopic photographs of molecular positions.[66,67]

When aqueous solutions are eventually studied by computer simulation, it will be interesting to see what characteristic hydrogen bond structures tend to form around chemically different types of solutes, in comparison with those in pure bulk water. Particular importance attaches to those nonpolar functional groups that engage in hydrophobic bonding.[93]

Up to the present only time-average properties have been calculated for the hydrogen bond network present in liquid water. However, there are analogous kinetic properties that could also be probed with the available simulation apparatus, whose understanding would unquestionably enrich our comprehension of the molecular nature of water. Using the same definition of hydrogen bonds as before in terms of a preassigned cutoff energy $V_{HB}$, we mention three distinct lifetime queries:

1. Identify the molecular pairs bonded at time $t = 0$. Define $P_1(t)$ to be the average fraction of pairs that remain bonded *without interruption* over the entire interval from 0 to $t$.

2. In terms of the same set of $t = 0$ pairs, let $P_2(t)$ be those that are bonded at later time $t$, *irrespective of intervening interruptions*.

3. Denote the total number of hydrogen bonds present in the system at time $t$ by $N_b(t)$, and set

$$P_3(t) = \frac{\langle [N_b(0) - \langle N_b \rangle][N_b(t) - \langle N_b \rangle] \rangle}{\langle [N_b - \langle N_b \rangle]^2 \rangle} \tag{8.10}$$

$P_1$, $P_2$, and $P_3$ are each equal to unity at $t = 0$, and in the infinite system limit they all approach zero as $t$ increases. Their long time behaviors ought to be roughly exponential with characteristic decay times $\tau_1, \tau_2$, and $\tau_3$. Naive considerations lead one to expect

$$\tau_1 < \tau_2 < \tau_3 \tag{8.11}$$

which could be checked for the molecular dynamics simulations.

The ability to pose and answer quantitative questions of the sort mentioned here illustrates the dramatic power of the simulation methods. No experimental techniques are known, or are ever likely to be developed, to determine the topological properties of the hydrogen bond network in liquid water and its solutions. Thus computer simulation has the dual tasks of reproducing experimental results as well as complementing them.

### E. Effect of Pressure

The crystallographic structures of the ice polymorphs[7] demonstrate that in the solid phases the response to increasing pressure involves the use of more and more efficient packing with little change in length of hydrogen bonds. In some of the ices, packings denser than ice Ih and Ic are achieved by deformation of angles between successive hydrogen bonds from the ideal $\theta_t$, so as to move second- and higher-order neighbors of any given molecule inward. But interpenetration of networks also accomplishes the same end and, as we have previously noted, the interpenetration of two equivalent ice Ic networks appears in the very-high-pressure ices VII and VIII, with the effect of doubling density and number of nearest neighbors (from four to eight).

One would expect similar considerations to apply in the liquid state. Although experiments to determine structure in very highly compressed liquid water, such as x-ray and neutron diffraction, are probably impractical to execute, molecular dynamics or Monte Carlo simulations are no more difficult to perform with extremes of temperature and pressure than under ordinary conditions.

Figure 20 shows an oxygen–oxygen pair correlation function $g_{OO}(r)$ computed for 97°C and mass density 1.346 g/cm³. This is the liquid density that exists at the experimental triple point (81.6°C, 22.0 kbars) for liquid, ice VI, and ice VII. Once again the ST2 interaction was employed for the molecular dynamics simulation.

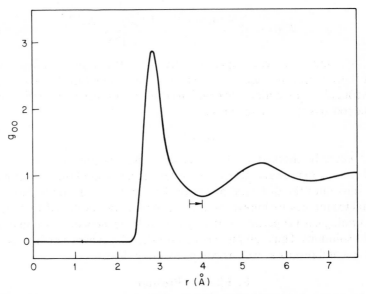

**Fig. 20** Oxygen-oxygen pair correlation function for 97°C, 1.346 g/cm³, based on the ST2 molecular dynamics simulation. The horizontal arrow shows how far the minimum displaces outward as a result of isothermal compression from 1 g/cm³.

The first maximum of $g_{OO}(r)$ occurs at 2.81 Å, only slightly less than its position at 1 g/cm³, namely, 2.86 Å. However, the subsequent minimum shifts outward to 4.00 Å, from the 1-g/cm³ distance 3.68 Å, as shown explicitly in Fig. 20. The mean number of neighbors out to the 4.00-Å minimum is 11.8. In order to make a fair comparison with the low-density number (5.8 neighbors at 97°C), the distance 3.68 Å should instead be used as the upper cutoff, and in this event the high-compression structure possesses

10.0 neighbors. From any point of view, the packing of molecules has been dramatically altered by compressing the liquid from 1 to 1.346 g/cm$^3$.

The four tetrahedrally disposed directions pointing from the oxygen nucleus to point charges $\pm q$ in definition of the ST2 potential bear an intimate geometric relation to the regular octahedron. In particular, by placing the oxygen at the center of the octahedron, the four tetrahedral directions may be oriented so as simultaneously to pierce the centers of four of the eight triangular faces which share vertices but not edges. This relationship is shown in Fig. 21.

The four pierced octahedron faces in Fig. 21 are faces through which first-neighbor oxygens in an unstrained hydrogen bond network would be seen from the position of the central oxygen, as in the ice crystal. The solid angles described by the unpierced triangular faces would be devoid of first neighbors. The octahedron thus provides a useful way to resolve $g_{OO}(r)$ into angular components:

$$g_{OO}(r) = g_{IV}(r) + g_V(r) \tag{8.12}$$

where the established convention[67] denotes pierced and unpierced faces by subscripts IV and V, respectively. The extent to which ice rules on first neighbors are violated in the liquid indicates either angular strains in the local network structure, or interstitial molecules (due possibly to network interpenetration).

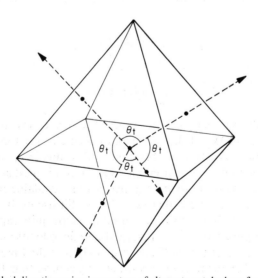

**Fig. 21**   Tetrahedral directions piercing centers of alternate octahedron faces.

In the low-pressure case (1 g/cm$^3$), most of the nearest neighbors are known to occur in $g_{IV}$.[67] But when $g_{IV}$ and $g_V$ are evaluated for high-compression molecular dynamics runs (1.346 g/cm$^3$), nearly equal numbers of nearest neighbors appear in $g_{IV}$ and $g_V$, undoubtedly indicating substantial network penetration.

The interpenetrating cubic ice networks forming ice VII place four nearest neighbors in each of $g_{IV}$ and $g_V$, not enough to make up the observed 10 to 12 in the compressed liquid simulation. Furthermore, the neighbor distance[7] in ice VII is 2.86 Å, significantly larger than the 2.81 Å found in the simulation for the liquid. Finally, the second maximum of $g_{OO}$ in Fig. 20 is too far out compared to the first maximum to be consistent with successive hydrogen bonds at an average angle $\theta_t$. Evidently, it cannot be valid to view the highly compressed liquid as dominated by ice VII structures.

Similar considerations seem to rule out ice VI as a major contributor to the liquid structure. In particular, a well resolved second-neighbor peak[7] would have to appear around 3.51 Å; clearly, it does not in Fig. 20.

The situation concerning the highly compressed liquid is similar to the liquid at atmospheric pressure. A much more diverse set of local structures is represented than can reasonably be generated by slight deformations of the crystal forms present at the respective pressures.

### F.  Molecular Motions

One aspect of molecular motion in liquids is revealed by the magnitude of the self-diffusion constant $D$. In an infinite system $D$ may be written in the alternative forms

$$D = \lim_{t \to \infty} \left\{ \frac{\langle [\Delta \mathbf{R}_j(t)]^2 \rangle}{6t} \right\}$$

$$= \tfrac{1}{3} \int_0^\infty \langle \mathbf{v}_j(0) \cdot \mathbf{v}_j(t) \rangle \, dt \tag{8.13}$$

in terms of equilibrium ensemble averages. The former involves the mean-square displacement $\Delta \mathbf{R}_j$ of the center of molecule $j$ over time $t$, while the second utilizes the velocity $\mathbf{v}_j$ of this molecule's center. For molecules that do not dissociate, it does not matter what fixed point is used as the center.

Neither infinite systems nor infinite periods are available in simulations by molecular dynamics. Nevertheless, (8.13) is still useful. In the first case $\langle [\Delta \mathbf{R}_j]^2 \rangle$ versus $t$ tends to approach a limiting slope quite rapidly, and this slope can be identified with $6D$. In the second case the velocity autocorrelation function decays towards zero sufficiently quickly that the time integral may normally be cut off at an upper limit considerably shorter than the dynamical run interval.

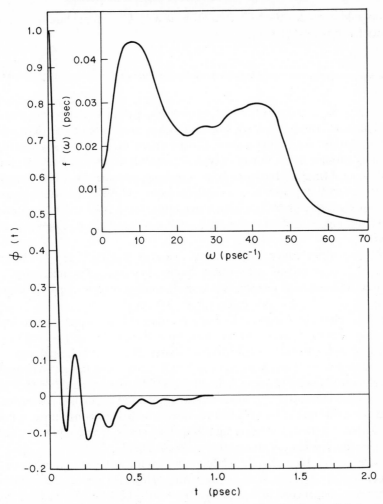

**Fig. 22** Normalized center-of-mass velocity autocorrelation function for the ST2 simulation of liquid water at 10°C, 1 g/cm³.

Figure 22 exhibits a normalized velocity autocorrelation function for center-of-mass motion:

$$\varphi(t) = \frac{\langle \mathbf{v}_j(0) \cdot \mathbf{v}_j(t) \rangle}{\langle v_j^2 \rangle}$$

$$\langle \mathbf{v}_j^2 \rangle = \frac{3k_B T}{m}$$

(8.14)

It was calculated for the ST2 simulation of a 10°C, 1-g/cm$^3$ liquid. The inset shows the power spectrum:

$$f(\omega) = \int_0^\infty \varphi(t) \cos(\omega t) \, dt \qquad (8.15)$$

Evidently, $\varphi(t)$ is essentially zero for times larger than a picosecond, whereas the dynamical run from which Fig. 22 was prepared lasted more than 8 psec.

The marked oscillatory nature of $\varphi(t)$ arises from the strong hydrogen bonding present in the liquid. A typical molecule tends to vibrate back and forth several times in the force field of its neighbors before it breaks free or the neighbors shift positions to modify the force field. Velocity autocorrelation functions computed by molecular dynamics for liquid argon[81] approach zero from below as does the water case, but without the distinctive oscillations.

The two broad maxima in $f(\omega)$ are centered at 44 and 215 cm$^{-1}$. These can probably be identified with broad bands observed experimentally[94] by infrared, Raman, and inelastic neutron-scattering spectroscopy to occur at about 60 cm$^{-1}$, and in the region 150 to 200 cm$^{-1}$.

As might have been expected, raising the temperature causes the oscillatory nature of $\varphi(t)$ to diminish. At the same time, the two broad maxima in $f(\omega)$ lose their distinctiveness, while drifting somewhat to lower frequencies.[82]

In principle the velocity autocorrelation function must have an asymptotic tail, of hydrodynamical origin, behaving at $t^{-3/2}$ at long times.[95,96] However, none of the molecular dynamics investigations for water has succeeded in identifying this tail. It is likely that the strong hydrogen bonds present in the liquid create a rigidity toward high-frequency stress, which tends to quench the hydrodynamical tail compared to simple liquids.

The values of $D$ found in the ST2 simulations for water at 1 g/cm$^3$ show the proper rapid increase with temperature. However, they are uniformly higher than experimental values by about 30% between the melting point and about 40°C.[82] Comparisons are not yet available beyond this temperature range. Although much better agreement could be obtained by increasing the strength of the ST2 interaction (by a simply multiplicative renormalization), such tampering would cause undesirable damage to other properties. Until quantum corrections can be appended to the manifestly classical simulation, it would be misdirected effort to insist on very precise agreement between predicted and measured $D$.

Rotational diffusion of water molecules has also been studied by the simulation method. Autocorrelation functions for angular momentum components about the three principal axes separately exhibit strong oscillatory character, showing that the molecules execute rapid librational motion under the directional hydrogen bond forces of neighbors. The quantities

$$\Gamma_n(t) = \langle P_n[\cos \theta(t)] \rangle \qquad (8.16)$$

measuring the distribution of angles $\theta(t)$ through which the dipole axis of a molecule turns in time $t$ have also been computed for $n = 1, 2$. The initial behavior of the $\Gamma_n$ reflects the high-frequency librational motion, but at long times the decay seems to involve a single dominant exponential function of time. Analogous to the center-of-mass velocity autocorrelation function, the presumed hydrodynamical nonexponential tails on the $\Gamma_n$ are not visible.[97] The decay time exhibited by $\Gamma_1$ is closely connected to the macroscopic dielectric relaxation time, and indeed compares reasonably well in magnitude with measured values. However, the exact connection between the microscopic and macroscopic rates is still controversial,[98] so that a decisive comparison cannot yet be made.

The ST2 simulation has also provided the basis for a molecular dynamics calculation of neutron inelastic scattering from liquid water.[99] Protons act as strong incoherent scatterers for neutrons, so observed scattering cross sections yield information about proton motions. In particular, the cross section for scattering with momentum change $\hbar\mathbf{k}$ and energy change $\hbar\omega$ gives the function

$$S_{\text{inc}}(k, \omega) = \int_0^\infty dt \cos(\omega t) \langle \exp[i\mathbf{k} \cdot \Delta\mathbf{r}_j(t)] \rangle \qquad (8.17)$$

where $\Delta\mathbf{r}_j(t)$ is the displacement of a typical proton $j$ over time interval $t$. It is known that

$$\lim_{k \to 0} \frac{\omega^2}{k^2} S_{\text{inc}}(k, \omega) = h(\omega) \qquad (8.18)$$

is, aside from trivial factors, the power spectrum of the proton velocity autocorrelation function. The molecular dynamics calculations explicitly show that for all intents and purposes the limiting behavior shown in (8.18) is achieved for $\omega^2 S_{\text{inc}}/k^2$ at $k = 1$ Å$^{-1}$, which ought to be directly obtainable in real experiments.

Careful studies of both proton and center-of-mass motions by molecular dynamics give no support to quasi-crystalline "jump-and-wait" descriptions for molecular motions in liquid water.[100]

## IX. WEAK-ELECTROLYTE MODEL

### A. Central Interactions

None of the statistical models mentioned above (lattice models, cell model, molecular dynamics simulations) has permitted water molecules to dissociate. Yet in real water this occurs to produce the chemically important hydrated $H^+$ and $OH^-$ ions. Unless it can eventually describe the structure and kinetics of these species in water, the theory of this liquid must be regarded as partial failure. To help avoid this shortcoming, Lemberg and Stillinger have proposed a weak-electrolyte model for water[101] in which the separate H and O particles become the basic dynamical entities, and $H_2O$ molecules spontaneously form as "ion triplets" on account of the specific forces attributed to these particles. The dissociation process thus becomes a natural thermally activated process in the model.

At the heart of the weak-electrolyte model is the selection of three central potentials $V_{HH}(r)$, $V_{OH}(r)$, and $V_{OO}(r)$ which must adequately represent in additive form, for all pairs of nuclei, the totality of interactions in an arbitrary collection of molecules and ions. The first two of these functions are constrained by the requirement that the energy minimum they give for an isolated $H_2O$ possesses the accurately known nonlinear triatomic geometry discussed in Section II. Furthermore, $V_{HH}$, $V_{OH}$, and $V_{OO}$ should each have Coulombic forms at large $r$ characteristic of electrostatic charges on Hs and Os which produce the correct dipole moment for the isolated molecules.

Beyond the capacity to describe dissociation and the nature of the resulting solvated ions, the weak-electrolyte model presents other attractive features. Owing to the fact that only central forces are used, it is much easier than otherwise to construct formal expressions for quantum corrections to the classical limit formulas both for equilibrium and for transport properties. Also, it is important that the three normal vibrations have been restored to each molecule, with frequencies determined by the curvatures of $V_{HH}$ and $V_{OH}$. As a result, it should be possible to examine the broadening of vibrational bands in condensed phases due to interactions between molecules which are themselves described by the three potential functions $V_{HH}$, $V_{OH}$, and $V_{OO}$. Finally, the use of only central pair interactions makes it feasible to examine binary mixture versions of standard integral equations (e.g., BGY, PY, CHNC equations[42]) for prediction of the three nuclear pair correlation functions $g_{OO}(r)$, $g_{OH}(r)$, and $g_{HH}(r)$, which is not possible with rigid-molecule models generally.

Continuing study of the weak-electrolyte model will be necessary to identify the optimal set of central interactions; however, we can display a tentative set which clearly illustrates the main ideas involved. First, the effective charges on the hydrogens ($Q^*$) and the oxygens ($-2Q^*$) must be

$$Q^* = 0.32983e$$
$$= 1.5841 \times 10^{-10} \text{ esu} \qquad (9.1)$$

to conform to the known molecular dipole moment. We then demand that $V_{OH}(r)$ consist just of two inverse power terms, $-2Q^{*2}/r$ for the Coulombic attraction between the O and H, plus a strong repulsion at short range $(n > 1)$:

$$V_{OH}(r) = \frac{2(Q^*)^2}{r_e}\left[\frac{1}{n}\left(\frac{r_e}{r}\right)^n - \frac{r_e}{r}\right] \tag{9.2}$$

This function passes through its minimum when $r = r_e$, the equilibrium bond length in the water molecule. In the present central force context, the asymmetric stretch normal mode frequency depends only on the curvature of $V_{OH}$ at $r_e$, so $n$ was determined to be

$$n = 14.9797 \tag{9.3}$$

from the observed $D_2O$ frequency. In terms of the convenient units angstroms for length and kilocalories per mole for energy, $V_{OH}$ has the specific expression

$$V_{OH}(r) = \frac{2.66366}{r^{14.9797}} - \frac{72.269}{r} \tag{9.4}$$

Two protons simultaneously present in the minimum of $V_{OH}$ at $r_e$ can be constrained to form a nonlinear triatomic molecule only if $V_{HH}$ possesses a relative minimum at the correct proton–proton distance $r_p = 1.5151$ Å. Energetic stability of the $H_2O$ molecule of course requires

$$2V_{OH}(r_e) + V_{HH}(r_p) < 0 \tag{9.5}$$

but $V_{HH}(r_p)$ can be positive. The symmetric stretch and symmetric bend normal mode frequencies are determined by both curvatures $V''_{OH}(r_e)$ and $V''_{HH}(r_p)$ but, since the first has already been fixed, the second must be chosen to achieve the best simultaneous fit to the symmetric mode frequencies. With this understanding the following specific function was constructed for $V_{HH}$ (in angstroms and kilocalories per mole):

$$V_{HH}(r) = \frac{36.1345}{r} + \frac{30}{1 + \exp\left[21.9722(r - 2.125)\right]}$$
$$- 26.51983 \exp\left[-4.728281(r - 1.4511)^2\right] \tag{9.6}$$

With this potential, and that shown for $V_{OH}$ in Eq. (9.4), the symmetric bend frequency is 14.4% too low and the symmetric stretch frequency is 14.4% too high, compared to observed values, all for $D_2O$.

The primary task assigned to $V_{OO}$ is to keep oxygens in different molecules apart. Otherwise the strong attraction between oxygen and hydrogen leads to linear hydrogen bonds of length about $2r_e$, with the bridging hydrogen half way between. One function found to produce proper asymmetric linear hydrogen bonds with the correct length and strength is

$$V_{OO}(r) = \frac{1697116}{r^{12}} - \frac{4039.394}{r^6} + \frac{144.538}{r} \qquad (9.7)$$

expressed as before in angstroms and kilocalories per mole. This and the other two potentials (9.4) and (9.6) are plotted in Fig. 23.

The mechanically stable structure for the hydronium cation $H_3O^+$ is pyramidal, with all three OH bonds having length $r_e$, and the apex angles at the oxygen the same as the bond angle in the water molecule. This arises because of the possibility that all six pairs of particles in $H_3O^+$ can exist simultaneously at the minima of their respective central potentials. It is relevant to note in this connection that the real $H_3O^+$ is also pyramidal, with apex angles and bond lengths only slightly larger than those predicted by the weak-electrolyte model.[102]

The disproportionation reaction

$$H_2O + H_2O \rightarrow H_3O^+ + OH^- \qquad (9.8)$$

with widely separated reactants and products all at their respective mechanical equilibria, is energetically unfavorable. Counting pairs, it becomes clear that the energy must increase by $V_{HH}(r_p)$ to produce the ionic products, which

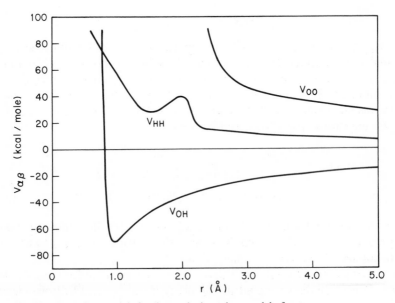

**Fig. 23** Central potentials for the weak electrolyte model of water.

Fig. 23 shows to be about 28 kcal/mole. A positive energy of this magnitude is necessary to prevent gross ionization in water molecule aggregates.

Although it is true that the weak-electrolyte model utilizes only interactions for pairs of ionic particles, the molecules out of which they are composed do not necessarily behave as though they have experienced additive *molecular* interactions. Neighboring molecules tend to perturb one another's normal models of vibration, and (4.7) to (4.10) show that vibrationally averaged component potentials $\overline{V}^{(n)}$ of all orders should in principle arise. The automatic incorporation of these many-molecule effects ought to be listed as one of the weak-electrolyte model's virtues.

The $H^+$ and $OH^-$ ions in the present model do not carry the full charges $\pm e$, but only the partial charges $\pm 0.32983e$. However, the electric fields of the real ions in liquid water are diminished, or shielded, by the electronic polarizability of neighboring molecules. This form of polarizability is not present in the weak-electrolyte model, but the reduced ionic charges offer partial compensation for this omission.

### B. Some Classical Formulas

At sufficiently high temperatures the partition function $Q$ for the weak-electrolyte model (with $N$ oxygens and $2N$ hydrogens) takes the classical form:

$$
\begin{aligned}
Q = \frac{1}{N!(2N)!\lambda_O^{3N}\lambda_H^{6N}} \int d\mathbf{r}_1 \cdots \int d\mathbf{r}_{3N} \Bigg\{ \exp\Bigg[ -\beta \sum_{i<j=1}^{N} V_{OO}(r_{ij}) \\
+ \sum_{i=1}^{N}\sum_{j=N+1}^{3N} V_{OH}(r_{ij}) + \sum_{i<j=N+1}^{3N} V_{HH}(r_{ij}) \Bigg] \Bigg\}
\end{aligned}
\tag{9.9}
$$

where $\mathbf{r}_1, \ldots, \mathbf{r}_N$ locate oxygens, and $\mathbf{r}_{N+1}, \ldots, \mathbf{r}_{3N}$ locate hydrogens. The integrations span the allowed volume $\mathscr{V}$, and

$$
\lambda_\alpha = h(2\pi m_\alpha k_B T)^{-1/2}, \qquad \alpha = O, H
\tag{9.10}
$$

are the mean thermal de Broglie wavelengths. Denoting the potential energy in (9.9) by $V_{N,2N}$ for compactness, the correlation functions in this classical limit may next be introduced:

$$
g_{OO}(r_{12}) = \frac{\mathscr{V}^2 \int d\mathbf{r}_3 \cdots \int d\mathbf{r}_{3N} \exp\left[-\beta V_{N,2N}(\mathbf{r}_1 \cdots \mathbf{r}_{3N})\right]}{\int d\mathbf{r}_1 \cdots \int d\mathbf{r}_{3N} \exp\left[-\beta V_{N,2N}(\mathbf{r}_1 \cdots \mathbf{r}_{3N})\right]}
$$

$$
g_{OH}(r_{1,N+1}) = \frac{\mathscr{V}^2 \int d\mathbf{r}_2 \cdots \int d\mathbf{r}_N \int d\mathbf{r}_{N+2} \cdots \int d\mathbf{r}_{3N} \exp\left[-\beta V_{N,2N}(\mathbf{r}_1 \cdots \mathbf{r}_{3N})\right]}{\int d\mathbf{r}_1 \cdots \int d\mathbf{r}_{3N} \exp\left[-\beta V_{N,2N}(\mathbf{r}_1 \cdots \mathbf{r}_{3N})\right]}
$$

$$
\begin{aligned}
&g_{HH}(r_{N+1,N+2}) \\
&= \frac{\mathscr{V}^2 \int d\mathbf{r}_1 \cdots \int d\mathbf{r}_N \int d\mathbf{r}_{N+3} \cdots \int d\mathbf{r}_{3N} \exp\left[-\beta V_{N,2N}(\mathbf{r}_1 \cdots \mathbf{r}_{3N})\right]}{\int d\mathbf{r}_1 \cdots \int d\mathbf{r}_{3N} \exp\left[-\beta V_{N,2N}(\mathbf{r}_1 \cdots \mathbf{r}_{3N})\right]}
\end{aligned}
\tag{9.11}
$$

In the infinite system limit, with fixed ionic densities, these correlation functions each approach unity at a large separation. $g_{OH}$ and $g_{HH}$ have sharp peaks at $r_e$ and $r_p$, respectively, corresponding to the intramolecular pairs.

Setting $\rho = N/\mathscr{V}$, the local electroneutrality conditions on the infinite system limit correlation functions may be written

$$\rho \int [g_{OH}(r) - g_{OO}(r)] \, d\mathbf{r} = 1 \tag{9.12}$$

and

$$2\rho \int [g_{OH}(r) - g_{HH}(r)] \, d\mathbf{r} = 1 \tag{9.13}$$

These can be supplemented by a "second moment condition" which must always exist for an ionic fluid:[103]

$$\int [2g_{OH}(r) - g_{OO}(r) - g_{HH}(r)]r^2 \, d\mathbf{r} = \frac{3k_B T}{8\pi\rho^2 (Q^*)^2} \tag{9.14}$$

The fact that interactions in the weak-electrolyte model are central and additive allows the thermodynamic energy $E$ to have an especially simple form:

$$\frac{E}{N} = \frac{9k_B T}{2} + \rho \int d\mathbf{r} \, [\tfrac{1}{2} V_{OO}(r)g_{OO}(r) + 2V_{OH}(r)g_{OH}(r) + 2V_{HH}(r)g_{HH}(r)] \tag{9.15}$$

The virial equation of state undergoes analogous simplifications:

$$\beta p = 3\rho - \frac{\beta\rho^2}{6} \int d\mathbf{r} \, \mathbf{r} \cdot [g_{OO}(r)\nabla V_{OO}(r) + 4g_{OH}(r)\nabla V_{OH}(r) + 4g_{HH}(r)\nabla V_{HH}(r)] \tag{9.16}$$

Although the ideal gas term $3\rho$ corresponds to $3N$ independent particles, one can show for the dilute vapor of undissociated molecules that the integral term equals $-2\rho$, leaving

$$\beta p = \rho \tag{9.17}$$

the appropriate ideal gas form for $N$ independent (but composite) particles.

The isothermal compressibility $\kappa_T$ may be expressed in terms of the oxygen–oxygen pair correlation function:

$$\rho k_B T \kappa_T = 1 + \rho \int d\mathbf{r} \, [g_{OO}(r) - 1] \tag{9.18}$$

The local electroneutrality conditions (9.12) and (9.13) allow this to be converted to the alternate forms

$$2\rho k_B T \kappa_T = 1 + 2\rho \int d\mathbf{r} \, [g_{HH}(r) - 1] \qquad (9.19)$$

and

$$k_B T \kappa_T = \int d\mathbf{r} \, [g_{OH}(r) - 1] \qquad (9.20)$$

Since large-scale charge separation is not possible in the collection of O and H ions forming the weak-electrolyte model, these two species are forced to execute density fluctuations of long wavelength together, thus producing the variety (9.18) to (9.20) of fluctuation-compressibility theorems. It should be stressed that their validity depends in no way on the extent of water molecule dissociation being small.

By using standard methods of statistical mechanics, one can easily express the static linear response of the ionic fluid to application of an external electrostatic potential in terms of $g_{OO}$, $g_{OH}$, and $g_{HH}$. We consider just a single Fourier component, for which the applied potential is

$$\Psi_{ap}(\mathbf{r}) = \Psi_0 \sin (\mathbf{k} \cdot \mathbf{r}) \qquad (9.21)$$

with $\Psi_0$ small. The H and O ions will rearrange under the influence of $\Psi_{ap}$ so as partially to shield it. The average potential $\overline{\Psi}$ will also be spatially sinusoidal, and we write it:

$$\overline{\Psi}(\mathbf{r}) = \frac{\Psi_0}{\varepsilon(k)} \sin (\mathbf{k} \cdot \mathbf{r}) \qquad (9.22)$$

thereby defining the wavelength-dependent dielectric constant $\varepsilon(k)$.[22] The requisite calculation yields the formula

$$\frac{1}{\varepsilon(k)} = 1 - \frac{16\pi\beta\rho(Q^*)^2}{k^2} \left\{ \frac{3}{2} + \rho \int d\mathbf{r} \left[ \frac{\sin (kr)}{kr} \right] [g_{OO}(r) - 2g_{OH}(r) + g_{HH}(r)] \right\} \qquad (9.23)$$

At any finite temperature some of the water molecules will have dissociated, rendering the pure liquid slightly conducting. Because it is a conductor, the water will be able to shield very long-wavelength external fields completely:

$$\lim_{k \to 0} \frac{1}{\varepsilon(k)} = 0 \qquad (9.24)$$

The electroneutrality and second moment conditions (9.12) to (9.14), when applied to (9.23), assure that this limit is obeyed. However, with a very small degree of dissociation, it is only for very small $k$ (comparable to, or less than, Debye's $\kappa$ for the dilute ionic solution) that $\varepsilon(k)$ begins to rise to infinity. For $k$ larger than these tiny values, but still reasonably small, $\varepsilon(k)$ should equal the static dielectric constant ($\approx 80$ at room temperature) normally quoted for water. At very large $k$, one sees from (9.23) that $\varepsilon(k)$ approaches unity. Figure 24 schematically illustrates the expected behavior.

Amplitudes $\sigma(\mathbf{k})$ for fluctuating charge density waves may be introduced by the definition

$$\sigma(\mathbf{k}) = \frac{Q^*}{\sqrt{N}} \left\{ -2 \sum_{j=1}^{N} \exp(i\mathbf{k} \cdot \mathbf{r}_j) + \sum_{j=N+1}^{3N} \exp(i\mathbf{k} \cdot \mathbf{r}_j) \right\} \qquad (9.25)$$

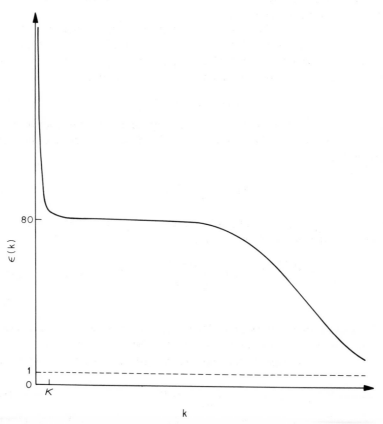

**Fig. 24** Schematic graph for $\varepsilon(k)$ in liquid water at room temperature. The Debye $\kappa$ refers to the very dilute solution of ions formed by molecule dissociation.

It is straightforward to compute the equal-time quadratic fluctuation average for the $\sigma$s, with the result

$$\langle \sigma^*(\mathbf{k})\sigma(\mathbf{k}) \rangle = 4(Q^*)^2 \left\{ \frac{3}{2} + \rho \int d\mathbf{r} \left[ \frac{\sin(kr)}{kr} \right] [g_{OO}(r) - 2g_{OH}(r) + g_{HH}(r)] \right\}$$

(9.26)

By comparing this expression (9.23), one obtains an alternate $\varepsilon(k)$ formula:

$$\frac{1}{\varepsilon(k)} = 1 - \frac{4\pi\beta\rho}{k^2} \langle \sigma^*(\mathbf{k})\sigma(\mathbf{k}) \rangle$$

(9.27)

The quadratic fluctuation is always positive for finite $k$, so $\varepsilon(k)$ cannot be unity or less, thus

$$-\frac{3}{2\rho} < \int d\mathbf{r} \left[ \frac{\sin(kr)}{kr} \right] [g_{OO}(r) - 2g_{OH}(r) + g_{HH}(r)]$$

(9.28)

for all $k$, irrespective of temperature and density.

Whereas equal-time fluctuations in $\sigma$s suffice to give static dielectric response, unequal-time correlation functions are related to dielectric response at a finite frequency $\omega$. Yet another advantage of the weak-electrolyte model for water therefore resides in its capacity to describe dielectric relaxation and conduction by providing a means to evaluate real and imaginary parts of $\varepsilon(k, \omega)$.

### C. Quantum Corrections

In applying the weak-electrolyte model to water at about room temperature, quantum corrections to the classical statistical mechanics just outlined are unavoidable. Fortunately, there is a simple prescription for generating leading quantum corrections to

$$\rho^{(3N)}(\mathbf{r}_1 \cdots \mathbf{r}_{3N} | \mathbf{r}_1 \cdots \mathbf{r}_{3N}) = \langle \mathbf{r}_1 \cdots \mathbf{r}_{3N} | \exp(-\beta \mathcal{H}) | \mathbf{r}_1 \cdots \mathbf{r}_{3N} \rangle$$

(9.29)

the diagonal elements of the density matrix for the $3N$ ionic particles. This permits one in principle to calculate corrected averages for operators involving position only, specific examples being the correlation functions $g_{OO}$, $g_{OH}$, and $g_{HH}$.

This simple prescription disregards spin and statistics, which should be valid for condensed phases. It states that the classical format applies, provided that the potential is replaced by a temperature-dependent apparent potential.[104] For the weak-electrolyte model, $V_{N,2N}$ is replaced by

$$W_{N,2N}(\mathbf{r}_1 \cdots \mathbf{r}_{3N}) = \sum_{i<j=1}^{N} W_{OO}(r_{ij}, \beta) + \sum_{i=1}^{N} \sum_{j=N+1}^{3N} W_{OH}(r_{ij}, \beta)$$

$$+ \sum_{i<j=N+1}^{3N} W_{HH}(r_{ij}, \beta) \qquad (9.30)$$

where $(\gamma, \delta = O, H)$.

$$W_{\gamma\delta}(r) = V_{\gamma\delta}(r) + \frac{\hbar^2}{m_{\gamma\delta}} \left\{ \frac{1}{12} \nabla^2 \beta V_{\gamma\delta}(r) - \frac{1}{24} [\nabla \beta V_{\gamma\delta}(r)]^2 \right\} + O(\hbar^4) \qquad (9.31)$$

$m_{\gamma\delta}$ is the reduced mass:

$$\frac{1}{m_{\gamma\delta}} = \frac{1}{m_\gamma} + \frac{1}{m_\delta} \qquad (9.32)$$

In particular this result assures that computer simulations (Monte Carlo or molecular dynamics) can be carried out as usual, using $W_{N,2N}$ in place of $V_{N,2N}$, to produce corrected static correlation functions.

It is easy to see from (9.31) that the leading quantum corrections (order $\hbar^2$ terms) have the following shape-changing effects on $V_{\gamma\delta}(r)$.

1. Regions of positive curvature (as in the vicinity of minima) are raised, while regions of negative curvature (such as maxima) are lowered, to produce diminished variation.

2. Minima are broadened; maxima are narrowed.

The net effect of these modifications at a given temperature is that the potential energy is less effective in localizing particles relative to one another. Consequently, the correlation function peaks should be somewhat broader and less distinct than if $V_{N,2N}$ alone were used.

Off-diagonal elements of the density matrix can also be displayed in a systematic expansion about the classical limit.[101] However, the results are more elaborate in appearance and interpretation. They are nevertheless required to calculate a wide variety of interesting equilibrium properties, such as the momentum distributions for the two types of atoms present.

Although a molecular dynamics simulation formally carried out with $W_{N,2N}$ in place of the interaction $V_{N,2N}$ is a valid procedure for generating equilibrium structural properties through order $\hbar^2$, one must be careful not to overinterpret the apparent transport properties emanating from such a simulation. As an example, the quantity

$$\frac{1}{3} \int_0^\infty \langle \mathbf{v}_j(0) \cdot \mathbf{v}_j(t) \rangle_W \, dt \qquad (9.33)$$

involving the classical velocity autocorrelation function for molecule $j$ in the presence of potential $W$ is not necessarily the correct expression for the self-diffusion constant $D$ through order $\hbar^2$. Instead, one would generally expect additional quantum corrections as well. The development of such quantum corrections for each of the transport coefficients (self-diffusion constant, shear and bulk viscosities, thermal conductivity, complex dielectric constant at arbitrary frequency) to be appended to the classical $W$-dynamics expression is an important problem whose solution would dramatically extend the power of the molecular dynamics technique.

## X. CONCLUSION

The major topics covered in this article show that the theory of water now has a well-established rational basis. Yet it is clear that considerable work remains to be done both on the quantum theory of water molecule interactions and on statistical mechanical techniques for utilizing these interactions in many-body calculations. Presuming that the present high level of interest in the subject persists, the prognosis nevertheless is good for rapid maturing of the field. Furthermore, the special techniques and insights developed for water should encourage similar activity for other polyatomic liquids.

Aqueous solutions have purposely been excluded from this article. However, each of the four statistical mechanical approaches discussed (lattice and cell theories, computer simulation, weak-electrolyte model) has obvious application to solutions. Quantum mechanical studies of solute–water and solute–solute potential energy functions are necessary prerequisites to quantitative statistical theory, and fortunately these calculations have been initiated for noble gases,[105,106] monatomic ions,[107,108] and small organic molecules.[109,110] Certainly, far more needs to be done, especially in examining characteristic functional groups which are included in large molecules of biological interest. A systematic study of the interactions between water molecules and hydrocarbons needs to be carried out, so that realistic models for hydrophobic bonding[93] can be constructed and investigated by the available statistical methods. With aggressive effort along these lines, it is possible within the foreseeable future that conformational kinetics of biopolymers will be simulated via computer, with full accounting of biopolymer intramolecular degrees of freedom, of ambient water molecule motions, and of interactions between them.

The theoretical study of chemical reactions in aqueous solution is another extremely important subject to which research attention needs to be directed. Fast reactions involving hydrogen ions[111] could probably be simulated by an extension of the weak-electrolyte model described in Section IX, since it already incorporates dissociation and proton transfer. In any case suitable energy surfaces would have to be constructed to describe the interaction of reactants. Furthermore, the precise distinction between reactants and products in terms of separate regions of configuration space leads to the same situation encountered above in defining hydrogen bonds. An element of arbitrariness intrudes, but prediction of measurable properties for the chemically reacting solution must be demonstrably independent of the criterion used to define reactants and products.

The importance of these extensions for a deeper understanding of chemistry cannot be ignored.

## References

1. G. Wald, *Proc. Natl. Acad. Sci. U.S.*, **52**, 595 (1964).
2. R. J. Curran, B. J. Conrath, R. A. Hanel, V. G. Kunde, and J. C. Pearl, *Science*, **182**, 381 (1973).
3. A. C. Cheung, D. M. Rank, C. H. Townes, D. D. Thornton, and W. J. Welch, *Nature*, **221**, 626 (1969).
4. I. S. Shklovskii and C. Sagan, *Intelligent Life in the Universe*, Holden–Day, San Francisco, 1966.
5. J. L. Kavenau, *Water and Solute-Water Interactions*, Holden–Day, San Francisco, 1964.
6. R. Brückner, *J. Non-Cryst. Solids*, **5**, 123 (1970).
7. D. Eisenberg and W, Kauzmann, *The Structure and Properties of Water*, Oxford University Press, New York, 1969, Chap. 3.
8. J. Padova, in R. A. Horne, Ed., *Water and Aqueous Solutions*, Wiley-Interscience, New York, 1972, Chap. 4.
9. S. B. Brummer and A. B. Gancy, Ref. 8, Chap. 19.
10. W. Kauzmann, *Advan. Protein Chem.*, **14**, 1 (Table III) (1959).
11. W. S. Benedict, N. Gailar, and E. K. Plyler, *J. Chem. Phys.*, **24**, 1139 (1956).
12. C. A. Coulson, *Valence*, 2nd ed., Oxford University Press, New York, 1961, Chap. VIII.
13. R. P. Feynman, *Phys. Rev.*, **56**, 340 (1939).
14. R. F. W. Bader, *J. Am. Chem. Soc.*, **86**, 5070 (1964).
15. T. R. Dyke and J. S. Muenter, *J. Chem. Phys.*, **59**, 3125 (1973).
16. J. Verhoeven and A. Dymanus, *J. Chem. Phys.*, **52**, 3222 (1970).
17. E. A. Moelwyn-Hughes, *Physical Chemistry*, 2nd ed. Pergamon, New York, 1961, p. 383.
18. S. P. Liebmann and J. W. Moskowitz, *J. Chem. Phys.*, **54**, 3622 (1971).
19. O. Crawford and A. Dalgarno, *Chem. Phys. Lett.*, **1**, 23 (1967).
20. K. Kuchitsu and Y. Morino, *Bull. Chem. Soc. Jap.*, **38**, 814 (1965).
21. C. W. Kern and M. Karplus, in F. Franks, Ed., *Water, A Comprehensive Treatise*, Vol. 1, Plenum, New York, 1972, p. 67.
22. F. H. Stillinger and R. Lovett, *J. Chem. Phys.*, **48**, 3858 (1968).

23. K. Morokuma and L. Pedersen, *J. Chem. Phys.*, **48**, 3275 (1968).
24. J. Del Bene and J. A. Pople, *J. Chem. Phys.*, **52**, 4858 (1970).
25. K. Morokuma and J. R. Winick, *J. Chem. Phys.*, **52**, 1301 (1970).
26. P. A. Kollman and L. C. Allen, *J. Chem. Phys.*, **51**, 3286 (1969).
27. D. Hankins, J. W. Moskowitz, and F. H. Stillinger, *J. Chem. Phys.*, **53**, 4544 (1970); Erratum, **59**, 995 (1973).
28. G. H. F. Diercksen, *Theor. Chim. Acta*, **21**, 335 (1971).
29. H. Popkie, H. Kistenmacher, and E. Clementi, *J. Chem. Phys.*, **59**, 1325 (1973).
30. J. C. Slater and J. G. Kirkwood, *Phys. Rev.*, **37**, 682 (1931).
31. J. G. Kirkwood, *Phys. Z.* **33**, 57 (1932).
32. E. U. Condon and H. Odishaw, Eds., *Handbook of Physics*, McGraw-Hill, New York, York, 1967, p. 7–40.
33. A. Dalgarno, *Advan. Chem. Phys.* **12**, 164 (1967).
34. J. Corner, *Trans. Faraday Soc.*, **44**, 914 (1948).
35. S. W. Peterson and H. A. Levy, *Acta Crystallogr.*, **10**, 70 (1957).
36. L. Pauling, *The Nature of the Chemical Bond* 3rd ed., Cornell University Press, Ithaca, 1960, p. 469.
37. B. R. Lentz and H. A. Scheraga, *J. Chem. Phys.*, **58**, 5296 (1973).
38. H. S. Frank and W. Y. Wen, *Discuss. Faraday Soc.*, **24**, 133 (1957).
39. H. S. Frank, *Proc. Roy. Soc.*, **A247**, 481 (1958).
40. A. Ben-Naim and F. H. Stillinger, in R. A. Horne, Ed., *Water and Aqueous* Solutions, Wiley-Interscience, New York, 1972, Chap. 8.
41. M. E. Fisher, *Rep. Prog. Phys.*, **XXX** (II), 615 (1967).
42. S. A. Rice and P. Gray, *The Statistical Mechanics of Simple Liquids*, Wiley-Interscience, New York, 1965, Sect. 2.6.
43. H. S. Green, *Proc. Roy. Soc.*, **A189**, 103 (1947).
44. T. L. Hill, *Statistical Mechanics*, McGraw-Hill, New York, 1956, p. 105.
45. W. G. McMillan, Jr., and J. E. Mayer, *J. Chem. Phys.*, **13**, 276 (1945).
46. F. H. Stillinger, *J. Phys. Chem.*, **74**, 3677 (1970).
47. F. H. Stillinger, *J. Chem. Phys.*, **57**, 1780 (1972).
48. F. Riesz and B. Sz.-Nagy, *Functional Analysis*, Frederick Ungar, New York, 1955, p. 40.
49. T. Halcioğlu and O. Sinanoğlu, *J. Chem. Phys.*, **49**, 996 (1968).
50. G. S. Rushbrooke and M. Silbert, *Mol. Phys.*, **12**, 505 (1967).
51. J. M. Levelt Sengers and S. C. Greer, *Int. J. Heat Mass Transfer*, **15**, 1865 (1972).
52. C. A. Rogers, *Packing and Covering*, Cambridge University Press, Cambridge, 1964, p. 74.
53. T. D. Lee and C. N. Yang, *Phys. Rev.*, **87**, 410 (1952).
54. B. Widom and F. H. Stillinger, *J. Chem. Phys.*, **58**, 616 (1973).
55. P. D. Fleming, III, and J. H. Gibbs, *J. Stat. Phys.*, **10**, 157 (1974).
56. P. D. Fleming, III, and J. H. Gibbs, *J. Stat. Phys.*, **10**, 351 (1974).
57. B. Kamb, *J. Chem. Phys.*, **43**, 3917 (1965).
58. A. H. Narten, M. D. Danford, and H. A. Levy, *Discuss. Faraday Soc.*, **43**, 97 (1967).
59. G. M. Bell, *J. Phys. C: Solid State Phys.*, **5**, 889 (1972).
60. E. A. Guggenheim and M. C. McGlashan, *Proc. Roy. Soc.*, **A206**, 335 (1951).
61. R. T. Whitlock and P. R. Zilsel, *Phys. Rev.*, **131**, 2409 (1963).
62. I. Prigogine, *The Molecular Theory of Solutions*, North-Holland, Amsterdam, 1957, Chaps. VII–X.
63. M. Weissmann and L. Blum, *Trans. Faraday Soc.*, **64**, 2605 (1968).
64. O. Weres and S. A. Rice, *J. Am. Chem. Soc.*, **94**, 8984 (1972).
65. N. Bjerrum, *Science*, **115**, 385 (1952).
66. A. Rahman and F. H. Stillinger, *J. Chem. Phys.*, **55**, 3336 (1971).

67. F. H. Stillinger and A. Rahman, *J. Chem. Phys.*, **57**, 1281 (1972).
68. D. P. Stevenson, *J. Phys. Chem.*, **69**, 2145 (1965).
69. L. Pauling, *J. Am. Chem. Soc.*, **57**, 2680 (1935).
70. H. M. J. Smith, *Phil. Trans. Roy. Soc.*, **241**, 105 (1948).
71. O. Weres, *Chem. Phys. Lett.*, **14**, 155 (1972).
72. G. E. Walrafen, in F. Franks, Ed., *Water, A Comprehensive Treatise*, Vol. 1, Plenum, New York, 1972, p. 170.
73. J. M. Hammersley and D. C. Handscomb, *Monte Carlo Methods*, Methuen-Wiley, New York, 1964.
74. D. Denley and S. A. Rice, *J. Am. Chem. Soc.*, **96**, 4369 (1974).
75. A. Ben-Naim, *J. Chem. Phys.*, **52**, 5531 (1970).
76. N. Metropolis, A. W. Rosenbluth, M. N. Rosenbluth, A. H. Teller, and E. Teller, *J. Chem. Phys.*, **21**, 1087 (1953).
77. J. A. Barker and R. O. Watts, *Chem. Phys. Lett.*, **3**, 144 (1969).
78. J. A. Barker and R. O. Watts, *Mol. Phys.*, **26**, 789 (1973).
79. B. J. Alder and T. Wainwright in I. Prigogine, Ed., *Transport Processes in Statistical Mechanics*, Interscience, New York, 1958, p. 97.
80. H. Goldstein, *Classical Mechanics*, Addison-Wesley, Reading, Mass., 1953, p. 158.
81. A. Rahman, *Phys. Rev.*, **136**, A405 (1964).
82. F. H. Stillinger and A. Rahman, *J. Chem. Phys.*, **60**, 1545 (1974).
83. A. H. Narten, *X-ray Diffraction Data on Liquid Water in the Temperature Range 4°C– 200°C*, Oak Ridge National Laboratory Technical Report ONRL-4578, July 1970.
84. A. H. Narten, *J. Chem. Phys.*, **56**, 5681 (1972).
85. S. N. Vinogradov and R. H. Linnell, *Hydrogen Bonding*, Van Nostrand Reinhold, New York, 1971.
86. A. Rahman and F. H. Stillinger, *J. Am. Chem. Soc.*, **95**, 7943 (1973).
87. F. H. Stillinger and A. Rahman, *J. Chem. Phys.*, **61**, 4973 (1974).
88. L. Pauling, in D. Hadzi and H. W. Thompson, Eds., *Hydrogen Bonding*, Pergamon, New York, 1959, pp. 1–5.
89. M. D. Danford and H. A. Levy, *J. Am. Chem. Soc.*, **84**, 3965 (1962).
90. O. Ya. Samoilov, *Structure of Aqueous Electrolyte Solutions and the Hydration of Ions*, Consultants Bureau, New York, 1965.
91. J. W. Perram, *Mol. Phys.*, **20**, 1077 (1971).
92. G. Némethy and H. A. Scheraga, *J. Chem. Phys.*, **36**, 3382 (1962).
93. W. Kauzmann, *Advan. Protein Chem.*, **14**, 1 (1959).
94. Ref. 7, Table 4.10.
95. B. J. Alder and T. E. Wainwright, *Phys. Rev.*, **A1**, 18 (1970).
96. M. H. Ernst, E. H. Hauge, and J. M. J. van Leeuwen, *Phys. Rev. Lett.*, **25**, 1254 (1970).
97. N. K. Ailawadi and B. J. Berne, *J. Chem. Phys.*, **54**, 3569 (1971).
98. J. M. Deutch, *Ann. Rev. Phys. Chem.* **24**, 319–22 (1973).
99. F. H. Stillinger and A. Rahman, in *Molecular Motion in Liquids, Proceedings of the 24th Annual Meeting*, Paris, July 3–6, 1973, Société de Chimie Physique, 1974, pp. 479–492.
100. K. S. Singwi and A. Sjölander, *Phys. Rev.*, **119**, 863 (1960).
101. H. L. Lemberg and F. H. Stillinger, *J. Chem. Phys.*, **62**, 1677 (1975).
102. P. A. Kollman and C. F. Bender, *Chem. Phys. Lett.*, **21**, 271 (1973).
103. F. H. Stillinger and R. Lovett, *J. Chem. Phys.*, **49**, 1991 (1968).
104. L. D. Landau and E. M. Lifshitz, *Statistical Physics*, Addison-Wesley, Reading, Mass., 1958, p. 103.
105. M. Losonczy, J. W. Moskowitz, and F. H. Stillinger, *J. Chem. Phys.*, **59**, 3264 (1973).
106. H. Lischka, *Chem. Phys. Lett.*, **20**, 448 (1973).

107. G. H. F. Diercksen and W. P. Kraemer, *Theor. Chim. Acta*, **23**, 387 (1972).
108. W. P. Kraemer and G. H. F. Diercksen, *Theor. Chim. Acta*, **27**, 265 (1972).
109. J. E. Del Bene, *J. Chem. Phys.*, **55**, 4633 (1971).
110. K. Morokuma, *J. Chem. Phys.*, **55**, 1236 (1971).
111. M. Eigen, in S. Claesson, Ed., *Fast Reactions and Primary Processes in Chemical Kinetics, Proceedings of the 5th Nobel Symposium*, Interscience-Wiley, New York, 1967, p. 245.

# THE STRUCTURE OF POLAR FLUIDS*

S. A. ADELMAN AND J. M. DEUTCH

*Department of Chemistry, Massachusetts Institute of Technology,
Cambridge, Massachusetts*

## CONTENTS

## I. INTRODUCTION

In the past few years progress in understanding the structure of polar liquids has been substantial. Indeed, interesting developments occur so rapidly that even recent reviews[1] of this topic are badly out of date. Our purpose in this article is to present a survey of this recent progress and to communicate to the reader a status report of our knowledge about this important class of fluids.

The progress we have enjoyed is the result of contributions from several groups. Each of these groups has approached the problem of polar liquids in a different manner with a distinctive perspective. Happily, however, the work from these different groups has been complementary, and it is possible with a little effort to discern the relationships between the various approaches. We have attempted to emphasize these relationships in this article in order to convey the attractive sense of unity we perceive in the present status of knowledge on polar liquids.

The topics in which important progress has been made in the past few years, which are the subject of this chapter, include:

* Supported in part by the National Science Foundation.

1. An exact result for the structure of rigid polar liquids, including the molecular nature of the dielectric constant and properties of the two-particle distribution function. These developments are reported in Section II.

2. The construction and solution of explicit models for polar fluids, polar fluid mixtures, polar-ionic solutions based on the mean spherical model (MSM). This topic is the subject of Section III.A. While the MSM is an approximate model which suffers important limitations, it does have the virtue of providing the opportunity for explicit calculation and thus permits one to examine various physical ideas with some precision.

3. The use of various models and procedures to obtain explicit approximate expressions for the thermodynamical properties and other physical quantities of interest for polar fluids. A variety of models and results is presented in Section III.B.

4. An improved understanding of the molecular basis of dielectric relaxation. In Section IV we review arguments which we believe establish the proper relationship between the frequency-dependent dielectric constant and various equilibrium time correlation function expressions.

We speculate that prospects are excellent for continued rapid progress on the structure of polar fluids. This progress is likely to occur in two directions. First, we anticipate that detailed numerical work on the models discussed in this article, with perhaps some modifications, will provide much better insight into the behavior at a molecular level of complex liquids and solutions. For example, we are in a position to compute the activity coefficients or the potential of mean force between ions within a model for concentrated ionic solutions in molecular polar solvents. Second, the influence of polar forces on transport coefficients, of which the frequency-dependent dielectric constant is only the simplest example, remains to be fully explored. Progress in this area is possible because of the long-range nature of dipole forces, and conceivably models can be constructed that will lead to numerical determinations of transport coefficients.

Polar liquids are an important class of fluids which deserve the attention of both theorists and experimentalists. There is every reason to expect continued progress in this area during the next few years.

## A. Local Structure and the Dielectric Constant

In this section we discuss two closely related issues: the nature of the orientational order in polar fluids, and the molecular basis for a sample-independent dielectric constant. The microscopic theory dealing with these issues has been developed in three stages. First, the theory was developed for the rigid polar fluid model;[2] that is, for fluids composed of hypothetical polar molecules with zero polarizability. Next, the theory was extended to the case of real nonpolar fluids;[3] that is, fluids composed of polarizable but non-polar molecules. Finally, very recently a general theory for polarizable and polar fluids was presented.[4]

A major conclusion of all these theories is that fluids are endowed with a sample-independent dielectric constant. That is, each sample of a fluid has a dielectric constant characteristic of its molecular composition but independent of its shape or surroundings. This sample independence of the dielectric constant has long been assumed, but its deduction from statistical mechanics is far from trivial and is one of the main concerns of this section. The theory has also yielded several useful formal expressions for the dielectric constant.

In addition the theory has provided valuable insights into the nature of the orientational correlations in dielectric fluids. These correlations, as we have mentioned, are closely connected to the existence of a sample-independent dielectric constant. In fact, simple arguments (see Section I.B) show that the existence of the dielectric constant implies that the pair correlation function of a polar fluid depends on sample shape and surroundings. For the case of rigid polar fluids,[2] the sample-dependent part of the correlation function has been obtained in explicit form; thus a very detailed picture of the structure of the rigid polar model is now available.

For this reason, as well as others, our plan for this section is to present only the theory for rigid polar fluids in detail. We then discuss the results of the more sophisticated analysis for fluids with polarizability mainly by analogy with the rigid polar case.

## B. Preliminary Survey of the Problem

The purpose of this section is threefold. First, we present those results from the classical electrostatics of polar media needed to interpret the microscopic calculations of Section I.C. Second, we see why it is necessary to confirm the existence of a sample-independent dielectric constant. Third, we give a qualitative description of the microscopic theories in order that the reader may see clearly what motivates the quantitative treatment of Section I.C.

Suppose we consider an arbitrarily shaped molecular dielectric sample with volume $V$ embedded in surroundings of volume $W$ composed of a continuum dielectric[2] with dielectric constant $\varepsilon_0$. (see Fig. 1). Then, if we subject the sample to an applied electric field $\mathbf{E}_0(\mathbf{r})$, a polarization $\mathbf{P}(\mathbf{r})$ will be induced in the sample. The applied field $\mathbf{E}_0(\mathbf{r})$ is defined as the electric field at point $\mathbf{r}$ in the absence of the sample. The actual macroscopic field $\mathbf{E}(r)$ which the sample experiences is a superposition of the applied field, the fields produced by the polarization in the sample, and the fields produced by the polarization in $W$, which is induced by the polarization in the sample.

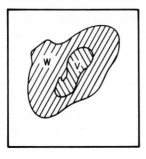

**Fig. 1** Illustrative dielectric geometry. The dielectric sample of volume $V$ is embedded in a dielectric continuum of volume $W$ with dielectric constant $\varepsilon_0$. This is the geometry employed in the analysis of Section II.C

First, consider the case $\varepsilon_0 = 1$. Then no polarization occurs in $W$ and the macroscopic field is

$$E(r) = E_0(r) + \int_V T(r, r') \cdot P(r') \, dr' \tag{1.1a}$$

where $T(r, r')$ is the dipole–dipole tensor

$$T(r, r') = \frac{3(r - r')(r - r')}{|r - r'|^5} - \frac{1}{|r - r'|^3}$$

The term $T(r, r') \cdot P(r')$ gives the contribution to the macroscopic field at $r$ from the polarization density at $r'$.

For the case $\varepsilon_0 \neq 1$, (1.1a) must be replaced by

$$E(r) = E_0(r) + \int_V D(r, r') \cdot P(r') \, dr' \tag{1.1b}$$

where

$$D(r, r') = T(r, r') + R_W(r, r'; \varepsilon_0) \tag{1.2}$$

The reaction-field tensor $R_W(r, r'; \varepsilon_0)$ depends on the shape of region $W$, as well as its dielectric constant $\varepsilon_0$. It gives the contribution to the macroscopic field at $r$ caused by that part of the polarization in $W$ induced by the polarization at $r'$. For $\varepsilon_0 = 1$, $R_W(r, r'; \varepsilon_0)$ vanishes and $D(r, r') = T(r, r')$.

Because of the long range of $D(r, r')$ and the dependence of $R_W(r, r'; \varepsilon_0)$ on surroundings, (1.1b) shows that $E(r)$ depends on both sample shape and surroundings. Also, the dielectric polarization at point $r$ should result from the actual macroscopic field $E(r)$ rather than from $E_0(r)$. Thus we expect that $P(r)$ also depends on sample shape and surroundings. Consequently, the relationship between $P(r)$ and $E_0(r)$ is sample-dependent and for arbitrary sample geometries can be very complicated. In classical electrostatics, however, it is assumed that the relationship between $\mathbf{P(r)}$ and $\mathbf{E(r)}$ is simple and independent of sample shape or surroundings. One assumes

$$\mathbf{P(r)} = \frac{\varepsilon - 1}{4\pi} \mathbf{E(r)} \tag{1.3}$$

where $\varepsilon$ is a sample-independent dielectric constant. An important result of the macroscopic theory we will describe is the deduction of the assumed form (1.3) from statistical mechanics.

For now, we accept the validity of (1.3) and examine its consequences. Combining equations (1.1) and (1.3) gives an integral equation for $\mathbf{P(r)}$, which can be explicitly solved for certain sample geometries and surroundings. This gives a sample-dependent relationship between $\mathbf{P(r)}$ and $\mathbf{E_0(r)}$. A relationship between $\mathbf{P(r)}$ and $\mathbf{E_0(r)}$ can also be found from a microscopic calculation.

To carry out the microscopic calculation, we require the statistical definition of the polarization $\mathbf{P(r)}$. This is

$$\mathbf{P(r)} = \left\langle \sum_{i=1}^{N} \boldsymbol{\mu}(\omega_i) \delta(\mathbf{r} - \mathbf{r}_i) \right\rangle_{E_0} \tag{1.4}$$

In the above equation $(\mathbf{r}_i, \omega_i)$ describes the position $\mathbf{r}_i$ and orientation $\omega_i$ of molecule $i$, while $\boldsymbol{\mu}(\omega_i)\delta(\mathbf{r} - \mathbf{r}_i)$ is the contribution of molecule $i$ to the dipole moment density at point $\mathbf{r}$. The sum runs over all $N$ molecules in the fluid, and the average is taken over the canonical distribution function in the presence of the field $\mathbf{E_0(r)}$.

To order $\mathbf{E_0(r)}$ the polarization can readily be computed in terms of $g(\mathbf{r}_1 \omega_1; \mathbf{r}_2 \omega_2)$, the pair distribution function of the fluid in the absence of the field. One finds

$$\mathbf{P(r)} = \frac{\beta\rho}{\Omega} \int_V d\omega \, d\omega' \, d\mathbf{r}' \, \boldsymbol{\mu}(\omega) \left[ \delta(\mathbf{r} - \mathbf{r}')\delta(\omega - \omega') + \frac{\rho}{\Omega} g(\mathbf{r}\omega; \mathbf{r}'\omega') \right] \boldsymbol{\mu}(\omega') \cdot \mathbf{E_0(r')} \tag{1.5a}$$

where $\rho$ is the density, and $\Omega$ is an angular-phase space volume equal to $4\pi$ for linear molecules and $(8\pi)^2$ for nonlinear molecules. Equation (1.5) can be further reduced to give an expression for $\mathbf{P(r)}$ in terms of $\langle \mathbf{M}^2 \rangle$ the mean-square total dipole moment of the fluid in the absence of the field. This in turn is related to $\langle \cos \gamma_{12} \rangle$ which is the average cosine of the angle between two representative molecular dipoles. The explicit relation is

$$\mathbf{P(r)} = \frac{\beta}{3V} \langle \mathbf{M}^2 \rangle \mathbf{E_0(r)} = \tfrac{1}{3}\beta\rho\mu^2(1 + \rho\langle \cos \gamma_{12} \rangle)\mathbf{E_0(r)} \tag{1.5b}$$

In the preceding analysis no mention is made of sample surroundings or geometry, and a general microscopic relation between $\mathbf{P(r)}$ and $\mathbf{E_0(r)}$ is found. The macroscopic relationship between $\mathbf{P(r)}$ and $\mathbf{E_0(r)}$, however, is, as we have discussed, sample-dependent. For example, for a spherical sample in vacuum, one finds from electrostatics:

$$\mathbf{P(r)} = \frac{3}{4\pi} \frac{\varepsilon - 1}{\varepsilon + 2} \mathbf{E_0(r)} \tag{1.6}$$

Comparison of (1.5b) and (1.6) then gives the following expression for $\langle \mathbf{M}^2 \rangle$:

$$\frac{3}{4\pi} \frac{\varepsilon - 1}{\varepsilon + 2} = \frac{\beta}{3V} \langle \mathbf{M}^2 \rangle_0 = \tfrac{1}{3}\beta\rho\mu^2(1 + \rho\langle \cos \gamma_{12} \rangle_0) \tag{1.7}$$

The subscript zero denotes a spherical sample in vacuum.

Similarly, for a spherical sample embedded in an infinite-continuum dielectric (subscript $\infty$) also with dielectric constant $\varepsilon$, one finds

$$\frac{\varepsilon - 1}{4\pi} \frac{2\varepsilon + 1}{3\varepsilon} = \frac{\beta}{3V} \langle \mathbf{M}^2 \rangle_\infty = \tfrac{1}{3}\beta\rho\mu^2(1 + \rho\langle \cos \gamma_{12} \rangle_\infty) \tag{1.8}$$

For samples with the shape of an infinite slab (subscript $l$), comparison of the microscopic and macroscopic calculations gives

$$\frac{\varepsilon - 1}{4\pi} \frac{1}{\varepsilon} = \frac{\beta}{V} \langle M_z^2 \rangle_l = \tfrac{1}{3}\beta\rho\mu^2(1 + \rho\langle \cos \gamma_{12} \rangle_l) \tag{1.9}$$

In the above equation $z$ is the direction perpendicular to the slab.

The conclusion to be drawn from this discussion is that $\langle M^2 \rangle$, $\langle \cos \gamma_{12} \rangle$, and therefore the pair distribution function must depend on sample shape and surroundings. That this shape dependence is nonnegligible can be seen by considering the ratio

$$\frac{\langle \mathbf{M}^2 \rangle_\infty}{\langle \mathbf{M}^2 \rangle_0} = \frac{(2\varepsilon + 1)(\varepsilon + 2)}{9\varepsilon}$$

which is about 20 for water.

This result may appear remarkable, since it seems to imply that the fluid structure inside the sample is strongly dependent on the boundaries and surroundings *in the absence of an external field*. Analysis shows that part of the pair distribution function *does* depend on shape and surroundings. This sample-dependent part, however, is inversely proportional to the sample volume$^2$ and thus does not affect the local structure in the interior of the fluid. Nevertheless the shape-dependent part of the distribution function can contribute significantly to expectation values like $\langle M^2 \rangle$ and $\langle \cos \gamma_{12} \rangle$, since it is of macroscopic range.[2]

This result raises the possibility that expectation values of *all* properties may depend on sample shape, since these expectation values are volume integrals over the pair distribution function. However, the statistical theory reviewed in Section II.C shows that this is not the case. In particular, it reveals that the dielectric constant and thermodynamic functions are *independent* of sample shape and surroundings even though they are defined microscopically as volume integrals over the pair distribution function.

Before proceeding to the statistical theory, it is necessary to introduce a bit more background. Combining (1.1) and (1.3) yields the macroscopic relation

$$\mathbf{P(r)} = \frac{\varepsilon - 1}{4\pi} \left[ \mathbf{E_0(r)} + \int_V \mathsf{D}(\mathbf{r}, \mathbf{r'}) \cdot \mathbf{P(r')} \, dr' \right] \qquad (1.10)$$

The strategy of the Nienhuis–Deutch formulation of the rigid polar fluid theory[2] is to cast the microscopic expression for $\mathbf{P(r)}$ in the form of (1.10). This then yields a microscopic formula for $\varepsilon - 1/4\pi$ which not surprisingly is a volume integral over the pair distribution function. What is very interesting is that the expression for $\varepsilon - 1/4\pi$ is an integral over only *part* of the pair distribution function and that this part is independent of sample shape and surroundings and also is of microscopic range. Thus the Nienhuis–Deutch theory provides a statistical formula for the dielectric constant $\varepsilon$, which shows explicitly that $\varepsilon$ is independent of sample shape and surroundings.

Two other related formulations of the theory of the dielectric constant for rigid polar fluids are available. Like the Nienhuis–Deutch method described above, these alternative formulations involve careful comparison of the microscopic and macroscopic theories. One approach is due to Ramshaw.[5] His idea is to invert the microscopic relationship (1.5) between $\mathbf{P(r)}$ and $\mathbf{E_0(r)}$ and thus obtain $\mathbf{E_0(r)}$ as a functional of $\mathbf{P(r)}$. Then $\mathbf{E_0(r)}$ is eliminated from this relation using (1.1). The resulting relationship is compared with the constitutive formula (1.3). The existence of the dielectric constant is verified, and an expression for the dielectric constant in terms of the Ornstein–Zernike direct correlation function is obtained. Ramshaw's motivation for developing the theory in this manner is that the key quantity that enters his formulation is the direct correlation function which is of simpler structure than the pair correlation function.

The Nienhuis–Deutch and the Ramshaw theories have proceeded from the microscopic relation (1.5) for the linear response of the polar fluid to an applied field. One can also probe this response by introducing impurities that interact with the polar fluid through electrostatic forces. Comparison of the predictions of the microscopic and macroscopic theories for the interaction energy of the impurities then gives a statistical formula for $\varepsilon$ which can be shown to be shape-independent.

For example, classical electrostatics predicts that the potential energy of interaction from a pair of point charges in a continuous medium of dielectric constant $\varepsilon$ is

$$\omega(\mathbf{r}) = \frac{z_1 z_2 e^2}{\varepsilon r}$$

That is, $\omega(r)$ is the vacuum Coulomb energy reduced by a factor $\varepsilon^{-1}$.

The interaction of energies of impurities can also be computed from statistical mechanics. Asymptotically, we expect that the potential of mean force should approach the classical value; structure in the potential of mean force due to the granularity of the polar solvent appears only at small inter-particle separations. Thus by computing the asymptotic potential of mean force and comparing the result with the electrostatic prediction, one can obtain a formal expression for the dielectric constant. It is important to verify that the same result for the dielectric constant emerges for all test impurities, and that this result is shape-independent. Further, one must show that this expression is identical to that found by computing the polar-ization in an applied field. For the case of a rigid polar fluid, this program[6] was carried out by Høye and Stell.[7] We also discuss the related work of Nienhuis and Deutch.[2]

### C. Theories of the Dielectric Constant

In this section we discuss the various theories of the rigid polar fluid dielectric constant mentioned in the Section I.B.

We begin with the Nienhuis–Deutch theory[2] based on calculation of the polarization. We rewrite (1.5) in the form

$$\mathbf{P}(\mathbf{r}) = \frac{\beta\rho}{\Omega} \left( \int_V d\boldsymbol{\omega}\, d\boldsymbol{\omega}'\, d\mathbf{r}' \right.$$

$$\left. \times \boldsymbol{\mu}(\boldsymbol{\omega}) \left[ \delta(\boldsymbol{\omega} - \boldsymbol{\omega}')\delta(\mathbf{r} - \mathbf{r}') + \frac{\rho}{\Omega} h(\mathbf{r}\boldsymbol{\omega}; \mathbf{r}'\boldsymbol{\omega}') \right] \boldsymbol{\mu}(\boldsymbol{\omega}') \cdot \mathbf{E}_0(\mathbf{r}') \right) \quad (1.11)$$

Note that we have replaced the pair distribution function $g(\mathbf{r}, \boldsymbol{\omega}; \mathbf{r}', \boldsymbol{\omega}')$ by the pair correlation function

$$h(\mathbf{r}\boldsymbol{\omega}; \mathbf{r}'\boldsymbol{\omega}') = g(\mathbf{r}\boldsymbol{\omega}; \mathbf{r}'\boldsymbol{\omega}') - 1$$

This is permissible, since the extra contribution so introduced vanishes when we integrate over angles.

We rewrite (1.11) in a shorthand operator notation as

$$\mathbf{P} = \mathsf{K} \cdot \mathbf{E}_0 \quad (1.12a)$$

where

$$K = \frac{\beta \rho}{\Omega} \boldsymbol{\mu} \left( 1 + \frac{\rho}{\Omega} h \right) \boldsymbol{\mu} \qquad (1.12b)$$

The quantities in equations (1.11) and (1.12) are now interpreted to be operators. Here and below we use the convention that the operators representing functions of the variables $(\mathbf{r}, \boldsymbol{\omega})$ and $(\mathbf{r}', \boldsymbol{\omega}')$ are denoted by the same symbols as the functions but without any functional dependence indicated.

The idea of the Nienhuis–Deutch theory is to express $\mathbf{P}$ in the form of (1.10) which we rewrite as

$$\mathbf{P} = \frac{\varepsilon - 1}{4\pi} (\mathbf{E}_0 + \mathbf{D} \cdot \mathbf{P}) \qquad (1.13)$$

To do this one introduces the Ornstein–Zernike relation between the pair correlation function and the direct correlation function $c(\mathbf{r}, \boldsymbol{\omega}; \mathbf{r}', \boldsymbol{\omega}')$. This relation is

$$h(\mathbf{r}\boldsymbol{\omega}; \mathbf{r}'\boldsymbol{\omega}') = c(\mathbf{r}\boldsymbol{\omega}; \mathbf{r}'\boldsymbol{\omega}) + \frac{\rho}{\Omega} \int_V c(\mathbf{r}\boldsymbol{\omega}; \mathbf{r}''\boldsymbol{\omega}'') h(\mathbf{r}''\boldsymbol{\omega}; \mathbf{r}'\boldsymbol{\omega}') \, d\mathbf{r}'' \, d\boldsymbol{\omega}'' \quad (1.14)$$

or in operator notation

$$h = c + \frac{\rho}{\Omega} ch \qquad (1.15)$$

The direct correlation function $c$ can be expressed as a sum of Mayer graphs.[8] By analyzing the range and possible shape dependence of these graphs, Nienhuis and Deutch show that $c$ can be written in the form

$$c = c_0 + \beta \boldsymbol{\mu} \cdot \mathbf{D} \cdot \boldsymbol{\mu} \qquad (1.16)$$

where $c_0$ is of microscopic range and is independent of sample shape and surroundings. Thus the only term in the direct correlation function that can contribute to the shape dependence of the pair correlation function is $-\beta$ times the effective dipole–dipole potential $-\boldsymbol{\mu} \cdot \mathbf{D} \cdot \boldsymbol{\mu}$ in the sample.

We now define a short-range part of the correlation function $h_0$ by the relation

$$h_0 = c_0 + \frac{\rho}{\Omega} c_0 h_0 \qquad (1.17)$$

The quantity $h_0$ is independent of sample geometry and is of microscopic range. All shape dependence is in the long-range part of the correlation function defined by Nienhuis and Deutch as $h_1 = h - h_0$.

We now develop an integral equation for $h_1$. We actually develop a more general integral equation which is of frequent use in the theory of fluids. The general problem can be stated as follows. Given an arbitrary division of the direct correlation function $c = c_0 + c_1$, and given an $h_0$ related to $c_0$ by (1.17), express $h_1 = h - h_0$ in terms of $h_0$ and $c_1$; that is, eliminate $c_0$ from the final result.

This problem was solved by Lebowitz, Stell, and Baer[9] using a graphical method. For completeness we present here a short algebraic derivation.

We rewrite (1.15) as

$$1 + \frac{\rho}{\Omega} h = \left(1 - \frac{\rho}{\Omega} c\right)^{-1} = \left[1 - \frac{\rho}{\Omega}(c_0 + c_1)\right]^{-1} \tag{1.18}$$

We have analogously for (1.17):

$$1 + \frac{\rho}{\Omega} h_0 = \left(1 - \frac{\rho}{\Omega} c_0\right)^{-1} \tag{1.19}$$

Now, introducing the operator identity

$$(1 + X + Y)^{-1} = (1 + X)^{-1} - (1 + X)^{-1} Y (1 + X + Y)^{-1}$$

into the rhs of (1.18) and then using (1.18) and (1.19), we find

$$1 + \frac{\rho}{\Omega} h = \left(1 + \frac{\rho}{\Omega} h_0\right) + \frac{\rho}{\Omega}\left(1 + \frac{\rho}{\Omega} h_0\right) c_1 \left(1 + \frac{\rho}{\Omega} h\right)$$

In terms of $h_1 = h - h_0$, this equation becomes

$$h_1 = \left(1 + \frac{\rho}{\Omega} h_0\right) c_1 \left(1 + \frac{\rho}{\Omega} h_0\right) + \frac{\rho}{\Omega}\left(1 + \frac{\rho}{\Omega} h_0\right) c_1 h_1 \tag{1.20a}$$

Equation (1.20) is our final result. Note that it is just as rearrangement of the Ornstein–Zernike (OZ) equation and thus is completely general. No assumption of pairwise additivity of the intermolecular forces is invoked; nor is the validity of (1.20) restricted to infinite-volume systems.

At this time a discussion of our operator notation may be clarifying, since this notation is not conventional in the theory of classical fluids. Similar notation, however, is commonplace in other areas of physics, for example, in collision theory.

The operators 1 and $h_0$, for example, have the following explicit "matrix" representations in coordinate space:

$$\langle X_1 | 1 | X_2 \rangle = \delta(X_1 - X_2)$$
$$\langle X_1 | h_0 | X_2 \rangle = h(X_1 X_2)$$

The operator $[1 + (\rho/\Omega)h_0]c_1h_1$ consequently has the coordinate space representation

$$\left\langle X_1 \left| \left(1 + \frac{\rho}{\Omega}h_0\right)c_1h_1 \right| X_2 \right\rangle$$

$$= \int_V \int_V \langle X_1|1|X_3\rangle\langle X_3|c_1|X_4\rangle\langle X_4|h_1|X_2\rangle \, dX_3 \, dX_4$$

$$+ \frac{\rho}{\Omega} \int_V \int_V \langle X_1|h_0|X_3\rangle\langle X_3|c_1|X_4\rangle\langle X_4|h_1|X_2\rangle \, dX_3 \, dX_4$$

$$= \int_V c(X_1X_3)h(X_3X_2) \, dX_3$$

$$+ \frac{\rho}{\Omega} \int_V \int_V h_0(X_1X_3)c(X_3X_4)h_1(X_4X_2) \, dX_3 \, dX_4$$

The operator

$$\left(1 - \frac{\rho}{\Omega}c\right)^{-1} = 1 + \frac{\rho}{\Omega}c + \left(\frac{\rho}{\Omega}\right)^2 c^2 + \cdots$$

and this has the coordinate representation

$$\left\langle X_1 \left| 1 - \frac{\rho}{\Omega}c \right| X_2 \right\rangle = \delta(X_1X_2)$$

$$+ \frac{\rho}{\Omega}c(X_1X_3) + \left(\frac{\rho}{\Omega}\right)^2 \int_V dX_3 \, c(X_1X_3)c(X_3X_2) + \cdots$$

With this discussion the nature and use of our shorthand notation should be clear, and we can return to the Nienhuis–Deutch theory.

Using (1.16) in (1.20) gives the integral equation for $h_1$, the shape- and surroundings-dependent part of the pair correlation function:

$$h_1 = \left(1 + \frac{\rho}{\Omega}h_0\right)(\beta\boldsymbol{\mu} \cdot \mathbf{D} \cdot \boldsymbol{\mu})\left(1 + \frac{\rho}{\Omega}h_0\right) + \frac{\rho}{\Omega}\left(1 + \frac{\rho}{\Omega}h_0\right)(\beta\boldsymbol{\mu} \cdot \mathbf{D} \cdot \boldsymbol{\mu})h_1$$

$$(1.20b)$$

Inserting this result in (1.12b) gives an integral equation for the polarization kernel $\mathbf{K}$. Then combining this integral equation with (1.12a) gives the integral equation for $\mathbf{P}$:

$$\mathbf{P} = \frac{\beta\rho}{\Omega}\left[\boldsymbol{\mu}\left(1 + \frac{\rho}{\Omega}h_0\right)\boldsymbol{\mu} \cdot \mathbf{E}_0 + \boldsymbol{\mu}\left(1 + \frac{\rho}{\Omega}h_0\right)\boldsymbol{\mu} \cdot \mathbf{D} \cdot \mathbf{P}\right] \qquad (1.21)$$

Because $h_0$ is of the molecular range, and because $\mathbf{E}_0$ and $\mathbf{P}$ vary negligibly over microscopic distances, (1.21) can be written in the more useful form:

$$\mathbf{P} = \frac{1}{3}\frac{\beta\rho}{\Omega}\left[\boldsymbol{\mu}\left(1 + \frac{\rho}{\Omega}h_0\right)\cdot\boldsymbol{\mu}\right](\mathbf{E}_0 + \mathbf{D}\cdot\mathbf{P}) \qquad (1.22)$$

Comparison with (1.13) yields

$$\frac{\varepsilon - 1}{4\pi} = \frac{\beta\rho}{3}\boldsymbol{\mu}\left(1 + \frac{\rho}{\Omega}h_0\right)\cdot\boldsymbol{\mu} \qquad (1.23)$$

This expression for the dielectric constant is clearly independent of shape and surroundings, because it involves only the short-range correlation function $h_0$.

Equation (1.23) can be written in terms of an effective dipole moment $\boldsymbol{\mu}_{\text{eff}} = \boldsymbol{\mu} + \boldsymbol{\kappa}$, where

$$\boldsymbol{\kappa} = \frac{\rho}{\Omega}h_0\boldsymbol{\mu}$$

More explicitly, we have

$$\boldsymbol{\kappa}(\omega) = \frac{\rho}{\Omega}\int_V h_0(\mathbf{r}\,\omega;\mathbf{r}'\,\omega')\,\boldsymbol{\mu}(\omega')\,d\omega'\,d\mathbf{r}' \qquad (1.24)$$

and thus

$$\frac{\varepsilon - 1}{4\pi} = \tfrac{1}{3}\beta\rho\boldsymbol{\mu}\cdot[\boldsymbol{\mu} + \boldsymbol{\kappa}] = \tfrac{1}{3}\beta\rho\boldsymbol{\mu}\cdot\boldsymbol{\mu}_{\text{eff}} \qquad (1.25)$$

which is the form for the dielectric constant given by Nienhuis and Deutch.[2]

Except for the use of (1.16), nowhere in this calculation have we invoked pairwise additivity of the intermolecular forces. Although (1.16) was derived by Nienhuis and Deutch with the assumption of pairwise additivity, we do not believe that this assumption is necessary. Recent work by Wertheim[4] appears to give a method for removing this assumption. Consequently, we believe that the existence of a shape-independent dielectric constant for general rigid polar fluids is established and that (1.25) is valid for rigid polar fluids with nonadditive as well as pairwise additive forces.

We now discuss another valuable result of the Nienhuis–Deutch analysis. For the case of large intermolecular separations, $h \cong h_1$ is proportional to the potential of mean force between a pair of dipoles in the fluid. For large separations we expect that this potential of mean force will *resemble* the form predicted by classical electrostatics. We have here emphasized the word resemble, since the asymptotic potential of a pair of molecular dipoles in a molecular medium depends on short-range anisotropic interactions with neighboring molecules as well as long-range dipolar interactions with the whole medium.[2] Only the latter type of interaction enters a continuum electrostatic calculation. Thus we cannot expect exact agreement with classical electrostatics. We return to this point below.

Still one expects simple behavior in the asymptotic limit, and so it is not surprising that Nienhuis and Deutch were able to obtain an asymptotic solution to (1.20) for the particularly simple case of samples embedded in continua of identical dielectric constant. They find asymptotically

$$h(\mathbf{r}_1\boldsymbol{\omega}_1; \mathbf{r}_2\boldsymbol{\omega}_2) = \frac{\beta}{\varepsilon}\,\boldsymbol{\mu}_{\mathrm{eff}}(\boldsymbol{\omega}_1) \cdot [\mathsf{T}(\mathbf{r}_1, \mathbf{r}_2) + \mathsf{R}^*_{V+W}(\mathbf{r}_1, \mathbf{r}_2; \varepsilon^{-1})] \cdot \boldsymbol{\mu}_{\mathrm{eff}}(\boldsymbol{\omega}_2) \quad (1.26)$$

Here $\mathsf{R}^*_{V+W}$ is the reaction-field tensor for a region *external* to $V + W$ with dielectric constant $\varepsilon^{-1}$. While not explicitly noted by Nienhuis and Deutch, it is possible to show[10] that the reaction-field tensor $\mathsf{R}^*_{V+W}(\mathbf{r}, \mathbf{r}'; \varepsilon^{-1})$ is identical to the reaction-field tensor $\mathsf{R}_{V+W}(\mathbf{r}, \mathbf{r}'; \varepsilon)$. The physical interpretation of the latter reaction field, in analogy to (1.2), is the field produced at $\mathbf{r}'$ by a unit dipole at $\mathbf{r}$ when the volume $V + W$ is filled with a dielectric of dielectric constant $\varepsilon$ from the surface polarization at the boundary of the sample $V + W$.

Equation (1.26) has several interesting features. The quantity $\mathsf{R}^*_{V+W}$ depends on the shape of $V + W$ and is the only sample-dependent part of the pair distribution function. $\mathsf{R}^*_{V+W}$ is proportional to $(V + W)^{-1}$ and thus does not significantly affect the fluid structure inside the sample. However, it is of macroscopic range and thus can contribute in a shape-dependent way to average values when the function to be averaged extends over the entire volume. The quantity $\mathsf{T}$ is not itself shape-dependent. However, it also is of macroscopic range and also can contribute to the shape dependence of properties.

The tensor $\mathsf{R}^*_{V+W}$ can be found explicitly for certain geometries. This calculation is feasible, in particular, for the cases discussed in Section I.B, the sphere in vacuum, the sphere embedded in a continuum dielectric of identical dielectric constant, and the infinite slab.

For these geometries an explicit form for $h_1$ is therefore available. This expression can then be employed to compute the shape-dependent property $\langle \mathbf{M}^2 \rangle$ or, alternatively, $\langle \cos \gamma_{12} \rangle$.

Nienhuis and Deutch carried out this calculation. In all three cases they found complete agreement between their theory and the results of Section I.B.

The Nienhuis–Deutch analysis thus provides a satisfying quantitative explanation for the apparently anomalous behavior described in Section I.B.

We now turn to Ramshaw's formulation[5] of the dielectric constant for rigid polar fluids. Ramshaw has presented an interesting new formal expression for the dielectric constant in terms of the direct correlation function for a fluid of polar *axially symmetric* molecules. His expression is

$$\frac{3}{4\pi}\frac{\varepsilon - 1}{\varepsilon + 2} = \frac{1}{3}\beta\rho\mu^2\left(1 - \frac{\rho}{4\pi}A\right)^{-1} \tag{1.27}$$

where

$$A = \frac{1}{4\pi}\int d\mathbf{r}_{12}\int d\omega_1\, d\omega_2\, c(r_{12}; \omega_1\omega_2)\cos\gamma_{12} \tag{1.28}$$

where $\gamma_{12}$ is the angle between dipoles 1 and 2. Ramshaw arrived at this expression based on the assumption that the direct correlation function was of the form

$$c(1, 2) = c_s(r_{12}) + c_\Delta(r_{12})\cos\gamma_{12} + c_0(r_{12})F(\mathbf{r}_1\omega_1; \mathbf{r}_2\omega_2) \tag{1.29a}$$

where

$$F(\mathbf{r}_1\omega_1; \mathbf{r}_2\omega_2) = r_{12}^3\mu^{-2}\boldsymbol{\mu}(\omega_1)\cdot\mathbf{T}(\mathbf{r}_{12})\cdot\boldsymbol{\mu}(\omega_2) \tag{1.29b}$$

According to (1.28) and (1.29) $A$ may be reexpressed as

$$A = \frac{(4\pi)^2}{3}\int_0^\infty r^2 c_\Delta(r)\, dr \tag{1.30}$$

Deutch[1] has shown that for this special form of $c$, for axially symmetric system, it is possible to compute $\boldsymbol{\mu}_{eff}$ [see (1.24)], hence $\varepsilon$ in the Nienhuis–Deutch theory and to demonstrate the equivalence of the two expressions for $\varepsilon$, (1.25) and (1.30). More recently, Høye and Stell[6] provided justification for Ramshaw's result for axially asymmetric systems which avoids Ramshaw's assumption about the special form of the direct correlation function (1.29). However, for the case of potentials of arbitrary orientation dependence, the Ramshaw expression for $\varepsilon$ cannot be expected to be correct.

Now we turn to the test impurity method of determining the dielectric constant. We recall that this procedure involves examining the asymptotic potential between test impurities and then comparing the result with classical electrostatics.[6]

We only present an outline of the method, since the detailed calculations are quite lengthy.

Although in the test impurity we assume that the impurity is infinitely dilute, it is interesting to first assume finite dilution[11] and then pass to the limit of infinite dilution. Thus we consider a two-component mixture composed of our polar fluid and a solute impurity.

The Ornstein–Zernike matrix equation for the mixture is

$$\begin{pmatrix} h_{XX} - c_{XX} & h_{XD} - c_{XD} \\ h_{DX} - c_{DX} & h_{DD} - c_{DD} \end{pmatrix} = \Omega^{-1} \begin{pmatrix} c_{XX} & c_{XD} \\ c_{DX} & c_{DD} \end{pmatrix} \begin{pmatrix} \rho_X & 0 \\ 0 & \rho_D \end{pmatrix} \begin{pmatrix} h_{XX} & h_{XD} \\ h_{DX} & h_{DD} \end{pmatrix} \quad (1.31)$$

Here $X$ denotes impurity and $D$ denotes dipole. In the absence of the impurity, or at infinite dilution, $h_{DD}$, $c_{DD}$, and $\rho_D$ reduce to $h$, $c$, and $\rho$, the pure polar fluid correlation functions, and density. We write out two of the components of (1.31) explicitly:

$$h_{XX} = c_{XX} + \frac{\rho_X}{\Omega} c_{XX} h_{XX} + \frac{\rho_D}{\Omega} c_{XD} h_{DX}$$

$$h_{DX} = c_{DX} + \frac{\rho_X}{\Omega} c_{DX} h_{XX} + \frac{\rho_D}{\Omega} c_{DD} h_{DX}$$

Formally solving the second of these equations for $h_{DX}$ and inserting the result into the first equation gives an effective OZ equation for the solute correlation formation. We find

$$h_{XX} = c_{XX}^{\text{eff}} + \frac{\rho_X}{\Omega} c_{XX}^{\text{eff}} h_{XX} \quad (1.32)$$

Note that (1.32) is of one-component form with the effective correlation function

$$c_{XX}^{\text{eff}} = c_{XX} + \frac{\rho_D}{\Omega} c_{XD} \left( 1 - \frac{\rho_D}{\Omega} c_{DD} \right)^{-1} c_{DX} \quad (1.33)$$

We can easily extend our analysis to the general case of an $m$-component fluid with $m$ solvents and $m - n$ solutes by means of a matrix partitioning technique. This then can be made the basis for a general theory of solutions.[11] For the present case we only require the results for three-component solutions composed of our solvent and two impurities $X$ and $Y$. This allows us to treat interactions between unlike impurities, for example, a charge–quadrupole interaction.

We find for this case that $h_{XX}$, $h_{XY}$, and $h_{YY}$ satisfy a two-component Ornstein–Zernike equation of the form of (1.31), but with effective direct correlation functions of the form of (1.33), that is,

$$c_{ij}^{\text{eff}} = c_{ij} + \frac{\rho_D}{\Omega} c_{iD} \left( 1 - \frac{\rho_D}{\Omega} c_{DD} \right)^{-1} c_{Dj} \quad (1.34)$$

where $i = X$ or $Y$.

Equation (1.34) is exact for all solute concentrations, and we illustrate its utility in Section I.C when we discuss the mean spherical model (MSM) for an ionic solute in a polar solvent.

However, in the present discussion we only require results valid at infinite dilution. In this limit,

$$h_{ij}(\text{infinite dilution}) = c_{ij}^{\text{eff}}(\text{infinite dilution}) = c_{ij} + \frac{\rho}{\Omega} c_{iD}\left(1 + \frac{\rho_D}{\Omega} h\right)c_{Dj}$$

(1.35)

where all quantities on the rhs of (1.35) are to be evaluated at infinite dilution. Note that we have used (1.18).

To obtain the asymptotic potentials of mean force $h_{ij}$ between impurities $i$ and $j$, one may evaluate the $r$-space representation of (1.35) explicitly for large interparticle separations. This is in essence the method of Nienhuis and Deutch.[2] Alternatively, one can examine the Fourier space representation (1.35) for the case of small wave vectors. This is essentially the procedure of Høye and Stell.[6]

The calculations of Nienhuis and Deutch and Høye and Stell are not, however, equivalent. Nienhuis and Deutch examine the asymptotic potential of mean force for realistic impurity molecules which interact with the solvent through anistropic short-range forces as well as through the dipole–impurity electrostatic interactions. Their calculation cannot be compared with continuum electrostatics, since there is no electrostatic analog of the orienting effects of the short-range torques. Høye and Stell study the case of fictitous impurities that interact only with the solvent via long-range electrostatic forces. Their calculation can be directly compared with electrostatics and provides a rigorous alternative method of obtaining the dielectric constant.

Clearly, the Nienhuis–Deutch and Høye–Stell calculations are complementary. The former gives information about the real interactions in the fluid, while the latter provides a rigorous method for obtaining the dielectric constant.

The results of the Høye–Stell theory are completely consistent with classical electrostatics; the theory yields the same formula for the dielectric constant for all impurity combinations studied. Moreover, this formula is equivalent to the Nienhuis–Deutch formula, and for the special case of axially symmetric molecules is also equivalent to Ramshaw's result. Thus Høye–Stell results are consistent with those obtained by calculating the response of the fluid to an external field.

The result of the Nienhuis–Deutch calculation is also very interesting. They find agreement with classical electrostatics for the interaction between impurity charges. But for the interaction of higher multipolar impurities, Nienhuis and Deutch find deviations from classical electrostatics.

These deviations arise since the anistropic short-range forces between the solvent and impurity can align the solvent molecules in the vicinity of the impurity. The effect of this alignment on the potential of mean force is to make the impurity dipole moment appear different from its true value.

For impurities with zero dipole moment, for example, quadrupolar impurities, this is particularly striking, since for short-range torques of the appropriate symmetry such molecules appear to acquire dipole moments. Nienhuis and Deutch give expressions for the effective interactions of impurities in a polar fluid at large separations in terms of the solvent dielectric constant and the effective dipole moment. In particular, they give a formal expression for the effective dipole moment $\mathbf{P}_Q(\omega)$ of a quadrupolar impurity as

$$\frac{1}{V\Omega} \mathbf{P}_Q(\omega) = \int d\mathbf{r}' \, d\omega' \, G_{ab}^{(0)}(\mathbf{r}'\omega'; \mathbf{r}\omega)\mu(\omega') \tag{1.36}$$

The quantity $G_{ab}^{(0)}(\mathbf{r}'\omega'; \mathbf{r}\omega)$ is the short-range part of the quadrupole–dipole interaction and is closely related to $c_{iD}$ (where $i$ refers to quadrupole) in (1.34). The exact definition is given by Nienhuis–Deutch.[2] The integration is over solvent molecule coordinates, and $\mu(\omega')$ is the dipole moment of a solvent molecule.

The asymptotic potential of mean force between an impurity quadrupole and a solvent dipole or another impurity is nearly identical in form to the interaction of a solvent molecule with a second solvent molecule or with an impurity. One simply replaces $\mu_{\text{eff}}(\omega)$ by $\mathbf{P}_Q(\omega)$. For example, the interaction between a solvent molecule $(\mathbf{r}_1\omega_1)$ and an impurity quadrupole $(\mathbf{r}_2\omega_2)$ is [cf. (1.20)]

$$\lim_{r_{12} \to \infty} h_{DQ}(\mathbf{r}_1\omega_1; \mathbf{r}_2\omega_2) = \beta\varepsilon^{-1}\mu_{\text{eff}}(\omega_1) \cdot [\mathbf{T}(\mathbf{r}_1\mathbf{r}_2) + \mathbf{R}^*_{V+W}(\mathbf{r}_1\mathbf{r}_2; \varepsilon^{-1})] \cdot \mathbf{P}_Q(\omega_2)$$
$$\tag{1.37}$$

## D. Polar and Polarizable Fluids

We now give a brief review of Wertheim's[4] general theory of the dielectric constant for polar and polarizable fluids.

For such fluids the instantaneous molecular dipole moment depends on the configuration of all the other molecules. This is because the field that polarizes a given molecule is the microscopic field (not an average local field) which includes the superposition of the electric fields from all the molecular dipole moments in the sample.

Clearly, the existence of a many-body molecular dipole moment greatly complicates the statistical mechanics of fluids with polarizability, that is, real fluids. Despite this, Wertheim has succeeded in formulating a remarkably elegant theory of the dielectric constant for these systems. The results of his theory provide the natural generalizations of the rigid polar fluid results.

Wertheim's analysis begins with the expression for the total molecular dipole moment $\mathbf{M}_i$ $(1, 2, \ldots, N)$ of molecule $i$ which is part of a fluid subjected to an applied field $\mathbf{E}_0(\mathbf{r})$. This moment is given in terms of an approximate one-body polarizability tensor $\boldsymbol{\alpha}$ as

$$\mathbf{M}_i(\omega_i) = \boldsymbol{\mu}(\omega) + \boldsymbol{\alpha} \cdot \mathbf{E}_i \qquad (1.38)$$

where $\mathbf{E}_i$ is the actual microscopic field at molecule $i$ given by

$$\mathbf{E}_i = \mathbf{E}_0(\mathbf{r}_i) + \sum_{\substack{j=1 \\ j \neq i}}^{N} \mathbf{T}_{ij} \cdot \mathbf{M}_j \qquad (1.39)$$

Wertheim next calculates the electrostatic energy for an arbitrary configuration of the fluid. This is simply expressed in terms of $\mathbf{M}_i$, $\mathbf{E}_i$, and $\boldsymbol{\alpha}$ as a sum of dipole–dipole, field–dipole, and electronic polarization energies. This expression cannot be directly used in a statistical calculation; the instantaneous dipole moments $\mathbf{M}_i$ and instantaneous fields $\mathbf{E}_i$ must first be eliminated in favor of the permanent moments $\boldsymbol{\mu}_i$ and the applied field $\mathbf{E}_0(\mathbf{r})$ using (1.38) and (1.39). This leads to a complicated form for the electrostatic energy involving sums over polarizability tensor and dipole–dipole tensor chains.

Wertheim then constructs a grand canonical distribution function from the electrostatic and nonelectrostatic contributions to the potential energy and proceeds to calculate the polarization $\mathbf{P}(\mathbf{r})$. To linear order in $\mathbf{E}_0(\mathbf{r})$, it can be written:

$$\mathbf{P}(\mathbf{r}) = \int_V d\omega \, d\omega' \, d\mathbf{r}' \, [\mathbf{B}(\mathbf{r}'\omega')\delta(\mathbf{r}\omega; \mathbf{r}'\omega') + \mathbf{A}(\mathbf{r}\omega; \mathbf{r}'\omega')] \cdot \mathbf{E}_0(\mathbf{r}') \qquad (1.40)$$

The quantities $\mathbf{A}$ and $\mathbf{B}$ at first appear very complicated, because of the complicated form of the electrostatic interaction energy. But the formal resemblance of (1.11) and (1.40) suggests that, if carefully analyzed, $\mathbf{A}$ and $\mathbf{B}$ might be found to be simple in form. This is because we expect that the dielectric constant exists and that a correspondence of (1.40) with the simple results of classical electrostatics can be made.

After a sophisticated graphical analysis, Wertheim is indeed able to show that $\mathbf{A}$ and $\mathbf{B}$ have simple structures. A key step in this analysis is the introduction of a renormalized polarizability $\boldsymbol{\alpha}'$ and a renormalized dipole moment $\boldsymbol{\mu}'$.

The renormalized dipole moment $\boldsymbol{\mu}'$ turns out to be the average value of the molecular dipole moment $\mathbf{M}_i$ $(1, 2, \ldots, N)$ in zero applied field. The renormalized polarizability $\boldsymbol{\alpha}'$ is related to $\boldsymbol{\mu}'$ by the functional derivative relationship:

$$\boldsymbol{\alpha}'(\omega) = \frac{\delta\boldsymbol{\mu}'(\omega)}{\delta\boldsymbol{\mu}(\omega)} \cdot \boldsymbol{\alpha}(\omega) \qquad (1.41)$$

In terms of these renormalized quantities, Wertheim finds for A and B

$$B(1) = \rho\Omega^{-1}[\beta\mu'(\omega_1)\mu(\omega_1) + \alpha'(\omega_1)]$$ (1.42)

and

$$A(1, 2) = (\rho\Omega^{-1})^2[\beta\mu'(\omega_1)h(1, 2)\mu'(\omega_2)$$
$$+ \mu'(\omega_1)h(1, 2) \cdot \alpha'(\omega_2) + \alpha'(\omega_1) \cdot h(2, 1)\mu'(\omega_2)$$
$$+ \beta^{-1}\alpha'(\omega_1) \cdot H(1, 2) \cdot \alpha'(\omega_2)]$$ (1.43)

In (1.43), $h(1, 2)$ is the pair correlation function and

$$\mathbf{h}(1, 2) = \frac{\delta h(1, 2)}{\delta\mu'(\omega_2)} \qquad \mathbf{H}(1, 2) = \frac{\delta^2 h(1, 2)}{\delta\mu'(\omega_1)\delta\mu'(\omega_2)}$$ (1.44)

For the rigid polar fluid $\alpha' = \alpha = 0$ and $\mu' = \mu$ and (1.40), (1.42), and (1.43) reduce to (1.11). For the nonpolar but polarizable case, Wertheim similarly finds agreement with the polarization formula obtained in his earlier analysis for nonpolar fluids.[3]

From (1.40), (1.42), and (1.43), and the graphical definitions of the quantities in these formulas, Wertheim establishes the existence of a shape-independent dielectric constant for the general polar and polarizable fluid case.

Wertheim's analysis leads to other significant results. Perhaps most important is his realization that the direct correlation function, heretofore regarded as the fundamental building block quantity in the theory of fluids, loses its central role when one includes polarizability. Analysis leads Wertheim to define new basic quantities $w(1, 2)$, $\mathbf{w}(1, 2)$, and $\mathbf{W}(1, 2)$. The function $w(1, 2)$ is defined by Wertheim in terms of its constituent graphs, while $\mathbf{w}(1, 2)$ and $\mathbf{W}(1, 2)$ are derived from $w(1, 2)$ through relationships identical in form to (1.44). For the rigid polar case $w(1, 2)$ is identical to the direct correlation function $c(1, 2)$. However, in the general case $c(1, 2)$ is no longer simple, in fact it is shape-dependent, and $w(1, 2)$ emerges as the fundamental quantity.

This is clearly a very important result. What Wertheim has done, in effect, is to generalize the OZ theory to fluids composed of polarizable molecules. The OZ equation, supplemented by the graphical definition of the direct correlation function, provides the essential structure of the theory of classical fluids *for rigid molecules*. Wertheim has shown how this structure must be modified when one includes polarizability, and how $w(1, 2)$ replaces $c(1, 2)$ as the basic building block of the theory.

Although we have not done justice to Wertheim's theory in this brief review, it is clear that his formalism is an important development in the theory of fluids. We expect that it will eventually have wide application outside the context of dielectric theory.

To obtain explicit results from the new formalism, one must make approximations. Wertheim has fortunately developed approximation methods for his theory, including an analytically tractable scheme analogous to the MSM. We review this model and other approximate theories in Section II.

## II. APPROXIMATE THEORIES FOR POLAR FLUIDS

### A. The Dielectric Constant

To implement the general formalism of Section I and obtain explicit results for polar fluids, one must turn to approximation methods. We are mainly concerned here with one approximate theory; the MSM introduced by Lebowitz and Percus[12] as an extension of the spherical model for Ising spin systems to fluids. The MSM has the virtue of yielding analytical results for fluids composed of hard-sphere molecules interacting via electrostatic forces.

This model was first solved explicitly for the case of charged hard spheres by Waisman and Lebowitz.[13] Shortly after, Wertheim[14] produced a solution for hard spheres with embedded dipoles. His calculation yielded an explicit formula for the dielectric constant. Later, Wertheim's solution was extended[15] to the case of equal-radius polar mixtures in a very simple manner.

Blum[16] extended Wertheim's work in a different direction. He showed that the MSM could be solved for neutral hard spheres with arbitrary multipolar interactions. We recently extended Blum's work to the case of charged hard spheres with arbitrary multipolar interactions and studied the case of charged hard spheres in a polar solvent in detail.[17] Our motivation for studying this system is that it provides a model for ionic solutions that includes the feature of a molecular solvent and thus goes beyond the continuum dielectric solvent model of conventional Debye-Hückel electrolyte theory.

Our plan for this section is to review the MSM theory described above, and also the results of Wertheim's analogous theory for polarizable fluids.[3] In this section our main focus is on the dielectric constant, and in Section II.B we discuss thermodynamical properties.

We begin with Wertheim's MSM theory for pure polar fluids. First we describe the results, and then we sketch the general method for this and other MSM problems.

The MSM is defined generally by the OZ equation [see (1.14)] along with the closure relations[12]

$$h(\mathbf{r}_1\boldsymbol{\omega}_1; \mathbf{r}_2\boldsymbol{\omega}_2) = -1 \qquad \text{for } \mathbf{r}_{12} \equiv |\mathbf{r}_2 - \mathbf{r}_1| < d \qquad (2.1)$$

$$c(\mathbf{r}_1\boldsymbol{\omega}_1; \mathbf{r}_2\boldsymbol{\omega}_2) = -\beta v(\mathbf{r}_1\boldsymbol{\omega}_1; \mathbf{r}_2\boldsymbol{\omega}_2) \qquad \text{for } \mathbf{r}_{12} > d \qquad (2.2)$$

where $v(\mathbf{r}_1\omega_1; \mathbf{r}_2\omega_2)$ is the potential over the hard-sphere interaction. Equation (2.1) is just the condition of nonoverlap for hard spheres and therefore is exact for hard spheres. Equation (2.2) is the essence of the MSM and also of the more general theory of dense fluids due to Anderson and Chandler.[18] These authors discuss under what conditions (2.2) might be a good approximation.[18]

For an MSM polar fluid, (2.3) becomes

$$c(\mathbf{r}_1, \omega_1; \mathbf{r}_2, \omega_2) = \beta\boldsymbol{\mu}(\omega_1) \cdot \mathbf{T}(\mathbf{r}_1\mathbf{r}_2) \cdot \boldsymbol{\mu}(\omega_2) \qquad (2.3)$$

For the special case $\mu = 0$ [or more generally $v(\mathbf{r}_1\omega_1; \mathbf{r}_2\omega_2) = 0$], the MSM closure conditions are

$$h_s(r, \rho) = -1 \qquad \text{for } r < d \qquad (2.4a)$$

$$c_s(r, \rho) = 0 \qquad \text{for } r > d \qquad (2.4b)$$

with $h_s(r, \rho)$ and $c_s(r, \rho)$ satisfying the OZ relation

$$h_s(r_{12}, \rho) = c_s(r_{12}, \rho) + \rho \int c_s(r_{13}, \rho)h_s(r_{32}, \rho)\, d\mathbf{r}_3 \qquad (2.5)$$

where $\rho$ is the density. Equations (2.4) and (2.5) define the Percus–Yevick (PY) hard-sphere fluid problem which was solved exactly by Wertheim[19] and Thiele.[19]

For the limiting case $d = 0$, the MSM polar fluid problem reduces to the problem of summing dipole chains. This approximation, which is the dipolar analog of Mayer's[20] formulation of Debye–Hückel theory, can also be carried through analytically.[21] It is therefore not wholly surprising that Wertheim succeeded in solving the MSM for polar fluids. For a polar fluid characterized by a hard-sphere radius $d$, a density $\rho$, and a dipole moment $\mu$, Wertheim finds that the correlation function $f(\mathbf{r}_1\omega_1; \mathbf{r}_2\omega_2) = f(1, 2)$ (here and below $f$ denotes either $c$ or $h$) can be expressed as

$$f(1, 2) = f_s(r_{12}, \rho) + f_\Delta(r_{12}; \rho, \mu)\boldsymbol{\mu}(\omega_1) \cdot \boldsymbol{\mu}(\omega_2)\mu^{-2} + f_D(r_{12}; \rho, \mu)F(1, 2)\mu^{-2} \qquad (2.6)$$

where $F(1, 2)$ is defined in (1.29b).

The quantities $f_s(r, \rho)$ are the PY hard-sphere correlation functions which solve (2.4) and (2.5). The quantity $f_\Delta$ is also expressable in terms of PY functions. The analysis yields

$$f_\Delta(r; \rho, \mu) = 2\kappa[f_s(r; 2\kappa\rho) - f_s(r; -\kappa\rho)] \qquad (2.7)$$

Note that the quantity $f_\Delta$ is of molecular range, since $f_s$ is of molecular range.

The parameter $\kappa$ is given implicitly as the solution of the equation

$$\frac{4\pi}{3} \beta \rho \mu^2 = q_s{}^+ - q_s{}^- = q_s(2\kappa\eta) - q_s(-\kappa\eta) \tag{2.8}$$

where $\eta$ is the packing fraction $(\pi \, d^3/6)\rho$, and

$$q_s(\eta) = 1 - 4\pi\rho \int_0^d r^2 c_s(r, \rho) \, dr = \frac{(1 + 2\eta)^2}{(1 - \eta)^4} \tag{2.9}$$

is the hard-sphere inverse compressibility in the PY approximation.[19] The functions $f_D$ have a component proportional to $r^{-3}$ and are thus of macroscopic range. They are determined, however, from short-range functions $\bar{f}_D$ by

$$f_D(r; \rho, \mu) = \bar{f}_D(r; \rho, \mu) - 3r^{-3} \int_0^R s^2 \bar{f}_D(s; \rho, \mu) \, ds \tag{2.10a}$$

where

$$\bar{f}_D(r; \rho, \mu) = \kappa[2f_s(r; \partial\kappa\rho) + f_s(r; -\kappa\rho)] \tag{2.10b}$$

Thus the MSM gives a remarkably simple solution. All correlation functions are expressed in terms of PY functions at three densities; the true density $\rho$ and the densities $2\kappa\rho$ and $-\kappa\rho$. The negative density causes no problem.

The PY hard-sphere direct correlation function $c_s(r, \rho)$ is a cubic polynomial for $r < d$ and vanishes for $r > d$. The MSM direct correlation function consequently can be obtained in explicit form. The PY pair correlation function $h_s(r, \rho)$ is more complicated, and some numerical work is necessary to calculate it. This is not surprising, since $h_s(r, \rho)$ must show the oscillations in local density characteristic of dense fluid structure. Calculation of the MSM $h(1, 2)$ therefore requires a substantial amount of numerical work. This calculation is worth performing, since it would give a valuable insight into polar fluid structure. Fortunately, much information can be obtained without $h(1, 2)$; the dielectric constant, asymptotic potential of mean force, and thermodynamic properties can be extracted from information obtained during the calculation of $c(1, 2)$. The dielectric constant $\varepsilon$, for example, is easily computed from $c_A(1, 2)$ using Ramshaw's formula [see (1.27) and (2.7)]. One finds

$$\varepsilon = \frac{q_s{}^+}{q_s{}^-} = \frac{(1 + 4\kappa\rho)^2(1 + \kappa\rho)^2}{(1 - 2\kappa\rho)^6} \tag{2.11}$$

Note that this formula is independent of the hard-sphere diameter $d$.

Using the identity

$$1 + 4\pi\rho \int_0^\infty r^2 h_s(r, \rho)\, dr = \left[1 - 4\pi\rho \int_0^d r^2 c_s(r, \rho)\, dr\right]^{-1}$$

which is a special case of (1.18), and also (2.9) and (2.10), one finds the following result for the asymptotic potential of mean force:

$$\omega^\infty(1, 2) = - (q_s^- q_s^+)^{-1} \boldsymbol{\mu}(\omega_1) \cdot \mathbf{T}(\mathbf{r}_1\mathbf{r}_2) \cdot \boldsymbol{\mu}(\omega_2) \qquad (2.12)$$

These results obtained by Wertheim can be compared with Nienhuis and Deutch's general theory of $\varepsilon$ and $w^\infty(1, 2)$ discussed in Section I.C. This has been done,[22] and the consistency of the MSM and the general theory has been demonstrated.

We have just seen that analysis of the MSM pure polar fluid problem yields simple explicit results. The model has the additional welcome feature of yielding equally simple results for equal-diameter polar fluid mixtures.[15]

We consider an $m$-component mixture. Each component is characterized by a density $\rho_i$ a dipole moment $\mu_i$. All have a common diameter $d$. We define an effective dipole moment:

$$\hat{\mu} = \left(m^{-1} \sum_{i=1}^m \mu_i^2\right)^{1/2} \qquad (2.13)$$

and an effective density:

$$\hat{\rho} = \sum_{i=1}^m \rho_i \mu_i^2 \qquad (2.14)$$

We also require the true total density:

$$\rho = \sum_{i=1}^m \rho_i \qquad (2.15)$$

Then, analysis shows[15] that the correlation functions $f_{ij}(1, 2)$ for species $i$ and $j$ have the form

$$f_{ij}(1, 2) = f_s(r_{12}, \rho) + f_\Delta(r_{12}; \hat{\rho}, \hat{\mu}) \boldsymbol{\mu}_1(\omega_1) \cdot \boldsymbol{\mu}_j(\omega_2) \hat{\mu}^{-2}$$
$$+ f_D(r_{12}; \hat{\rho}, \hat{\mu}) F_{ij}(1, 2) \hat{\mu}^{-2} \qquad (2.16)$$

where $f_\Delta$ and $f_D$ are the correlation functions found by Wertheim and given above, and $F_{ij}(1, 2)$ is defined analogously to $F(1, 2)$ as

$$F_{ij}(1, 2) = r_{12}^3 \boldsymbol{\mu}_i(\omega_i) \cdot \mathbf{T}(\mathbf{r}_1\mathbf{r}_2) \cdot \boldsymbol{\mu}_j(\omega_2)$$

The spherically symmetric component of the two-particle correlation functions is that of the reference PY hard-sphere fluid just as in the pure polar fluid case. The functions $f_\Delta$ and $f_D$ are those for an effective pure polar fluid with dipole moment $\hat{\mu}$, density $\hat{\rho}$, and diameter $d$. From (2.16) it is readily established that the dielectric constant is that of the effective pure polar fluid. The asymptotic potential of mean force between species $i$ and $j$ is

$$w_{ij}^{\infty}(1, 2) = \hat{\mu}^{-2}\mu_i\mu_j w^{\infty}(1, 2)$$

where $w^{\infty}(1, 2)$ is the asymptotic potential of mean force in the effective pure fluid.

The mixture model displays phase separations with classical exponents but has the unsatisfactory feature [see (2.16)] of allowing no correlation between nonpolar ($\mu_i$ is set equal to zero) and polar species beyond that included in $f_s(r, \rho)$. Also, in this model the potential of mean force between the nonpolar species (hard spheres without dipole moments) is uninfluenced by the strength of the dipole moments of the polar components of the solution. Thus the model does not display the local segregation or "hydrophobic bonding" expected of nonpolar species in a polar fluid.

To understand how the results discussed above are obtained unfortunately involves working through a lengthy calculation. We present only the most important features of this calculation here.

The analysis begins with the observation that in infinite fluids all spatial directions are equivalent, and consequently the correlation functions $f(1, 2)$ are unchanged if we rotate the pair with respect to the fluid keeping the relative pair position and orientation fixed. Thus it is natural to expand $f(1, 2)$ in terms of a complete set of rotationally invariant angular functions.[23] For the general case of nonlinear molecules, the invariants are constructed from triple products of Wigner $D$ functions. For linear molecules they are formed from the corresponding products of spherical harmonics. For the sake of simplicity we discuss only the case of fluids composed of linear molecules, and this restriction should be kept in mind.

For linear molecules we can specify the relative position and orientation of a molecular pair by $r_{12}, \omega_1, \omega_2, \omega_{12}$, where the $\omega_i$ are the spherical polar angles describing, respectively, the axis of molecule 1, the axis of molecule 2, and the orientation of the interparticle vector $\mathbf{r}_{12}$.

The rotational invariants $\phi_{l_1l_2l_3}(\omega_1, \omega_2, \omega_{12})$ are[24]

$$\phi_{l_1l_2l_3}(\omega_1, \omega_2, \omega_{12}) = \sum_{m_1m_2m_3} (-)^{m_1} \begin{pmatrix} l_1 & l_2 & l_3 \\ -m_1 & m_2 & m_3 \end{pmatrix}$$
$$\times Y_{l_1m_1}^*(\omega_1)Y_{l_2m_2}(\omega_2)Y_{l_3m_3}(\omega_{12}) \qquad (2.17)$$

The form of (2.17) is identical to the form one finds when coupling three quantum mechanical angular momenta $l_1$, $l_2$, and $l_3$ to obtain a zero resultant.

We now expand $f(X_1 X_2)$ in terms of the complete set of functions $\phi_{l_1 l_2 l_3}$. Thus

$$f(X_1 X_2) = \sum_{l_1 l_2 l_3} f(l_1 l_2 l_3; r_{12}) \phi_{l_1 l_2 l_3}(\omega_1, \omega_2, \omega_{12}) \qquad (2.18)$$

The correspondence between the notation used to discuss Wertheim's results and our new notation is

$$f(r) = f(000; r) \qquad (2.19a)$$

$$f_\Delta(r) = -\sqrt{3}\, f(110; r) \qquad (2.19b)$$

$$f_D(r) = \sqrt{\tfrac{3}{2}}\, f(112; r) \qquad (2.19c)$$

We now insert (2.18) into the OZ equation (1.14). A straightforward but lengthy calculation gives the following infinite set of coupled Fourier space equations:

$$h^{(m)}(l_1 l_2; k) = c^{(m)}(l_1 l_2; k) + (-)^m (2\pi)^{3/2} \rho \sum_{L_2=0}^{\infty} c^{(m)}(l_1 L_2; k) h^{(m)}(L_2 l_2; k) \qquad (2.20)$$

where the functions $f^{(m)}(l_1 l_2; k)$ are related to the Fourier transforms

$$f(l_1 l_2 l_3; k) = \frac{4\pi}{(2\pi)^{3/2}} i^{l_3} \int_0^{\infty} r^2 j_{l_3}(kr) f(l_1 l_2 l_3; r)\, dr \qquad (2.21)$$

by

$$f^{(m)}(l_1 l_2; k) = \sum_{l_3} (2l_3 + 1)^{1/2} \begin{pmatrix} l_1 & l_2 & l_3 \\ m & -m & 0 \end{pmatrix} f(l_1 l_2 l_3; k) \qquad (2.22)$$

In (2.21), $j_l(kr)$ is the spherical Bessel function of order $l$.

Equation (2.20) is an infinite matrix set of equations which is equivalent to the original OZ equation (1.14).

Equation (2.20), however, has two important advantages over the original OZ equation. It is free of angular variables and can be approximated in an obvious way, that is, by truncation. We will see that the MSM closure relations for polar fluids [see (2.1) and (2.3)] yield a very simple truncation for (2.20). In fact, we obtain three uncoupled one-dimensional equations from (2.20). Moreover, the remarkable fact is that the closure relations for these three equations are of the PY hard-sphere type. The solutions of the three uncoupled equations are in fact $f_s(r, \rho)$, $f_s(r, 2\kappa\rho)$, and $f_s(r, -\kappa\rho)$, the three PY functions that entered Wertheim's final result.

To see this, we must examine the closure relations for the quantities $f^{(m)}(l_1 l_2; r)$ defined below [see (2.30), (2.42), and (2.57a)] as appropriate Fourier transforms of the corresponding functions $f^{(m)}(l_1 l_2; k)$.

We first consider the transforms of $f(l_1 l_2 l_3; k)$, $\bar{f}(l_1 l_2 l_3; r)$ defined as

$$\bar{f}(l_1 l_2 l_3; r) = \frac{4\pi r^{-1}}{(2\pi)^{3/2}} \int_0^\infty k \sin krf(l_1 l_2 l_3; k)\, dk \qquad \text{for } l_3 \text{ even} \qquad (2.23a)$$

$$\bar{f}(l_1 l_2 l_3; r) = \frac{4\pi i r^{-1}}{(2\pi)^{3/2}} \int_0^\infty k \cos krf(l_1 l_2 l_3; k)\, dk \qquad \text{for } l_3 \text{ odd} \qquad (2.23b)$$

Eliminating $f(l_1 l_2 l_3; k)$ between (2.21) and (2.23) gives after an integration:

$$\bar{f}(l_1 l_2 l_3; r) = f(l_1 l_2 l_3; r) - r^{-1} \int_r^\infty P'_{l_3}\left(\frac{r}{y}\right) f(l_1 l_2 l_3; y)\, dy \qquad (2.24)$$

for both $l_3$ even and $l_3$ odd. In (2.24), $P'_l(X)$ is the derivative of the $l$th Legendre polynomial.

We also require the inverse of (2.24), which is similarly found to be

$$f(l_1 l_2 l_3; r) = \bar{f}(l_1 l_2 l_3; r) - r^{-2} \int_0^r y P'_{l_3}\left(\frac{y}{r}\right) \bar{f}(l_1 l_2 l_3; y)\, dy \qquad (2.25)$$

So far our treatment has been free of approximations. Before specializing to the MSM, we discuss several additional exact results which can be obtained from the formalism.

To do this we write (2.24) and (2.25) out explicitly for $f(112; r)$ which is proportional to $f_D(r)$. We find

$$\bar{f}(112; r) = f(112; r) - 3 \int_r^\infty y^{-1} f(112; y)\, dy \qquad (2.26)$$

$$f(112; r) = \bar{f}(112; r) - 3r^{-3} \int_0^r y^2 \bar{f}(112; y)\, dy \qquad (2.27)$$

The functions $\bar{f}(112; r)$ are in general of microscopic range. This follows from the result that $f(112; r)$ can always be within the form

$$f(112; r) = f_{SR}(112; r) + Ar^{-3} \qquad (2.28)$$

where $A$ is independent of $r$, and $f_{SR}(112; r)$ is of short range. Equation (2.28) follows by specializing the general equations (1.16) and (1.26) to the present case of an infinite fluid composed of cylindrically symmetric molecules.

Then, inserting (2.28) into (2.26) gives the important result that the $Ar^{-3}$ term cancels, that is, that

$$\bar{f}(112; r) = f_{SR}(112; r) - 3 \int_r^\infty y^{-1} f_{SR}(112; y)\, dy \qquad (2.29)$$

which clearly shows that $\bar{f}(112; r)$ is of short range.

We will see below that the cancellation of the $Ar^{-3}$ term is the essential reason why the MSM can be solved.

We can now easily obtain formal expressions for the dielectric constant and asymptotic potential of mean force, which are exact for linear molecules and which also are similar in form to the MSM results (2.11) and (2.12).

We first introduce the Fourier transform:

$$f^{(m)}(11;r) = \frac{4\pi r^{-1}}{(2\pi)^{3/2}} \int_0^\infty k \sin kr f^{(m)}(11;k)\, dk \qquad m = 0, \pm 1 \qquad (2.30)$$

which has the inverse

$$f^{(m)}(11;k) = \frac{4\pi k^{-1}}{(2\pi)^{3/2}} \int_0^\infty r \sin kr f^{(m)}(11;r)\, dr \qquad (2.31)$$

For zero wave vector this inverse becomes

$$\lim_{k\to 0} f^{(m)}(11;k) = \frac{4\pi}{(2\pi)^{3/2}} \int_0^\infty r^2 f^{(m)}(11;r)\, dr \qquad (2.32)$$

Since $\bar{f}(112;r)$ is of short range, we also have, from (2.27),

$$\lim_{r\to\infty} h(112;r) = -3r^{-3} \int_0^\infty y^2 \bar{h}(112;y)\, dy \qquad (2.33)$$

and

$$\lim_{r\to\infty} c(112;r) = -3r^{-3} \int_0^\infty y^2 \bar{c}(112;y)\, dy \qquad (2.34)$$

From the analysis of Nienhuis and Deutch, we know that the asymptotic form of $c(X_1, X_2)$ is $\beta\boldsymbol{\mu}(\omega_1) \cdot \mathsf{T}(r_1, r_2) \cdot \boldsymbol{\mu}(\omega_2)$. This is (1.16). We can use this result to obtain the asymptotic form of $c(112;r)$, required to evaluate the lhs of (2.34). Further, by taking the Fourier transform of the inverse of (2.22), we can express $\bar{c}(112;y)$ in terms of $c^{(\pm 1)}(11;y)$ and $c^{(0)}(11;y)$.

We then eventually find that (2.34) gives the relation [cf. (2.8) and (2.9)]

$$\frac{4\pi}{3}\beta\rho\mu^2 = q^+ - q^- \qquad (2.35)$$

where

$$q^+ = 1 - 4\pi\rho \int_0^\infty y^2 c^{(0)}(11;y)\, dy \qquad (2.36)$$

$$q^- = 1 + 4\pi\rho \int_0^\infty y^2 c^{(\pm 1)}(11;y)\, dy \qquad (2.37)$$

We can similarly express $\bar{h}(112; y)$ in terms of $h^{(0)}(11; y)$ and $h^{(\pm 1)}(11; y)$. This then gives for (2.33)

$$\lim_{r \to \infty} h(112; r) = -\frac{\sqrt{6}}{4\pi\rho r^3}\left[1 + 4\pi\rho \int_0^\infty y^2 h^{(0)}(11; y)\, dy \right.$$

$$\left. - \left(1 - 4\pi\rho \int_0^\infty y^2 h^{(11)}(11; y)\, dy\right)\right] \qquad (2.38)$$

Now we use the result that the coupled matrix set (2.20) becomes uncoupled for $f^{(m)}(11; k)$ in the zero wave vector limit; that is,

$$\lim_{k \to 0}[h^{(m)}(11; k) - c^{(m)}(11; k)] = (-)^m(2\pi)^{3/2}\rho \lim_{k \to 0} c^{(m)}(11; k)h^{(m)}(11; k) \qquad (2.39)$$

This is an exact result for nonionic fluids composed of linear molecules. From it one can prove that Ramshaw's formula [see (1.27)] is exact for rigid polar fluids composed of linear molecules.

Equation (2.39) is established by examining the small wave vector limits of the functions $f(l_1 l_2 l_3; k)$, taking proper cognizance of the long range of some of the functions $f(l_1 l_2 l_3; r)$. One finds that only $f(110; k)$ and $f(112; k)$ are of order $k^0$ for small $k$; all other functions that can occur for nonionic fluids vanish for small $k$. Using this result in conjunction with (2.20) and (2.22) yields (2.39).

Using (2.39) to express $\lim_{k \to 0} h^{(m)}(11; k)$ in terms of $\lim_{k \to 0} c^{(m)}(11; k)$, and further using (2.32) and (2.35) in (2.38), one can readily establish

$$\omega^\infty(1, 2) = -(q^- q^+)^{-1}\mathbf{\mu}(\mathbf{\omega}_1) \cdot \mathbf{T}(\mathbf{r}_1\mathbf{r}_2) \cdot \mathbf{\mu}(\mathbf{\omega}_2) \qquad (2.40)$$

Equation (2.40), which is exact within the restrictions mentioned, is of the same form as the MSM result (2.12).

Using similar arguments we can calculate the dielectric constant from Ramshaw's equation [see (1.27)]. We find [cf. (2.11)]

$$\varepsilon = \frac{q^+}{q^-} \qquad (2.41)$$

In order to obtain explicit results from our formal discussion, we now turn to the MSM. The key point is that we can deduce from (2.1) and (2.3) that only $f(000; r)$, $f(110; r)$, and $f(112; r)$ are nonvanishing. This further implies from (2.22) that only

$$f^{(0)}(00; r) = \frac{4\pi r^{-1}}{(2\pi)^{3/2}} \int_0^\infty k \sin kr f^{(0)}(00; k)\, dk \qquad (2.42)$$

$f^{(0)}(11; r)$, and $f^{(1)}(11; r)$ are nonvanishing. Thus Fourier transforming equation (2.20) gives us three uncoupled OZ equations. We find

$$h^{(0)}(11; r_{12}) = c^{(0)}(11; r_{12}) + \rho \int c^{(0)}(11; r_{13}) h^{(0)}(11; r_{32}) \, d\mathbf{r}_3 \qquad (2.43a)$$

$$h^{(\pm 1)}(11; r_{12}) = c^{(\pm 1)}(11; r_{12}) - \rho \int c^{(\pm 1)}(11; r_{13}) h^{(\pm 1)}(11; r_{32}) \, d\mathbf{r}_3 \qquad (2.43b)$$

$$h^{(0)}(00; r_{12}) = c^{(0)}(00; r_{12}) + \rho \int c^{(0)}(00; r_{13}) h^{(0)}(00; r_{32}) \qquad (2.43c)$$

Further, the MSM closure conditions can be expressed in terms of the expansion coefficients $f(l_1 l_2 l_3; r)$ as

$$h(000; r) = -1 \qquad \text{for} \quad r < d \qquad (2.44a)$$

$$c(000; r) = 0 \qquad \text{for} \quad r > d \qquad (2.44b)$$

$$h(110; r) = 0 \qquad \text{for} \quad r < d \qquad (2.45a)$$

$$c(110; r) = 0 \qquad \text{for} \quad r > d \qquad (2.45b)$$

$$h(112; r) = 0 \qquad \text{for} \quad r < d \qquad (2.46a)$$

$$c(112; r) = (\tfrac{2}{3})^{1/2} \beta \mu^2 r^{-3} \qquad \text{for} \quad r > d \qquad (2.46b)$$

The above closure relations for $f(l_1 l_2 l_3; r)$ can be transformed to closure relations for $\bar{f}(l_1 l_2 l_3; r)$. The functions $\bar{f}(112; r)$ are found using (2.26). An important point is that, since (2.26) transforms $r^{-3}$ into zero, the function $\bar{c}(112; r)$ vanishes for $r > d$. This is why the polar fluid problem reduces to the PY hard-sphere problem. The functions $\bar{f}(110; r)$ and $\bar{f}(000; r)$ are by (2.24) identical to $f(110; r)$ and $f(000; r)$. Thus the closure relations for all the required functions $\bar{f}(l_1 l_2 l_3; r)$ are easily obtained. From these results, the closure relations for the functions $f^{(m)}(l_1 l_2; r)$ are easily constructed. They are

$$h^{(0)}(00; r) = -1 \qquad \text{for} \quad r < d \qquad (2.47a)$$

$$c^{(0)}(00; r) = 0 \qquad \text{for} \quad r > d \qquad (2.47b)$$

$$h^{(0)}(11; r) = -2\kappa \qquad \text{for} \quad r < d \qquad (2.48a)$$

$$c^{(0)}(11; r) = 0 \qquad \text{for} \quad r > d \qquad (2.48b)$$

$$h^{(\pm 1)}(11; r) = -\kappa \qquad \text{for} \quad r < d \qquad (2.49a)$$

$$c^{(\pm 1)}(11; r) = 0 \qquad \text{for} \quad r > d \qquad (2.49b)$$

Comparing the OZ equations (2.43) and the above closure relations with the definition of the PY hard-sphere problem [see (2.4) and (2.5)] then yields

$$f^{(0)}(00; r) = f_s(r; \rho) \tag{2.50}$$

$$f^{(0)}(11; r) = 2\kappa f_s(r; 2\kappa\rho) \tag{2.51}$$

$$f^{(\pm 1)}(11; r) = \kappa f_s(r; -\kappa\rho) \tag{2.52}$$

Therefore, as we have discussed earlier, the MSM polar fluid problem can be solved in terms of three PY hard-sphere correlation functions $f_s(r, \rho)$, $f_s(r, 2\kappa\rho)$, and $f_s(r, -\kappa\rho)$. From this result and from the other results we have derived, particularly (2.35) to (2.37) and (2.40) and (2.41), equations (2.6) to (2.12) can be easily verified.

Next we turn to another physically interesting problem which can be treated explicitly within the MSM. We refer to the polar ion mixture[17] which, as mentioned earlier, provides a model for ionic solutions which includes the feature of solvent granularity. We consider a general $m$-component mixture. The $i$th component is characterized by a charge $e_i$, a dipole moment $\mu_i$, and a hard-sphere diameter $d$ which is the same for all species. The dipole moments $\mu_i$ can assume any nonnegative value including zero. The charges can have any value within the constraint of overall electro-neutrality of the fluid.

As in the case of the polar mixture previously discussed, our analysis leads us to define an effective dipole moment $\hat{\mu}$ and an effective density of dipoles $\hat{\rho}_1$:

$$\hat{\mu} = \left( m^{-1} \sum_{i=1}^{m} \mu_i^2 \right)^{1/2} \tag{2.53}$$

$$\hat{\rho}_1 = \hat{\mu}^{-2} \sum_{i=1}^{m} \rho_i \mu_i^2 \tag{2.54}$$

We also require an effective charge $\hat{e}$ and an effective density of charges $\hat{\rho}_0$ defined analogously as

$$\hat{e} = \left( m^{-1} \sum_{i=1}^{m} e_i^2 \right)^{1/2} \tag{2.55}$$

$$\hat{\rho}_0 = \hat{e}^{-2} \sum_{i=1}^{m} \rho_i e_i^2 \tag{2.56}$$

As in the case of the MSM polar fluid problem, analysis leads us to solve a truncated set of OZ equations derived from (2.20). Because of the MSM closure conditions, only the functions $f^{(0)}(00; k)$ $f^{(0)}(10; k) = -f^{(0)}(01; k)$, $f^{(0)}(11; k)$, and $f^{(\pm 1)}(11; k)$ are nonvanishing. We will later also require the

$r$-space transforms $f^{(0)}(00; r)$, $f^{(0)}(10; r)$, $f^{(0)}(11; r)$, and $f^{(\pm 1)}(11; r)$. The transforms for all but $f^{(0)}(10; r)$ have been defined in the discussion of the polar fluid problem. The function $f^{(0)}(10; r)$ is defined by

$$f^{(0)}(10; r) = \frac{4\pi i r^{-1}}{(2\pi)^{3/2}} \int_0^\infty k \cos kr f^{(0)}(10; k) \, dk \qquad (2.57a)$$

with the inverse

$$f^{(0)}(10; k) = -\frac{4\pi i k^{-1}}{(2\pi)^{3/2}} \int_0^\infty r \cos kr f^{(0)}(10; r) \, dr \qquad (2.57b)$$

For small $k$ this last relation becomes

$$\lim_{k \to 0} f^{(0)}(10; k) = -\frac{4\pi i k^{-1}}{(2\pi)^{3/2}} \int_0^\infty r f^{(0)}(10; r) \, dr \qquad (2.57c)$$

The function $f^{(\pm 1)}(11; r)$ can be expressed in terms of a hard-sphere PY function just as in the polar fluid case [cf. (2.52)]

$$f^{(\pm 1)}(11; r) = \hat{\kappa}_e f_s(r; -\hat{\kappa}_e \hat{\rho}_1) \qquad (2.58)$$

The quantity $\hat{\kappa}_e$, however, is no longer determined as the solution to a single algebraic equation like (2.8). Instead it, along with two other quantities, are the solutions to a set of three coupled nonlinear algebraic equations. We require only two of these equations here. The first is [cf. (2.8)]

$$\frac{4\pi\beta\hat{\rho}_1\hat{\mu}^2}{3} = q_e^+ - q_e^- \qquad (2.59)$$

and the second is

$$\frac{1}{\sqrt{3}} 4\pi\beta(\hat{\rho}_0\hat{\rho}_1)^{1/2}\hat{e}\hat{\mu} = 4\pi\sqrt{\hat{\rho}_0\hat{\rho}_1} \int_0^d r c^{(0)}(10; r) \, dr \qquad (2.60)$$

where

$$q_e^+ = 1 - 4\pi\hat{\rho}_1 \int_0^d r^2 c^{(0)}(11; r) \, dr \qquad (2.61)$$

$$q_e^- = 1 + 4\pi\hat{\rho}_1 \int_0^d r^2 c^{(\pm 1)}(11; r) \, dr \qquad (2.62)$$

Equations (2.59) and (2.60) are derived from an argument similar to that used to obtain (2.35).

The quantity $q_e^-$, by (2.58), is simply expressed in terms of the PY inverse compressibility (2.9). The quantity $q_e^+$, however, is of much more complicated form. This is because $f^{(0)}(11; r)$ does not satisfy a simple one-dimensional PY equation but is coupled to the functions $f^{(0)}(10; r)$ and $f^{(0)}(00; r)$.

Actually, it is only coupled to part of $f^{(0)}(00; r)$, which we denote by $f_e^{(0)}(00; r)$. This function is the contribution to the spherically symmetric part of the correlation function from the charge–charge interaction. The contribution from the hard-sphere interaction completely decouples from the rest of the problem.

The functions just discussed satisfy the following matrix OZ equation which we write in $k$ space as (suppressing the superscript zero)

$$
\begin{pmatrix} h_e(00;k) - c_e(00;k) & h(01;k) - c(01;k) \\ h(10;k) - c(10;k) & h(11;k) - c(11;k) \end{pmatrix}
$$
$$
= (2\pi)^{3/2} \begin{pmatrix} c_e(00;k) & c(01;k) \\ c(10;k) & c(11;k) \end{pmatrix} \begin{pmatrix} \hat{\rho}_0 & 0 \\ 0 & \hat{\rho}_1 \end{pmatrix} \begin{pmatrix} h_e(00;k) & h(01;k) \\ h(10;k) & h(11;k) \end{pmatrix} \qquad (2.63)
$$

Note that the above OZ equation describes an effective two-component fluid containing a charged component with charge $\hat{e}$ at an effective density $\hat{\rho}_0$ and an effective polar component with dipole moment $\hat{\mu}$ at density $\hat{\rho}_1$. The quantities $\hat{e}$ and $\hat{\mu}$ enter the closure equations as well as (2.59) and (2.60). We require only one closure relation for the limited analysis to be presented here:

$$
c_e(00; r) = -\frac{\beta \hat{e}^2}{r} \qquad \text{for} \quad r > d \qquad (2.64)
$$

From (2.64) one obtains the Fourier space relation

$$
(2\pi)^{3/2} \lim_{k \to 0} c_e(00; k) = -\frac{\beta \hat{e}^2}{k^2} \qquad (2.65)
$$

We should mention that the effective two-component fluid does not correspond to any real ionic solution. For example, it contains charges of only one sign. It is, however, a convenient mathematical device for solving the MSM and discussing the results.

We next obtain an effective OZ equation for $f_e(00; k)$ of one-component form. The motivation for this approach is the fact that only the functions $f_e(00; k)$ are necessary to discuss the osmotic properties of ionic solutions. This result is in the spirit of the well-known solution theory of MacMillan and Mayer.[25]

To obtain an effective one-component problem for the ions, we recall the technique, equations (1.31) to (1.34), employed in Section I.C. We find the result:

$$
h_e(00; k) = c_e^{\text{eff}}(00; k) + (2\pi)^{3/2} \hat{\rho}_0 c_e^{\text{eff}}(00; k) h_e(00; k) \qquad (2.66)
$$

where the effective direct correlation function is

$$c_e^{\text{eff}}(00;k) = c_e(00;k) + (2\pi)^{3/2}\hat{\rho}_1 \frac{c(01;k)c(10;k)}{1 - (2\pi)^{3/2}\hat{\rho}_1 c(11;k)} \quad (2.67)$$

We can explicitly evaluate (2.67) for small $k$ using (2.32), (2.57c), (2.60), (2.61), and (2.65). We find

$$(2\pi)^{3/2} \lim_{k\to 0} c_e^{\text{eff}}(00;k) = \frac{\beta\hat{e}^2}{k^2}\left(1 - \frac{1}{3}\frac{4\pi\hat{\rho}_1\beta\hat{\mu}^2}{q_e^+}\right) \quad (2.68)$$

Combining this expression with (2.59) then yields the important result [cf. (2.65)]

$$(2\pi)^{3/2} \lim_{k\to 0} c_e^{\text{eff}}(00;k) = -\frac{\beta\hat{e}^2}{k^2\mathscr{E}} \quad (2.69)$$

where

$$\mathscr{E} = \frac{q_e^+}{q_e^-} \quad (2.70)$$

is an effective *solute-dependent* dielectric constant which screens the "bare" direct correlation function $c_e(00;k)$.

In $r$ space (2.69) yields

$$\lim_{r\to\infty} c_e^{\text{eff}}(00;r) = -\frac{\beta\hat{e}^2}{\mathscr{E}r} \quad (2.71)$$

which clearly shows that $\mathscr{E}$ plays the role of a dielectric constant. For sufficiently small solute concentration,

$$\mathscr{E} \cong \varepsilon = \frac{q_s^+}{q_s^-} \quad (2.72)$$

where $\varepsilon$ is the MSM dielectric constant for the polar fluid mixture that results if we set $\hat{\rho}_0 = 0$.

In this limit we can set

$$\lim_{\hat{\rho}_0\to 0} c_e^{\text{eff}}(00;r) = -\frac{\beta\hat{e}^2}{\varepsilon r} \quad (2.73)$$

for all $r$, since the charges are nearly always very far apart. In $k$ space we then have

$$\lim_{\hat{\rho}_0\to 0} c_e^{\text{eff}}(00;k) = -\frac{\beta\hat{e}^2}{\varepsilon k^2} \quad (2.74)$$

Inserting this result into (2.66), solving for $h_e(00; k)$, and then transforming to $r$ space gives

$$\lim_{\hat{\rho}_0 \to 0} h_e(00; r) = - \frac{\beta \hat{e}^2}{\varepsilon r} e^{-\kappa_0 r} \tag{2.75}$$

where

$$\kappa_0 = \left( \frac{4\pi \beta \hat{\rho}_0 \hat{e}^2}{\varepsilon} \right)^{1/2} \tag{2.76}$$

which is the inverse Debye length for an electrolyte of charge $\hat{e}$ and density $\hat{\rho}_0$ in a dielectric continuum with dielectric constant $\varepsilon$.

Thus we have derived the Debye–Hückel distribution function (2.75) in the low-ionic-strength limit without invoking the prior assumption of a continuum dielectric.

From (2.75) it is easy to recover the Debye–Hückel limiting thermodynamics. This was done by us[17] previously, using a different method by which we obtained a formal solution of the MSM for arbitrary ionic concentrations including expressions for the thermodynamic properties at high ionic strength. More precisely, we obtained our solution in term of three parameters which are the roots of the set of three coupled nonlinear algebraic equations mentioned previously. An important additional qualitative conclusion that emerges from the analysis is that the potential of mean force between the solvent dipoles in the ionic solution is shielded by Debye screening from the charges in a way analogous to the shielding the charges experience. The charge–dipole interaction also is modified by Debye shielding. The MSM model for ionic-polar mixtures has also recently been reported by Blum.[26]

We are currently attempting to obtain explicit results at arbitrary solute concentrations. One simple way to do this is to assume

$$c_e^{\text{eff}}(00; k) = - \frac{\beta \hat{e}^2}{\mathscr{E} r} \tag{2.77}$$

where $\mathscr{E}$ is the solute-dependent dielectric constant of (2.70). That is, we assume that the asymptotic relation (2.71) holds for all $r > d$.

This type of approximation is in the spirit of the MSM. We recall that the essence of the MSM is to replace the *true* direct correlation function by its asymptotic limit for $r > d$. Here we replace the *effective* direct correlation function by its asymptotic limit for $r > d$.

The assumed closure equation (2.77), which has the virtue of yielding a very simple MSM theory for solvent effects in ionic solutions, ignores certain correlations, and we are currently investigating the importance of the neglected correlations. The MSM charge-dipole model has the limitation implied by the approximation of replacing the true direct correlation function outside the hard core by the potential. But the MSM has the virtue of permitting explicit calculation of several quantities of physical interest. For example, the model yields values for the thermodynamical properties and activity coefficients of concentrated solutions, as well as quantitative expressions for the potential of mean force between two ions at close separation in a "molecular" polar solvent. These results, while only approximations to the behavior of actual solutions, go beyond the primitive electrolyte model in which the solvent is treated as a dielectric continuum. Thus the MSM may be expected to improve qualitative understanding of solution behavior. The MSM may well be the only analytic model that permits practical calculation of these interesting physical quantities.

Summarizing, we believe that the statistical methods discussed in this article, which have proved so powerful in developing the theory of pure polar fluids, may soon lead to considerable progress in the important area of ionic solution theory.

Before we conclude this section, we present a very brief discussion of Wertheim's MSM-like theory for fluids with polarizability.[3,4] We restrict ourselves to the nonpolar put polarizable case.[3] Wertheim's result for the dielectric constant of polarizable fluids shows a strong similarity to the rigid polar fluid result. He finds for the case of zero permanent moment:[3]

$$\varepsilon = \frac{q^+}{q^-} \tag{2.78}$$

with

$$q^+ = \frac{(1 + 4\xi)^2}{(1 - 2\xi)^4} \qquad q^- = \frac{(1 - 2\xi)^2}{(1 + \xi)^4} \tag{2.79}$$

and with $\xi$ determined by the relation

$$4\pi\rho\alpha' = q^+ - q^-$$

where $\alpha'$ is a renormalized polarizability expressed in terms of the true polarizability $\alpha$ by

$$\alpha' = \frac{\alpha}{1 - 16\xi\alpha\,d^{-3}} \tag{2.80}$$

As before, $d$ is the hard-sphere radius. If we make the replacement $\alpha' \to \frac{1}{3}\mu^2\beta$, the polarizable fluid results and the rigid polar fluid results become identical.

Stell and Rushbrooke[27] have examined Wertheim's result [see (2.78)] in the following manner. They compare the Kirkwood–Yvon[28] perturbation theory valid for small $\alpha$ with the small-$\alpha$ limit of Wertheim's theory. Comparison is made for the Clausius–Mossotti function $(4\pi/3)(\varepsilon - 1)(\varepsilon + 2)^{-1}$ to order $\alpha^3$. Stell and Rushbrooke find that the perturbation limit of Wertheim's theory is a reasonable approximation to his full result for a wide range of polarizabilities and densities. They thus argue that the Kirkwood–Yvon theory should provide a moderately good description of the true dielectric constant over the same range of polarizabilities and densities. Thus by comparison of the small-polarizability expansion of the Wertheim theory for $\varepsilon$ with the corresponding Kirkwood–Yvon expansion of the exact result, one might hope to assess the validity of the Wertheim result for a wide range of polarizabilities and densities.

Unfortunately, the results of this comparison show that the Wertheim theory is inadequate except at low density. The discrepancy between the Kirkwood–Yvon results and the small-$\alpha$ expansion of Wertheim's theory is much greater than the discrepancy between Wertheim's full theory and its small-polarizability limit. Stell and Rushbrooke also show that the expanded form of Wertheim's theory is the low-density limit of the Kirkwood–Yvon theory. That this is the case is also apparent from Wertheim's article.

In concluding this section we wish to emphasize that the MSM suffers from some severe limitations as a result of the approximation of the direct correlation function by the pair potential outside the hard core. These limitations may be traced to an unphysical behavior of MSM pair correlation functions at close separations. These difficulties may be improved by use of the exponential approximation of Anderson and Chandler,[18] which they present as a simple method for improving the MSM, and as well as their own more sophisticated theory of fluids. The importance of the MSM, as we have stressed, is that it is the only model available that permits detailed calculation of many molecular and thermodynamical properties of interest.

## B. Thermodynamics of Polar Fluids

We now turn to thermodynamical predictions of some of the microscopic theories we have discussed. Thermodynamic results are not yet available for fluids composed of polarizable molecules, so our discussion is restricted to rigid polar fluids.

First, we dispose of the possibility of shape dependence of the thermodynamic functions. Although this possibility is clearly excluded on physical grounds, a microscopic proof is useful as a consistency check of the theory.

We begin by computing the dipolar contribution to the Helmholtz free energy of the fluid. If this quantity is shape-independent, all the other thermodynamic properties will also be shape-independent.

The dipolar Helmholtz free energy $\Delta A$ is easily computed in terms of the pair correlation function $h(\mathbf{r}_1 \boldsymbol{\omega}_1; \mathbf{r}_2 \boldsymbol{\omega}_2; \lambda)$ of a fictitious fluid with nondipolar forces which are identical to the corresponding forces in the true fluid but with all dipoles $\mu$ scaled down to $\lambda^{1/2} \mu$, $0 \leq \lambda \leq 1$. This is the familiar charging process method. We find

$$V^{-1}\Delta A = -\tfrac{1}{2}\rho^2 \int_0^1 d\lambda \int_V d\mathbf{r}_1 \, d\boldsymbol{\omega}_1 \, d\mathbf{r}_2 \, d\boldsymbol{\omega}_2$$

$$\times h(\mathbf{r}_1 \boldsymbol{\omega}_1; \mathbf{r}_2 \boldsymbol{\omega}_2; \lambda)\boldsymbol{\mu}(\boldsymbol{\omega}_1) \cdot \mathsf{T}(\mathbf{r}_1 \mathbf{r}_2) \cdot \boldsymbol{\mu}(\boldsymbol{\omega}_2) \tag{2.81}$$

From the Nienhuis–Deutch analysis[2] [see (1.24) and (1.26)], one sees

$$\lim_{r_{12} \to \infty} h(\mathbf{r}_1 \boldsymbol{\omega}_1; \mathbf{r}_2 \boldsymbol{\omega}_2; \lambda) = \beta\varepsilon^{-1}(\lambda)\, \boldsymbol{\mu}_{\text{eff}}(\boldsymbol{\omega}_1; \lambda)$$

$$\cdot [\mathsf{T}(\mathbf{r}_1 \mathbf{r}_2) + \mathsf{R}_{V+W}^*[\mathbf{r}_1 \mathbf{r}_2; \varepsilon^{-1}(\lambda)]\} \cdot \boldsymbol{\mu}_{\text{eff}}(\boldsymbol{\omega}_2; \lambda)$$

$$\tag{2.82}$$

where

$$\boldsymbol{\mu}_{\text{eff}}(\boldsymbol{\omega}; \lambda) = \lambda^{1/2}\boldsymbol{\mu}(\boldsymbol{\omega}_1) + \boldsymbol{\kappa}(\boldsymbol{\omega}_1; \lambda) \tag{2.83}$$

with

$$\boldsymbol{\kappa}(\boldsymbol{\omega}; \lambda) = \int_V \lambda^{1/2} h_0(\mathbf{r}\boldsymbol{\omega}; \mathbf{r}'\boldsymbol{\omega}'; \lambda)\boldsymbol{\mu}(\boldsymbol{\omega}') \, d\mathbf{r}' \, d\boldsymbol{\omega}' \tag{2.84}$$

The long-range limit of $h(\mathbf{r}_1 \boldsymbol{\omega}_1; \mathbf{r}_2 \boldsymbol{\omega}_2; \lambda)$ given in (2.82) is the only part of the pair correlation function that can conceivably contribute shape dependence to $\Delta A$. However since $\mathsf{T} \cdot \mathsf{T} \sim r_{12}^{-6}$ is of short range, and since $\int_V \mathsf{T} \cdot \mathsf{R}^*$ is negligible because $\mathsf{R}^*$ is proportional to $(V + W)^{-1}$, the long-range part of the pair correlation function does not contribute in a shape-dependent way. The thermodynamic properties of the fluid are consequently independent of sample shape and surroundings.

Now we turn to the predictions of three approximate theories for the thermodynamic properties of polar fluids. These are the Onsager model proposed by Nienhuis and Deutch,[2] the MSM[14] discussed in Section II.A, and the thermodynamic perturbation theory of Stell, Rasaiah, and Narang (SRN)[29] as discussed by Rushbrooke, Stell and Høye.[30]

The Onsager model[2] is a translation of Onsager's theory of the dielectric constant into the pair correlation function language.

One assumes that the pair correlation function has the form

$$h(\mathbf{r}_1 \boldsymbol{\omega}_1; \mathbf{r}_2 \boldsymbol{\omega}_2; \lambda) = -1 \qquad \text{for } r_{12} < d$$

$$h(\mathbf{r}_1 \boldsymbol{\omega}_1; \mathbf{r}_2 \boldsymbol{\omega}_2; \lambda) = h_1(\mathbf{r}_1 \boldsymbol{\omega}_1; \mathbf{r}_2 \boldsymbol{\omega}_2; \lambda)$$

$$= \beta\varepsilon^{-1}(\lambda)\boldsymbol{\mu}_{\text{eff}}(\boldsymbol{\omega}_1; \lambda) \cdot [\mathsf{T}(\mathbf{r}_1 \mathbf{r}_2) + \mathsf{R}_{V+W}^*(\mathbf{r}_1 \mathbf{r}_2; \varepsilon^{-1}(\lambda))]$$

$$\cdot \boldsymbol{\mu}_{\text{eff}}(\boldsymbol{\omega}_2; \lambda) \qquad \text{for } r_{12} > d$$

$$\tag{2.85}$$

From this assumed form for $h$, for $r_{12} > d$, the short-range part of the pair correlation function $h_0 = h - h_1$ is seen to be

$$h_0(\mathbf{r}_1\omega_1; \mathbf{r}_2\omega_2; \lambda) = -1 - h_1(\mathbf{r}_1\omega_1; \mathbf{r}_2\omega_2; \lambda) \qquad \text{for } r_{12} < d$$

$$h_0(\mathbf{r}_1\omega_1; \mathbf{r}_2\omega_2; \lambda) = 0 \qquad \text{for } r_{12} > d \tag{2.86}$$

From the assumed form of $h$ in (2.85), one can determine $\mu_{\text{eff}}(\omega, \lambda)$ from (2.83) to (2.86) as

$$\mu_{\text{eff}}(\omega; \lambda) = \frac{3\varepsilon(\lambda)}{2\varepsilon(\lambda) + 1} \lambda^{1/2}\mu(\omega) \tag{2.87}$$

with [cf. (1.25)]

$$\frac{1}{4\pi} [\varepsilon(\lambda) - 1] = \tfrac{1}{3}\beta\rho\lambda^{1/2}\mu \cdot \mu_{\text{eff}}(\lambda) \tag{2.88}$$

Combining the last two equations yields

$$\tfrac{1}{3}\beta\rho(\lambda^{1/2}\mu)^2 = \frac{[\varepsilon(\lambda) - 1][2\varepsilon(\lambda) + 1]}{12\pi\varepsilon(\lambda)} \tag{2.89}$$

which when $\lambda = 1$ is Onsager's formula for the dielectric constant. This confirms that (2.85) is a pair correlation function consistent with Onsager's theory for $\varepsilon$, but the uniqueness of (2.85) has not been proved.

The pair correlation formulation of Onsager's theory has the advantage that it provides a solvable model for the thermodynamic properties of the hard-sphere polar fluid. For example, from (2.81), (2.85), (2.87), and (2.88) we can easily compute the Onsager model dipolar Helmholtz energy $\Delta A_{\text{ons}}$. This is expressed by Sutherland, Nienhuis, and Deutch[31] as

$$V^{-1}\beta\Delta A_{\text{ons}} = -\frac{18}{\pi d^3} \int_0^y dX\, F_{\text{ons}}(X) \tag{2.90}$$

where

$$y = \frac{4\pi\beta\rho\mu^2}{9} \qquad F_{\text{ons}}(X) = \frac{1}{8}\frac{\varepsilon(X) - 1}{2\varepsilon(X) + 1} \tag{2.91}$$

and

$$X = \frac{\varepsilon(X) - 1}{9\varepsilon(X)}[2\varepsilon(X) + 1] \tag{2.92}$$

where $\varepsilon(X)$ is the Onsager dielectric constant given in (2.89).

The dipolar Helmholtz energy for the MSM polar fluid discussed in Section II.A can also be readily evaluated. One finds[22]

$$V^{-1}\beta\Delta A_{MSM} = -\frac{18}{\pi d^3} \int_0^y dX \, F_{MSM}(X) \tag{2.93}$$

where $F_{MSM}(X)$ is given by the solution to the implicit equation [cf. (2.8)]

$$X = \tfrac{1}{3}\{q_s[2F_{MSM}(X)] - q_s[-F_{MSM}(X)]\} \tag{2.94}$$

where, as before [see (2.9)],

$$q_s(\eta) = \frac{(1 + 2\eta)^2}{(1 - \eta)^4} \tag{2.95}$$

is the hard-sphere PY inverse compressibility.

Also recall that the MSM dielectric constant [see (2.11)] is

$$\varepsilon_{MSM}(y) = \frac{q_s[2F_{MSM}(y)]}{q_s[-F_{MSM}(y)]} \tag{2.96}$$

As is conventional, we write $\varepsilon_{MSM}(y)$ in a form similar to (2.89), the Onsager formula, by introducing the MSM Kirkwood "$g$" factor $g_{MSM}$. This gives

$$\frac{[\varepsilon_{MSM}(y) - 1][2\varepsilon_{MSM}(y) + 1]}{9\varepsilon_{MSM}(y)} = yg_{MSM} \tag{2.97}$$

We can easily evaluate $g_{MSM}$ from (2.94) (2.96)–(2.97) as

$$g_{MSM} = \frac{1}{3}\left\{\frac{2}{q[-F(y)]} + \frac{1}{q[2F(y)]}\right\} \tag{2.98}$$

The Kirkwood $g$ factor is a measure of the short-range orientational order neglected in the Onsager theory. The MSM is the only theory of polar fluids now available that gives an explicit form for the $g$ factor.

With this introduction to the Onsager model of Nienhuis and Deutch[2] and with this brief review of the MSM,[14,22] we can now turn to a detailed study of the thermodynamical predictions of these approximate theories and also to the predictions of the SRN perturbation theory[29] to be discussed later. The thermodynamics of the Onsager model and the MSM have been studied by Sutherland, Nienhuis, and Deutch.[31] Rushbrooke, Stell, and Høye[30] have also studied MSM thermodynamics and in addition have compared these results to SRN perturbation theory predictions.

Both the Onsager[31] [see (2.90)] and MSM results[32] [see (2.93)] for the dipolar free energy can be evaluated explicitly. One finds

$$V^{-1}\beta\Delta A_{ons} = (4\pi\ d^3)^{-1}\left\{\frac{[\varepsilon(y)-1]^2}{\varepsilon(y)} - \frac{3}{2}\ln\left[\left(\frac{2\varepsilon(y)+1}{3}\right)^3(\varepsilon(y))^{-2}\right]\right\} \quad (2.99)$$

$$V^{-1}\beta\Delta A_{MSM} = -\frac{48}{\pi\ d^3}\left\{\frac{(1+F(y))^2}{(1-2F(y))^4} + \frac{[2-F(y)]^2}{8[1+F(y)]^4}\right\}[F(y)]^2 \quad (2.100)$$

In (2.99), $\varepsilon(y)$ is the Onsager dielectric constant, while in (2.100) $F(y) = F_{MSM}(y)$ which is defined implicitly in (2.94). The dipolar internal energy for both models can be expressed as

$$V^{-1}\Delta E = \left(\frac{\partial\beta V^{-1}\Delta A}{\partial\beta}\right)_{N_1V} = -\frac{18y}{\pi\ d^3\beta}F(y) \quad (2.101)$$

where $F(y) = F_{ons}(y)$ for the Onsager case and $F_{MSM}(y)$ for the MSM. For both models the reference integral energy is $\frac{3}{2}NkT$, the hard-sphere energy. So the total energy is

$$V^{-1}E = \frac{3}{2}\rho kT - \frac{8\rho\mu^2}{d^3}F(y) \quad (2.102)$$

Thus $c_V$, the specific heat at constant volume, is

$$V^{-1}c_V = \frac{3}{2}\rho k + \frac{18k}{\pi\ d^3}y^2 F'(y) \quad (2.103)$$

For both models the excess pressure takes the very simple form

$$\Delta P = -\frac{\partial\Delta A}{\partial V} = V^{-1}[\Delta E - \Delta A] = V^{-1}T\Delta S \quad (2.104)$$

From this result and from an expression for the pressure of the reference hard-sphere fluid, one can readily obtain the equation of state for the polar fluid. Two convenient choices for the reference fluid equation of state are available, the PY hard-sphere result[19] and the more accurate Carnahan–Starling[32] formula. The former was employed in the calculations of Sutherland, Nienhuis, and Deutch, while the latter expression was used by Rushbrooke, Stell, and Høye. Only minor differences in the MSM equation of state result from these differing choices.

The equations of state for the SRN, MSM, and Onsager theories show van der Waals loops. One can thus identify the two-phase region, equilibrium vapor pressure curve, and critical point by the Maxwell construction method.

The equation of state is conveniently expressed in terms of the reduced variables:

$$v^* = \frac{6}{\pi \, d^3 \rho} \qquad (2.105a)$$

$$T^* = \frac{3k_B \, d^3}{8\mu^3} \, T \qquad (2.105b)$$

$$P^* = \frac{\pi \, d^3}{16\mu^2} \, P \qquad (2.105c)$$

Values for the critical parameters $v_c^*$, $T_c^*$, and $P_c^*$ are given in Table I for the Onsager model and the MSM, as well as for the SRN perturbation theory. The critical behaviors of all three theories are classical. Thus the compressibility diverges near the critical point as $|T - T_c|^{-1}$ and the specific heat remains finite. The detailed predictions of the three theories, however, are very different, as can be seen from Table I. The phase boundary, critical point, and vapor pressure curve of the Onsager model are located at lower reduced temperatures and pressures than the MSM results, which in turn lie at lower reduced temperatures and pressures than the perturbation results found by Rushbrooke, Stell, and Høye.

TABLE I

Critical Parameters for Model Polar Fluids

| | $P_c^* \times 10^3$ | $T_c^*$ | $\eta_c$ | $\varepsilon_c$ | $\alpha_c$ |
|---|---|---|---|---|---|
| MSM | 1.382 | 0.0844 | 0.0555 | 4.232 | 0.6581 |
| Onsager | 0.1937 | 0.0457 | 0.0151 | 2.210 | 0.3297 |
| RSH[a] | 2.773 | 0.1047 | 0.0833 | — | 0.7956 |

[a] Perturbation theory results of RSH.

Recently, Patey and Valleau[33] performed Monte Carlo calculations for the dipolar hard-sphere system. They obtain the dipolar free energy at two liquid densities and several values of the parameter $\beta\mu^2$. They compare their results with the MSM and SRN theory predictions and find the SRN results are in better agreement with the Monte Carlo calculations.

We now give a brief outline of the SRN perturbation theory. One expands the dipolar free energy of (2.81) in the parameter $\rho\beta\mu^2$ to obtain a series of the form:

$$\beta V^{-1} \Delta A = [\rho\beta\mu^2]^2 A_2 + [\rho\beta\mu^2]^3 A_3 + \cdots \qquad (2.106)$$

where

$$A_2 = -\tfrac{1}{6} \int g_0(r) r^{-6} \, d\mathbf{r} \tag{2.107}$$

$$A_3 = \tfrac{1}{54} \int g_0(1, 2, 3) u(1, 2, 3) \, d\mathbf{r}_1 \, d\mathbf{r}_3 \tag{2.108}$$

where $g_0(r)$ is the pair distribution function, $g_0(1, 2, 3)$ is the triplet distribution function of the reference hard-sphere fluid, and

$$u(1, 2, 3) = \frac{1 + 3 \cos \alpha_1 \cos \alpha_2 \cos \alpha_3}{[r_{12} r_{23} r_{13}]^3} \tag{2.109}$$

Here $\alpha_1$, $\alpha_2$, and $\alpha_3$ are the interior angles of the triangle connecting points 1, 2, and 3.

Equation (2.106), as it stands, is not a satisfactory approximation. To improve (2.106) SRN form the simple Pade approximant

$$V^{-1}\beta \Delta A_\rho = \frac{(\rho \beta \mu^2)^2 A_2}{1 - (\rho \beta \mu^2) A_2^{-1} A_3} \tag{2.110}$$

which is equivalent to summing (2.106) as a geometric series. The thermodynamic properties in the SRN perturbation theory are computed from (2.110).

As a justification for this procedure, Rushbrooke, Stell, and Høye explore an analogous approximation within the MSM. That is, they expand the MSM free energy [see (2.100)] in a series analogous to (2.106), perform an approximate resummation of the form of (2.110), and then numerically compare the Pade MSM free energy with (2.100). They find close agreement. Rushbrooke, Stell, and Høye also observe that the MSM coefficients analogous to $A_2$ and $A_3$ are the low-density limits of $A_2$ and $A_3$. They attribute the limitations of the MSM as a theory for the thermodynamic properties of polar fluids to this result.

The SRN resummed perturbation theory is attractive because of its simplicity and because of the good agreement it gives with the Monte Carlo calculations. However, recent results of Sullivan and Deutch[34] on rigid polar lattices indicate that the success of the simple Pade approximant may be fortuitous. Further work aimed at clarifying our understanding of (2.110) would be welcome.

Before concluding this section we briefly discuss the thermodynamics of polar mixtures. Sutherland Nienhuis, and Deutch[31] have analyzed the MSM polar mixture model[15] discussed in Section II.A and have established that it displays regions of thermodynamic instability.

They show that the MSM mixture model exhibits a variety of liquid–liquid phase separations as well as liquid–vapor regions. They do not, however, map out coexistence regions in detail. Sutherland Nienhuis, and Deutch[31] also show that the multicomponent Onsager model can be solved in terms of an effective one-component Onsager fluid analogous to the effective one-component MSM fluid discussed earlier. The multicomponent Onsager mixture displays instabilities and critical behavior completely analogous to that expected in the MSM mixture. For both models the interesting special case of nonpolar–polar mixtures also shows a variety of phase separations. This is despite the limitations of the MSM for the potential of mean force in this case, as discussed earlier in Section II.A.

## III. DIELECTRIC RELAXATION

Thus far this article has been exclusively concerned with polar fluids at equilibrium. One might well suspect that long-range dipole–dipole forces would be manifest in nonequilibrium behavior of simple polar fluids and that issues such as shape dependence would arise in a manner not encountered for short-range fluids. Dipole–dipole forces may have a qualitative effect on all transport coefficients, but it is probable that the effects are most pronounced in dielectric relaxation, that is, on the complex frequency-dependent dielectric constant $\varepsilon(\omega)$. In fact, dielectric relaxation is the only nonequilibrium property that has been examined in any detail. This section is devoted to presenting an assessment of the present state of knowledge on dielectric relaxation.

The general procedure one should follow to investigate the influence of dipole–dipole forces on nonequilibrium properties is to relate the transport coefficient of interest to an appropriate equilibrium time correlation function and then examine the influence of dipole–dipole forces compared to the short-range forces present on the temporal behavior of the correlation function. For fluid systems this program has not been completely accomplished for any transport coefficient, although some progress has been made on dielectric relaxation both for fluid systems in general and for specific relaxation models. The investigation of the effect of dipole–dipole forces on other transport coefficients, such as the self-diffusion coefficient and viscosity, remains an outstanding problem of some interest, and there are excellent prospects for progress on this topic. We note that an analogous problem exists in the case of plasma systems[35] in which one seeks to disentangle the influence of Coulomb and short-range forces on transport coefficients.

Our discussion is restricted to samples of uniform polarization geometry. For this case, under the influence of a spatially homogeneous oscillatory *external* electric field $\mathbf{E}_e(\omega) = \mathbf{E}_e \exp(i\omega t)$, the constitutive relation, which is assumed to hold, is

$$\mathbf{P}(\omega) = \frac{\varepsilon(\omega) - 1}{4\pi} \mathbf{E}(\omega) \qquad (3.1)$$

where $\mathbf{P}(\omega)$ is the polarization and $\mathbf{E}(\omega)$ is the *macroscopic* electric field in the sample.

If we employ the familiar algorithm of linear response theory, we obtain a linear relationship between $\mathbf{P}(\omega)$ and the external field $\mathbf{E}_e(\omega)$ of the form

$$\mathbf{P}(\omega) = \chi(\omega)\mathbf{E}_e(\omega) \qquad (3.2)$$

where $\mathbf{P}(\omega)$ is identified with the nonequilibrium steady-state response of the average of the net dipole moment per unit volume of the system to $\mathbf{E}_e$:

$$\mathbf{P}(\omega) \exp(i\omega t) = V^{-1} \left\langle \sum_{i=1}^{N} \boldsymbol{\mu}_i(t) \right\rangle_{\mathbf{E}_e} \qquad (3.3)$$

In (3.2) $\chi(\omega)$ is the susceptibility which is related to the one-sided Fourier transform of the equilibrium time autocorrelation of the net dipole moment $\mathbf{m}(t) = \sum_i \boldsymbol{\mu}_i(t)$ according to

$$\chi(\omega) = -\frac{B}{V} \int_0^\infty \exp(-i\omega t) \frac{d}{dt} \langle m_z(0)m_z(t) \rangle \, dt \qquad (3.4)$$

Comparison of (3.1) and (3.2) immediately permits one to conclude that $\chi(\omega)$, hence the autocorrelation function $\langle m_z(0)m_z(t) \rangle$ must depend on sample shape and surroundings. The reasoning is as follows. Since the relation between $\mathbf{E}(\omega)$ and $\mathbf{E}_e(\omega)$ depends on sample shape, it is necessary for $\chi(\omega)$ to contain geometry-dependent factors to compensate for this dependence if the computed $\varepsilon(\omega)$ is to be a single function independent of the sample configuration. A discussion of the influence of sample shape has been given by Mazur.[36] For example, in the case of a spherical sample in vacuum, the relation between $\mathbf{E}_e(\omega)$ and $\mathbf{E}(\omega)$ is

$$\mathbf{E}_e(\omega) = \frac{\varepsilon(\omega) + 2}{3} \mathbf{E}(\omega) \qquad (3.5)$$

so that

$$\frac{\varepsilon(\omega) - 1}{4\pi} \frac{3}{\varepsilon(\omega) + 2} = \chi_0(\omega) \qquad (3.6)$$

where

$$\chi_0(\omega) = -\frac{\beta}{V} \langle m_z^2(0) \rangle_0 \int_0^\infty e^{-i\omega t} \frac{d}{dt} \Phi(t) \, dt \qquad (3.7)$$

with

$$\Phi(t) = \frac{\langle m_z(0)m_z(t)\rangle_0}{\langle m_z^2(0)\rangle_0} \tag{3.8}$$

In these expressions a subscript zero has been added to emphasize that the relations are only valid for the particular sample geometry of a sphere in vacuum.

The first application of linear response theory to dielectric relaxation is due to Glarum.[37] Glarum obtained the result for a sphere in vacuum noted above, but also presented an expression for $\varepsilon(\omega)$ in terms of the auto-correlation function of the moment of spherical region embedded in an infinite dielectric medium, $\phi(t)$:

$$\phi(t) = \frac{\langle m_z(0)m_z(t)\rangle_\infty}{\langle m_z^2(0)\rangle_\infty} \tag{3.9}$$

where the subscript denotes the particular sample geometry of a sphere embedded in an infinite medium. Glarum transformed the geometry from a sphere in vacuum to a sphere in an infinite dielectric medium, employing macroscopic reasoning analogous to that used by Kirkwood[38] for the static dielectric constant. Glarum's result, which was later extended by Cole,[39] is

$$\frac{\varepsilon(\omega) - 1}{4\pi} \frac{2\varepsilon_0 + 1}{2\varepsilon_0 + \varepsilon(\omega)} = \chi_\infty(\omega) \tag{3.10}$$

where

$$\chi_\infty(\omega) = -\frac{\beta}{V} \langle m_z^2(0)\rangle_\infty \int_0^\infty e^{-i\omega t} \frac{d}{dt} \phi(t)\, dt \tag{3.11}$$

and $\varepsilon_0 = \varepsilon(0)$ is the static dielectric constant.

It should be noted that the equilibrium averages $\langle m_z^2(0)\rangle_0$ and $\langle m_z^2(0)\rangle_\infty$ appearing in $\chi_0(\omega)$ and $\chi_\infty(\omega)$, respectively, can be eliminated in terms of the static dielectric constant $\varepsilon_0$ according to

$$\frac{3}{4\pi} \frac{\varepsilon_0 - 1}{\varepsilon_0 + 2} = \frac{\beta}{V} \langle m_z^2(0)\rangle_0 \tag{3.12}$$

and

$$\frac{\varepsilon_0 - 1}{4\pi} \frac{2\varepsilon_0 + 1}{3\varepsilon_0} = \frac{\beta}{V} \langle m_z^2(0)\rangle_\infty \tag{3.13}$$

Thus the expression for $\varepsilon(\omega)$ arising for a sphere in vacuum [see (3.6)] is

$$\frac{\varepsilon(\omega) - 1}{\varepsilon(\omega) + 2}\left[\frac{\varepsilon_0 - 1}{\varepsilon_0 + 2}\right]^{-1} = \mathscr{L}[-\dot{\Phi}(t)] \tag{3.14}$$

and for a sphere in an infinite medium [see (3.10)]

$$\frac{\varepsilon(\omega) - 1}{\varepsilon_0 - 1}\frac{3\varepsilon_0}{2\varepsilon_0 + \varepsilon(\omega)} = \mathscr{L}[-\dot{\phi}(t)] \tag{3.15}$$

where the operator $\mathscr{L}$ indicates the one-sided Fourier transform

$$\mathscr{L}[f(t)] = \int_0^\infty e^{-i\omega t} f(t)\, dt \tag{3.16}$$

Subsequently, an alternative formula relating $\varepsilon(\omega)$ to $\phi(t)$ was put forward by Fatuzzo and Mason,[40] Klug, Kranbuehl, and Vaughn,[41] and Scaife.[42] Their result may be expressed as

$$\frac{\varepsilon(\omega) - 1}{4\pi}\frac{2\varepsilon(\omega) + 1}{3\varepsilon(\omega)} = \chi_\infty(\omega)$$

or, alternatively,

$$\frac{[\varepsilon(\omega) - 1][2\varepsilon(\omega) + 1]\varepsilon_0}{(\varepsilon_0 - 1)(2\varepsilon_0 + 1)\varepsilon(\omega)} = \mathscr{L}[-\dot{\phi}(t)] \tag{3.17}$$

The difference between the Glarum expression [see (3.15)] and the Fatuzzo–Mason expression [see (3.17)] has prompted much controversy. As Cole has pointed out, the significance of this difference is that, if one wishes to infer from a measured $\varepsilon(\omega)$ some information about the correlation function $\phi(t)$ in order to assess various molecular models, it is important to know if (3.15) or (3.17) is correct. These expressions have been generalized by Cole[39] and Hill[43] to include the presence of induced as well as permanent dipole moments, a complication that is not treated here.

Nee and Zwanzig[44] presented an argument based on a rotational Brownian motion model of dipoles in a polar fluid, which employed the Fatuzzo–Mason result. Glarum[41] criticized the derivations leading to the Fatuzzo–Mason results, and recently Cole presented two arguments, one based on a rotational Brownian motion model for a high-temperature lattice,[46] and the other based on a "molecular" analysis[47] of the correlation functions $\Phi(t)$ and $\phi(t)$, which support the Glarum result.

Recently Titulaer and Deutch[48] presented an analysis of these conflicting theories of dielectric relaxation. Their principal result was to find that the Fatuzzo–Mason result [see (3.17)] is correct for the case of a spherical sample embedded in its own medium, and that the Glarum expression (3.15) is incorrect for this case. The Glarum expression refers to the case of a spherical sample embedded in a medium with a frequency-independent dielectric constant equal to the static dielectric constant of the sample.

We present here a brief summary of the salient features of the analysis presented by Titulaer and Deutch.[48] These investigators considered an isotropic molecular sample in a spherical volume of radius $a$ consisting of a material characterized by a frequency-dependent dielectric constant $\varepsilon_2(\omega)$. This molecular sample is embedded in an infinite medium, which by assumption consists of a macroscopic dielectric continuum characterized by a dielectric constant $\varepsilon_1(\omega)$. This physical situation is depicted in Fig. 2.

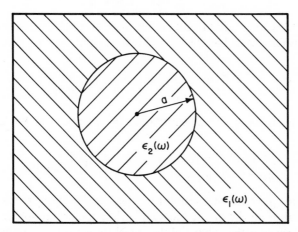

**Fig. 2**   The system discussed in Section III. A sphere of radius $a$ with dielectric constant $\varepsilon_2(\omega)$ is embedded in an infinite dielectric continuum with dielectric constant $\varepsilon_1(\omega)$.

In linear response formalism the external field $\mathbf{E}_e$ is the field present in the cavity of radius $a$ when the molecular sample is removed from the surrounding medium. Titulaer and Deutch[48] present further arguments supporting this identification of the cavity field with the external field. If an applied electric field is present in the medium which approaches the constant value $\mathbf{E}^\infty(t) = \mathbf{E}^\infty \exp(i\omega t)$ at large distances from the cavity then, by means of a simple electrostatic boundary[49] value calculation, we obtain for the external field:

$$\mathbf{E}_e(\omega) = \frac{3\varepsilon_1(\omega)}{2\varepsilon_1(\omega) + 1}\, \mathbf{E}^\infty(t) \qquad (3.18)$$

Use of (3.2) gives the relation between the polarization of the molecular sample and this external field, where the susceptibility $\chi(\omega)$ is given in (3.4). For uniform polarization geometries such as the one discussed here, this polarization is related to the macroscopic field $\mathbf{E}(\omega)$ present inside the spherical sample by

$$\mathbf{P}(\omega) = \frac{\varepsilon_2(\omega) - 1}{4\pi} \mathbf{E}(\omega) \tag{3.19}$$

The macroscopic field $\mathbf{E}(\omega)$ may be expressed in terms of $\mathbf{E}^\infty(t)$ by a macroscopic calculation[49]

$$\mathbf{E}(\omega) = \frac{3\varepsilon_1(\omega)}{2\varepsilon_1(\omega) + \varepsilon_2(\omega)} \mathbf{E}^\infty(t) \tag{3.20}$$

and (3.2), (3.18), (3.19), and (3.20) may be used to obtain a relation between the susceptibility $\chi(\omega)$ [see (3.4)] and $\varepsilon_1(\omega)$, $\varepsilon_2(\omega)$ for this geometry. The result is

$$\frac{3}{4\pi} \frac{[\varepsilon_2(\omega) - 1][2\varepsilon_1(\omega) + 1]}{2\varepsilon_1(\omega) + \varepsilon_2(\omega)} = \chi(\omega) \tag{3.21}$$

An analogous linear response calculation for a static field $\mathbf{E}^\infty$ leads to the expression

$$\frac{3}{4\pi} \frac{[\varepsilon_2(0) - 1][2\varepsilon_1(0) + 1]}{2\varepsilon_1(0) + \varepsilon_2(0)} = \frac{\beta}{V} \langle m_z^2(0)m_z(0) \rangle \tag{3.22}$$

Here, and in the defining relation for $\chi(\omega)$ [see (3.4)], $m_z$ refers to the net $z$ component of all the dipole moments in the molecular spherical sample, and $V = (4\pi a^3/3)$ is the volume of this inner region. If we combine the result of (3.4), (3.21), and (3.22), we derive the important relation[48]

$$\mathscr{L}[-\dot{f}(t)] = \frac{[\varepsilon_2(\omega) - 1][2\varepsilon_1(\omega) + 1][2\varepsilon_1(0) + \varepsilon_2(0)]}{[\varepsilon_2(0) - 1][2\varepsilon_1(0) + 1][2\varepsilon_1(\omega) + \varepsilon_2(\omega)]} \tag{3.23}$$

where we have defined $f(t)$ the normalized autocorrelation function for the net dipole moment of the molecular sample as

$$f(t) = \frac{\langle m_z(0)m_z(t) \rangle}{\langle m_z(0)m_z(0) \rangle} \tag{3.24}$$

The controversy concerning the various relationships between correlation functions and dielectric constants may be clarified by considering various special cases of (3.24).

1. First, we set $\varepsilon_1(\omega) = 1$, which corresponds to eliminating the dielectric in the external region. In this case $f(t) = \Phi(t)$, and we recover the result for

the sphere in vacuum [see (3.14)] when we omit the subscript 2 and write $\varepsilon_0$ for $\varepsilon_2(0)$.

2. If we set $\varepsilon_1(\omega) = \varepsilon_2(\omega) = \varepsilon(\omega)$, the physical situation corresponds to the molecular sample embedded in an infinite macroscopic medium characterized by the same dielectric constant. For this case $f(t) = \phi(t)$, and we recover the Fatuzzo-Mason result[40] displayed in (3.17). The Glarum result [see (3.15)] is incorrect for this physical situation.

3. However, if we consider the special case in which $\varepsilon_2(\omega) = \varepsilon(\omega)$ while $\varepsilon_1(\omega) = \varepsilon_2(0) = \varepsilon_0$ for all frequencies, we obtain exactly the Glarum result [see (3.15)]. Accordingly, we denote $f(t)$ for this case by $\phi_G(t)$. Note that this physical situation corresponds to a molecular medium characterized by dielectric constant $\varepsilon(\omega)$ embedded in an infinite macroscopic medium which responds instantaneously to fields of all frequencies as a static dielectric with a value of the static dielectric constant equal to $\varepsilon(0)$ of the molecular sample.

Confusion between the second case in which $\varepsilon_1(\omega) = \varepsilon_2(\omega) = \varepsilon(\omega)$ and the third case in which $\varepsilon_1(\omega) = \varepsilon_2(0) = \varepsilon_0$ is responsible for part of the controversy on dielectric relaxation. Both results are correct, however, they correspond to entirely different physical situations. The second case $f(t) = \phi(t)$ corresponds to the fluctuating moment of a spherical molecular sample embedded in a medium of its own frequency-dependent dielectric constant, while the third case $f(t) = \phi_G(t)$ corresponds to the fluctuating moment of a spherical molecular sample embedded in a medium characterized by a frequency-independent dielectric constant equal to the static dielectric constant of the molecular sample. Of course, on physical grounds we expect that an exact calculation, if carried to completion, would yield identical results for $\varepsilon(\omega)$. However, for models or approximate calculations, the third case may prove awkward.

The remainder of the article by Titulaer and Deutch[48] is devoted to a careful discussion of where previous analyses[44,45,46] of the question of the correlation function expression for the dielectric constant have gone astray.

An important additional point to appreciate about the result in (3.23) is that it clearly illustrates the dependence of the equilibrium time correlation functions $f(t)$ on the nature of the external region, even in the absence of fields and irrespective of the size of the molecular sample. This situation is similar to the situation arising in the case of equilibrium polar fluids, which was discussed earlier in this article. Titulaer and Deutch[50] examined, by use of a macroscopic theory based on linear response and the fluctuation-dissipation theorem, the nature of equilibrium correlation functions of dipole moments at different times and at different widely separated points in space for general sample geometries. The dependence on the shape of the sample and on the nature of the surroundings is clearly exhibited. Of course, for equal time correlation functions the results agree with the microscopic theory of Nienhuis and Deutch.

The next few years should see considerable progress in understanding the nonequilibrium behavior of polar fluids. This progress should include development of a rigorous molecular theory of the frequency-dependent dielectric constant and of other transport properties of polar fluids. Such a theory would isolate the effect of long-range dipolar forces and should indicate possible methods for developing perturbation theories for these fluids that expand about a reference fluid in which only short-range forces are present. There certainly will be further efforts in developing models of dielectric relaxation in fluids that go beyond the classical stochastic relaxation[44] models available today.

## References

1. J. M. Deutch, *Ann. Rev. Phys. Chem.*, **24**, 301 (1974).
2. G. Nienhuis and J. M. Deutch, *J. Chem. Phys.*, **55**, 4213 (1971); **56**, 235 (1972); **56**, 1819 (1972).
3. M. S. Wertheim, *Mol. Phys.*, **25**, 211 (1973).
4. M. S. Wertheim, *Mol. Phys.*, **26**, 1425 (1973).
5. J. D. Ramshaw, *J. Chem. Phys.*, **57**, 2684 (1972).
6. J. S. Høye and G. Stell, preprint.
7. D. W. Jepsen, *J. Chem. Phys.*, **45**, 709 (1966).
8. See, for example, G. Stell, in H. L. Frisch and J. L. Lebowitz, Eds., *The Equilibrium Theory of Classical Fluids*, Benjamin, New York, 1964.
9. J. L. Lebowitz, G. Stell, and S. Baer, *J. Math. Phys.*, **6**, 1282 (1964).
10. U. Titulaer and J. M. Deutch, unpublished work.
11. S. A. Adelman and J. M. Deutch, work in progress.
12. J. L. Lebowitz and J. K. Percus, *Phys. Rev.*, **144**, 251 (1966).
13. F. Waisman and J. L. Lebowitz, *J. Chem. Phys.*, **52**, 4307 (1970); **56**, 3093 (1972).
14. M. S. Wertheim, *J. Chem. Phys.*, **55**, 4291 (1971).
15. S. A. Adelman and J. M. Deutch, *J. Chem. Phys.*, **59**, 3971 (1973).
16. L. Blum, *J. Chem. Phys.*, **57**, 1862 (1972); **58**, 3295 (1973).
17. S. A. Adelman and J. M. Deutch, *J. Chem. Phys.*, **60**, 3935 (1974).
18. H. C. Andersen and D. Chandler, *J. Chem. Phys.*, **57**, 1918 (1972).
19. M. S. Wertheim, *Phys. Rev. Lett.*, **10**, 321 (1963); E. Thiele, *J. Chem. Phys.*, **39**, 374 (1963).
20. J. E. Mayer, *J. Chem. Phys.*, **18**, 1426 (1950).
21. D. W. Jepsen and H. L. Friedman, *J. Chem. Phys.*, **38**, 846 (1963).
22. G. Nienhuis and J. M. Deutch, *J. Chem. Phys.*, **56**, 5511 (1972).
23. The invariant expansion method was developed by L. Blum in Ref. 16. The notation and some of the specific results employed in the following discussion are in Refs. 15 and 17.
24. Our notation and conventions for angular functions follows A. R. Edmonds, *Angular Momentum in Quantum Mechanics*, Princeton University Press, Princeton, N.J., 1960.
25. W. G. MacMillan and J. F. Mayer, *J. Chem. Phys.*, **13**, 276 (1945).
26. L. Blum, Chem. Phys. Letters, **26**, 200 (1974).
27. G. Stell and G. S. Rushbrooke, preprint.

28. J. Kirkwood, *J. Chem. Phys.*, **4**, 592 (1936); J. Yvon, *Actualités Scientifiques et Industrielles*, No. 543, Hermann, Paris, 1937.
29. G. S. Stell, J. C. Rasaiah, and H. Narang, *Mol. Phys.*, **23**, 393 (1972).
30. G. S. Rushbrooke, G. Stell, and J. S. Høye, *Mol. Phys.*, **26**, 1199 (1973).
31. W. Sutherland, G. Nienhuis, and J. M. Deutch, Mol. Phys, **27**, 721 (1974).
32. N. F. Carnahan and K. E. Starling, *J. Chem. Phys.*, **51**, 635 (1969).
33. G. N. Patey and J. P. Valleau, *Chem. Phys. Lett.*, **21**, 297 (1973).
34. D. E. Sullivan, J. M. Deutch, and G. Stell, Mol. Phys., **28**, 1359 (1974).
35. J. T. Bartis and I. Oppenheim, *Phys. Rev.*, **A8**, 3174 (1973).
36. P. Mazur in B. Jancovici, Ed., *Cargèse Lectures in Theoretical Physics, 1964, Statistical Mechanics*, Gordon and Breach, New York, 1966.
37. S. H. Glarum, *J. Chem. Phys.*, **33**, 1371 (1960).
38. J. G. Kirkwood, *J. Chem. Phys.*, **4**, 592 (1936).
39. R. H. Cole, *J. Chem. Phys.*, **42**, 637 (1965).
40. E. Fatuzzo and P. R. Mason, *Proc. Phys. Soc.*, **90**, 741 (1967).
41. D. D. Klug, D. E. Kranbuehl, and W. E. Vaughn, *J. Chem. Phys.*, **50**, 3904 (1969).
42. B. K. P. Scaife in J. H. Calderwood, Ed., *Complex Permittivity*, English University Press, London, 1971.
43. N. E. Hill, *J. Phys. C: Solid State*, **5**, 415 (1972).
44. T. W. Nee and R. Zwanzig, *J. Chem. Phys.*, 6353 (1970).
45. S. H. Glarum, *Mol. Phys.*, **24**, 1327 (1972).
46. R. G. Cole, *Mol. Phys.*, in press.
47. R. G. Cole, *Mol. Phys.*, in press.
48. U. M. Titulaer and J. M. Deutch, *J. Chem. Phys.*, **60**, 1502 (1974).
49. H. Frohlich, *Theory of Dielectrics*, 2nd ed., Clarendon Press, Oxford, 1958, Appendix B.
50. U. M. Titulaer and J. M. Deutch, *J. Chem. Phys.*, **60**, 2703 (1974).

# THE KINETIC THEORY OF DENSE POLYATOMIC FLUIDS*

## JOHN S. DAHLER

*Departments of Chemical Engineering and Chemistry
University of Minnesota, Minneapolis, Minnesota*

## MARY THEODOSOPULU

*Faculté des Sciences
Université Libre de Bruxelles, Belgique*

## TABLE OF CONTENTS

## I. INTRODUCTION

The kinetic theory of dense polyatomic fluids is in its infancy. The literature on this topic is very nearly nonexistent, and the few results that have been reported are the products of quite recent research.[1,2] As such, these theories have yet to be thoroughly tested numerically or to be as critically and carefully examined as have been the older and better known theories of dense monoatomic fluids and of dilute polyatomic gases.

* This research was supported in part by a grant from the National Science Foundation.

Within the last 10 to 15 years, there has been much progress in the non-equilibrium statistical mechanics of dense monatomic fluids and of dilute polyatomic gases. It is natural to build on these foundations as one attempts to fashion a theory of dense polyatomic fluids. This is what we strive to do in this chapter, to present a theory combining the best features of theories originally developed for rather different purposes. We cannot do this without exercising considerable bias regarding which of several options we judge to be the best. In general, we place high priority on computational practicability. Other criteria we adopt will emerge in the course of the development. Indeed, the pattern of our presentation is to describe what we perceive to be the basic problems, to indicate how similar difficulties have been dealt with in related areas of research, and finally to apply some variant or generalization of one of these solutions to the problem at hand.

Only recently has it become clear how one can generalize the kinetic theory of dilute gases to include molecular internal degrees of freedom.[3] At the time of this writing, very few calculations based on this extended theory have been made. Thus, even in the case of extreme dilution where theory faces less severe problems, the task of constructing numerical estimates for the coefficients characteristic of the nonequilibrium behavior of a polyatomic fluid is still largely unfinished. Viewed in this light it scarcely is surprising to find that the kinetic theory of dense polyatomic fluids is in a rather primitive stage. Foremost among the difficulties associated with transport and relaxation in polyatomic fluids is that of dealing with the collisional events between constituent molecules. In addition to collisional transfers of linear momentum and translational energy, one also must concern oneself with alterations in angular momentum and in rotational and vibrational energy. The forces of interaction are noncentral. Therefore, even at low densities where only binary encounters are of importance, one is faced with dynamical problems which cannot be dealt with by analytical means. It was this as much as the lack of a truly satisfactory statistical mechanics formalism that delayed for so long the emergence of generally acceptable kinetic theories of dilute polyatomic gases. It is this alone that currently blocks the way to a thorough numerical test of the dilute gas theories that now at last are available.

The problem before us here is that of constructing a kinetic theory of dense polyatomic fluids. To solve this we must cope in a satisfactory fashion with all the problems encountered in the dilute gas, and then more. Especially, must we learn to live with the fact that no one has yet derived a kinetic equation for dense fluids that enjoys the general acceptance or claims for rigor accorded to the Boltzmann equation of the dilute gas theory. However, what *has* been learned is a way of generalizing to polyatomic species the methods that have been used for so long and with such notable success in treating monatomic gases. It is on this pattern that we shall build. What we relate in this article is the story of how one goes about the job of extending to dense polyatomic fluids the currently existing kinetic theories of dense monatomic fluids. Because these theories of dense monatomic fluids are not yet comparable in quality to the Boltzmann theory of gases, we scarcely can expect to do better when dealing with the polyatomic case. But what we can do is develop our theory in such a way that it will be readily adaptable to the incorporation of improvements in the monatomic theory whenever they become available.

The program of research we have just mentioned very well might begin with the generalization to polyatomic species of the pioneering studies conducted by John Kirkwood and his co-workers, starting with Irving and Kirkwood's[4] derivation of formally exact molecular expressions for quantities such as the pressure tensor and heat flux and then proceeding through those of his works aimed at the actual evaluation of liquid transport coefficients.[5] The first of these steps already took place long ago.[6] The second, were we to attempt it now, surely should be done with full awareness of the further developments of the theory of dense monatomic fluids due to Rice and his co-workers and to the Brussels school. Although this is not precisely the way in which we shall proceed, it is close to the mark.

Also, if one intends—as we do—to borrow from the theory of monatomic fluids, it would seem natural to construct polyatomic versions of one or more of the several available (and quite closely related) nonequilibrium ensemble theories and from these to extract correlation integral formulas for the relevant transport coefficients and relaxation times. Indeed, this too already has been done.[7] These formal results are helpful in guiding molecular dynamics investigators to the correct identification of relevant correlation integrals. However, the difficulty with this elegant and currently fashionable approach to the theory of irreversible processes is not so much that of generating formal results, but of proceeding beyond them to the development of practical computational procedures. The formalism is excellently adapted to the stochastic modeling of irreversible processes, as exemplified by the popular and successful continuous- and jump-diffusion theories of orientational relaxation. In principle, this approach should be just as valuable a tool for the analysis of other nonequilibrium processes. However, despite numerous specific calculations that have been reported, it appears that no general procedure has yet been developed for evaluating correlation integrals of dense fluids. Therefore, since this approach has produced little we can find to borrow, our pragmatic doctrine dictates that it be discarded in favor of more attractive alternatives such as the more immediately adaptable kinetic theories of Enskog,[9] of Rice and Allnatt,[10] of Prigogine, Misguich, and Nicolis,[11] and of Davis.[12]

A final item we must contend with in these introductory remarks is the Chapman–Enskog (CE) method for solving kinetic equations. It was clear from the earliest writings of these authors (see especially the treatise by Chapman and Cowling[9]) that their procedure selected from all possible solutions of the Boltzmann equation one specially related to the hydro-(thermo)dynamical quasi-steady state in the sense that it was characterized by secular variations with time of the hydro(thermo)dynamical local variables, concentration, temperature, and fluid velocity. Thanks to the investigations of Grad[13] and of those who followed in his path, the CE "normal solution" is now much better understood. Its identification as a specially restricted asymptotic solution of the (Boltzmann) kinetic equation is thoroughly consistent with the less rigorously defined but no less physically understandable characteristics originally ascribed to it.

The spirit and methodology of the CE procedure has been extended to dense gases and liquids, first by Enskog in his theory of a dense gas of rigid spheres, and then more recently to a host of other systems, a few of the most important being (1) the Kirkwood theory of liquid transport; (2) the square-well generalization of the Enskog dense gas theory;[14] (3) the Rice–Allnatt theory and its generalizations and improvements;[12] (4) the Prigogine–Misguich–Nicolis and Davis theories; and (5) quasi-steady state treatments based on the Bogoliubov theory of dense gases and general nonequilibrium ensemble theories such as those of Green,[15] McLennan,[16] and Kirkwood and Fitts.[17] In each of these the primary concern was with deriving formulas for the phenomenological coefficients descriptive of the quasi-steady state, that is, coefficients of viscosity, of thermal conduction, and of mutual and thermal diffusion. However, when molecular internal degrees of freedom were introduced, difficulties arose not because the method was inapplicable but because it failed to provide means for dealing with all the questions one wanted to ask. How, for example, was one to extract from the CE normal solution information about the rate of interconversion of kinetic and internal (rotational or vibrational) energy or of the coupling between the fluid vorticity and the average rotational or "spin" velocity of nonspherical species? Attempts to obtain this sort of information from the normal solution began with Wang Chang and Uhlenbeck[18] and later were pursued by several others.[19] In no case can it be said that the attempts really were successful.

Because it is descriptive of the quasi-steady state, the CE normal solution simply is inappropriate for dealing with relaxation processes characterized by time scales far shorter than the intervals required to establish the hydrothermal steady state. This inherent limitation of the method is particularly severe in the case of polyatomic fluids. However, the need for an alternative is also clearly dictated by the demands placed on the kinetic theory (of monatomic as well as polyatomic fluids) by the current period of experimentation that deals less often with the steady state than (in the case of ultra-sonics, high-frequency field induction effects, and the scattering of light and neutrons) with the rate of approach or relaxation to this steady state. In the case of a gas, such an alternative has been known for over a century. Thus, long before the discovery of the CE method, Maxwell,[20] extracted many useful results from kinetic theory by studying the equations of change for various velocity moments of the singlet distribution function. Then, 25 years ago this almost forgotten approach was reinstated by Grad[21] who argued that the equations of change for the "13 (velocity) moments" $\langle 1 \rangle$, $\langle \mathbf{c} \rangle$, $\langle \mathbf{cc} \rangle$ and $\langle \mathbf{cc}^2 \rangle$ provided a better basis for the analysis of many problems in gas dynamics than did the more traditional Navier–Stokes and heat-balance equations. During the past decade the great utility of moment equations has been demonstrated by Waldmann and Hess[22] through their many studies of polyatomic gas transport and relaxation phenomena, by Chen and Snider[23] in their kinetic theory analysis of gaseous NMR, and by Yip and his co-workers[24] in their numerous calculations of van Hove functions for dilute gases.

The point of all this is to establish that a real need exists for the generalization to dense fluids of the familiar method of moments. However, if the moment equations were to be derived on the basis of specific and necessarily approximate kinetic equations, the entire theory would be in constant need of revision. Thus it is far more sensible to derive the generalized moment equations directly from particle mechanics, and then use the approximations specific to a particular statistical theory only to evaluate the quantities that occur in these equations. These moment equations are generalizations of the equations of motion Irving and Kirkwood derived long ago for the specific examples of density, fluid velocity, and energy. As such, they are exact consequences of Hamilton's equations, and of the model Hamiltonian function we choose to represent the physical system under consideration. These equations can provide us with useful insights about the ways in which specific functions of dynamical variables are coupled to others. However, we shall be able to proceed further only if means are devised for constructing realistic estimates of the singlet and pair-space distribution functions. In the case of a dilute gas, this essential ingredient is provided by the functional relationship between the pair and singlet distribution functions implicit in the Boltzmann equation. The moment method for a dense polyatomic fluid depends for its success on the discovery of analogous functional relationships. Because our numerical predictions depend so sensitively on this relationship, the quantitative success of the whole scheme requires that careful attention be devoted to this part of the theory.

Now that all the components of the theory have been identified, we can set about the task of putting them together. Our first step is to approximate the nonequilibrium singlet distribution function $f$ by a finite number of its moments, taken with respect to the mechanical variables (exclusive of position) descriptive of the dynamical state of an individual molecule. Next comes the construction of the exact equations of change for these moments. The final ingredient is a functional relationship that permits construction of the pair-space distribution function $f_2$ from a knowledge of $f_1$. Through the use of this relationship, the equations of change for the moments are transformed into a closed set of equations from which the moments themselves then can be determined.

## II. MOMENT EQUATIONS

Although it may be necessary to adopt a semiclassical representation for situations in which vibrational degrees of freedom are of particular interest and/or when the fluid temperature is low enough to warrant a quantal treatment of the rotational motions, we have temporarily assigned a rather low priority to these refinements. Accordingly, we neglect molecular vibrations and also adopt a classical mechanical description of the fluid. Finally, we deal specifically in these pages with only two kinematic models, the spherical top and the linear rotor.

### A. Moment Expansion of the Distribution Function

The dynamical state of a single molecule can be fully specified by the position $\mathbf{x}$ and velocity $\mathbf{c} = \mathbf{p}/m$ of the center of mass, by the angular momentum $\mathbf{L}$, and by the three Euler angles $\boldsymbol{\varepsilon} = (\theta, \phi, \psi)$. The molecular peculiar velocity will be denoted by the symbol $\mathbf{C} = \mathbf{c} - \mathbf{u}$. Furthermore, it is convenient to introduce the dimensionless peculiar velocity and angular momentum $\mathbf{W} = (m/2kT)^{1/2}\mathbf{C}$ and $\boldsymbol{\Omega} = (2IkT)^{-1/2}\mathbf{L}$, respectively.

We now elect to write the singlet distribution function in the form $f = f^0(1 + \phi)$, where $f^0$ is the Maxwell–Boltzmann function associated with the *local* thermomechanical state of the fluid. Because $f^0$ has been chosen in this way, the distortion $\phi$ must be orthogonal to each of $\psi_1 = 1$, $\psi_2 = \mathbf{W}$ and $\psi_3 = W^2 + \Omega^2$, that is, $\langle \phi, \psi_k \rangle^0 \equiv \int d\boldsymbol{\varepsilon}\, d\mathbf{c}\, d\mathbf{L}\, f^0 \phi \psi_k = 0$ for $k = 1$, 2, 3. The first and second of these constraints imply that the expansion of $\phi$ includes neither a constant term nor one proportional to $\mathbf{W}$ alone. The third imposes the restriction $3T_t + rT_r = (3 + r)T$ among the translational, rotational, and "average" kinetic temperatures. Here the symbol $r$ denotes the number (two or three) of rotational degrees of freedom per molecule. The subscripts $t$ and $r$ refer to translational and rotational, respectively.

The distortion $\phi$ can be expanded in a complete set of irreducible tensor functions of $\boldsymbol{\varepsilon}$, $\mathbf{W}$ and $\boldsymbol{\Omega}$, with coefficients that depend on $\mathbf{x}$ and $t$. Indeed, the first few of the expansion coefficients that appear in the series

$$\phi = (\Theta_t - 1)(W^2 - \tfrac{3}{2}) + (\Theta_r - 1)\left(\Omega^2 - \frac{r}{2}\right) + \left(\frac{2m}{kT}\right)^{1/2} \frac{2}{5p} \mathbf{Q}_K{}^t \cdot \mathbf{W}(W^2 - \tfrac{5}{2})$$

$$+ \left(\frac{2m}{kT}\right)^{1/2} \frac{2}{rp} \mathbf{Q}_K{}^r \cdot \mathbf{W}\left(\Omega^2 - \frac{r}{2}\right) + \frac{1}{p} \, \mathsf{P}_K : \mathbf{WW}$$

$$+ \frac{3m}{r}\left(\frac{2}{IkT}\right)^{1/2} \boldsymbol{\ell} \cdot \boldsymbol{\Omega} + \frac{6}{rp}\left(\frac{m}{I}\right)^{1/2} \mathsf{C}_K : \mathbf{W}\boldsymbol{\Omega} + \cdots \tag{2.1}$$

(with $p = nkT$) can be identified with quantities that are familiar and of immediate interest to those concerned with the description of nonequilibrium processes. Thus for example, $\Theta_t \equiv T_t/T$ and $\Theta_r \equiv T_r/T$. It also can be verified that $\mathbf{Q}_K{}^t$ and $\mathbf{Q}_K{}^r$ are the "kinetic" or diffusional contributions to the fluxes of translational and rotational energy, that $\mathsf{P}_K$ is the kinetic contribution to the viscous portion of the pressure tensor, and that $\mathsf{C}_K$ is a similarly defined contribution to the flux of the molecular angular momentum or "spin" $\boldsymbol{\ell}$.

The few terms explicitly displayed here do of course constitute only a very primitive approximation to the distribution function. However, theoretical investigations of dilute gases have established that much of value can be extracted from truncated approximations to the exact distribution function. For example, the complete series contains, in addition to the terms proportional to $\mathbf{W}(W^2 - \tfrac{5}{2})$ and $\mathbf{W}(\Omega^2 - r/2)$, an infinite number of the form $\mathbf{W}S_{5/2}^{(n)}(W^2)S_{r/2}^{(n')}(\Omega^2)$ with $S$ a Sonine (associated Laguerre) polynomial. It is a well-documented matter of experience[25] that these terms alter by only a few percent the numerical value of the coefficient of thermal conductivity. The story is similar for terms proportional to $\mathbf{WW}$ and the coefficient of gas viscosity. And finally, analogous conclusions have been drawn regarding the rapid convergence of the moment expansion when applied to the calculation of the van Hove function of monatomic gases.[24] In general, one expects (or hopes) that the higher moments will have little effect on the behavior of the lower moments. And it usually turns out to be a moment of low order that is related directly to a measurable property of the fluid. For dilute gases these statements definitely are true. Although we anticipate that they will apply as well to dense gases and liquids, the rate of convergence of the moment expansion—as gauged by numerical estimates of directly measurable fluid properties—really has to be determined empirically.

Here we retain only those terms of the moment expansion that appear explicitly in (2.1). They are sufficient to provide a sort of minimal, and hopefully rather realistic, description of the approach to a steady state of transport of momentum, energy, and spin. In the present case our selection of these few terms has been specially tailored to the description of transport processes. Had our concern instead been with the theory of depolarized Rayleigh scattering, we then might have chosen from the infinite series a quite different "minimal basis set."

## B. Derivation of the Moment Equations

Each moment of the singlet distribution function (with respect to the variables $\varepsilon$, $C$, and $L$) is the ensemble average of the sum of one or more many-body functions of the form $\sum_{i=1}^{N} G_i \delta_i = \sum_{i=1}^{N} G(x_i, p_i, \varepsilon_i, L_i)\delta(x_i - x)$. We now introduce the shorthand notation

$$\langle (\cdots) f\delta_1 \rangle = \int d\tau_1 \, (\cdots) f(\tau_1; t)\delta(x_1 - x)$$

$$\langle (\cdots) f_2 \delta_1 \rangle = \int d\tau_1 \int d\tau_2 \, (\cdots) f_2(\tau_1, \tau_2; t)\delta(x_1 - x)$$

where $\tau_K$ and $d\tau_K$, respectively, denote a point and the differential of extension in the phase space of molecule $k$. One then obtains directly from the Liouville equation the equation of change

$$\partial_t \langle G_1 f\delta_1 \rangle + \mathbf{V} \cdot \langle c_1 G_1 f\delta_1 \rangle + \left\langle \left( \mathbf{V}_1 V_{12} \cdot \partial_1 G_1 - \mathbf{N}_{21} \cdot \frac{\partial G_1}{\partial \mathbf{L}_1} \right) f_2 \delta_1 \right\rangle$$

$$= \left\langle \left( \mathbf{c}_1 \cdot \mathbf{V}_1 G_1 + \dot{\varepsilon}_1 \cdot \frac{\partial G_1}{\partial \varepsilon_1} \right) \delta_1 \right\rangle$$

for the ensemble average of $\Sigma G_i \delta_i$. In this equation $V_{12} = V(x_{12}, \varepsilon_1, \varepsilon_2)$ refers to the potential of interaction between molecules 1 and 2, and

$$\mathbf{N}_{21} = -\mathbf{e}_1 \times \frac{\partial V_{12}}{\partial \mathbf{e}_1} - (\mathbf{\theta}_1 \cot \theta_1 + \mathbf{e}_1) \frac{\partial V_{12}}{\partial \psi_1}$$

with

$$\frac{\partial}{\partial \mathbf{e}} = \mathbf{\theta} \frac{\partial}{\partial \theta} + \mathbf{\phi} \csc \theta \frac{\partial}{\partial \phi}$$

to the torque acting on molecule 1 because of its interaction with molecule 2. In these formulas $\mathbf{V}_k = \mathbf{V}_{x_k}$, $\partial_k = \mathbf{V}_{p_k}$, and $\mathbf{e}_k$ is a unit vector directed along the polar axis of the body-fixed frame for molecule $k$; $(\mathbf{e}_k, \mathbf{\theta}_k, \mathbf{\phi}_k)$ is an associated (right-handed) set of orthonormal vectors. And finally, the derivative $\partial/\partial\varepsilon$ is performed holding constant the space-fixed components of angular momentum.

Next we write the third term from the equation of change in the form

$$\langle(\cdots)f_2\delta_1\rangle = \langle(\cdots)f_2(\tau_1, \tau_2; t)\delta(\mathbf{x}_1 - \mathbf{x})\rangle$$

$$= \int d\mathbf{y} \langle(\cdots)f_2(\tau_1, \tau_2; t)\delta(\mathbf{x}_1 - \mathbf{x})\delta(\mathbf{x}_2 - \mathbf{y})\rangle$$

$$\equiv \int d\mathbf{y}\, F(\mathbf{x}, \mathbf{y})$$

and then resolve the function $F(\mathbf{x}, \mathbf{y})$ into the sum of its symmetric and antisymmetric parts $F^S(\mathbf{x}, \mathbf{y})$ and $F^A(\mathbf{x}, \mathbf{y})$. The motivation for this is the fact that associated with each integral $\int d\mathbf{y}\, F(\mathbf{x}, \mathbf{y})$ is the unique "flux contribution" given by[26]

$$\int d\mathbf{y}\, F^A(\mathbf{x}, \mathbf{y}) = \mathbf{V}_\mathbf{x} \cdot \tfrac{1}{2} \int d\mathbf{z}\mathbf{z} \int_0^1 da\, F[\mathbf{x} - a\mathbf{z}, \mathbf{x} + (1 - a)\mathbf{z}]$$

From this it follows that the equation of change for $G = \langle G_1 f\delta_1\rangle$ can be written as $\partial_t G + \mathbf{V}_\mathbf{x} \cdot j_G = s_G$, where the objects

$$j_G = \langle \mathbf{c}_1 G_1 f\delta_1\rangle + \tfrac{1}{2} \int d\mathbf{z}\mathbf{z} \int_0^1 da \left\langle \left(\mathbf{V}_1 V_{12} \cdot \partial_1 G_1 \right.\right.$$

$$\left.\left. - \mathbf{N}_{21} \cdot \frac{\partial G_1}{\partial \mathbf{L}_1}\right) f_2\delta[\mathbf{x}_1 - (\mathbf{x} - a\mathbf{z})]\delta[\mathbf{x}_2 - \{\mathbf{x} + (1 - a)\mathbf{z}\}]\right\rangle \qquad (2.2)$$

and

$$s_G = \left\langle \left(\mathbf{c}_1 \cdot \mathbf{V}_1 G_1 + \dot{\boldsymbol{\varepsilon}}_1 \cdot \frac{\partial G_1}{\partial \boldsymbol{\varepsilon}_1}\right) f\delta_1 \right\rangle - \tfrac{1}{2}\left\langle \left[\mathbf{V}_1 V_{12} \cdot (\partial_1 G_1 - \partial_2 G_2)\right.\right.$$

$$\left.\left. - \left(\mathbf{N}_{21} \cdot \frac{\partial G_1}{\partial \mathbf{L}_1} + \mathbf{N}_{12} \cdot \frac{\partial G_2}{\partial \mathbf{L}_2}\right)\right] f_2\delta_1 \right\rangle \qquad (2.3)$$

respectively, may be interpreted as the flux and source strength (or source density) of the quantity $G$. These formulas for $j_G$ and $s_G$ can be applied to a fluid of (linear) rigid rotors provided that the proper expressions, $\mathbf{N} = -\mathbf{e} \times \partial V/\partial \mathbf{e}$, $\dot{\theta} = L_\phi/I$, and $\dot{\phi} = -L_\theta/I \sin\theta$, are used for the torque and for the rates of change of the angular variables.

It is now a routine task to construct the desired equations of change. For mass, momentum, and energy one obtains

$$d_t\rho + \rho\mathbf{V} \cdot \mathbf{u} = 0$$

$$\rho\, d_t\mathbf{u} + \mathbf{V} \cdot \mathbf{p} = 0$$

$$\rho\, d_t e_t + \mathbf{V} \cdot \mathbf{Q}^t + \mathbf{p}:\mathbf{V}\mathbf{u} = -\tfrac{1}{2}\langle(\mathbf{V}_1 V_{12}) \cdot \mathbf{C}_{12} f_2\delta_1\rangle$$

$$\rho\, d_t e_r + \mathbf{V} \cdot \mathbf{Q}^r = +\tfrac{1}{2}\langle(\mathbf{N}_{21} \cdot \boldsymbol{\omega}_1 + \mathbf{N}_{12} \cdot \boldsymbol{\omega}_2)f_2\delta_1\rangle$$

where $\boldsymbol{\omega}_k = I_k^{-1} \cdot \mathbf{L}_k$, $d_t = \partial_t + \mathbf{u} \cdot \mathbf{V}$, $\rho = \langle mf\delta_1 \rangle$, and $\rho\mathbf{u} = \langle \mathbf{c}f\delta_1 \rangle$. In these equations and elsewhere in this chapter, we use the "serial convention" $\mathbf{a} : \mathbf{b} = a_{ij}b_{ij}$, $\mathbf{a} \vdots \mathbf{b} = a_{ijk}b_{ijk}$, and so on, when forming products of tensors. The energy densities and translational ($t$) and rotational ($r$) temperatures are defined by $\rho e_t = \frac{3}{2}nkT_t = \langle \frac{1}{2}mC_1{}^2 f\delta_1 \rangle$ and $\rho e_r = (r/2)nkT_r = \langle \frac{1}{2}\mathbf{L}_1 \cdot \boldsymbol{\omega}_1 f\delta_1 \rangle$. Finally, the fluxes of momentum $\mathbf{p} = nkT_t\boldsymbol{\delta} + \mathbf{P}'$ and of translational and rotational energy are given by

$$\mathbf{P}' = \mathbf{P}_K + \mathbf{P}_V = \langle m[\mathbf{C}_1\mathbf{C}_1]^{(2)}f\delta_1 \rangle - \tfrac{1}{2}\langle \mathbf{x}_{12}\mathbf{V}_1 V_{12}(0_{21}f_2)\delta_1 \rangle$$

$$\mathbf{Q}^t = \mathbf{Q}_K{}^t + \mathbf{Q}_V{}^t = \langle \mathbf{C}_1 \tfrac{1}{2}mC_1{}^2 f\delta_1 \rangle - \tfrac{1}{2}\langle \mathbf{x}_{12}\mathbf{V}_1 V_{12} \cdot \mathbf{C}_1(0_{21}f_2)\delta_1 \rangle$$

$$\mathbf{Q}^r = \mathbf{Q}_K{}^r + \mathbf{Q}_V{}^r = \langle \mathbf{C}_1 \tfrac{1}{2}(\mathbf{L}_1 \cdot \boldsymbol{\omega}_1)f\delta_1 \rangle + \tfrac{1}{2}\langle \mathbf{x}_{12}\mathbf{N}_{21} \cdot \boldsymbol{\omega}_1(0_{21}f_2)\delta_1 \rangle$$

wherein

$$0_{21} = \sum_{n \geq 0} \frac{1}{(n+1)!} (\mathbf{x}_{12} \cdot \mathbf{V}_x)^n.$$

Here and henceforth we denote by $[\mathbf{a}]^{(2)}$ the traceless and symmetric part of the second-rank tensor $\mathbf{a}$. The total flux of energy equals the sum $\mathbf{Q} = \mathbf{Q}^t + \mathbf{Q}^r + \mathbf{Q}_\phi$, where $\mathbf{Q}_\phi = \tfrac{1}{2}\langle \mathbf{C}_1 V_{12}f_2\delta_1 \rangle$ is the diffusional flow of potential energy.

The other moments retained in our truncated expansion of $f$ are $\mathbf{P}_K$, $\mathbf{Q}_K{}^t$, $\mathbf{Q}_K{}^r$, $\boldsymbol{\ell} = \rho^{-1}\langle \mathbf{L}_1 f\delta_1 \rangle$ and $\mathbf{C}_K = \langle \mathbf{C}_1\mathbf{L}_1 f\delta_1 \rangle$. For these the equations of change are

$$\partial_t \mathbf{P}_K + \mathbf{V} \cdot (\mathbf{u}\mathbf{P}_K) + \mathbf{V} \cdot \mathbf{J}_{\mathbf{P}_K} + 2[\mathbf{p}^T \cdot \mathbf{V}\mathbf{u}]^{(2)} = \boldsymbol{\sigma}_{\mathbf{P}_K}$$

$$\partial_t \mathbf{Q}_K{}^t + \mathbf{V} \cdot (\mathbf{u}\mathbf{Q}_K{}^t) + \mathbf{V} \cdot \mathbf{J}_{\mathbf{Q}_K{}^t} + \mathbf{F}_K : \mathbf{V}\mathbf{u} + \mathbf{Q}^t \cdot \mathbf{V}\mathbf{u} + d_t\mathbf{u}(3\rho e_t\boldsymbol{\delta} + \mathbf{P}_K) = \boldsymbol{\sigma}_{\mathbf{Q}_K{}^t}$$

$$\partial_t \mathbf{Q}_K{}^r + \mathbf{V} \cdot (\mathbf{u}\mathbf{Q}_K{}^r) + \mathbf{V} \cdot \mathbf{J}_{\mathbf{Q}_K{}^r} + d_t\mathbf{u}(\rho e_r) + \mathbf{Q}^r \cdot \mathbf{V}\mathbf{u} = \boldsymbol{\sigma}_{\mathbf{Q}_K{}^r}$$

$$\partial_t \mathbf{C}_K + \mathbf{V} \cdot (\mathbf{u}\mathbf{C}_K) + \mathbf{V} \cdot \mathbf{J}_{\mathbf{C}_K} + d_t\mathbf{u}(\rho\boldsymbol{\ell}) + (\mathbf{V}\mathbf{u})^T \cdot \mathbf{C} = \underline{\boldsymbol{\sigma}}_{\mathbf{C}_K}$$

and

$$\rho d_t \boldsymbol{\ell} + \mathbf{V} \cdot \mathbf{C} = \boldsymbol{\sigma}_\ell$$

The flux and source terms appearing in these equations are defined by

$$\boldsymbol{\sigma}_{\mathbf{P}_K} = -\tfrac{1}{2}\langle [(\mathbf{V}_1 V_{12})\mathbf{C}_{12}]^{(2)}f_2\delta_1 \rangle$$

$$\boldsymbol{\sigma}_{\mathbf{Q}_K{}^t} = -\tfrac{1}{2}\langle [\mathbf{V}_1 V_{12} \cdot (\mathbf{C}_1\mathbf{C}_1 - \mathbf{C}_2\mathbf{C}_2) + \tfrac{1}{2}\mathbf{V}_1 V_{12}(C_1{}^2 - C_2{}^2)]f_2\delta_1 \rangle$$

$$\boldsymbol{\sigma}_{\mathbf{Q}_K{}^r} = -\tfrac{1}{2}\left\langle \left[ \frac{\mathbf{V}_1 V_{12}(\mathbf{L}_1 \cdot \boldsymbol{\omega}_1 - \mathbf{L}_2 \cdot \boldsymbol{\omega}_2)}{2m} - \mathbf{C}_1\mathbf{N}_{21} \cdot \boldsymbol{\omega}_1 - \mathbf{C}_2\mathbf{N}_{12} \cdot \boldsymbol{\omega}_2 \right]f_2\delta_1 \right\rangle$$

$$\underline{\boldsymbol{\sigma}}_{\mathbf{C}_K} = -\tfrac{1}{2}\left\langle \left[ \frac{(\mathbf{V}_1 V_{12})(\mathbf{L}_1 - \mathbf{L}_2)}{m} - \mathbf{C}_1\mathbf{N}_{21} - \mathbf{C}_2\mathbf{N}_{12} \right]f_2\delta_1 \right\rangle$$

$$\boldsymbol{\sigma}_l = \tfrac{1}{2}\langle (\mathbf{N}_{12} + \mathbf{N}_{21})f_2\delta_1 \rangle$$

$$\mathbf{J}_{\mathbf{Q}_{K'}} = \langle [\mathbf{C}_1 \mathbf{C}_1 \tfrac{1}{2} m C_1{}^2] f \delta_1 \rangle - \tfrac{1}{2} \langle x_{12} (\mathbf{V}_1 V_{12} \cdot \mathbf{C}_1 \mathbf{C}_1 + \tfrac{1}{2} \mathbf{V}_1 V_{12} C_1{}^2) (0_{21} f_2) \delta_1 \rangle$$

$$\mathbf{J}_{\mathbf{Q}_{K'}} = \langle (\mathbf{C}_1 \mathbf{C}_1 \tfrac{1}{2} \mathbf{L}_1 \cdot \boldsymbol{\omega}_1) f \delta_1 \rangle$$

$$- \tfrac{1}{2} \left\langle x_{12} \left( \mathbf{V}_1 V_{12} \frac{\mathbf{L}_1 \cdot \boldsymbol{\omega}_1}{2m} - \mathbf{C}_1 \mathbf{N}_{21} \cdot \boldsymbol{\omega}_1 \right) (0_{21} f_2) \delta_1 \right\rangle$$

$$\mathbf{J}_{\mathbf{C}_K} = \langle \mathbf{C}_1 \mathbf{C}_1 \mathbf{L}_1 f \delta_1 \rangle - \tfrac{1}{2} \left\langle x_{12} \left( \mathbf{V}_1 V_{12} \frac{\mathbf{L}_1}{m} - \mathbf{C}_1 \mathbf{N}_{21} \right) (0_{21} f_2) \delta_1 \right\rangle$$

$$\mathbf{C} = \mathbf{C}_K + \mathbf{C}_V = \langle \mathbf{C}_1 \mathbf{L}_1 f \delta_1 \rangle + \tfrac{1}{2} \langle x_{12} \mathbf{N}_{21} (0_{21} f_2) \delta_1 \rangle$$

and finally, $\mathbf{J}_{\mathbf{P}_K}$ is the portion of the third-rank tensor

$$\mathbf{F}_K = \langle m \mathbf{C}_1 \mathbf{C}_1 \mathbf{C}_1 f \delta_1 \rangle - \tfrac{1}{2} \langle x_{12} [(\mathbf{V}_1 V_{12}) \mathbf{C}_1 + \mathbf{C}_1 (\mathbf{V}_1 V_{12})] (0_{21} f_2) \delta_1 \rangle$$

which is traceless and symmetric in the last pair of indices.

## C. Impulsive Interactions

It was previously mentioned that one of the greatest difficulties we face is that of analyzing the collisions among polyatomic molecules. The expense of performing dynamical calculations for polyatomic species is considerable, and it is also far from easy to adapt the results of such calculations to the problems of kinetic theory. One strategy for avoiding these difficulties is to adopt some sort of stochastic model for molecular-scale events. However, this is not the approach we have decided to employ. We prefer instead to retain as much as we can of the flavor of a theory based on the detailed analysis of molecular collisions. To accomplish this we are prepared to accept models for the molecular interactions that sacrifice some elements of reality in exchange for computational tractability. Specifically, we intend to exploit models that already have proved useful in theories of dense monatomic fluids and of dilute polyatomic gases. The common feature in both these categories is a reliance on impulsive interactions, for example, the hard-sphere model of the Enskog dense gas theory, the impulsive forces of attraction and repulsion of the square-well model for liquids, the hard core of the Rice–Allnatt interaction, and the rigid nonspherical models used in the classical kinetic theory of polyatomic gases. Calculations using simple models such as these are the indispensible preliminaries to studies based on more realistic interactions.

Therefore the task immediately before us is that of adapting the previously derived moment equations to models involving impulsive molecular inter-actions. By the term impulsive we mean to characterize a contribution to the intermolecular force which is operative only over a very restricted range of molecular separations (and/or relative orientations), but which within this range is of much greater magnitude than any other contribution. The effective ranges of these forces are assumed to be so small that their effects can be described in terms of binary collisional events regardless of the value of the fluid density. Interactions that fall into this category are steeply sloped potentials similar to the rigid-sphere and square-well models.

Now the forces of interaction appear explicitly only in the potential contributions to the fluxes and source terms of the moment equations. They occur in factors of the form

$$-\frac{\partial V_{12}}{\partial \mathbf{x}_1} \cdot \frac{\partial G_1}{\partial \mathbf{p}_1} + \mathbf{N}_{21} \cdot \frac{\partial G_1}{\partial \mathbf{L}_1} = \left(\frac{d\mathbf{p}_1}{dt}\right)_2 \cdot \frac{\partial G_1}{\partial \mathbf{p}_1} + \left(\frac{d\mathbf{L}_1}{dt}\right)_2 \cdot \frac{\partial G_1}{\partial \mathbf{L}_1} = \left(\frac{dG_1}{dt}\right)_2$$

where the symbol $(d\psi_1/dt)_2$ denotes the rate of change in $\psi_1$ due to the inter-action between molecules 1 and 2. Dynamical events associated with impulsive forces are completed within intervals of time briefer by far than those characteristic of transport and relaxation processes. Therefore it is reasonable to average the moment equations over the durations of these impulsive events. For this we employ the time-smoothing operation

$$\alpha_\tau \langle \psi(q, p) f_2(t) \delta_1 \rangle = \left\langle \frac{1}{\tau} \int_t^{t+\tau} ds \, \psi[q(s), p(s)] f_2(t) \delta_1 \right\rangle$$

defined by Kirkwood.[27] When $\psi(q, p) = (dG_1/dt)_2$, this can be written in the form $\langle (1/\tau)(\Delta_2 G_1) f_2 \delta_1 \rangle$, where $\Delta_2 G_1 = G_1(t + \tau) - G_1(t)$. If the interval $\tau$ is chosen to be greater than the duration of the impulsive event, $\Delta_2 G_1$ can be identified with the change in $G_1$ due to a "completed" binary (impulsive) encounter between molecules 1 and 2. For this interpretation to be valid, such an event must actually occur within the interval $(t, t + \tau)$. This in turn imposes restrictions on the initial value of the separation $\mathbf{x}_{21}$. In particular, $d\mathbf{x}_{21}$ or $dz$ must be equal to $-S \, \mathbf{k} \cdot \mathbf{g}_{21} \tau \, d\mathbf{k}$, with $\mathbf{k} \cdot \mathbf{g}_{21} < 0$, where $\mathbf{k}$ is the outward-directed unit normal to the surface of molecule 1 at its point of contact with molecule 2, and where $S \, d\mathbf{k}$ denotes the differential surface element of the "excluded volume" associated with the impulsive force. Finally, we denote by $\mathbf{g}_{21} = \mathbf{c}_2 - \mathbf{c}_1 + \mathbf{\omega}_2 \times \mathbf{\sigma}_2 - \mathbf{\omega} \times \mathbf{\sigma}_1$ the relative velocity at the moment of impact of the points of contact on the surfaces of the two molecules. Here $\mathbf{\sigma}_i$ is the vector that extends from the center of molecule $i$ to this point of contact.

The impulsive contributions to the flux and source terms (for linear rotors) then become

$$-\tfrac{1}{2} \int d\mathbf{k} \int d\mathbf{c}_1 \, d\mathbf{c}_2 \, d\mathbf{L}_1 \, d\mathbf{L}_2 \, d\mathbf{e}_1 \, d\mathbf{e}_2 \, S \, (\mathbf{k} \cdot \mathbf{g}_{21}) \mathbf{x}_{12} (\Delta_2 G_1)(0_{21} f_2) \quad (2.4)$$

and

$$-\tfrac{1}{2} \int d\mathbf{k} \int d\mathbf{c}_1 \cdots d\mathbf{e}_2 \, S \, (\mathbf{k} \cdot \mathbf{g}_{21})(\Delta_1 G_2 + \Delta_2 G_1) f_2$$

respectively. In these formulas $x_{12}$, $0_{21} f_2$, and $f_2$ are to be evaluated at contact, that is, for the configuration in which the impulse occurs, and the integrals are to be performed only over the hemisphere for which $k \cdot g_{21} < 0$.

## III. MODELS OF DENSE POLYATOMIC FLUIDS

The preceding section was devoted to the presentation of two of the three components of our theory, the truncated moment approximation to the singlet distribution function and the derivation of the formally exact equations of change for these moments. It also included brief discussions of the kinematic and collisional models we intend to study. The remaining step is the introduction of a functional relationship between the pair and singlet distribution function. If we had for dense fluids a theory with the generality of the Boltzmann theory for dilute gases, this final portion of the development could be carried out once and for all. But there is no such theory. There exist instead several model theories such as those of Enskog, of Rice and Allnatt, of Severne, Prigogine, Nicolis, and Misguich, and of Davis, each enjoying some degree of success but also exhibiting certain defects and limitations. Some of these appear to be more fundamentally sound than others, and some are much less demanding from a computational point of view. Those in the latter category have their obvious attractions. The former presumably point the way to the better theories of tomorrow and also serve to introduce us to the new computational problems that must be surmounted before those better theories can be implemented numerically. Both of these categories (and others that fit comfortably into neither) are deserving of our attention, since they can contribute to an increased understanding of the requirements for a truly comprehensive theory of transport and relaxation in polyatomic fluids.

Therefore our objective here is to illustrate how various models or approximate theories can be used to complete the program of the moment method. Regardless of which model we select, the algebra involved turns out to be disconcertingly complex. However, there is partial compensation for this in the fact that the structure of the formal manipulations and of the final results themselves is much the same for each. Because of this we outline the general procedure but once, and then subsequently focus on those features that distinguish one model from another. First on our agenda come generalizations of the Enskog dense gas theory. From these we proceed to the Rice–Allnatt model, and then to Davis' adaptation of the theory of Severne[28] and of Prigogine, Nicolis, and Misguich. This is followed by the derivation of a theory very similar to that of Davis. The section concludes with a discussion of Brinser and Condiff's[29] generalization of the Enskog dense gas theory and a demonstration of how this theory can be adapted to the description of polyatomic fluids.

We mention in passing that our first test of the moment method was to compute the first- and second-order transport coefficients for a dense gas of rigid spheres.[30] In this case we adopted for $f_2$ the Enskog approximation that $f_2(\mathbf{x}_1 \mathbf{c}_1, \mathbf{x}_2 \mathbf{c}_2; t) = g(\sigma) f(\mathbf{x}_1 \mathbf{c}_1; t) f(\mathbf{x}_1 + \mathbf{k}\sigma, \mathbf{c}_2; t)$ just prior to collision. Here $\sigma$ is the diameter of a sphere $\mathbf{k} = \mathbf{x}_{21}/x_{21}$, and $g(\sigma)$ denotes the equilibrium radial distribution function evaluated for two spheres in contact. Since our interest was not only in testing the method but also in determining the importance of stress tensor contributions proportional to the square of the velocity gradient, we selected the single-term expansion $\phi = (1/p)\mathbf{P}_K : \mathbf{WW}$. The equation of change for $\mathbf{P}_K$ then was solved iteratively. The first and second iterates yielded the very same formulas for the equation of state and Newtonian stress tensor that had previously been generated by the CE method. From the next higher iterate we obtained the set of phenomenological coefficients associated with stress tensor terms proportional to the square of the velocity gradient. The same method was extended to a dense fluid of structureless molecules which interact with the square-well potential. Again it was found that the equation of state and linear phenomenological coefficients agreed with those derived previously from the CE method. These calculations are, to the best of our knowledge, the only detailed theoretical estimates ever made for the second-order transport coefficients of a dense fluid.

## A. Enskog Theory for Rough Spheres

This is the most thoroughly studied of the collision models used to mimic real polyatomic fluids.[3c,19] Its popularity stems from the relative simplicity of the computations based on it; all the integrals occurring in the theory can be evaluated by elementary methods and expressed in compact analytical form. The dynamical state of a rough sphere is fully specified by the location of its center of mass $\mathbf{x}_i$, by its linear momentum, $\mathbf{p}_i = m\mathbf{c}_i$, and by its internal angular momentum $\mathbf{L}_i = I\boldsymbol{\omega}_i$. The isotropic internal distribution of mass is characterized by the moment of inertia $I$. It is convenient to introduce the dimensionless moment $\kappa = 4I/m\sigma^2$ which can vary in value from zero, when the mass is localized at the center of the sphere, to $\frac{2}{3}$ when the mass is uniformly distributed over the surface of the sphere.

A collision between two rough spheres occurs when the distance between their centers diminishes in value to the diameter $\sigma$. At the instant before collision, the center of one lies at $\mathbf{x}_1$ and that of the other at $\mathbf{x}_1 + \sigma\mathbf{k}$. The condition $\mathbf{k} \cdot \mathbf{c}_{21} < 0$ ensures that they are about to collide. The collisional alterations of the linear momenta, angular momenta, and of the energy are given by $\Delta\mathbf{p}_1 = -\Delta\mathbf{p}_2 = \mathbf{K}$, $\Delta\mathbf{L}_1 = \Delta\mathbf{L}_2 = \frac{1}{2}\sigma\mathbf{k} \times \mathbf{K}$, and $\Delta E = -\mathbf{K} \cdot (\mathbf{g}'_{21} + \mathbf{g}_{21}) = 0$, respectively. Here $\mathbf{K}$ is the collisional impulse imparted to sphere 1, and $\mathbf{g}_{21} = \mathbf{c}_{21} + \frac{1}{2}\sigma\mathbf{k} \times (\boldsymbol{\omega}_1 + \boldsymbol{\omega}_2)$ is the relative velocity just before collision of the points of contact on the two spheres. The rough-sphere model is defined by the requirement that $\mathbf{g}'_{21}$, the postcollisional value of the relative velocity, be the negative of $\mathbf{g}_{21}$. From this and the equations of impact, it follows that $\mathbf{K} = m\kappa(\kappa + 1)^{-1}(\mathbf{g}_{21} + \mathbf{kk} \cdot \mathbf{c}_{21})$.

The principal shortcoming of this model is its failure to provide a realistic description of inelastic collisions. However, because it so drastically over-emphasizes the importance of noncentral forces, the model does provide useful upper limits on the contributions stemming from inelastic events. There is no value of $\kappa$ for which the collisions are truly elastic. Even when $\kappa \to 0$ and no transfer between rotational and translational degrees of freedom can occur, the mechanism for collisional exchange of spin or angular momentum still remains.

For the purposes of the moment method, the essential features of this model are: (1) the formulas it provides for the collisional changes in linear and angular momentum, and (2) the Enskog assumption that

$$f_2(\mathbf{x}_1, \mathbf{c}_1, \mathbf{L}_1; \mathbf{x}_2, \mathbf{c}_2, \mathbf{L}_2) = g(\sigma)f(\mathbf{x}_1, \mathbf{c}_1, \mathbf{L}_1)f(\mathbf{x}_1 + \sigma\mathbf{k}, \mathbf{c}_2, \mathbf{L}_2)$$

just prior to impact. [The equilibrium radial distribution function

$$g(\sigma) = g[\mathbf{x}_{21} = \mathbf{k}\sigma, \tfrac{1}{2}(\mathbf{x}_1 + \mathbf{x}_2)]$$

is the same for rough and smooth spheres, and so can be obtained either from experiment (Monte Carlo or molecular dynamics) or from theory (Percus–Yevick).] From the first of these we are able to construct the integrand factors $\Delta_j G_i$ which occur in the formulas for the potential contributions to the fluxes and to the source terms. The second allows us to express the integrand factors of $f_2$ in terms of (the moments of) the singlet distribution function $f$. Taken together these are the means by which the exact (but useless) moment equations of Section II can be transformed into a closed set of equations for the moments.

The procedure is straightforward but tedious to carry out. By means of the Enskog approximation, one is able to express $f_2$ in terms of $g(\sigma)$ and the moments of $f$. We then discard the quadratic terms in this expression and so obtain a closed set of linear, first-order partial differential equations for the moments. The coefficients that occur in the flux and source terms of these equations can be expressed in terms of the collision integrals

$$\{\alpha, \beta\} = -\frac{g\sigma^3}{\pi^6}$$

$$\times \int_{\mathbf{k} \cdot \boldsymbol{\gamma}_T > 0} d\boldsymbol{\gamma} \, d\Gamma \, d\boldsymbol{\Omega}_1 \, d\boldsymbol{\Omega}_2 \, d\mathbf{k}(\mathbf{k} \cdot \boldsymbol{\gamma}_T)e^{-(1/2)\gamma^2 - 2\Gamma^2 - \Omega_1^2 - \Omega_2^2}\mathbf{k}(\Delta_2\alpha_1)(\beta_1 + \beta_2)$$

and

$$[\alpha, \beta] = +\frac{g\sigma^2}{\pi^6}\int_{\mathbf{k}\cdot\gamma_T>0}\cdots(\Delta_2\alpha_1 + \Delta_1\alpha_2)(\beta_1 + \beta_2)$$

respectively. Here $\gamma_T = 2^{-1/2}(\mathbf{W}_1 - \mathbf{W}_2) - (2\kappa)^{-1/2}\mathbf{k}(\Omega_1 + \Omega_2)$, $\gamma = \mathbf{W}_1 - \mathbf{W}_2$, and $\Gamma = \frac{1}{2}(\mathbf{W}_1 + \mathbf{W}_2)$; $\Delta_i\alpha_j$ denotes the difference in values of the function $\alpha_j$ evaluated after and before a binary collision with molecule $i$. By symbols such as $\{\alpha; \beta\}$ and $[\underline{\alpha}\,\vdots\,\underline{\beta}]$, we indicate the tensor contractions $\Sigma_i\{\alpha_i, \beta_i\}$ and $\Sigma_i\Sigma_j(\alpha_{ij}, \beta_{ij})$, respectively. When an element of one of these integrals is a second-rank tensor, it is understood that what is intended is the traceless and symmetric part of that tensor.

The square-bracket integrals appear in the kinetic theory of the dilute gas as well as in the dense gas theory. Most that we need here already have been evaluated by Condiff, Lu, and Dahler;[31] those that were not are given in Ref. 2. The integrals $\{\alpha, \beta\}$ occur only in the dense gas theory. They arise not only from the terms $\langle\mathbf{x}_{12}(\Delta_2 G_1)0_{21}f_2\rangle$ of the moment equations, but also from the potential or interactional parts of the flux tensors. These integrals have been evaluated in Ref. 2 using the very efficient procedures proposed by Montgomery and Hoffman.[32]

The equation of change for the traceless and symmetric portion of the kinetic flux of momentum is

$$\tau_P\,\partial_t\,\mathbf{P}_K + \mathbf{P}_K = -2\eta_K[\mathbf{V}u]^{(2)} \tag{3.1}$$

Here $[\mathbf{V}u]^{(2)}$ denotes the traceless and symmetric part of the tensor $\mathbf{V}u$, and the momentum relaxation time $\tau_P$ is given by

$$\tau_P^{-1} = \frac{8}{15}\,ng(\sigma)\sigma^2\left(\frac{\pi kT}{m}\right)^{1/2}\frac{13\kappa + 6}{(\kappa + 1)^{1/2}}$$

The kinetic part of the shear viscosity coefficient is related by the formula

$$\eta_K = \eta^{(K)}\left[1 + \frac{5\kappa + 2}{5(\kappa + 1)}bng\right]$$

to the Pidduck approximation $\eta^{(K)} = (15/8\sigma^2 g)(mkT/\pi)^{1/2}(\kappa + 1)^2(13\kappa + 6)^{-1}$ to the coefficient of shear viscosity for a *dilute* gas of rough spheres. [Actually one must set $g$ in $\eta^{(K)}$ equal to unity in order to obtain the usual dilute gas result.]

For the translational and rotational temperatures we obtain an indeterminate system of two equations plus the subsidiary condition $(\theta_t - 1) = -(\theta_r - 1)$. To resolve this redundancy we combine one of these equations with the subsidiary condition to obtain the single equation of change:

$$\tau_T\partial_t(\theta_t - 1) + (\theta_t - 1) = -\mathscr{K}\mathbf{V}\cdot\mathbf{u} \tag{3.2}$$

with

$$\tau_T^{-1} = \frac{32}{3} n\sigma^2 g \left(\frac{\pi k T}{m}\right)^{1/2} \frac{\kappa}{(\kappa + 1)^2}$$

$$\mathscr{K} = \frac{1}{32} \frac{1 + bng}{n\sigma^2 g} \left(\frac{m}{\pi k T}\right)^{1/2} \frac{(\kappa + 1)^2}{\kappa}$$

and where $b = \frac{2}{3}\pi\sigma^3$ is the second virial coefficient. The time Fourier transforms of (3.1) and (3.2) may be written in the forms

$$\tilde{P}_K = -\frac{2\eta_K}{1 + i\omega\tau_p} [\mathbf{V}\tilde{\mathbf{u}}]^{(2)}$$

$$\tilde{\theta}_t - 1 = -\frac{\kappa}{1 + i\omega\tau_T} \mathbf{V} \cdot \tilde{\mathbf{u}}$$

When the Enskog approximation for $f_2$ is substituted into the defining formula for $\mathbf{P}_V$, we obtain, to first order in the moments, the expression

$$\mathbf{P}_V = (\mathscr{P} - p)\underline{\delta} + \alpha_p \mathbf{P}_K + \alpha_\theta(\theta_t - 1)\underline{\delta}$$
$$+ 2\zeta\underline{\varepsilon} \cdot (\mathbf{\omega} - \tfrac{1}{2}\,\text{curl }\mathbf{u}) + \alpha_u[\mathbf{V}\mathbf{u}]^{(2)} + \bar{\alpha}_u \mathbf{V} \cdot \mathbf{u}\underline{\delta}$$

where $\mathscr{P} = p(1 + bng)$, with $p = nkT$ and $b = \frac{2}{3}\pi\sigma^3$, is the equation of state for rigid spheres. The quantities $\alpha_p \propto \{\mathbf{x}_{12}\mathbf{W} \,\vdots\, \mathbf{W}\mathbf{W}\}$ and $\alpha_\theta \propto \{\mathbf{x}_{12}\mathbf{W}, W^2 - \Omega^2\}$ are the coefficients of proportionality between $\mathbf{P}_V$ and the kinetic fluxes $\mathbf{P}_K$ and $\theta_t - 1$, respectively, whereas $\zeta \propto \{\mathbf{x}_{12} \times \mathbf{W};$ $\mathbf{x}_{2i} \times \mathbf{W}_2\}$, $\alpha_u \propto \{\mathbf{x}_{12}\mathbf{W} \,\vdots\, \mathbf{x}_{21}\mathbf{W}_2\}$ and $\bar{\alpha}_u \propto \{\mathbf{x}_{12} \cdot \mathbf{W}, \mathbf{x}_{21} \cdot \mathbf{W}_2\}$ are proportionality coefficients between $\mathbf{P}_V$ and the "driving forces" $2\mathbf{\omega} - \text{curl }\mathbf{u}$, $[\mathbf{V}\mathbf{u}]^{(2)}$, and $(\mathbf{V} \cdot \mathbf{u})$. Here $\mathbf{\omega}$ refers to the local spin velocity which, for spherical-top rough spheres, is equal to $(m/I)\ell$.

By combining these results we obtain for the Fourier transform (indicated by a tilde) of the pressure tensor the formula

$$\tilde{\mathbf{p}} = \mathscr{P}\underline{\delta} - 2\eta(\omega)[\mathbf{V}\tilde{\mathbf{u}}]^{(2)} - \eta_b(\omega)\mathbf{V} \cdot \tilde{\mathbf{u}}\underline{\delta} + 2\zeta\underline{\varepsilon} \cdot (\tilde{\mathbf{\omega}} - \tfrac{1}{2}\,\text{curl }\tilde{\mathbf{u}})$$

with

$$\eta(\omega) = \frac{\eta}{1 + i\omega\tau_p} - \frac{1}{2}\alpha_u \frac{i\omega\tau_p}{1 + i\omega\tau_p}$$

$$\eta_b(\omega) = \frac{\eta_b}{1 + i\omega\tau_T} - \bar{\alpha}_u \frac{i\omega\tau_T}{1 + i\omega\tau_T}$$

and

$$\zeta = \frac{1}{2}\rho\sigma\left(\frac{kT}{\pi m}\right)^{1/2} bng \frac{\kappa}{\kappa + 1}$$

The two quantities

$$\eta = \eta^{(K)}\left[1 + \frac{5\kappa + 2}{5(\kappa + 1)} bng\right]^2 + \frac{1}{10}\rho\sigma\left(\frac{kT}{\pi m}\right)^{1/2} bng \frac{7\kappa + 4}{\kappa + 1}$$

and

$$\eta_b = \eta_b^{(K)}(1 + bng) + \frac{2}{3}\rho\sigma\left(\frac{kT}{\pi m}\right)^{1/2} bng$$

where $\eta_n^{(K)} = p\mathscr{K}$, are the zero frequency limits of the coefficients of shear and bulk viscosity. Finally, $\alpha_u$ and $\bar{\alpha}_u$ are given by

$$\alpha_u = -\frac{1}{5}\rho\sigma\left(\frac{m}{\pi kT}\right)^{1/2} bng \frac{7\kappa + 4}{\kappa + 1}$$

and

$$\bar{\alpha}_u = -\frac{2}{3}\rho\sigma\left(\frac{m}{\pi kT}\right)^{1/2} bng$$

respectively.

To the present approximation the equation of change for the local spin velocity is

$$\partial_t \boldsymbol{\omega} = -\tau_{\boldsymbol{\omega}}^{-1}(\boldsymbol{\omega} - \tfrac{1}{2}\nabla \times \mathbf{u})$$

with a spin relaxation time $\tau_{\boldsymbol{\omega}}$ given by

$$\tau_{\boldsymbol{\omega}}^{-1} = \frac{4\zeta}{nI} = \frac{16}{3}ng\sigma^2\left(\frac{\pi kT}{m}\right)^{1/2}\frac{1}{\kappa + 1}$$

The kinetic part of the spin flux can be written as the sum $\frac{1}{3}C_K\underline{\boldsymbol{\delta}} + \frac{1}{2}\boldsymbol{\varepsilon}\cdot\mathbf{C}_K$ + $[\mathbf{C}_K]^{(2)}$ of three linearly independent contributions, one proportional to the trace $C_K$, one to the dual vector $\mathbf{C}_K$, and one to the traceless symmetric part $[\mathbf{C}_K]^{(2)}$ of $\mathbf{C}_K$. Henceforth we shall write $\mathbf{C}_K$ in place of $[\mathbf{C}_K]^{(2)}$. The equations of change for $\mathbf{C}_K$ and $C_K$ are

$$\tau_C\,\partial_t \mathbf{C}_K + \mathbf{C}_K = -\nu_{K2}[\nabla\boldsymbol{\ell}]^{(2)}$$
$$\tau_C\partial_t C_K + C_K = -\nu_{K1}\nabla\cdot\boldsymbol{\ell}$$

where the two relaxation times are given by $\tau_C^{-1} = \tau_C^{-1} = \frac{16}{3}ng\sigma^2(\pi kT/m)^{1/2}$, and where

$$\nu_{K1} = \frac{3}{16g\sigma^2}\left(\frac{mkT}{\pi}\right)^{1/2}$$

$$\nu_{K2} = \frac{3}{32g\sigma^2}\left(\frac{mkT}{m}\right)^{1/2}\left(1 - \frac{1}{5}bng\frac{5\kappa + 3}{\kappa + 1}\right)$$

The moment equation for the vector dual of the spin flux is coupled to the equations for the fluxes of translational and rotational energy. These can be solved to yield constitutive relations of the form

$$\tilde{\mathbf{Q}}_K{}^t = -\lambda_K{}^t(\omega)\mathbf{V}\tilde{T} - \pi_K{}^t(\omega)\mathbf{V}\times\tilde{\ell}$$
$$\tilde{\mathbf{Q}}_K{}^r = -\lambda_K{}^r(\omega)\mathbf{V}\tilde{T} - \pi_K{}^r(\omega)\mathbf{V}\times\tilde{\ell}$$
$$\tilde{\mathbf{C}}_K = -\mu_K(\omega)\mathbf{V}\tilde{T} - \nu_{K3}(\omega)\mathbf{V}\times\tilde{\ell}$$

with phenomenological coefficients which are simply related to the coefficients appearing in the three coupled moment equations.

To complete the picture we must add to these the potential or interactional contributions

$$\mathbf{Q}_V{}^t = \alpha_{\mathbf{Q}^t}\mathbf{Q}_K{}^t + \gamma_c\mathbf{C}_K + \gamma_{\mathbf{Q}^r}\mathbf{Q}_K{}^r + c_T\mathbf{V}T + c_l\mathbf{V}\times\ell + c_n\mathbf{V}n$$
$$\mathbf{Q}_V{}^r = \alpha_{\mathbf{Q}^r}\mathbf{Q}_K{}^r + \gamma_c\mathbf{C}_K + d_T\mathbf{V}T + d_l\mathbf{V}\times\ell$$
$$\mathbf{C}_V = \alpha_c\mathbf{C}_K + \beta_{\mathbf{Q}^r}\mathbf{Q}_K{}^r + b_T\mathbf{V}T + b_l\mathbf{V}\times\ell$$

In the present situation it is found that $c_n = c_l = d_l = 0$ and (a special property of the rough sphere model) also $\mathbf{C}_V = 0$. Therefore, the constitutive relations for $\mathbf{Q}^t$, $\mathbf{Q}^r$, and $\mathbf{C}$ are of the same forms as those for $\mathbf{Q}_K{}^t$, $\mathbf{Q}_K{}^r$ and $\mathbf{C}_K$ with coefficients to which we assign the same symbols but with the subscripts $K$ omitted. Finally, since $\mathbf{Q}_\phi = 0$ for this particular choice of the molecular interactions, it follows that $\mathbf{Q} = \mathbf{Q}^t + \mathbf{Q}^r$.

Our numerical studies for rough spheres are summarized in Figs. 1 to 5 where $bn = \frac{2}{3}\pi\sigma^3 n$ denotes the dimensionless density. The values of $g(\sigma)$ used in these calculations are tabulated in Ref. 19. The theory of Sather and Dahler[37] to which reference is made only included the potential contributions to the fluxes and neglected nonequilibrium distortions of the singlet distribution function. The McCoy–Sandler–Dahler calculations of thermal conductivity and those computed here both exhibit a remarkable independence from the value of $\kappa$: $\lambda^t$ and $\lambda^r$ vary considerably, but their sum does not. Indeed, the $\kappa$ dependence of $\lambda$ we found in this latest calculation is too slight to be recorded on the graph. We are surprised by the significant difference between ours and the McCoy–Sandler–Dahler (McCSD) results; the only difference in the two is that our singlet distribution function contains a term proportional to $\mathbf{\Omega W}$, whereas that of McCSD does not. Although the possibility of numerical errors can not be ignored, we have been unable to discover any.

In Fig. 3 values of shear viscosity have been plotted along with those of $\zeta$ in order to facilitate comparison. The variation of $\zeta$ with $\kappa$ is very similar to the dependence on eccentricity (see Fig. 11 and Section III.B) that $\zeta$ exhibits in the case of smooth ellipsoids. There is a corresponding similarity in the $\kappa$ and eccentricity dependencies of the coefficient $\pi$; compare Figs. 4 and 14.

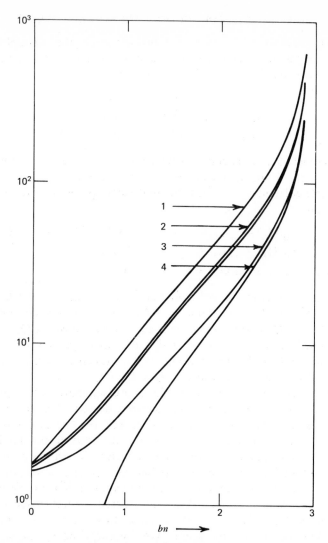

**Fig. 1** Thermal conductivity [in units of $\sigma^{-2}(k^3 T/\pi m)^{1/2}$] of rough spheres. Curve 1 is the result of the present calculation, whereas those labeled 2 are the curves of McCoy, Sandler, and Dahler (as corrected by Klein, Hoffman, and Dahler, cf. Ref. 19); the uppermost of the two is for $\kappa = 0.7$ and the lower for $\kappa = 0.0$. Finally, curves 3 and 4 are the results of the Enskog dense-gas theory (for smooth spheres) with the Eucken correction and the Sather-Dahler theory, respectively.

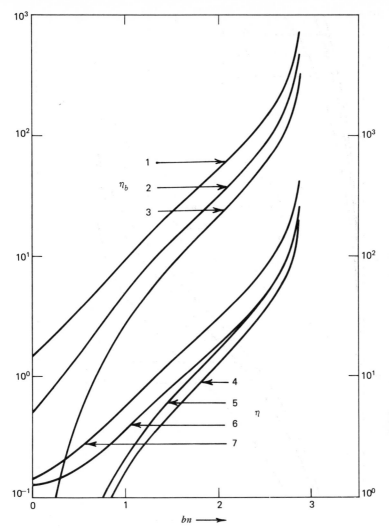

**Fig. 2** Coefficients of shear and bulk viscosity [in units of $(4\sigma^2)^{-1}(mkT/\pi)^{1/2}$] for rough spheres. Curves 1, 2, and 3 are referred to the left-hand ordinate scale; curves 4 to 7 to the right-hand scale. Curves 1 and 2 are the results for $\eta_b$ of the present calculation (and of Ref. 19 as well) for $\kappa$ equal to 0.1 and 0.7, respectively. Curve 3 is Sather and Dahler's estimate of $\eta_b$. Curves 4 and 5 are the Sather-Dahler predictions for $\eta$ with $\kappa$ equal to 0.0 and 0.7, respectively. Finally, curves 6 and 7 are values of $\eta$ for $\kappa$ equal to 0.0 and 0.7 given by the present theory and by that of Ref. 19 as well.

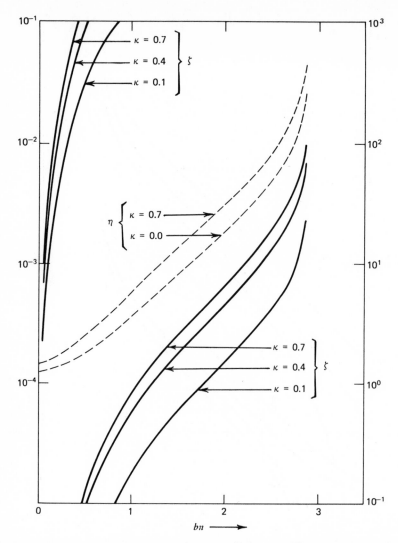

**Fig. 3** Coefficient of vortex viscosity [in units of $(4\sigma^2)^{-1}(mkT/\pi)^{1/2}$] for rough spheres. All except the three curves in the upper left-hand corner are referred to the right-hand ordinate scale. For purposes of comparison the coefficient of shear viscosity is shown by the dashed curves.

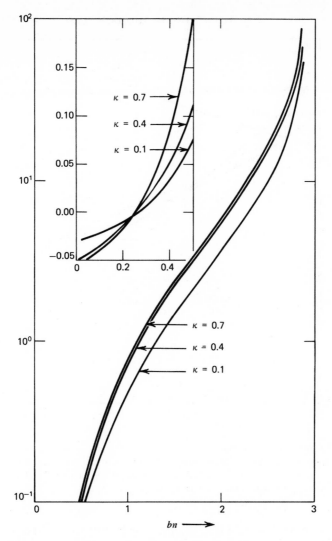

**Fig. 4** Coefficient $-\pi$ [in units of $(m/I)^{1/2}(kT/\sigma^2)$] for rough spheres. The inset shows the behavior near $bn = 0.2$, where the coefficient changes algebraic sign.

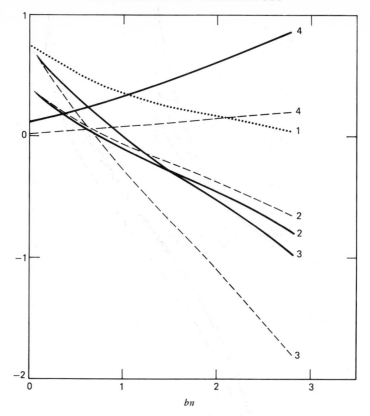

**Fig. 5**   The coefficients $v_i$ [in units of $(4\sigma^2)^{-1}(mkT/\pi)^{1/2}$] and $-\mu$ [in units of $(k/2\sigma)(2\pi)^{-1/2}$] for rough spheres. The labels 1, 2 and 3 refer to the coefficients $v_1$, $v_2$, and $v_3$, respectively, and the label 4 refers to $-\mu$. The solid and dashed curves are for $\kappa$ equal to 0.7 and 0.1, respectively. To the present approximation the coefficient $v_1$ is independent of $\kappa$.

Finally, we have devoted Fig. 5 to those coefficients that behave very differently in the case of rough spheres than they do for the other models we later examine. These four coefficients, $v_1$, $v_2$, $v_3$, and $\mu$, are all associated with the spin flux **C**. It may be that the differences are due to the fact that $C_V = 0$ for rough spheres. This unique characteristic of rough spheres is a consequence of the collisional invariance of the difference $\mathbf{L}_1 - \mathbf{L}_2$ of molecular angular momenta.

## B.  Enskog Theory for Rigid and Square-Well Ellipsoids

In applications to polyatomic fluids, the most notable advantage of an impulsive collision model is that the changes in linear and angular momentum take place so suddenly there are no associated changes in molecular orientations or locations. Provided one neglects the occurrence of multiple or successive impacts between a single pair of particles, it then is a very simple task to determine the consequences of a binary collision. (In the especially simple case of rough spheres, there are none of these successive impacts.) Although calculations based on these models are not trivial to perform, they nevertheless are far less difficult to do for these than for models involving continuous noncentral forces.

Next after rough spheres come the more complicated loaded-sphere and rigid (but smooth) convex-body models. Dilute gas theories for these already have been developed. Because of the great success of the Enskog dense gas theory for rigid spheres, it makes good sense to apply the same approximation to these two models for polyatomic fluids. (At sufficiently high density the collision frequency actually may become great enough to effectively exclude the possibility of multiple-impact events. However, we do not in any case consider our neglect of these processes to be a serious defect of the theory.) But there is more to it than this, because at high densities it is reasonable to apply the Enskog approximation to other impulsive interactions such as the square-well potential. In a "square-well fluid" the trajectory of an individual particle consists of a sequence of linear segments, each beginning and ending with an impulsive collisional event. There are four categories of these events: (I) core collisions; (II) one particle entering the potential well about another; (IIIa) one particle escaping from the potential well of another; (IIIb) one particle reflected from the edge of the well about another, that is, a particle unable to escape from a well because the binding energy is greater than the relative kinetic energy. According to the Enskog approximation, there is no dynamical correlation (other than that associated with the radial distribution function $g$) between the states of the two particles prior to the moment when they first come within range of their mutual interaction. This is the familiar "chaos" assumption which has been used so successfully in dilute gases and in the Enskog theory for a dense gas of rigid spheres. The rationale for applying this same assumption to events of the types II, IIIa, and IIIb is of a different sort and can be defended only for quite high densities. Specifically, it is argued that, at high density, two particles separated by a distance within the range $(\sigma, R\sigma)$ experience (on the average) several "randomizing" collisions before their separation reaches either of the critical values $\sigma$ or $R\sigma^-$. The lack of correlations implied by the Enskog approximation is a consequence of these third-particle events which interrupt the two-body collision in progress.

One can of course obtain the Enskog theory for rigid spheres from the Enskog square-well theory in the limit that the well depth tends to zero. In an analogous way we can obtain the Enskog theory for rigid convex bodies from the limit of the theory for a fluid composed of particles that interact with a square-well convex-body potential. The specific case considered here is that of square-well ellipsoids, that is, of rigid ellipsoidal molecules that interact with an additional square-well potential which is geometrically similar. This interaction is fully characterized by the well depth $\varepsilon$, by the eccentricity of the ellipsoids, and by $\sigma$ and $R\sigma$, the minor radii of the ellipsoidal core and well, respectively. Each molecule is taken to be a rigid rotor with a mass distribution coincident with the major axis of its ellipsoidal core and well. The ellipsoids and mass distributions are assumed to be invariant under reflection through their common center.

The collisional alterations of state accompanying the impulsive events I, II, IIIa, and IIIb are given by $\Delta \mathbf{p}_i = \mathbf{k}_i K$ and $\Delta \mathbf{L}_i = \boldsymbol{\sigma}_i \times \mathbf{k}_i K$, where $\mathbf{k}_i$ is the outward-directed unit normal to the surface of molecule $i$ at the point of common tangency with its collision partner. $\boldsymbol{\sigma}_i$ is the vector extending from the mass center of $i$ to this point of mutual contact. And finally $K$, the magnitude of the normally directed impulse, is related to the collisional changes in kinetic energy and in the normal component of relative velocity by $\Delta E_{\text{kin}} = -K(\mathbf{k} \cdot \mathbf{g}_{21}) + \frac{1}{2}Kh$ and $\Delta(\mathbf{k} \cdot \mathbf{g}_{12}) = -Kh$, respectively. Here $h = 2d^2/m$, $d^2 = 1 + (m/2I)[\mathbf{k} \times \boldsymbol{\sigma}_1)^2 + (\mathbf{k} \times \boldsymbol{\sigma}_2)^2]$ and $\mathbf{k} = \mathbf{k}_1$. For events of the types I, II, IIIa, and IIIb, the magnitude of the impulse is given by

$$2(\mathbf{k} \cdot \mathbf{g}_{21})h^{-1}, (\mathbf{k} \cdot \mathbf{g}_{21} + \sqrt{(\mathbf{k} \cdot \mathbf{g}_{21})^2 + 2\varepsilon h})\, h^{-1},$$

$$(\mathbf{k} \cdot \mathbf{g}_{21} - \sqrt{(\mathbf{k} \cdot \mathbf{g}_{21})^2 - 2\varepsilon h})\, h^{-1},$$

and $2(\mathbf{k} \cdot \mathbf{g}_{21})h^{-1}$, respectively.

The flux and source terms of the moment equations (and the collisional transfer portions of the fluxes as well) can be expressed in terms of the two sets of integrals $\{\alpha, \beta\} = \{\alpha, \beta\}^{(1)} + \{\alpha, \beta\}^{(2)}$ and $[\alpha, \beta] = [\alpha, \beta]^{(1)} + [\alpha, \beta]^{(2)}$. The superscript 1 refers to the hard-core contributions

$$\{\alpha, \beta\}^{(1)} = \int dV F \int d\mathbf{k}\, S^{(1)}g^{(1)}H(\mathbf{k} \cdot \boldsymbol{\gamma}_T)\mathbf{x}(\Delta_2{}^1\alpha_1)(\beta_1 + \beta_2)$$

$$[\alpha, \beta]^{(1)} = \int dV F \int d\mathbf{k}\, S^{(1)}g^{(1)}H(\mathbf{k} \cdot \boldsymbol{\gamma}_T)(\Delta_1{}^1\alpha_2 + \Delta_2{}^1\alpha_1)(\beta_1 + \beta_2)$$

where

$$F = (16\pi^7)^{-1}(\mathbf{k} \cdot \boldsymbol{\gamma}_T)\exp(-\tfrac{1}{2}\gamma^2 - 2\Gamma^2 - \Omega_1{}^2 - \Omega_2{}^2),$$

$$dV = de_1\, de_2\, d\gamma\, d\Gamma\, d\Omega_1\, d\Omega_2$$

and $\mathbf{x} = (m/I)^{1/2}\mathbf{x}_{12}$. The symbol $g^{(1)}$ denotes the value of the equilibrium "radial" distribution function $g(\mathbf{x}_1, \mathbf{x}_2, \mathbf{e}_1, \mathbf{e}_2)$ evaluated with the two molecular cores in contact, that is, with $\mathbf{x}_{12} = \boldsymbol{\sigma}_{12}(\mathbf{e}_1, \mathbf{e}_2, \mathbf{k})$. Finally, $S^{(1)}\, d\mathbf{k}$ is the differential surface element of the "excluded volume" associated with the two cores, $H(x)$ is the unit step function, equal to unity for $x > 0$ and to zero for $x < 0$; $\mathbf{g} = \mathbf{g}_{12}$ is the relative velocity of the points of contact at the moment of impact.

The superscript 2 refers to contributions from the three other types of impulsive events we have just described. The associated integrals are defined by

$$\{\alpha, \beta\}^{(2)} = \int dVF \int d\mathbf{k}\; S^{(2)}g^{(2)}\mathbf{x}[e^{-\varepsilon/kT}H(\mathbf{k}\cdot\mathbf{g})\Delta_2^{II}\alpha_1$$
$$- H(-\mathbf{k}\cdot\mathbf{g} - v)\Delta_2^{IIIa}\alpha_1 - H(\mathbf{k}\cdot\mathbf{g} + v)H(-\mathbf{k}\cdot\mathbf{g})\Delta_2^{IIIb}\alpha_1](\beta_1 + \beta_2)$$

and

$$[\alpha, \beta]^{(2)} = \int dVF \int d\mathbf{k}\; S^{(2)}g^{(2)}[e^{-\varepsilon/kT}H(\mathbf{k}\cdot\mathbf{g})(\Delta_1^{II}\alpha_2 + \Delta_2^{II}\alpha_1)$$
$$- H(-\mathbf{k}\cdot\mathbf{g} - v)(\Delta_1^{IIIa}\alpha_2 + \Delta_2^{IIIa}\alpha_1)$$
$$- H(\mathbf{k}\cdot\mathbf{g} + v)H(-\mathbf{k}\cdot\mathbf{g})(\Delta_1^{IIIb}\alpha_2 + \Delta_2^{IIIb}\alpha_1)](\beta_1 + \beta_2)$$

where $v = (4\varepsilon/m)^{1/2}d$, and $\Delta_2^{II}\alpha_1$, for example, denotes the change in the function $\alpha_1$ that results from a collisional event of type II with molecule 2. The symbol $S^{(2)}$ is defined analogously to $S^{(1)}$, and $g^{(2)}$ stands for the value of $g(\mathbf{x}_1, \mathbf{x}_2, \mathbf{e}_1, \mathbf{e}_2)$ evaluated just inside the lip of the potential well. We have eliminated from our formulas the value of $g$ just outside the well by making use of the relationship $g(\mathbf{x}_{12} + 0, \ldots) = g(\mathbf{x}_{12} - 0, \ldots) \exp(-\varepsilon/kT)$. All other factors in the integrand are to be similarly evaluated.

To handle the parts of these integrals associated with the lip of the well, it was necessary for us to extend to the square-well model Hoffman's techniques for dealing with the collision integrals of rigid nonspherical objects.[33] The critical element in Hoffman's scheme is that all exchanges of momentum be parallel to $\mathbf{k}$, the normal (at the point of contact of the two bodies) to the surface of the excluded volume. Since this same condition is satisfied at the lip of the well, we can carry Hoffman's analysis over in its entirety provided that the relation $\varepsilon'_n = -\varepsilon_n$ (for the collisional alteration of the normal component of the dimensionless relative momentum) is replaced with $\varepsilon'_n = -\sqrt{\varepsilon_n{}^2 + \varepsilon/kT}$, $\varepsilon'_n = \sqrt{\varepsilon_n{}^2 - \varepsilon/kT}$, and $\varepsilon'_n = -\varepsilon_n$ for events of the type II, IIIa, IIIb, respectively. Because these relationships depend on no characteristic of the noncentral potential other than the well depth, $\varepsilon$, the integrand contributions from all four of the collision types included in the integrals $\{\alpha, \beta\}$ and $[\alpha, \beta]$ can be collected into a single compact term which is then integrated over all directions of the unit normal $\mathbf{k}$ and all orientations of the two ellipsoids.

The integrations with respect to the momenta $\gamma$, $\mathbf{\Gamma}$, $\mathbf{\Omega}_1$, and $\mathbf{\Omega}_2$ can be performed analytically, so that the only remaining task is the evaluation of integrals of the form

$$J^{(i)}[\chi] = \frac{1}{16\pi^{5/2}} \int d\mathbf{e}_1 \, d\mathbf{e}_2 \, d\mathbf{k} \, S^{(i)} g^{(i)} \chi(\mathbf{e}_1, \mathbf{e}_2, \mathbf{k})$$

Here is where we encounter a computational hangup, for no one has yet evaluated the equilibrium pair correlation function either for rigid ellipsoids or for square-well ellipsoids. However, there has been a recent flurry of theoretical investigations concerning the equilibrium properties of dense polyatomic fluids. Some of these, particularly the studies by Chen, Steele and Sandler and those of Chandler and Lowden,[34] can be adapted to the needs of the theory of transport and relaxation presented here. But for present purposes, we disregard entirely the orientation dependence of the functions $g^{(i)}$. In the case of rigid ellipsoids, we set the well depth $\varepsilon$ equal to zero and so replace the integrals $\{\alpha, \beta\}$ and $[\alpha, \beta]$ with $\{\alpha, \beta\}^{(1)}$ and $[\alpha, \beta]^{(1)}$, respectively. In place of $g^{(1)}(\mathbf{x}_1, \mathbf{x}_2, \mathbf{e}_1, \mathbf{e}_2)$ we then substitute the orientation-independent values characteristic of a fluid composed of rigid spheres of diameter $\sigma = 2(A^2 B)^{1/3}$, where $A$ and $B$, respectively, denote the radii of the minor and major axes of an ellipsoid. The volume of the sphere with this diameter is equal to that of the ellipsoid. Although this choice for the sphere diameter certainly is not unique, we prefer it to the other possibilities, $2A$, $A + B$, and $B$, for which we also have performed calculations. For square-well ellipsoids we use for $g^{(1)}$ and $g^{(2)}$ values of the radial distribution function calculated (by the Monte Carlo method) by Rotenberg[35] for the square-well potential; the diameter of the spherical core is related to the ellipsoidal parameters in the same way as for rigid ellipsoids. The numerical values of the radial distribution functions used in the present calculations are tabulated in Refs. 19 and 30.

The moment equations for the rigid ellipsoid and square-well ellipsoid models are identical in form to those for rough spheres. The only significant differences between these models is that the potential contribution to the spin flux no longer vanishes identically and that $\boldsymbol{\omega} = (3m/2I)\boldsymbol{\ell}$ for linear rotors.

The various phenomenological coefficients depend not only on the value of well depth but also on the eccentricity $\alpha = A/B$ and the moment of inertia, which for a diatomic species is given by $I = mD^2/4$. From studies of the magnetic field dependence of the thermal conductivity,[36] it has been found that for gaseous nitrogen $\alpha = 1.9074/2.2002 \doteq 0.867$ and $\beta \equiv D/B = 1.098/2.2002 = 0.453$. This provides us with an idea of the magnitudes of $\alpha$ and $\beta$ for small, almost spherical molecules. The purpose of the graphs presented here is to give some indication of how the transport coefficients vary with density, well depth, eccentricity, and moment of inertia. However, because there are so many coefficients and so many independent variables, only a small sampling of our results can be presented. (See Figs. 6 to 16). In all cases the diameter $\sigma$ has been set equal to the "nitrogen value" of 4.001 Å.

The plots of $\eta$, $\eta_b$, and $\zeta$ are in units of $(5/16\sigma^2)(mkT/\pi)^{1/2}$, the CE first approximation to the shear viscosity for a dilute gas of rigid spheres. $\lambda$ and $\lambda^t$ are plotted in units of $(75/64\sigma^2)(k^3T/\pi m)^{1/2}$, the corresponding CE approximation to the thermal conductivity.

Because we have neglected the orientation dependence of the equilibrium pair correlation function, the reliability of these results diminishes as the value of the eccentricity declines. Furthermore, Rotenberg's radial distribution functions for the square-well potential vary with density in a rather unlikely manner. This is especially pronounced for $\varepsilon/kT = 0.1$ and is responsible for the sharp breaks in the plots of viscosity and thermal conductivity versus density. Also, it appears that as the well depth becomes smaller the square-well results do not approach those for rigid ellipsoids. We intend to repair these deficiencies by redoing our calculations with more accurate estimates for the square-well radial distribution functions.

Generally speaking, the value of the vortex viscosity coefficient $\zeta$ will be very much less than either the shear or bulk viscosity. Only at high densities and for small values of the eccentricity is it otherwise. Invariably, it is found that a coefficient depends much less sensitively on $\beta$ (and so on the moment of inertia) than on the eccentricity. And this makes good sense. Of special interest is the variation of $\lambda/\lambda^t$ with $\alpha$ and $\beta$; as $\alpha$ tends to unity $\lambda \to \lambda^t$, but the rotational contribution to the thermal conductivity generally turns out to be large.

As one examines these figures, it is important to bear in mind that the value of $\sigma = 2\sqrt[3]{A^2B}$ has been held constant and that $\alpha = A/B = (2A/\sigma)^3$ and $\beta \equiv D/B = (4I/m\sigma^2)^{1/2}\alpha^{2/3} = (4I/m\sigma^2)^{1/2}(2A/\sigma)^2$. Therefore $\alpha$ is proportional to the cube of the minor dimension of the ellipsoid and, for a fixed value of $\alpha$ (or of $A$), the parameter $\beta$ varies as the square root of the moment of inertia. The complexity of the variations with parameter values is apparent from the figures. One further example of this complexity is provided by a description of the density and moment of inertia ($\beta$) dependencies of $\lambda$, $\eta$ and $\eta_b$ for the two values 0.2 and 0.6 of the parameter $\alpha$; at very high densities $\lambda$ diminishes, but $\eta$ and $\eta_b$ both rise in value along with $\beta$. This behavior of $\lambda$ persists at the lower density $bn = 1.00$, but now $\eta$ first falls in value and then rises with $\beta$. At $bn = 0.1$ and $\alpha = 0.2$, $\lambda$, $\eta$, and $\eta_b$ decrease as $\beta$ increases; for $bn = 0.1$ and $\alpha = 0.6$, $\eta$ exhibits this same behavior, but $\lambda$ and $\eta_b$ both pass through minima as the value of $\beta$ rises. The situation is similar at the still lower density $bn = 0.01$. From Figs. 7 and 10 we see that, at least at high densities, the introduction of the square well does not alter these conclusions.

It is likely that one can fashion simple explanations for the parametric variations in the coefficients that emerge from our calculations, but we have not yet had sufficient opportunity to investigate this possibility.

The first application of the square-well ellipsoid model to the kinetic theory of dense polyatomic fluids was made by Sather and Dahler[37] who estimated the potential contributions to the fluxes of momentum and energy by using the Enskog approximation and by neglecting nonequilibrium

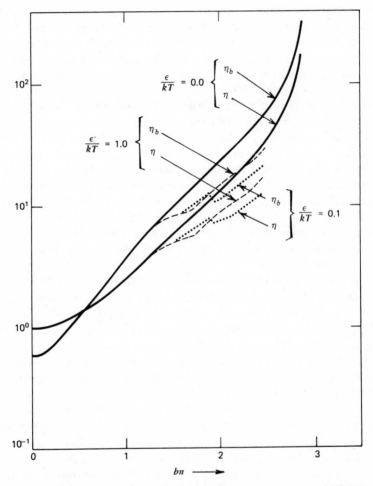

**Fig. 6**    The dependencies of $\eta$ and $\eta_b$ on the dimensionless density $bn = \frac{2}{3}\pi\sigma^3 n$. The solid curves are for rigid ellipsoids, the dashed for the square-well with $\varepsilon/kT = 1$, and the dotted for $\varepsilon/kT = 0.1$. In all cases $\alpha = 0.867$ and $\beta = 0.453$.

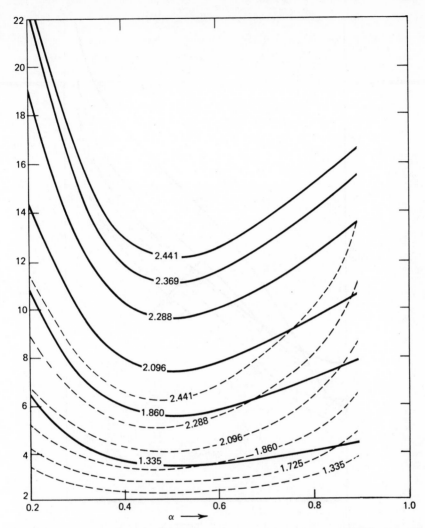

**Fig. 7** Eccentricity ($\alpha = A/B$) dependence of the shear viscosity for square-well ellipsoids with $\varepsilon/kT = 1$. The curves are labeled with values of the density $bn$. The solid curves are for $\beta = 0.5$, and the dashed for $\beta = 0.2$. As $\beta$ increases, so also does the moment of intertia $I \propto \beta^2$.

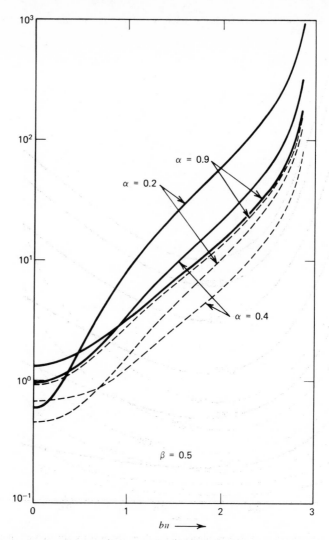

**Fig. 8a** Dependence on *bn* of the thermal conductivity $\lambda$ (solid curves) and its translational part $\lambda^t$ (dashed curves) for rigid ellipsoids. Here $\beta = 0.5$. Note that as $\alpha$ increases the curves of $\lambda$ and $\lambda^t$ draw together.

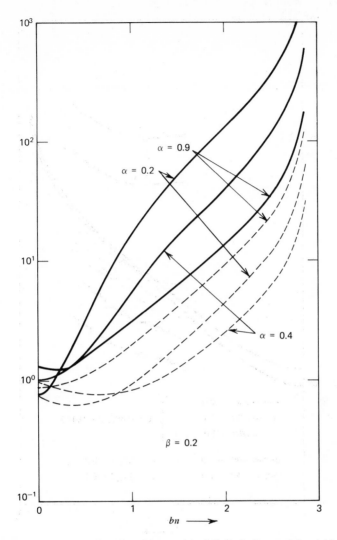

**Fig. 8b** Dependence on *bn* of the $\lambda$ (solid curves) and $\lambda^t$ (dashed curves) for rigid ellipsoids. Here $\beta = 0.2$.

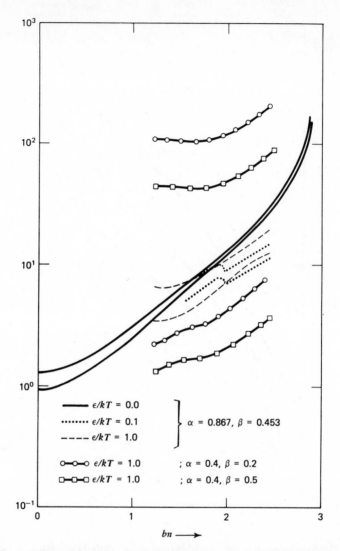

**Fig. 9** The variations of $\lambda$ and $\lambda^t$ with density. The upper curve of each pair (solid, dashed, and dotted) is of $\lambda$, and the lower of $\lambda^t$. These curves illustrate the parametric dependence of $\lambda$ and $\lambda^t$ on $\alpha$, $\beta$, and $\varepsilon/kT$. The three pairs of curves are computed for the nitrogen parameters $\alpha = 0.867$, $\beta = 0.453$.

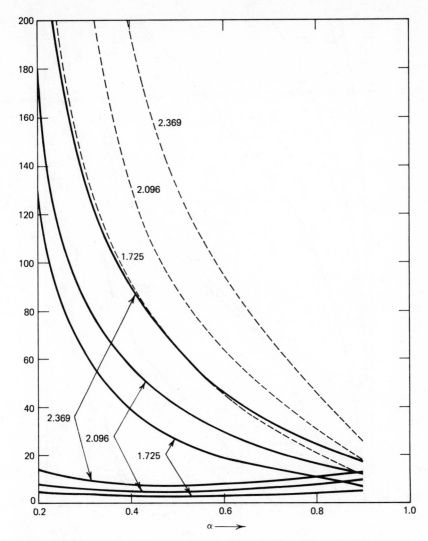

**Fig. 10** Eccentricity ($\alpha = A/B$) dependencies of $\lambda$ and $\lambda^t$ for $\varepsilon/kT = 1.0$. The upper three solid curves are of $\lambda$ for $\beta = 0.5$, and the lower three are of $\lambda^t$ for $\beta = 0.5$. The dashed curves show $\lambda$ for $\beta = 0.2$. Each curve is labeled with a value of the density $bn$.

189

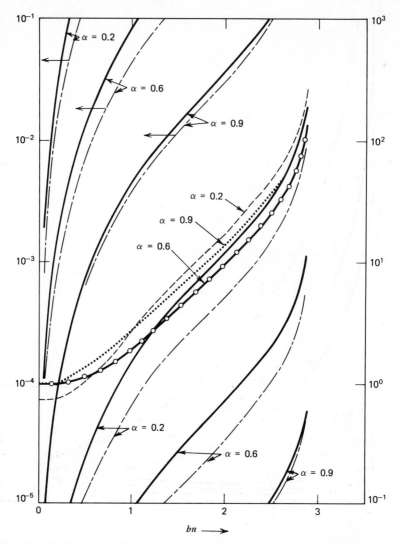

**Fig. 11** The density dependence of the vortex viscosity coefficient $\zeta$ for rigid ellipsoids. The solid curve of each pair is for $\beta = 0.5$, and the dash-dot curve is for $\beta = 0.2$. The three other curves, dashed, dotted, and wide-line are of $\eta$ for $\beta = 0.5$. $\eta$ values for other choices of $\beta$ are not shown; the dependence of $\eta$ on $\beta$ is even less than it is on $\alpha$. The ordinate scale is on the right except for the three curves in the upper left-hand corner marked with arrows.

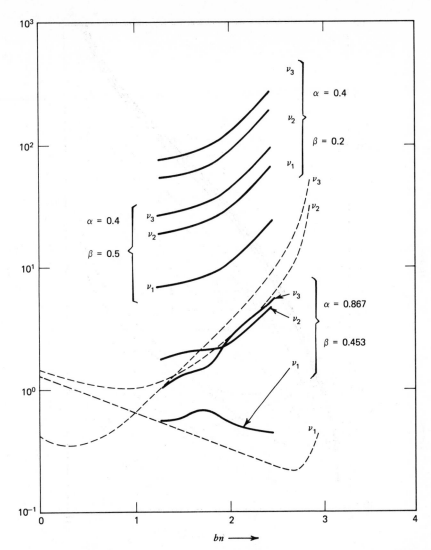

**Fig. 12** The variations with density, eccentricity, and moment of inertia of the coefficients $v_1$, $v_2$ and $v_3$ [in units of $(4\sigma^2)^{-1}(mkT/\pi)^{1/2}$]. The curves grouped together by braces are for the square-well model with $\varepsilon/kT = 1.0$. The dashed curves are for rigid ellipsoids. In all cases $\alpha = 0.867$ and $\beta = 0.453$. The curve of $v_2$ for $\alpha = 0.4$, $\beta = 0.5$ is coincident with that of $v_1$ for $\alpha = 0.4$, $\beta = 0.2$.

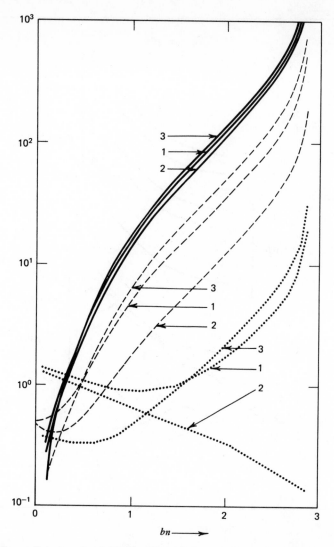

**Fig. 13** The coefficients $v_i$ for rigid ellipsoids with $\beta = 0.5$. The ordinate scale is in units of $(4\sigma^2)^{-1}(mkT/\pi)^{1/2}$. The curves labeled 1, 2, and 3 refer to the coefficients $v_1$, $v_2$, and $v_3$, respectively. The solid, dashed, and dotted curves are for $\alpha$ equal to 0.2, 0.4, and 0.9, respectively.

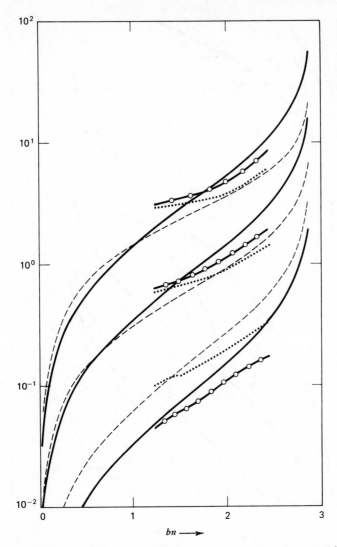

**Fig. 14** Parametric dependencies of the transport coefficient $-\pi$ [in units of $(m/I)^{1/2}\sigma^{-2}kT$]. The solid and dashed curves are for rigid ellipsoids with $\beta = 0.5$ and $\beta = 0.2$, respectively. The top pair of curves is for $\alpha = 0.2$, the middle pair for $\alpha = 0.6$, and the lower pair is for $\alpha = 0.9$. The associated pairs of open-circle and dotted curves are for $\varepsilon/kT = 1.0$, with $\beta = 0.5$ and 0.2, respectively.

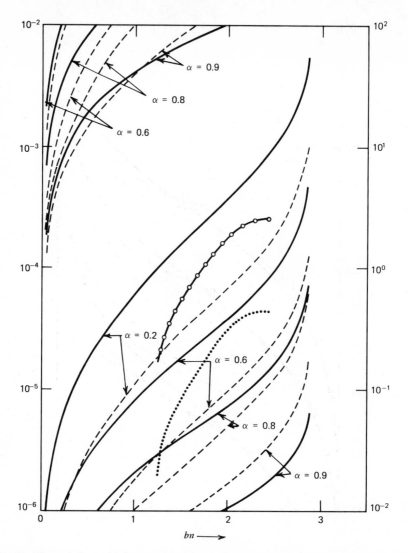

**Fig. 15** The variations with $bn$, $\alpha$, and $\beta$ of the coefficient $-\mu$. The ordinate scale is in units of $(k/2\sigma)(2\pi)^{-1/2}$. The curves in the upper left-hand corner are referred to the left ordinate scale; the right-hand scale is for all others. The solid and dashed curves are for rigid ellipsoids with $\beta = 0.5$ and 0.2, respectively. The open-circle and dotted curves are for the square-well potential with $\varepsilon/kT = 1$ and $\alpha = 0.2$, and refer to $\beta = 0.5$ and $\beta = 0.2$, respectively.

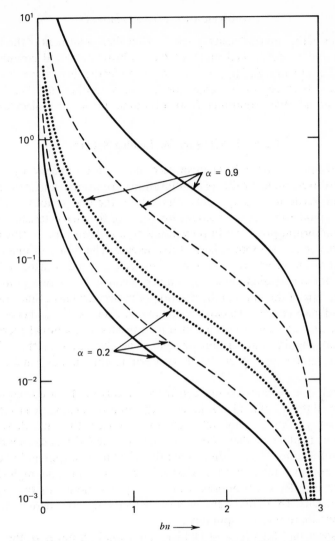

**Fig. 16** Variations with density and eccentricity of the rigid ellipsoid relaxation times $\tau_\omega$ (spin), $\tau_T$ (rotational energy), and $\tau_P$ (viscous stress). These relaxation times are plotted in units of $\sigma(m\pi/kT)^{1/2}$. In each grouping of three, the solid, dashed, and dotted curves are for $\tau_\omega$, $\tau_T$, and $\tau_P$, respectively. It can be seen that for nearly spherical molecules $\tau_\omega > \tau_T > \tau_P$, whereas for highly eccentric species $\tau_P > \tau_T > \tau_\omega$.

distortions of the singlet density function. Thus their treatment of the square-well ellipsoid model lacked much of the sophistication and completeness of the present theory. So far as we are aware, no further investigations of this model have been reported, except for its use by Sather and Dahler[38] to compute liquid-phase rotational relaxation times for linear molecules.

## C. Enskog Theory for Loaded Spheres

Loaded spheres were first introduced into kinetic theory by Jeans[20] in an effort to estimate the rate of equilibration of rotational and translational degrees of freedom. This appears to have been the first serious attempt to develop a model that could account for some of the characteristic features of nonequilibrium processes in polyatomic fluids. According to this model, each molecule is a smooth sphere of diameter $\sigma$. The center of mass is displaced from the geometric center by a distance $\varepsilon < \sigma/2$. Furthermore, we assume that the straight line that passes through these two points is a symmetry axis of the internal distribution of molecular mass. Since the geometric and mass centers of a particle do not coincide, a collision between two of these loaded spheres is accompanied by transfers of rotational energy and of internal angular momentum or spin. The dynamics of these binary encounters and many other details concerning this model can be found in Ref. 8.

Although one certainly could construct for this model a corresponding "square-well theory of loaded spheres"—the limit being a dense gas loaded-sphere theory—this has not yet been done. Indeed, the only dense fluid calculations made for either of these models are the CE results reported by Sandler and Dahler.[39] They discovered that the fluxes of spin and of energy for the repulsive interaction loaded-sphere model include high-density collisional transfer contributions which are proportional to the gradient of fluid density. (Actually, Sandler and Dahler only considered explicitly the $\mathbf{V}n$ contribution to the spin flux.)

We suspect that Sandler and Dahler's numerical estimates of the vortex viscosity $\zeta$ and of the coefficients of porportionality between the spin flux and the gradients of temperature and density are erroneous, not only because their formulas were obtained by the CE method but also because the results themselves depend on the model parameters in ways that we question. Although it would not be difficult to reexamine this model using the moment method, we have not had the time to do so. Consequently, the only results of Sandler and Dahler reported here (in Figs. 17 and 18) are their numerical findings for the coefficients of thermal conductivity and of shear and bulk viscosity. The CE values for these should agree with those of the moment method.

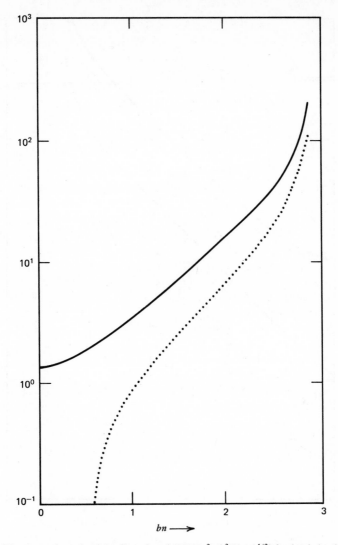

**Fig. 17** The thermal conductivity [in units of $(75/64\sigma^2)(k^3T/\pi m)^{1/2}$] for loaded spheres. The dotted curve is the nondistortional contribution to $\lambda$ computed by Sather and Dahler.[37]

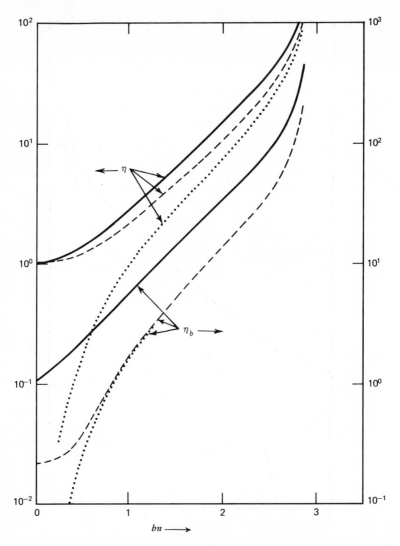

**Fig. 18** The dependence on density and $a = m\varepsilon^2/2I$ of the coefficients $\eta$ and $\eta_b$. The ordinate scale is in units of $(5/16\sigma^2)(mkT/\pi)^{1/2}$. In both cases the solid curve refers to $a = 0.05$ and the dashed to $a = 0.5$. The dotted curves are the non-distortional contributions of Sather and Dahler computed for $a = 0.5$. The $\eta$ curves are referred to the left hand ordinate scale and the $\eta_b$ curves to the right.

The most remarkable feature is how very slightly these coefficients depend on the values of the two parameters $\varepsilon/\sigma$ and $a = m\varepsilon^2/2I$, which occur separately in the formulas. Indeed, the dependence on the former is so weak as to be ignorable. The curve for $\lambda$ shown in Fig. 17 is the same to within a few percent for all $0 \le \varepsilon/\sigma \le 0.5$ and $0 \le a \le 0.5$. The dependence of $\eta_b$ on $a$ does of course become very pronounced as $a$ approaches zero; in the limit $\eta_b$ is unbounded. Since $a$ decreases as the value of $I$ rises, the trends of $\eta$ and $\eta_b$ with $I$ are in the same directions as were found above for ellipsoids. The numerical values of the radial distribution functions used in these calculations were the same as those we used previously for rigid ellipsoids.

In both figures we have included for comparison (the dotted curves) the earlier predictions due to Sather and Dahler[37] who completely neglected distortions of $f$ from the local Maxwell–Boltzmann function $f^0$. (The PNM theory, which we soon consider, includes a similar approximation.) Therefore the differences between the Sandler–Dahler and Sather–Dahler results provide indications of how important these distortions are. They illustrate that as the density rises the relative importance of distortions falls. For a detailed comparison of the distortional and nondistortional contributions to the Enskog theory for smooth spheres, the reader should consult Ref. 40.

## D. Rice–Allnatt Theory for Rough Spheres

Invariably it is possible to separate the molecular pair potential into a short-range exclusively repulsive interaction and a longer-range more slowly varying part; the square-well potential is a rather unphysical exception to this rule. Therefore, as a molecule moves through a dense fluid, it experiences many small alterations in momentum (both linear and angular) as a result of its interactions with relatively distant neighbors and, with much less frequency, it suffers occasional and violent, almost impulsive core collisions. The effect of the slowly varying "soft" part of the pair force is to produce a sort of high-frequency noise which is superimposed on the intermittent large changes in molecular momentum caused by the core or "hard" parts of the forces. It is sensible to assume that these rapidly fluctuating soft forces destroy the dynamical correlations which otherwise would persist from one core collision to the next. This argument is closely analogous to that which we already have used to justify our application of the chaos approximation to the dense gas theory for the square-well potential. The difference is that here we wish to take explicitly into account the changes in molecular state due to the intervening and randomizing (soft) collisional events. In the Rice–Allnatt theory[10] this is accomplished by using for the mathematical description of these events the weak-coupling approximation to the many-body collision operation. The kinetic equation for the singlet distribution function is written, accordingly, in the form $(\partial_t + \mathbf{c}_1 \cdot \mathbf{V})f = \hat{\Omega}_H[f] + \hat{\Omega}_S[f]$, where $\hat{\Omega}_H$ is an Enskog-like collisional operator which takes account of the short-range hard potential, and $\hat{\Omega}_S$ is a weak-coupling operator characteristic of the soft part of the pair interaction. In the work of Rice and Allnatt, the second of these is replaced by a simple Fokker–Planck (FP) operator, so that the effect of the soft potential on the evolution of the singlet distribution function is given exclusively by the contribution of that potential to the friction coefficient which occurs in the FP operator.

To complete the theory one also must compute the collisional transfer contributions to the fluxes, and these depend on the pair distribution function. For this Rice and Allnatt proposed a generalization of their singlet kinetic equation. Despite the realistic conceptual basis of this theory, it does suffer from several deficiencies, the most serious being the two very different ways $f_2$ is treated, first in the derivation of the kinetic equation for $f$, and then in the estimation of the potential or collisional transfer portions of the fluxes. This inconsistency and other defects of the RA theory have been studied carefully by Davis and his co-workers;[12] they and others, notably Misguich and Nicolis[41] and Allen, Cole, and Foster[42] have proposed alternative generalized Rice–Allnatt theories, the details of which have been summarized excellently by Davis.[12]

These theories all have played important roles in preparing the way for the more satisfactory effective potential theories to which we turn our attention in Sections III.E and F. But prior to these very recent developments, the theory of Rice and Allnatt was the best available, and so it seemed worthwhile to determine whether it could be applied to polyatomic fluids. The first step in this direction was taken by Condiff and Dahler[43] who derived a weak-coupling approximation to the kinetic equation for the singlet distribution function appropriate to a fluid of rigid rotors. From this they obtained as a further approximation the generalization to rotors of the FP operator, together with correlation integral expressions of the Kirkwood–Green variety for the friction tensors associated with translational and rotational motions. The program begun in this way was to have been completed by (1) including in the singlet kinetic equation an Enskog-like collision operator characteristic of the short-range central or noncentral forces, and (2) deriving a suitable generalization of the Rice–Allnatt kinetic equation for $f_2$. The theory never has been presented in this form. From the very beginning we were hampered by our inability to cope successfully with the inconsistencies of the Rice–Allnatt theory; these appeared to be of more consequence, or at least were more visible, for the polyatomic fluid than in the monatomic case. Then too, we could not convince ourselves of the adequacy of the Rice–Allnatt approximation of neglecting the momentum (and in our case angular momentum as well) dependence of the friction tensors that occur in the generalized FP operator. The second of these difficulties can be avoided by retaining the exact weak-coupling formula for the soft-potential collision operator, but the first problem is not so easily solved. In any case the effective potential theories presented in the following two sections appear to deal successfully with all these difficulties.

Despite our reservations about the Rice–Allnatt theory itself and the additional problems encountered in our attempt to construct a satisfactory generalization for polyatomic fluids, there is a restricted class of systems for which generalization can be accomplished with scarcely any trouble. These are situations in which the pair interactions consist of long-range central and short-range noncentral forces. For these the soft-potential part of the problem is the same as that dealt with by the original RA theory, and it is only the Enskog-like portion that need be modified. A theory of this sort could be developed for any choice of the short-range potential, but the only calculations that actually have been performed are those by McCoy and Dahler[44] for rough spheres. They combined the Enskog theory for rough spheres with the Rice–Allnatt treatment of a soft central potential, and then solved the resulting kinetic equations by the CE method. The really difficult part of the calculation was not that associated with the rough-sphere cores but the soft-potential portion; and this already had been done by Rice and co-workers (for argon) for a few values of density and temperature. The approximation used by McCoy and Dahler for the singlet distribution function was more elaborate than the truncated moment expansion (2.1) and included terms proportional to $\mathbf{W}\boldsymbol{\Omega}\boldsymbol{\Omega}$, $\boldsymbol{\Omega}\mathbf{W}$, and $\boldsymbol{\Omega}\boldsymbol{\Omega}$.

According to this theory the coefficient of vortex viscosity is given by the same formula as reported in Section III.A, but with the radial distribution function $g(\sigma)$ appropriate to the sum of rigid-sphere and soft (Lennard–Jones 12-6) potentials. Typical of McCoy and Dahler's results were the findings that, for $\varepsilon/kT = 0.952$ and $bn = 1.402$, the shear coefficient rose monotonically from 1.74 to 1.87 mp as the value of $\kappa = 4I/m\sigma^2$ climbed from 0 to 0.7. Under these same conditions Rice and co-workers computed for $\eta_b$ the value of 0.933 mp. McCoy and Dahler found that $\eta_b$ for the rough-sphere model was unbounded at $\kappa = 0$ and fell in value from 2.259 to 1.381 as $\kappa$ varied from 0.1 and 0.7. Thus $\eta$ increases and $\eta_b$ decreases with rising $I$, just as in the case of a dilute gas of rough spheres. However, it was found that $\lambda$ (which is equal to 169.2 $\mu$cal/(cm)(sec)($^\circ$K) in the Rice–Allnatt smooth-core theory) decreased in value from 292.6 to 266.8 as $\kappa$ rose from zero to 0.7. This is opposite the trend exhibited by the dilute gas.

## E. The Effective Potential Theory

It is clear that at high densities one can invoke the simple and appealing chaos approximation of the Enskog theory only if one is dealing with impulsive molecular interactions. If one wishes to consider more slowly varying, continuous potentials of interaction, it becomes necessary to abandon this approach and use other means to construct the desired functional relationship between the pair and singlet distribution functions. The scheme we consider next is a product of the theory of nonequilibrium processes due to Prigogine and his collaborators.

Severne[28] showed that in the long-time limit (when initial correlations may be neglected) the Fourier transform of the doublet correlation function $g_2 = f_2 - ff$ for a monatomic fluid is of the form

$$g_2(\ell, \mathbf{c}_1, \mathbf{c}_2; \mathbf{x}_1, t) = \sum_{s=2}^{N} \int \prod_{i=3}^{s} d\mathbf{c}_i \prod_{j=2}^{N} d\mathbf{x}_j \, \delta(\mathbf{x}_1 - \mathbf{x}_j)$$

$$\times \, C^{(s)}(\ell, \{\mathbf{V}\}; io + \partial_t) \prod_{i=1}^{s} f(\mathbf{x}_i, \mathbf{c}_i; t) \tag{3.3}$$

with "creation operations"

$$C^{(s)}(\ell, \{\mathbf{V}\}; io + \partial_t) = -\frac{\Omega}{8\pi^3} c^{-s+2} \sum_{n=1}^{\infty} e^{-i\mathbf{V}_2 \cdot (\partial/\partial\ell) \cdot i \, \Sigma_{j=v+1}^{N} \mathbf{V}_j \cdot (\partial/\partial\ell)}$$

$$\times \, \frac{1}{\ell \cdot \mathbf{c}_{12} - i(\mathbf{c}_1 \cdot \mathbf{V}_1 + \mathbf{c}_2 \cdot \mathbf{V}_2 + \sum_{j=3}^{N} \mathbf{c}_1 \cdot \mathbf{V}_j + \sum_{j=v+1}^{N} \mathbf{c}_2 \cdot \mathbf{V}_j) - io - i\partial_t}$$

$$\times \sum_{\{k''\}} \left\langle \ell + \mathbf{q}_1 + \sum_{j=3}^{v} \mathbf{q}_j, -\ell + \mathbf{q}_2 + \sum_{j=v+1}^{N} \mathbf{q}_j \middle| (-\delta L) \middle| \{k''\} \right\rangle$$

$$\times \, \langle \{k''\} | [(L_0 - z)^{-1}(-\delta L)]^{n-1} | \mathbf{q}_1 \cdots \mathbf{q}_N \rangle_s^{ir} \, |_{z \to io + i\partial_t}^{\mathbf{q}_j \to -i\mathbf{V}_j} \tag{3.4}$$

wherein $v$ may be selected equal to any integer greater than or equal to 3. The notation used here is that of Prigogine and co-workers; for definitions the reader is referred either to their original papers or to the recent review article by Davis.

In the paper in which he derived this expression, Severne also examined in detail the first term of the series, thereby obtaining the explicit formula

$$g_2^{WC}(\ell, \mathbf{c}_1, \mathbf{c}_2; \mathbf{x}_1, t) = \int d\mathbf{x}_2 \, (\mathbf{x}_1 - \mathbf{x}_2) e^{-i\mathbf{V}_2 \cdot (\partial/\partial\ell)}$$

$$\times \, \frac{1}{\ell \cdot \mathbf{c}_{12} - i(\mathbf{c}_1 \cdot \mathbf{V}_1 + \mathbf{c}_2 \cdot \mathbf{V}_2) - io - i\partial_t}$$

$$\times \, V_\ell \ell \cdot \partial_{12} f(\mathbf{x}_1, \mathbf{c}_1, t) f(\mathbf{x}_2, \mathbf{c}_2; t) \tag{3.5}$$

for the weak-coupling approximation to $g_2$. Here $V_\ell$ denotes the Fourier transform of the pair potential, and $\partial_{12} = \partial_1 - \partial_2$. Although this derivation was a very considerable achievement, the result itself was valid only in the limit of weak coupling and so could not be used to describe nonequilibrium processes in real fluids.

The second step in the development was taken by Prigogine, Nicolis, and Misguich (PNM)[11] who considered a potential similar to that used in the Rice–Allnatt theory. The contributions appearing in Severne's formula for $g_2$ were divided into three categories—those characteristic of the state of equilibrium, and two others labeled *sommet* and *propagateur*—each of the three being a functional of the singlet distribution function and of the pair interaction potential. They then established that the result of including the higher-order terms was to replace the pair potential in the *sommet* part with an effective many-body potential equal to the product of $-kT$ with the equilibrium pair correlation function $G(r) = g(r) - 1$. Also, in place of the exact singlet distribution functions, PNM inserted local Maxwell–Boltzmann functions. A concise description of this theory has been given by Davis;[12] for a more detailed account, one should consult the paper by Misguich.[45]

The next step was due to Davis[12] who proposed that the pair potential $V_l$ be replaced with this effective potential

$$-kT\left(\frac{\mathbf{x}_1 + \mathbf{x}_2}{2}, t\right)G_\ell\left(\frac{\mathbf{x}_1 + \mathbf{x}_2}{2}, t\right)$$

not only in the *sommet* part of $g_2$ but wherever it occurs. He then showed that this ansatz for $g_2$ is exact at equilibrium and that it generates the correct nondissipative hydrodynamical equations. This theory of $g_2$ includes a consideration of the finite duration of collisions and the finite extent of the region over which the interactions extend. Davis applied this theory to the case of a "Rice–Allnatt" interaction consisting of a short-range impulsive potential (rigid spheres) and a longer-range continuous potential (Lennard–Jones 12-6 function). This provides a description of nonequilibrium processes that incorporates all the desirable and intuitively appealing features of the RA theory but which is free from its several deficiencies. So far as we can determine, there presently is no theory better than this for describing non-equilibrium phenomena in dense monatomic fluids. Therefore it is natural that we attempt to generalize it to nonspherical molecular species.

Severne's formalism is cast in terms of action-angle variables. We denote by $a$ and $b$ the angle variables for a rigid rotor and by $L$ and $L_3$ the associated conjugate momenta. The definitions of these variables and their relationships to the usual polar angles and conjugate momenta are given by Cohen and Marcus[46] (who use the symbol $\psi$ in place of $a$).

The Hamiltonian function for a system of $N$ identical molecules is $H = \sum_{i=1}(p_i{}^2/2m + L_i{}^2/2I) + \frac{1}{2}\sum_i\sum_{j\neq i}\lambda V_{ij}$, with $V_{ij} = V(\mathbf{x}_{ij}, \mathbf{e}_i, \mathbf{e}_j)$ and where $\mathbf{e}_i$ depends on $a_i$, $b_i$, and $L_{3i}/L_i$. The parameter $\lambda(=1)$ is a marker used to order terms in powers of the pair potential. It is usual to write the Liouville equation in the form $i\partial_t f_N = (\mathfrak{L}_0 + \lambda\mathfrak{L})f_N$, where

$$\mathfrak{L}_0 = -i\sum_j\left[\left(\frac{\mathbf{p}_j}{m}\right)\left(\frac{\partial}{\partial\mathbf{x}_j}\right) + \left(\frac{L_j}{I}\right)\left(\frac{\partial}{\partial a_j}\right)\right]$$

and

$$\mathfrak{L} = \sum_j\sum_{n\neq j}\mathfrak{L}_{jn} = i\sum_j\sum_{n\neq j}\left[\frac{\partial V_{jn}}{\partial\mathbf{x}_j}\cdot\frac{\partial}{\partial\mathbf{p}_j} + \frac{\partial V_{jn}}{\partial a_j}\frac{\partial}{\partial L_j} + \frac{\partial V_{jn}}{\partial b_j}\frac{\partial}{\partial L_{3j}}\right.$$
$$\left. - \frac{\partial V_{jn}}{\partial L_j}\frac{\partial}{\partial a_j} - \frac{\partial V_{jn}}{\partial L_{3j}}\frac{\partial}{\partial b_j}\right]$$

The formal solution of the Liouville equation then can be expressed as

$$f_N(t) = \frac{1}{2\pi i}\int_C dz\,\frac{e^{-izt}}{\mathfrak{L}_0 + \lambda\mathfrak{L} - z}f_N(0); \quad t > 0$$

where $[\mathfrak{L}_0 + \lambda\mathfrak{L} - z]^{-1} = \sum_{n\geq 0}[-\lambda(\mathfrak{L}_0 - z)^{-1}\mathfrak{L}]^n(\mathfrak{L}_0 - z)^{-1}$, and $C$ is any contour that encircles the entire real axis.

Next we expand $f_N(t)$ as the Fourier series

$$f_N(t) = \Omega^{-N/2}(2\pi)^{-N}\sum_{\{\mathbf{k}lm\}}f_{N\{\mathbf{k}lm\}}(\mathbf{c}LL_3\,t)\phi_{\{\mathbf{k}lm\}}$$

in terms of the eigenfunctions

$$\phi_{\{\mathbf{k}lm\}} = \Omega^{-N/2}(2\pi)^{-N}\exp i(\mathbf{k}\cdot\mathbf{x} + la + mb)$$

of $\mathfrak{L}_0$. Here $\mathbf{k}$ is the ordered set of $N$ wave vectors which are the Fourier variables conjugate to the particle coordinates $\mathbf{x} = (\mathbf{x}_1, \ldots, \mathbf{x}_N)$. Similarly, $l$ and $m$ are the $N$-tuples of Fourier variables associated with the sets of coordinates $a$ and $b$, respectively. The eigenvalue of $\mathfrak{L}_0$ corresponding to $\phi_{\{\mathbf{k}lm\}}$ is $\lambda_{\{\mathbf{k}lm\}} = \mathbf{k}\cdot\mathbf{c} + (L/I)l$. Finally, the symbol $\Omega$ denotes the volume of the system. In terms of the Fourier coefficients $f_{N\{\mathbf{k}lm\}} = \langle\mathbf{k}lm|f_N\rangle$, the Liouville equation then becomes

$$f_{N\{\mathbf{k}lm\}}(\mathbf{c}LL_3\,t) = -\frac{1}{2\pi i}\int_C dz\,e^{-izt}\sum_{\{\mathbf{k}'l'm'\}}\sum_{n\geq 0}\langle\mathbf{k}lm|[-\lambda(\mathfrak{L}_0 - z)^{-1}\mathfrak{L}]^n$$
$$\times (\mathfrak{L}_0 - z)^{-1}|\mathbf{k}'l'm'\rangle f_{N\{\mathbf{k}'l'm'\}}(\mathbf{c}LL_3\,0)$$

To proceed further one must evaluate the matrix elements of the free propagator $(\mathfrak{L}_0 - z)^{-1}$ and of the interaction operator $\mathfrak{L}$. The first of these is given by

$$\langle klm|(\mathfrak{L}_0 - z)^{-1}|k'l'm'\rangle = \frac{1}{\mathbf{c}\cdot\mathbf{k} + (L/I)l - z}\,\delta_{\mathbf{kk'}}\,\delta_{ll'}\,\delta_{mm'}$$

where, for example, $\delta_{ll'} = \prod_1^N \delta_{l_j l'_j}$. To compute the second we expand the pair potential in the Fourier series

$$V_{jn} = \frac{8\pi^3}{\Omega}\sum_{\mathbf{k}''l''_j, n\, m''_j, n} V_{\mathbf{k}''l''_j, n\, m''_j, n}\left(\frac{L_{3j}}{L_j}, \frac{L_{3n}}{L_n}\right)\exp i(\mathbf{k}''\cdot\mathbf{x}_{jn} + l''_{j,n}a_{j,n} + m''_{j,n}b_{j,n})$$

where $l_{j,n} = (l_j, l_n)$. It then follows that

$$
\begin{aligned}
\langle klm|\mathfrak{L}_{jn}|k'l'm'\rangle = \frac{8\pi^3}{\Omega}&\left\{V_{(\mathbf{k}'_n - \mathbf{k}_n),\,(l_{j,n} - l'_{j,n}),\,(m_{j,n} - m'_{j,n})}\right.\\
&\times\left[(\mathbf{k}'_n - \mathbf{k}_n)\cdot\partial_{nj} + (l'_j - l_j)\frac{\partial}{\partial L_j} + (l'_n - l_n)\frac{\partial}{\partial L_n}\right.\\
&\left.+ (m'_j - m_j)\frac{\partial}{\partial L_{3j}} + (m'_n - m_n)\frac{\partial}{\partial L_{3n}}\right]\\
&+ \left[l'_j\frac{\partial}{\partial L_j} + l'_n\frac{\partial}{\partial L_n} + m'_j\frac{\partial}{\partial L_{3n}} + m'_n\frac{\partial}{\partial L_{3n}}\right]\\
&\left.\times V_{(\mathbf{k}'_n - \mathbf{k}_n),\,(l_{j,n} - l'_{j,n}),\,(m_{j,n} - m'_{j,n})}\right\}\\
&\times\delta_{(\mathbf{k}_j - \mathbf{k}'_j),\,(\mathbf{k}'_n - \mathbf{k}_n)}\prod_{\alpha\neq j,n}\delta_{\mathbf{k}_\alpha\mathbf{k}'_\alpha}\delta_{l_\alpha l'_\alpha}\delta_{m_\alpha m'_\alpha}
\end{aligned}
$$

These constitute the generalizations necessary in order to adapt Severne's theory to a fluid composed of rigid rotors. Beginning at this point one now can proceed through his development in step-by-step fashion to obtain the weak-coupling approximation,

$$
\begin{aligned}
g_{2(kl_3,4m_3,4)}(\mathbf{x}_1\sigma'_1\sigma'_2) &= \int d\mathbf{x}_2\,\delta(\mathbf{x}_1 - \mathbf{x}_2)e^{-i\mathbf{V}_2\cdot(\partial/\partial\mathbf{k})}\\
&\times\frac{1}{\mathbf{c}_{12}\cdot\mathbf{k} + (L_1/I)l_3 + (L_2/I)l_4 - i(\mathbf{c}_1\cdot\mathbf{V}_1 + \mathbf{c}_2\cdot\mathbf{V}_2 + \partial_t) - io}\\
&\times\mathscr{L}[f_{l_1m_1}(\mathbf{x}_1\sigma'_1t)f_{l_2m_2}(\mathbf{x}_2\sigma'_2t)]
\end{aligned}
\tag{3.6}
$$

to the Fourier transform of the doublet correlation function. Here $\sigma'_1 = (\mathbf{p}_i\mathbf{L}_i)$ and $(\phi - io)^{-1} = i\int_0^\infty d\tau\exp\left[-i\phi\tau\right] \equiv i\pi\delta_-(\phi)$. This singular quantity also can be expressed as the linear combination $\delta^-(\phi) = \delta(\phi) - (i/\pi)\mathscr{P}(1/\phi)$

of the Dirac delta and of the principal value of $1/\phi$. Finally, the linear operator $\hat{\mathscr{L}}$ is defined by

$$
\hat{\mathscr{L}}[\psi(\cdots)] = \sum_{l_{1,2}m_{1,2}} \left\{ V_{\mathbf{k},\,(l_{3,4}-l_{1,2}),\,(m_{3,4}-m_{1,2})} \right.
$$

$$
\times \left[ \mathbf{k} \cdot \partial_{12} + (l_3 - l_1)\frac{\partial}{\partial L_1} + (l_4 - l_2)\frac{\partial}{\partial L_2} \right.
$$

$$
\left. + (m_3 - m_1)\frac{\partial}{\partial L_{31}} + (m_4 - m_2)\frac{\partial}{\partial L_{32}} \right] \psi_{l_{1,2}m_{1,2}}(\cdots) - \psi_{l_{1,2}m_{1,2}}(\cdots)
$$

$$
\times \left. \left[ l_1\frac{\partial}{\partial L_1} + l_2\frac{\partial}{\partial L_2} + m_1\frac{\partial}{\partial L_{31}} + m_2\frac{\partial}{\partial L_{32}} \right] V_{\mathbf{k},\,(l_{3,4}-l_{1,2}),\,(m_{3,4}-m_{1,2})} \right\}
$$

To obtain the polyatomic version of Davis' theory we then have only to replace the potential $V$ in the operator $\hat{\mathscr{L}}$ with $\tilde{V} = -kTG$. Since our attention is limited here to the first order in gradients, we truncate the displacement operator $\exp\left[-i\mathbf{V}_2 \cdot (\partial/\partial \mathbf{k})\right]$ and the propagator $[\mathbf{k} \cdot \mathbf{c}_{12} + (L_1/I)l_3 + \cdots]^{-1}$ accordingly. It follows then that $g_2 \doteq \sum_{i=1}^{4} g_2^{(i)}$, where

$$
g_2^{(1)} = \frac{1}{\mathbf{k} \cdot \mathbf{c}_{12} + (L_1/I)l_3 + (L_2/I)l_4 - io} \hat{\mathscr{L}}[f(\mathbf{x}_1, \mathfrak{d}'_1, t)f(\mathbf{x}_1, \mathfrak{d}'_2, t)]
$$

$$
g_2^{(2)} = \int d\mathbf{x}_2\, \delta(\mathbf{x}_1 - \mathbf{x}_2)\left(-i\mathbf{V}_2 \cdot \frac{\partial}{\partial \mathbf{k}}\right)\frac{1}{\mathbf{k} \cdot \mathbf{c}_{12} + (L_1/I)l_3 + (L_2/I)l_4 - io}
$$

$$
\times \hat{\mathscr{L}}[f(\mathfrak{d}_1, t)f(\mathfrak{d}_2, t)] \tag{3.7}
$$

$$
g_2^{(3)} = \frac{io_t}{(\mathbf{k} \cdot \mathbf{c}_{12} + (L_1/I)l_3 + (L_2/I)l_4 - io)^2} \hat{\mathscr{L}}[f(\mathbf{x}_1, \mathfrak{d}'_1, t)f(\mathbf{x}_1, \mathfrak{d}'_2, t)]
$$

$$
g_2^{(4)} = \int d\mathbf{x}_2\, \delta(\mathbf{x}_1 - \mathbf{x}_2)\frac{i(\mathbf{c}_1 \cdot \mathbf{V}_1 + \mathbf{c}_2 \cdot \mathbf{V}_2)}{(\mathbf{k} \cdot \mathbf{c}_{12} + (L_1/I)l_3 + (L_2/I)l_4 - io)^2}
$$

$$
\times \hat{\mathscr{L}}[f(\mathfrak{d}_1, t)f(\mathfrak{d}_2, t)]
$$

Next we insert the expansion (2.1) into each of these formulas and so obtain for $g_2$ an expression in terms of moments of the singlet distribution function. In the cases of $g_2^{(2)}$ and $g_2^{(4)}$, one first acts with the operators $\hat{\mathscr{L}}$ and $\mathbf{V}_2$ and then conducts the $\mathbf{x}_2$ integration. The moments appearing in the final expressions are all evaluated at the point $\mathbf{x}_1$.

To construct $f_2$ we must add to $g_2$ the product $f(\mathfrak{d}_1)f(\mathfrak{d}_2)$ and expand $f(\mathfrak{d}_2)$ in a Taylor series about the point $\mathbf{x}_1$. The final formulas for $g_2$ and $f_2$ are complicated and are not given here.

Because of the rotational invariance of $\tilde{V}$ (or equivalently of the equilibrium pair correlation function $G$) it can be expressed as a function of only four independent variables, for example, $|\mathbf{x}_{21}|$, $\hat{\mathbf{x}}_{21} \cdot \mathbf{e}_1$, $\hat{\mathbf{x}}_{21} \cdot \mathbf{e}_2$, and $\mathbf{e}_1 \cdot \mathbf{e}_2$. A further consequence of this invariance is the identity

$$-\mathbf{x}_{12} \times \frac{\partial G}{\partial \mathbf{x}_{12}} + \tilde{\mathbf{N}}_{12} + \tilde{\mathbf{N}}_{21} = 0$$

where $\tilde{\mathbf{N}}_{21} = -\mathbf{e}_1 \times (\partial G / \partial \mathbf{e}_1)$. The use of this relationship leads to several simplifications in the final results of the theory.

Because we have selected for the intermolecular force a sum of impulsive and slowly varying portions, each flux and source term is the sum of an impulsive contribution given by (2.4) and a soft-potential part given by (2.2). Each quantity associated with the impulsive interactions can be separated into a part only involving the Enskog $(E)$ approximation to $f_2$ and a remainder which we label "additional to Enskog." To see how this division of $f_2$ is accomplished, let us focus on the two functions $g_2^{(1)}$ and $g_2^{(2)}$ given by (3.7). From the first of these we take only that part resulting from the action of $\mathscr{L}$ on the factor $f^0 f^0$ of

$$f(\mathbf{x}_1, \mathscr{d}'_1) f(\mathbf{x}_1, \mathscr{d}'_2) = f^0(\mathbf{x}_1, \mathscr{d}'_1) f^0(\mathbf{x}_1, \mathscr{d}'_2)[1 + \phi(\mathbf{x}_1, \mathscr{d}'_1) + \phi(\mathbf{x}_1, \mathscr{d}'_2)].$$

This is the Fourier transform of

$$g_2^{(1)E} = Gf(\mathbf{x}_1, \mathscr{d}'_1) f(\mathbf{x}_1, \mathscr{d}'_2) \tag{3.8}$$

In $g_2^{(2)}$ we only consider contributions from the portions

$$f^0 \left\{ 1 + \left( \frac{3m}{2} \right) \left( \frac{2}{kTI} \right)^{1/2} \mathbf{\Omega} \cdot \boldsymbol{\ell} \right\}$$

of the singlet distribution functions. Then we act with $\mathscr{L}$ on the product $f^0 f^0$ and with $\partial / \partial \mathbf{k}$ on the pair function $G$. Finally, we restrict the action of $\mathbf{V}_2$ to $f^0 f^0$ and to the spin density $\boldsymbol{\ell}$. The result of this is the Fourier transform of the function

$$g_2^{(2)E} = Gf^0(\mathbf{x}_1, \mathscr{d}'_1) f^0(\mathbf{x}_1, \mathscr{d}'_2) \left\{ \frac{\mathbf{x}_{21} \cdot \nabla \mathbf{x}_1 f^0(\mathbf{x}_1, \mathscr{d}'_2)}{f^0(\mathbf{x}_1, \mathscr{d}'_2)} + \left( \frac{2}{kTI} \right)^{1/2} \left( \frac{3m}{2} \right) \mathbf{x}_{21} \mathbf{\Omega}_2 : \nabla \boldsymbol{\ell} \right\} \tag{3.9}$$

with $\mathbf{x}_{21}$ and $G$ evaluated at contact. The two contributions (3.8) and (3.9), together with $f(\mathbf{x}_1, \mathscr{d}'_1) f(\mathbf{x}_2, \mathscr{d}'_2)$, give (to first order in gradients) the Enskog approximation

$$f_2^E(\mathscr{d}_1, \mathscr{d}_2) = g(\mathbf{x}_{21}, \mathbf{e}_1, \mathbf{e}_2) f(\mathbf{x}_1, \mathscr{d}'_1) f(\mathbf{x}_1 + \mathbf{x}_{21}, \mathscr{d}'_2)$$

to the pair distribution function. Were one to retain only this portion of $f_2$, one would obtain the Enskog dense gas theory (for a fluid composed of rigid ellipsoids), which we have already examined.

Next we substitute the expression obtained for $f_2$ into the source and flux terms of the moment equations. The Enskog contributions then can be expressed in terms of the same two sets of integrals, $\{\alpha, \beta\}^{(1)}$ and $[\alpha, \beta]^{(1)}$, occurring in the Enskog theory for ellipsoids. The contributions of the soft potential together with those arising from the additional-to-Enskog terms are combined in the integrals

$$(\alpha, \beta) = I^{-1} \int dV \left[ \left(\frac{m}{kT}\right)^{1/2} F' \int dx_{12}\, \alpha \int_0^\infty dt\, \beta(t) \right.$$
$$\left. + \int d\mathbf{k}\, SH(\mathbf{k} \cdot \boldsymbol{\gamma}_T) \bar\alpha \int_0^\infty dt\, \beta(t) \right]$$

where $F' = F(\mathbf{k} \cdot \boldsymbol{\gamma}_T)^{-1}$, $\beta(t) = \beta(\mathbf{x}_{12} - \mathbf{c}_{12}t, \mathbf{c}_{12}, \mathbf{e}_1 - \dot{\mathbf{e}}_1 t, \mathbf{e}_2 - \dot{\mathbf{e}}_2 t, \boldsymbol{\Omega}_1, \boldsymbol{\Omega}_2)$, and

$$\bar\alpha = \lim_{\tau \to 0^+} \int_t^{t+\tau} \alpha[q(s), p(s)]\, ds$$

is the integral of $\alpha$ over the brief duration of an impulsive collision, and where the arguments of $\beta$ in the second term are selected so that after a time $t$ they become $\boldsymbol{\sigma}_{12}$, $\mathbf{c}_{12}$, $\mathbf{e}_1$, $\mathbf{e}_2$, $\boldsymbol{\Omega}_1$, and $\boldsymbol{\Omega}_2$. The integrand of $(\alpha, \beta)$ can be written in the alternate form

$$\left[ \int dx_{12}\, \alpha \int_0^\infty dt\, e^{-it\mathfrak{L}_0^{(2)}} \beta + \left(\frac{kT}{m}\right)^{1/2} \int_{(\mathbf{k}\cdot\boldsymbol{\gamma}_T > 0)} d\mathbf{k}\, S\mathbf{k} \cdot \boldsymbol{\gamma}_T \bar\alpha \int_0^\infty dt\, e^{-it\mathfrak{L}_0^{(2)}} \beta \right]$$

where $\exp(-it\mathfrak{L}_0^{(2)})$ is the streaming operator which transforms the phases of the two molecules backward in time along force- and torque-free trajectories. Finally, in the second term $\beta$ is to be evaluated at contact.

Here again the moment equations are identical in form to those for rough spheres. However, the transport coefficients and relaxation times now are composed of the soft-potential and additional-to-Enskog contributions as well as the simpler Enskog parts. Thus the formulas for $\tau_p$ and $\eta_K$ are

$$\tau_p^{-1} = -\frac{2n}{5} \left(\frac{kT}{m}\right)^{1/2} \{[\mathbf{WW} \vdots \mathbf{WW}] + (\mathfrak{D} \vdots \tilde{\mathfrak{D}})\}$$

and

$$\eta_K = \tau_p \left\{ \mathscr{P} - \tfrac{1}{5}p_s + ([\mathbf{WW} \vdots \mathbf{xW}_2] - \tfrac{1}{2}(\mathfrak{D} \vdots \tilde{\mathfrak{E}} + \gamma\gamma G) \frac{pn}{5} \left(\frac{2I}{m}\right)^{1/2} \right\}$$

where the symbols $\mathfrak{D}$, $\overset{\rightharpoonup}{\mathfrak{D}}$, and $\overset{\circ}{\mathfrak{D}}$, respectively, stand for the trace, vector dual, and the traceless symmetric part of the tensor $\gamma(\partial V/\partial \mathbf{x})(I/m)^{1/2}$. A tilde above one of these symbols indicates that the potential $V$ is to be replaced with the equilibrium correlation function $G$. The symbols $\mathfrak{E}$, $\overset{\rightharpoonup}{\mathfrak{E}}$, $\overset{\circ}{\mathfrak{E}}$, $\tilde{\mathfrak{E}}$, $\overset{\approx}{\mathfrak{E}}$, and $\overset{\tilde{\circ}}{\mathfrak{E}}$ are analogously associated with the second-rank tensor $\mathbf{x}(\partial V/\partial \mathbf{x})$. Finally, $\mathscr{P}$ is the total pressure, and $p_s$ is the part due to the soft potential alone.

The most distinctive feature of this theory is the appearance in $\mathbf{Q}$ and $\mathbf{C}$ of terms proportional to the gradient of density (or pressure). Although there are no fundamental symmetry conditions (resulting from considerations of time reversal or coordinate frame invariance) that legislate against the existence of such terms, their appearance nevertheless comes as somewhat of a surprise. According to conventional formulations of the thermodynamics of irreversible processes, terms of this nature should not occur. It is interesting to note that in the theory presented here the occurrence of the "anomalous" terms is a dense gas phenomenon. Thus the coefficients $b(\omega)$ and $\delta(\omega)$ of

$$\tilde{\mathbf{Q}} = -\lambda(\omega)\nabla\tilde{T} - \pi(\omega)\nabla \times \tilde{\ell} - b(\omega)\nabla\tilde{n}$$

and

$$\tilde{\mathbf{C}} = -\mu(\omega)\nabla\tilde{T} - v_3(\omega)\nabla \times \tilde{\ell} - \delta(\omega)\nabla\tilde{n}$$

both vanish in the dilute gas limit, and it is in this same limit that the predictions of kinetic theory are well known to agree with those of irreversible thermodynamics. At this moment we still are unsure about these terms, but what we have learned is that they are not unique to our theory nor even to the theory of *poly*atomic fluids. They are an embarrassment which, it appears, others have discovered before us. Thus, in several papers on the kinetic theory of dense monatomic fluids, heat flux contributions proportional to $\nabla n$ or $\nabla p$ have gone unreported only because the authors chose to qualify their results with the provision that pressure or density be held constant. The most recent reference to this term of which we are aware, and one that certainly bears a closer relationship to the present investigation than any other, is that of Dowling and Davis.[47] These investigators found that a heat flux contribution proportional to $\nabla n$ arose quite naturally in their theoretical investigations of transport in weakly coupled systems. However, because they believed this should not be, they speculated that an exact cancelation might occur between the kinetic term they could have evaluated and the interactional term they had not examined. Our results do not appear to support this hypothesis. It is important to note that it is not the moment method that is responsible for these anomalous terms; in the steady state these same results can be obtained by application of the CE method. Further-more, since these terms persist in the limit of weak coupling, they can not be discounted as artefacts resulting from Davis' extension of the PNM theory.

There are further remarkable characteristics of the contribution to the energy flux. It is not specific to polyatomic fluids. The coefficient $b$ vanishes for *every* "hard" or impulsive interaction with center of symmetry coincident with the molecular center of mass, *provided* that one retains only the Enskog approximation to $f_2$. It is for this reason that no heat flux contribution of this form appears in the Enskog dense gas theory for smooth or rough spheres. Certainly, it is reasonable to suspect that these offending contributions to the fluxes of energy and angular momentum are consequences of some inadequacy of our theory. But if so, then several other theories of (monatomic) dense gases and fluids are similarly deficient. We are of course anxious to discover the resolution of this important and hitherto unreported difficulty and are striving in this direction.

This effective potential (EP) theory also can be applied to the square-well ellipsoid model, in which case there appear in addition to the integrals $\{\alpha, \beta\}$ and $[\alpha, \beta]$ others we denote by $\{\alpha, \beta\} = \{\alpha, \beta\}^{(1)} + \{\alpha, \beta\}^{(2)}$ and $[\alpha, \beta] = [\alpha, \beta]^{(1)} + [\alpha, \beta]^{(2)}$. The boldface symbols are defined in the same way as the corresponding Enskog integrals, *except* that the function $\beta = \beta(\mathbf{x}_1, \mathbf{c}_1, \mathbf{e}_1, \mathbf{\Omega}_1; \mathbf{x}_2, \mathbf{c}_2, \mathbf{e}_2, \mathbf{\Omega}_2)$ is replaced with

$$\left(\frac{2kT}{I}\right)^{1/2} \int_0^\infty dt\; \beta(\mathbf{x}_1 - \mathbf{c}_1 t, \mathbf{c}_1, \mathbf{e}_1 - \dot{\mathbf{e}}_1 t, \mathbf{\Omega}_1; \mathbf{x}_2 - \mathbf{c}_2 t, \mathbf{c}_2, \mathbf{e}_2 - \dot{\mathbf{e}}_2 t, \mathbf{\Omega}_2)$$

and the radial distribution function $g^{(1)}$ or $g^{(2)}$ does not appear. Typical of the results one obtains are the formulas

$$\tau_P^{-1} = -\frac{2n}{5}\left(\frac{kT}{m}\right)^{1/2} \{[\mathbf{WW}:\mathbf{WW}] - [\mathbf{WW}\; \vdots \; \tilde{\mathfrak{D}}]\}$$

and

$$\eta_K = \tau_P\left\{p + ([\mathbf{WW}\; \vdots\; \mathbf{xW}_2] + \tfrac{1}{2}[\mathbf{WW}\; \vdots\; \tilde{\mathfrak{C}} + \gamma\gamma G])\frac{pn}{\cdot 5}\left(\frac{2I}{m}\right)^{1/2}\right\}$$

As yet we have not succeeded in perfecting an inexpensive way of evaluating the soft-potential and additional-to-Enskog integrals for polyatomic species. However, we do have a satisfactory procedure for dealing with the corre-sponding integrals occurring in the EP theory for the square-well model of a monatomic fluid, and so we soon shall learn just how large these terms really are.

## F. Extended PNM Theory

Davis' ansatz for the nonequilibrium pair correlation function combines the best features of the RA and PNM theories and, as we just have demonstrated, is also computationally tractable. Indeed, the arguments given in support of this formula stress its desirable qualities rather than the existence of a precisely defined connection between it and Severne's exact relationship [see (3.3)]. It is lack of the latter that casts a semiempirical aura over Davis' EP theory and thus may leave one uncertain and perhaps even a bit uneasy about the nature of the approximations on which it rests. In an effort to remedy this we now try to derive from Severne's exact expression the closest possible equivalent we can to the formula postulated by Davis. To minimize notational complexity we only examine the monatomic theory. Although there are no fundamental obstacles that prevent generalization, the development of the polyatomic theory is much more complicated. Since the procedure followed here is modeled closely after the work of Prigogine, Nicolis, and Misguich, we confine most of our remarks to the new points that arise; for additional details the reader is referred to the paper by PNM.[11]

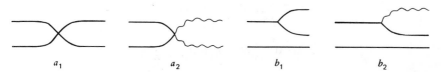

$a_1$ $\qquad$ $a_2$ $\qquad$ $b_1$ $\qquad$ $b_2$

**Fig. 19**  Diagrammatic illustrations of the four types of contributions to the left-most vertex of the creation operator $C^{(s)}$. Categories $a$ and $b$ involve the interaction of two and of three particles, respectively. The contributions labeled with subscripts 1 and 2 (and drawn with straight and wavy lines, respectively) are distinguished by different wave number dependencies. Thus, for example, the waves numbers associated with the upper portions of $b_1$ and $b_2$ may be written as $(\ell + \mathbf{q}_1 + \mathbf{q}_2; \ell - \ell' + \mathbf{q}_1, \ell' + \mathbf{q}_2)$ and as $(\ell + \mathbf{q}_1 + \mathbf{q}_2; \ell + \mathbf{q}_1, \mathbf{q}_2)$, respectively.

Our first step is to separate from the infinite summations (3.3) and (3.4) those features of the creation operator $C^{(s)}$ on which we wish to focus attention. These special items are (1) the inhomogeneities created by the first vertex of the diagonal operator $\psi^{(s)}$ (see Ref. 28) and transferred to $C^{(s)}$, (2) the inhomogeneities appearing in the first propagator of $C^{(s)}$, and (3) the leftmost vertex of $C^{(s)}$. Prigogine, Nicolis, and Misguich furnish an analysis of the various contributions that arise in connection with the leftmost vertex of $C^{(s)}$. These are illustrated in Fig. 19. By writing these terms explicitly and then combining them with the definitions for $f_2$ in terms of $g_2$ and

$$g_3(123) \equiv f_3(123) - f(1)f(2)f(3) + f(1)g_2(23) + f(2)g_2(13) + f(3)g_2(12) \quad \text{(for}$$

the latter use (3.3) but with the sum index $s$ beginning with the value 3 instead of 2) we obtain

$$
\begin{aligned}
g_2(\ell, \mathbf{c}_1, \mathbf{c}_2; \mathbf{x}_1, t) = \int \prod_{j=2}^{N} d\mathbf{x}_j \, \delta(\mathbf{x}_1 - \mathbf{x}_j) \hat{D}\hat{P} \int d\ell' \\
\times \left\{ V_{\ell-\ell'}(\ell - \ell') \cdot \partial_{12} f_2 \left( \ell' + \mathbf{q}_1 + \sum_{j=3}^{\nu} \mathbf{q}_j, \right. \right. \\
\left. - \ell' + \mathbf{q}_2 + \sum_{j=\nu+1}^{N} \mathbf{q}_j, \mathbf{c}_1, \mathbf{c}_2; \{\mathbf{x}_i\}, t \right) \\
+ 8\pi^3 \int dc_3 \left[ V_{\ell-\ell'+\mathbf{q}_1}(\ell - \ell' + \mathbf{q}_1) \cdot \partial_1 g_3 \left( \ell' + \sum_{j=3}^{\nu} \mathbf{q}_j, \right. \right. \\
\left. - \ell + \mathbf{q}_2 + \sum_{j=\nu+1}^{N} \mathbf{q}_j, \ell - \ell' + \mathbf{q}_1, \mathbf{c}_1, \mathbf{c}_2, \mathbf{c}_3; \{\mathbf{x}_i\}, t \right) \\
- V_{\ell-\ell'-\mathbf{q}_2}(\ell - \ell' - \mathbf{q}_2) \cdot \partial_2 g_3 \left( \ell' + \mathbf{q}_1 + \sum_{j=3}^{\nu} \mathbf{q}_j, \right. \\
\left. - \ell' + \sum_{j=\nu+1}^{N} \mathbf{q}_j, - \ell + \ell' + \mathbf{q}_2, \mathbf{c}_1, \mathbf{c}_2, \mathbf{c}_3; \{\mathbf{x}_i\}, t \right) \\
+ \delta(\ell') V_{\ell+\mathbf{q}_1}(\ell + \mathbf{q}_1) \cdot \partial_1 g_3 \left( \sum_{j=3}^{\nu} \mathbf{q}_j, \right. \\
\left. - \ell + \mathbf{q}_2 + \sum_{j=\nu+1}^{N} \mathbf{q}_j, \ell + \mathbf{q}_1, \mathbf{c}_1, \mathbf{c}_2, \mathbf{c}_3; \{\mathbf{x}_i\}, t \right) \\
- \delta(\ell') V_{\ell-\mathbf{q}_2}(\ell - \mathbf{q}_2) \cdot \partial_2 g_3 \left( \ell + \mathbf{q}_1 + \sum_{j=3}^{\nu} \mathbf{q}_j, \right. \\
\left. \left. \left. - \ell + \mathbf{q}_2, \mathbf{c}_1, \mathbf{c}_2, \mathbf{c}_3; \{\mathbf{x}_i\}, t \right) \right] \right\}^{\mathbf{q}_j \to -i\boldsymbol{\nabla}_j}
\end{aligned}
$$

$$ \tag{3.10} $$

where

$$
\hat{D} = \exp\left[ -i\boldsymbol{\nabla}_2 \cdot \frac{\partial}{\partial \ell} - i \sum_{j=\nu+1}^{N} \mathbf{V}_j \cdot \frac{\partial}{\partial \ell} \right],
$$

and

$$
\hat{P} = \left[ \ell \cdot \mathbf{c}_{12} - i\left( \mathbf{c}_1 \cdot \mathbf{V}_1 + \mathbf{c}_2 \cdot \mathbf{V}_2 + \sum_{j=3}^{\nu} \mathbf{c}_1 \cdot \mathbf{V}_j + \sum_{j=\nu+1}^{N} \mathbf{c}_2 \cdot \mathbf{V}_j \right) - io - i\partial_t \right]^{-1}
$$

Integration by parts with respect to the variable $\ell'$ provides a way of transferring the $\mathbf{q}$ wave vector dependence of the vertex to the subsequent infinite sum [this in the context of Severne's relationships; see (3.3) and (3.4)] and at the same time separates from each term of (3.10) the common multiplicative factor $V_{\ell-\ell'}(\ell - \ell')$. By the use of this simple device we obtain for $g_2$ the formula

$$g_2(\ell, \mathbf{c}_1, \mathbf{c}_2; \mathbf{x}_1 t) = \int \prod_{j=2}^{N} d\mathbf{x}_j \, \delta(\mathbf{x}_1 - \mathbf{x}_j) \hat{D}\hat{P} \int d\ell' \, V_{\ell-\ell'}(\ell - \ell')$$

$$\cdot \left\{ \partial_{12} f_2\left( \ell' + \mathbf{q}_1 + \sum_{j=3}^{v} \mathbf{q}_j, \right. \right.$$

$$- \ell' + \mathbf{q}_2 + \sum_{j=v+1}^{N} \mathbf{q}_j, \mathbf{c}_1, \mathbf{c}_2; \{\mathbf{x}_i\} t \right)$$

$$+ 8\pi^3 \int d\mathbf{c}_3 \left[ \partial_1 g_3\left( \ell' + \mathbf{q}_1 + \sum_{j=3}^{v} \mathbf{q}_j, \right. \right.$$

$$- \ell + \mathbf{q}_2 + \sum_{j=v+1}^{N} \mathbf{q}_j, \ell - \ell', \mathbf{c}_1, \mathbf{c}_2, \mathbf{c}_3; \{\mathbf{x}_i\}, t \right)$$

$$- \partial_2 g_3\left( \ell + \mathbf{q}_1 + \sum_{j=3}^{v} \mathbf{q}_j, \right.$$

$$- \ell' + \mathbf{q}_2 + \sum_{j=v+1}^{N} \mathbf{q}_j, - \ell + \ell', \mathbf{c}_1, \mathbf{c}_2, \mathbf{c}_3; \{\mathbf{x}_i\}, t \right)$$

$$+ \delta(\ell' + \mathbf{q}_1)\partial_1 g_3\left( \sum_{j=3}^{v} \mathbf{q}_j, \right.$$

$$- \ell + \mathbf{q}_2 + \sum_{j=v+1}^{N} \mathbf{q}_j, \ell + \mathbf{q}_1, \mathbf{c}_1, \mathbf{c}_2, \mathbf{c}_3; \{\mathbf{x}_i\}, t \right)$$

$$- \delta(\ell' - \mathbf{q}_2)\partial_2 g_3\left( \ell + \mathbf{q}_1 + \sum_{j=3}^{v} \mathbf{q}_j, \right.$$

$$\times \sum_{j=v+1}^{N} \mathbf{q}_j, - \ell + \mathbf{q}_2, \mathbf{c}_1, \mathbf{c}_2, \mathbf{c}_3; \{\mathbf{x}_i\}, t \right) \Bigg] \Bigg\}^{\mathbf{q}_j \to i\nabla_j} \quad (3.11)$$

which is equivalent to (3.3) and (3.4) but far better adapted to the approximation we now introduce.

The first of our two approximations is to assume that the two- and three-body correlation functions which appear on the rhs of (3.11) depend on the

particle coordinates and momenta as they would in a state of local equilibrium. This approximation is expressed by the relationships

$$f_2 \doteq n_2{}^e f(1) f(2) \quad g_2 \doteq h_2{}^e f(1) f(2) \quad \text{and} \quad g_3 \doteq h_3{}^e f(1) f(2) f(3) \quad (3.12)$$

where

$$n_2{}^e = 1 + h_2{}^e(12)$$

and

$$n_3{}^e = 1 + h_2{}^e(12) + h_2{}^e(13) + h_2{}^e(23) + h_3{}^e(123)$$

respectively, denote the local values of the equilibrium ($e$) radial distri-
bution function (to which we previously have assigned the symbol $g = g[\mathbf{x}_{12}, \tfrac{1}{2}(\mathbf{x}_1 + \mathbf{x}_2)]$) and the analogous three-particle function.

Before stating the second of our approximations, let us first distinguish
between two different kinds of position dependence. By definition, the
functions $n_2{}^e$ and $g_2{}^e$ (and similarly $n_3{}^e$ and $g_3{}^e$) exhibit the same dependence
on the interparticle separation $\mathbf{x}_{12}$ as do the radial and pair correlation
functions for a fluid at equilibrium. However, at equilibrium the density
and temperature are uniform, whereas in the present case these quantities
vary from point to point. Therefore, in addition to their microscopic dependence on the interparticle separations $\{\mathbf{x}_{ij}\}$, the correlation functions encountered here also display a macroscopic dependence on position. The
latter is a consequence of the functional dependence of the correlation
functions on the macroscopic variables $n = n(\mathbf{x}_i, t)$ and $T = T(\mathbf{x}_i, t)$. Furthermore, since the singlet distribution function can be expressed uniquely in
terms of moments, of which $n$ and $T$ are but two special examples, the position
dependence of this function can be classified as macroscopic.

With the help of this distinction, we now can describe the second of the
two approximations on which our theory rests. In particular, we disregard
in (3.11) all inhomogeneities connected with the *micro*scopic variations of
the correlation functions. This means that the only $\mathbf{q}$ (wave number) dependence retained is that arising from variations of $n$ and $T$—on which $n_r{}^e$
and $g_r{}^e$ are functionally dependent—and from variations of $f(\mathbf{x}_i\mathbf{c}_i; t)$.

Now we insert the Fourier transform of (3.12) into (3.11) and obtain the
approximate relationship

$$g_2(\ell, \mathbf{c}_1, \mathbf{c}_2; \mathbf{x}_1, t) = \int \prod_{j=2}^{N} d\mathbf{x}_j \, \delta(\mathbf{x}_1 - \mathbf{x}_j) \hat{D}\hat{P} \int d\ell' \, V_{\ell - \ell'}(\ell - \ell')$$

$$\cdot \{n^e_{2\ell'} \partial_{12} + 8\pi^3 n(\mathbf{x}_3)[n_3{}^e(\ell', -\ell, \ell - \ell'; \{\mathbf{x}_i\})\partial_1$$

$$- n_3{}^e(\ell, -\ell', \ell' - \ell; \{\mathbf{x}_i\})\partial_2]\} f(\mathbf{x}_1\mathbf{c}_1) f(\mathbf{x}_2\mathbf{c}_2) \quad (3.13)$$

To arrive at this result, use has been made of the identities

$$\int d\ell' \, V_{\ell-\ell'}(\ell - \ell') h^e_{2\ell'}(12)\delta(\ell - \ell') = 0$$

and

$$\int d\ell' \, V_{\ell'}\ell'[h^e_{2\ell'}(13)\partial_1 - h^e_{2\ell'}(23)\partial_2] = 0$$

The final step of our development requires us to express $n_3{}^e$ in terms of $n_2{}^e$. For this we cannot use the Born–Green–Yvon connection (as was done in Ref. 45), because the latter is specific to a system in a uniform state of thermodynamical equilibrium whereas we are concerned here with local equilibrium. To obtain the desired relationship, we follow closely the procedures familiar from ordinary equilibrium statistical mechanics but select as our starting point the local functions defined by

$$n_s{}^e(\{x_{ij}\}; \{x_i\}) = (N - s + 1)! \, \frac{\int dx^{N-s} \, e^{U_N} \prod_{i=1}^{N-s} n_i}{\int dx^N \, e^{U_N} \prod_{i=1}^{N} n_i} \qquad (3.14)$$

with

$$U_N = -\sum_i^N \sum_{j \neq i}^N \beta_{ij} V_{ij}(x_{ij}), \quad \beta_{ij} = \frac{1}{2}\left(\frac{1}{kT_i} + \frac{1}{kT_j}\right)$$

and where $n_i = n(x_i)$ and $T_i = T(x_i)$. To derive the connection between $n_3{}^e$ and $n_2{}^e$, we differentiate the latter with respect to the microscopic variables on which it depends. This operation does not introduce derivatives of $n$ and $T$, because these two functions only exhibit macroscopic variations with position. By proceeding in this manner one can obtain the identity

$$\ell n^e_{2\ell'} \cdot (-\beta_{13}^{-1}\partial_1 + \beta_{23}^{-1}\partial_2) + \int d\ell' \, V_{\ell-\ell'}(\ell - \ell')n^e_{2\ell'} \cdot$$

$$\cdot \left(-\frac{\beta_{12}}{\beta_{13}} \partial_1 + \frac{\beta_{12}}{\beta_{23}} \partial_2\right) = 8\pi^3 n(x_3) \int d\ell' \, V_{\ell-\ell'}(\ell - \ell')$$

$$\cdot [n_3{}^e(\ell', -\ell, \ell - \ell')\partial_1 - n_3{}^e(\ell, \ell', \ell' - \ell)\partial_2] \qquad (3.15)$$

which in the limit of absolute equilibrium (where $\beta_{ij} = \beta$; $V_{i,j}$) reduces to the first equation of the BGY hierarchy. The result of substituting this into (3.3) is the formula

$$g_2(\ell, \mathbf{c}_1, \mathbf{c}_2; \mathbf{x}_1, t) = \int \prod_{j=2}^{N} d\mathbf{x}_j \, \delta(\mathbf{x}_1 - \mathbf{x}_j) \hat{D}\hat{P}\Bigg\{ \ell n_{2\ell}^e \cdot (-\beta_{13}^{-1}\partial_1 + \beta_{23}^{-1}\partial_2)$$

$$+ \int d\ell' \, V_{\ell - \ell'}(\ell - \ell') n_{2\ell'}^e \Bigg[ \left( 1 - \frac{\beta_{12}}{\beta_{13}} \right)\partial_1$$

$$- \left( 1 - \frac{\beta_{12}}{\beta_{23}} \right)\partial_2 \Bigg] \Bigg\} f(\mathbf{x}_1 \mathbf{c}_1) f(\mathbf{x}_2 \mathbf{c}_2) \qquad (3.16)$$

where the dependence of $n_{2\ell}^e$ on the macroscopic variables $\{\mathbf{x}_i\}$ is shown explicitly by (3.14).

From this expression for $g_2$ we shall construct a much simpler function $g_2^{tr}$, which is suitable for evaluating the linear phenomenological coefficients. As a first step in this direction we observe that only the parts

$$\exp\left[ -i\mathbf{V}_2 \cdot \frac{\partial}{\partial \ell} \right]$$

and

$$\left[ \ell \cdot \mathbf{c}_{12} - i\left( \mathbf{c}_1 \cdot \mathbf{V}_1 + \mathbf{c}_2 \cdot \mathbf{V}_2 + \mathbf{G} \cdot \sum_{j=3}^{N} \mathbf{V}_j \right) - io - i\partial_t \right]^{-1},$$

where $\mathbf{G} = \frac{1}{2}(\mathbf{c}_1 + \mathbf{c}_2)$, of $\hat{D}$ and $\hat{P}$ will contribute to $g_2^{tr}$. Next we perform a linearization and write the result as the sum of three parts:

$$g_2^{tr} = g_2^{(1)} + g_{2s} + g_{2pr}$$

where $g_{2s}$ and $g_{2pr}$ are associated with inhomogeneities arising from the vertex $\hat{D}$ and the propagator $\hat{P}$, respectively; according to Davis' notation $g_{2s} = g_2^{(2)}$ and $g_{2pr} = g_2^{(3)} + g_2^{(4)}$. These three terms are given explicitly by

$$g_2^{(1)}(\ell, \mathbf{c}_1, \mathbf{c}_1; \mathbf{x}_1, t) = \frac{1}{\ell \cdot \mathbf{c}_{12} - io} \ell \cdot \partial_{12}(-kTn_{2\ell}^e)_{\mathbf{x}_1} f(\mathbf{x}_1, \mathbf{c}_1) f(\mathbf{x}_1, \mathbf{c}_2) \qquad (3.17)$$

$$g_{2s}(\ell, \mathbf{c}_1, \mathbf{c}_2; \mathbf{x}_1, t) = \int \prod_{j=2} d\mathbf{x}_j \, \delta(\mathbf{x}_1 - \mathbf{x}_j)$$

$$\left( -i\mathbf{V}_2 \cdot \frac{\partial}{\partial \ell} \right)\frac{1}{\ell \cdot \mathbf{c}_{12} - io} A_\ell(\mathbf{c}_1, \mathbf{c}_2; \mathbf{x}_1, \mathbf{x}_2, t)$$

and

$$g_{2pr}(\ell, \mathbf{c}_1, \mathbf{c}_2; \mathbf{x}_1, t) = \int \prod_{j=2} d\mathbf{x}_j \, \delta(\mathbf{x}_1 - \mathbf{x}_j)$$

$$\times i \frac{\mathbf{c}_1 \cdot \mathbf{V}_1 + \mathbf{c}_2 \cdot \mathbf{V}_2 + \partial_t + \mathbf{G} \cdot \sum_{j=3}^{N} \mathbf{V}_j}{(\ell \cdot \mathbf{c}_{12} - io)^2} A_\ell(\mathbf{c}_1, \mathbf{c}_2; \mathbf{x}_1, \mathbf{x}_2, t)$$

where

$$A_\ell(\mathbf{c}_1, \mathbf{c}_2; \mathbf{x}_1, \mathbf{x}_2, t) = \ell n_{2\ell}^e \cdot (-\beta_{13}^{-1}\partial_1 + \beta_{23}^{-1}\partial_2)$$

$$+ \int d\ell' \, V_{\ell - \ell'}(\ell - \ell') n_{2\ell'}^e \left[\left(1 - \frac{\beta_{12}}{\beta_{23}}\right)\partial_1 - \left(1 - \frac{\beta_{12}}{\beta_{23}}\right)\partial_2\right] f(\mathbf{x}_1, \mathbf{c}_1) f(\mathbf{x}_2, \mathbf{c}_2)$$

In $g_{2s}$ and $g_{2pr}$ we perform the indicated differentiations, retaining only those terms that are relevant to the transport problem. This results in the expressions

$$g_{2s} = -i\frac{\partial}{\partial \ell} \cdot \left[\frac{1}{\ell \cdot \mathbf{c}_{12} - io} (-kTn_{2\ell}^e)\ell \cdot \partial_{12}\right][\mathbf{V}_2 f(\mathbf{x}_1, \mathbf{c}_1) f(\mathbf{x}_2, \mathbf{c}_2)]_{\mathbf{x}_2 = \mathbf{x}_1}$$

$$+ i\frac{\partial}{\partial \ell} \cdot \left\{\frac{1}{\ell \cdot \mathbf{c}_{12} - io} \frac{1}{2}\left[(-kTn_{2\ell}^e)\ell - \int d\ell' \, V_{\ell - \ell'}(\ell - \ell') n_{2\ell'}^e\right]\right.$$

$$\left. \cdot \partial_1 f(\mathbf{x}_1 \mathbf{c}_1) f(\mathbf{x}_1 \mathbf{c}_2)\right\} \cdot \mathbf{V} \ln T \qquad (3.18)$$

and

$$g_{2pr} = \frac{i}{(\ell \cdot \mathbf{c}_{12} - io)^2} \mathbf{G} \cdot [-kT\mathbf{V}n_{2\ell}^e + (-kTn_{2\ell}^e)\mathbf{V} \ln T]$$

$$\ell \cdot \partial_{12} f(\mathbf{x}_1, \mathbf{c}_1) f(\mathbf{x}_2, \mathbf{c}_2)$$

$$+ \frac{i}{(\ell \cdot \mathbf{c}_{12} - io)^2} (-kTn_{2\ell}^e)[(\mathbf{c}_1 \cdot \mathbf{V}_1 + \mathbf{c}_2 \cdot \mathbf{V}_2 + \partial_t)]$$

$$\ell \cdot \partial_{12} f(\mathbf{x}_1 \mathbf{c}_1) f(\mathbf{x}_2 \mathbf{c}_2)]_{\mathbf{x}_2 = \mathbf{x}_1}$$

$$+ \frac{i}{(\ell \cdot \mathbf{c}_{12} - io)^2} \frac{1}{4}[(-kTn_{2\ell}^e)\ell - \int d\ell' \, V_{\ell - \ell'}(\ell - \ell') n_{2\ell'}^e]$$

$$\cdot (\mathbf{c}_{12} \cdot \mathbf{V} \ln T)(\partial_1 + \partial_2) f(\mathbf{x}_1 \mathbf{c}_1) f(\mathbf{x}_1 \mathbf{c}_2) \qquad (3.19)$$

where use has been made of the relationship

$$\left[\sum_{j=1}^{N} \mathbf{V}_j n_{2\ell}^e\right]_{\mathbf{x}_1 = \mathbf{x}_2 = \cdots = \mathbf{x}_N} = \left(\frac{\partial n_{2\ell}^e}{\partial T}\right)_n \mathbf{V}T + \left(\frac{\partial n_{2\ell}^e}{\partial n}\right)_T \mathbf{V}n = \mathbf{V}n_{2\ell}^e$$

The formulas (3.17), (3.18), (3.19) constitute our final expression for $g_2^{tr}$. They are analogous to the results (2.4–7) given by Davis and are related to them in the manner

$$g_2^{tr} = g_2^{D(tr)} + g_2^{cor}$$

with $g_2^{cor} = g_{2s}^{cor} + g_{2pr}^{cor}$,

$$g_{2s}^{cor} = i \frac{\partial}{\partial \ell} \cdot \left\{ \frac{1}{\ell \cdot \mathbf{c}_{12} - io} \frac{1}{2} \left[ (-kTn_{2\ell}^e)\ell - \int d\ell'\, V_{\ell-\ell'}(\ell - \ell')n_{2\ell'}^e \right] \right.$$
$$\left. \cdot \partial_1 f(\mathbf{x}_1 \mathbf{c}_1) f(\mathbf{x}_1 \mathbf{c}_2) \right\} \cdot \nabla \ln T$$

and

$$g_{2pr}^{cor} = \frac{i}{(\ell \cdot \mathbf{c}_{12} - io)^2} \frac{1}{4} \left[ (-kTn_{2\ell}^e)\ell - \int d\ell'\, V_{\ell-\ell'}(\ell - \ell')n_{2\ell'}^e \right]$$
$$\cdot (\mathbf{c}_{12} \cdot \nabla \ln T)(\partial_1 + \partial_2) f(\mathbf{x}_1 \mathbf{c}_1) f(\mathbf{x}_1 \mathbf{c}_2)$$

Here, $g_2^{D(tr)}$ denotes the portion of Davis' pair correlation function that contributes to the linearized description of transport phenomena.

Our theory differs from PNM's in that we have treated in a symmetric way the contributions arising from the vertex and the propagator, whereas PNM dealt only with the lowest-order approximation to the latter. But if this is so, then the term $g_{2s}^{cor}$ should appear in the theory of PNM. And this really is the case for, although this term was missing from the original derivation, it since has been discovered by Misguich.[12] It is interesting to note that, had we used in place of (3.15) the analogous relationship which is valid for the case of absolute equilibrium, we then would have found that $g_2^{cor}$ was zero and so obtained Davis' formula for $g_2$.

A final remark is in order. Silin[48] obtained the weak-coupling approximation to $g_2$ directly from the BBGKY hierarchy by retaining only the lowest-order terms in the pair potential. A similar derivation is possible here. We begin with the equation

$$(\partial_t + \mathbf{c}_1 \cdot \nabla_1 + \mathbf{c}_2 \cdot \nabla_2)g_2 = \mathbf{F}_{12} \cdot \partial_{12} f_2 + \int d\mathbf{x}_3\, d\mathbf{c}_3\, (\mathbf{F}_{13} \cdot \partial_1 + \mathbf{F}_{23} \cdot \partial_2) f_3$$
$$- \int d\mathbf{x}_3\, d\mathbf{c}_3\, [\mathbf{F}_{13} \cdot \partial_1 f_2(13)f(2) + \mathbf{F}_{23} \cdot \partial_2 f_2(23)f(1)]$$

which can be obtained from the first two members of the BBGKY hierarchy. Here, $\mathbf{F}_{ij} = -\partial V_{ij}/\partial \mathbf{x}_{ij}$ is the force on molecule $i$ due to its interaction with molecule $j$. We apply our two basic approximations to the rhs of this equation and obtain the result

$$(\partial_t + \mathbf{c}_1 \cdot \mathbf{V}_1 + \mathbf{c}_2 \cdot \mathbf{V}_2)g_2 = \mathbf{F}_{12} \cdot \partial_{12} n_2{}^e f(1)f(2)$$

$$+ \int d\mathbf{x}_3 \, d\mathbf{c}_3 \, (\mathbf{F}_{13} \cdot \partial_1 + \mathbf{F}_{23} \cdot \partial_2)n_3{}^e f(1)f(2)f(3)$$

From this we eliminate $n_3{}^e$ by the use of (3.15). The final step consists of integrating the resulting equation along a trajectory characteristic of two noninteracting particles, that is, by acting on the equation with the inverse of the operator $\partial_t + \mathbf{c}_1 \cdot \mathbf{V}_1 + \mathbf{c}_2 \cdot \mathbf{V}_2$. This leads to the formula

$$g_2(\mathbf{x}_1, \mathbf{x}_2, \mathbf{c}_1, \mathbf{c}_2, t) = \int_0^\infty d\tau \, e^{-\tau(\mathbf{c}_1 \cdot \mathbf{V}_1 + \mathbf{c}_2 \cdot \mathbf{V}_2 + \partial_t)} \left\{ \frac{\partial n_2{}^e}{\partial \mathbf{x}_{12}} \cdot (-\beta_{13}^{-1}\partial_1 + \beta_{23}^{-1}\partial_2) \right.$$

$$\left. + n_2{}^e \frac{\partial V_{12}}{\partial \mathbf{x}_{12}} \cdot \left[ \left(1 - \frac{\beta_{12}}{\beta_{13}}\right)\partial_1 - \left(1 - \frac{\beta_{12}}{\beta_{23}}\right)\partial_2 \right] \right\} f(\mathbf{x}_1 \mathbf{c}_1; t)f(\mathbf{x}_2 \mathbf{c}_2; t) \quad (3.20)$$

which reduces to Silin's result in the limit ($n_2{}^e \approx \exp - \beta_{12}V_{12} \approx 1 - \beta_{12}V_{12}$) of weak coupling. Beginning with (3.20) we now may invoke the simplifications appropriate to the linear transport problem and so rederive our earlier results in inverted form.

Although this extension of the PNM theory is certainly not exact, the approximations on which it is based seem to us to be eminently reasonable. This bolsters our confidence in Davis' effective potential theory and so also in the generalization of that theory which we have developed for polyatomic fluids. Until more has been learned about the numerical importance of $g_2^{\text{cor}}$, there is little to be gained from an examination of the extended PNM theory for polyatomic species. Therefore we refrain from doing so here.

### G.  Brinser and Condiff's Generalization of the Enskog Theory

This is the final example we consider of theories of the pair distribution function that have been specially designed to deal with the nonequilibrium properties of dense fluids. Although the Brinser–Condiff (BC)[29] theory is similar in several respects to the extended PNM theory—a resemblance that becomes particularly apparent when one compares BC's development with our Silin-type derivation of the preceding section—the two certainly are not equivalent. The originators of this theory have described it as a soft potential analog of the Enskog theory which is obtained by neglecting dynamical correlations. And the extended PNM theory could be described in much the same words. The difference lies in the precise manner by which this approximation is introduced into the analysis. BC draw attention to the calculations performed by Condiff and his colleagues[49] and by Gillespie and Sengers[50] for a dense gas of rigid spheres. These studies revealed that the difficult-to-calculate many-body dynamical correlations produced corrections to the Enskog values for the transport coefficients which amounted to only a few percent. They then conjectured that these same dynamical correlations would be no more important for soft-potential models of the liquid than they were for rigid spheres. Therefore they set out to develop a soft-potential theory which would incorporate the neglect of dynamical correlations in such a way that the theory would reduce to that of Enskog in the limit of rigid spheres. This seems a very sensible thing to do. Unlike the EP theory which is chosen to reduce properly in the (rigorous) limit of weak coupling, the BC theory is designed to converge on the approximate but highly successful Enskog theory for a strongly coupled system. In the following discussion we apply the methods of BC to a fluid of rigid rotors. How the theory can be generalized to other polyatomic models will become apparent.

The theory begins with the first and second members of the BBGKY hierarchy, which are written here in the forms

$$(\partial_t + \mathscr{L}_1)f(1) = J_{12}[f_2(12)]$$
$$(\partial_t + \mathscr{L}_{12})f_2(12) = J_{13}[f_3(123)] + J_{23}[f_3(123)]$$

(3.21)

where $\mathscr{L}_1 = \mathbf{c}_1 \cdot \mathbf{V}_1 + \mathscr{L}_1^r$, $\mathscr{L}_{12} = \mathscr{L}_1 + \mathscr{L}_2 + \hat{\theta}_{12}$, and $\hat{\theta}_{12} = -\mathbf{V}_1 V_{12} \cdot \partial_{12} + \hat{\theta}_{12}^r$. For the rotational ($r$) parts of these operators, we can use action-angle variables, in which case

$$\mathscr{L}_1^r = \frac{L_1}{I} \frac{\partial}{\partial a_1}$$

$$\hat{\theta}_{12}^r = \sum_{i=1,2} \left( \frac{\partial V_{12}}{\partial a_i} \frac{\partial}{\partial L_i} + \frac{\partial V_{12}}{\partial b_i} \frac{\partial}{\partial L_{3i}} - \frac{\partial V_{12}}{\partial L_i} \frac{\partial}{\partial a_i} - \frac{\partial V_{12}}{\partial L_{3i}} \frac{\partial}{\partial b_i} \right)$$

or polar spherical coordinates and angular momenta for which

$$\mathscr{L}_1^r = \boldsymbol{\omega}_1 \cdot \mathbf{e}_1 \times \frac{\partial}{\partial \mathbf{e}_1}, \quad \hat{\theta}_{12}^r = \mathbf{N}_{12} \cdot \frac{\partial}{\partial \mathbf{L}_2} + \mathbf{N}_{21} \cdot \frac{\partial}{\partial \mathbf{L}_1}$$

Depending on which of these choices is made, $J_{ij}[\psi]$ will be given either by $-\int d\mathbf{x}_{ij} \, d\mathbf{c}_j \, (da_j \, db_j \, dL_j \, dL_{3j}) \hat{\theta}_{ij}(\psi)$ or by $-\int d\mathbf{x}_{ij} \, d\mathbf{c}_j \, (d\mathbf{e}_j \, d\mathbf{L}_j) \hat{\theta}_{ij}(\psi)$.

BC write the nonequilibrium pair and triplet distribution functions in the forms

$$f_2(12) = \chi_2(12)[f(1)f(2) + \gamma_2(12)]$$
$$f_3(123) = \chi_3(123)[f(1)f(2)f(3) + f(1)\gamma_2(23) + f(2)\gamma_2(13) \quad (3.22)$$
$$+ f(3)\gamma_2(12) + \gamma_3(123)]$$

which, should be contrasted with the expressions

$$f_2(12) = f(1)f(2) + g_2(12)$$
$$f_3(123) = f(1)f(2)f(3) + f(1)g_2(23) + f(2)g_2(13)$$
$$+ f(3)g_2(12) + g_3(123)$$

of Severne, PNM, and Davis. The function $\chi_2(12)$ is selected by BC to equal the corresponding equilibrium function ($n_2^{\,e}$ of Section III.F) evaluated for the values of density and temperature associated with the point $\mathbf{R} = \frac{1}{2}(\mathbf{x}_1 + \mathbf{x}_2)$ midway between the two molecules. $\chi_3(123)$ is similarly defined. Because of this, the functions $\gamma_2$ and $\gamma_3$ vanish for a state of equilibrium.

By substituting (3.22) into (3.21) one obtains the relationship

$$\chi_2(\partial_t + \mathscr{L}_{12})\gamma_2 = -\chi_2 \hat{\theta}_{12} f(1)f(2) - [f(1)f(2) + \gamma_2](\partial_t + \mathscr{L}_{12})\chi_2$$
$$+ (K_{12} + K_{21}) \tag{3.23}$$

where

$$K_{ij} = J_{i3}[\chi_3(ij3)\{\gamma_3(ij3) + f(i)\gamma_2(j3) + f(3)\gamma_2(ij)\}$$
$$+ \{\chi_3(ij3) - \chi_2(ij)\chi_2(i3)\} \{f(i)f(3) + \gamma_2(i3)\} f(j)]$$

Next comes the assumption that $\chi_2(12) \equiv \rho_2(12) \exp(-\beta V_{12})$ [where $\beta = \beta(\mathbf{R})$ and $\rho_2 = \rho_2(\mathbf{R}, \mathbf{x}_{12}, \mathbf{e}_1, \mathbf{e}_2)$] and $\chi_3$ are related to one another by the expression

$$\left(\mathbf{c}_{12} \cdot \frac{\partial}{\partial \mathbf{x}_{12}} + \mathscr{L}_1^{\,r} + \mathscr{L}_2^{\,r}\right)\chi_2 = \rho_2\left(\mathbf{c}_{12} \cdot \frac{\partial}{\partial \mathbf{x}_{12}} + \mathscr{L}_1^{\,r} + \mathscr{L}_2^{\,r}\right)e^{-\beta V_{12}}$$
$$+ J_{13}[\chi_3 f^0(3)] + J_{23}[\chi_3 f^0(3)]$$

which is the form taken by the second of the BBGKY equations when the dynamical correlation functions $\gamma_2$ and $\gamma_3$ are set equal to zero, when the singlet distribution functions are replaced with Maxwell–Boltzmann functions, and when one ignores the space and time variations of density and temperature. We previously have envoked a strikingly similar assumption in our derivation of the extended PNM theory. When this is substituted into (3.23), the result can be expressed in the form

$$(\partial_t + \mathscr{L}_{12})\gamma_2 e^{-\beta V_{12}} = \gamma_2(\partial_t + \mathscr{L}_{12})e^{-\beta V_{12}} - \hat{\theta}_{12}f(1)f(2)$$

$$- [f(1)f(2) + \gamma_2]e^{-\beta V_{12}}\left(\partial_t + \mathbf{G} \cdot \frac{\partial}{\partial \mathbf{R}} + \hat{\theta}_{12}\right)\ln \chi_2$$

$$- [f(1)f(2) + \gamma_2]\left(-\mathbf{G} \cdot \frac{\partial}{\partial \mathbf{R}} + \mathscr{L}_1^{\,r} + \mathscr{L}_2^{\,r}\right)e^{-\beta V_{12}}$$

$$+ (C_{12} + C_{21}) \tag{3.24}$$

where $\mathbf{G} = \frac{1}{2}(\mathbf{c}_1 + \mathbf{c}_2)$ and

$$
\begin{aligned}
C_{ij} = \rho_2^{-1} J_{i3}\{\chi_3[\gamma_3 &+ f(i)\gamma(j3)] + [\chi_3 - \chi_2(ij)\chi_2(i3)]f(j)\gamma(i3) \\
&+ \chi_3[f(i)f(j) + \gamma_2(ij)][f(3) - f^0(3)] \\
&- \chi_2(ij)f(i)f(j) \cdot \chi_2(i3)f(3)\}
\end{aligned}
$$

In order to treat properly the explicit time dependence, we use the relationship

$$
\begin{aligned}
\hat{\theta}_{12}e^{-\beta V_{12}}f(1)f(2) &= [(\partial_t + \mathscr{L}_{12}) - (\partial_t + \mathscr{L}_{12} - \hat{\theta}_{12})]e^{-\beta V_{12}}f(1)f(2) \\
&= (\partial_t + \mathscr{L}_{12})e^{-\beta V_{12}}f(1)f(2) - f(1)f(2)(\partial_t + \mathscr{L}_{12} - \hat{\theta}_{12})e^{-\beta V_{12}} \\
&\qquad - e^{-\beta V_{12}}(\partial_t + \mathscr{L}_{12} - \hat{\theta}_{12})f(1)f(2)
\end{aligned}
$$

instead of the expression used by BC. With the aid of this we now obtain, in place of (3.24),

$$
(\partial_t + \mathscr{L}_{12})\gamma_2 e^{-\beta V_{12}} = \mathscr{G}_{12}(t) \tag{3.25}
$$

with

$$
\begin{aligned}
\mathscr{G}_{12}(t) = &-(\partial_t + \mathscr{L}_{12})f(1)f(2)\exp(-\beta V_{12}) \\
&+ \mathscr{H}_{12}(t) + C_{12}(t) + C_{12}(t) + C_{21}(t)
\end{aligned}
$$

and

$$
\begin{aligned}
\mathscr{H}_{12}(t) = &f(1)f(2)\left(\mathbf{G} \cdot \frac{\partial}{\partial \mathbf{R}}\right)e^{-\beta V_{12}} + e^{-\beta V_{12}}(\partial_t + \mathscr{L}_{12} - \hat{\theta}_{12})f(1)f(2) \\
&+ \gamma_2\left(\partial_t + \mathbf{G} \cdot \frac{\partial}{\partial \mathbf{R}}\right)e^{-\beta V_{12}} - [\gamma_2 + f(1)f(2)]e^{-\beta V_{12}}\left(\partial_t + \mathbf{G} \cdot \frac{\partial}{\partial \mathbf{R}}\right)\ln \chi_2
\end{aligned}
$$

From this and the first of the equations it follows that

$$
f_2 = \rho_2\left[f(1)f(2)e^{-\beta V_{12}} + \int_0^\infty dt'\, e^{-t'(\partial_t + \mathscr{L}_{12})}\mathscr{G}_{12}(t)\right]
$$

To proceed beyond this formally exact result BC discard, from the integrand factor $\mathscr{G}_{12}(t)$ the terms $C_{12}$ and $C_{21}$ which account for dynamical correlations. Then, since

$$
\begin{aligned}
e^{-t'(\partial_t + \mathscr{L}_{12})}f(1)f(2)e^{-\beta V_{12}} &= [f(1)f(2)e^{-\beta V_{12}}]_{t-t'} \\
&= f(\mathbf{x}_1', \boldsymbol{\sigma}_1'; t - t')f(\mathbf{x}_2', \boldsymbol{\sigma}_2'; t - t')e^{-\beta(\mathbf{R}_{12}')V_{12}(\mathbf{x}_{12}', \mathbf{e}_1', \mathbf{e}_2')} \\
&= f(1')f(2')e^{-\beta V_{12}} + e^{-\beta V_{12}}\{(\mathbf{x}_1' - \mathbf{x}) \cdot [\nabla f(1')]f(2') \\
&\qquad + (\mathbf{x}_2' - \mathbf{x}) \cdot [\nabla f(2')]f(1') - (\mathbf{R}' - \mathbf{x}) \cdot (\nabla\beta)V_{12}'f(1')f(2') + \cdots\}
\end{aligned}
$$

where $\mathbf{x}_1' = \mathbf{x}_1(t - t'), \ldots, f(i') = f(\mathbf{x}, \sigma_1'(t))$ and $\mathbf{x} = \mathbf{x}_1(t)$, we see that

$$\int_0^\infty dt' \, e^{-t'(\partial_t + \mathscr{L}_{12})} \mathscr{G}_{12}(t) \simeq f'(1')f(2')e^{-\beta V_{12}} \big|_{t'=0}^{t'=\infty}$$

$$+ \int_0^\infty dt' \left( \frac{D}{Dt'} \{ e^{-\beta V_{12}} [(\mathbf{x}_1' - \mathbf{x}) \cdot f(2')\nabla f(1') + (\mathbf{x}_2' - \mathbf{x}) \cdot f(1')\nabla f(2') \right.$$

$$\left. + (t' - t)\partial_t f(1')f(2') - (\mathbf{R}' - \mathbf{x}) \cdot (\nabla \beta) V_{12}' f(1')f(2') \} + \mathscr{H}_{12}(t') \right)$$

The symbol $D\phi/Dt'$ denotes the total derivative of a function $\phi$ which depends on time and the dynamical states of two molecules. Here and henceforth all singlet distribution functions that appear are evaluated with their spatial coordinate set equal to $\mathbf{x}$.

The first term of this last expression may be written as

$$f(1^*)f(2^*) - f(1)f(2) \exp(-\beta V_{12}),$$

where the asterisks refer to values of the dynamical variables prior to the commencement of a hypothetical isolated binary collision. The remainder of the manipulations involved in the BC theory have to do with identifying and discarding from this expression contributions of second or higher order in the derivatives with respect to time and position. We shall not concern ourselves with the details of this but give only the final result,

$$f_2(1, 2; t) = \rho_2 f(1^*)f(2^*) + \mathbf{x}_{21} \cdot [\tfrac{1}{2} f(1)f(2)\nabla \chi_2 + \chi_{12} f(1)\nabla f(2)]$$

$$+ \rho_2 \int_0^\infty dt' \, \frac{1}{2} \mathbf{x}_{21}' \cdot \frac{D}{Dt'} \{ e^{-\beta V_{12}} [f(1')\nabla f(2') - f(2')\nabla f(1')] \}$$

$$- \rho_2 \int_0^\infty dt' \, f(1')f(2') e^{-\beta V_{12}} \left( \partial_t + \mathbf{G}_{12}' \cdot \frac{\partial}{\partial \mathbf{R}_{12}'} \right) \ln \chi_2$$

$$+ \rho_2 \int_0^\infty dt' \, e^{-\beta V_{12}} (\mathscr{L}_1^{r} + \mathscr{L}_2^{r}) f(1')f(2') \qquad (3.26)$$

This is the generalization of the BC theory to a dense fluid of rigid rotors. Therefore we expect (3.26) to reduce exactly to the Enskog dense gas theory for rigid ellipsoids. And, indeed, by performing manipulations analogous to those of BC, we find that the contribution of the first three terms of (3.26) to the impulsive interaction limit of $J_{12}[f_2]$ is the (linearized approximation to) Enskog result. Next, just as in BC, the fourth term of (3.26) is found to contribute nothing to the transport theory. The last term of (3.26) provides a contribution to $J_{12}[f_2]$ which does *not* vanish identically in the rigid ellipsoid limit but does for rigid spheres. However, if one is interested only in the usual transport theory and uses for $f$ the truncated moment expansion (2.1), this term is of no consequence. However, if one includes in the expansion a term such as $\mathbf{ee} \times \boldsymbol{\Omega}$, one then will obtain a contribution from the last term of (3.26).

If we select for $f$ the truncated expansion (2.1), the moment equations are the same as those presented previously for rough spheres. The coefficients appearing in the equations for $\mathsf{P}_K$, $\mathbf{C}_K$ and $C_K$ are given by

$$\tau_P^{-1} = \frac{2n}{5}[\mathbf{WW} \vdots \mathbf{WW}]_{BC}$$

$$\eta_K = \tau_P\left(\mathscr{P} - p_V - \frac{n^2}{2}[\mathbf{WW} \vdots \mathbf{x}_{12}\,\mathbf{F}_{12}]_{BC}\right)$$

$$\tau_c^{-1} = 2n[\mathbf{W} \cdot \boldsymbol{\Omega}, \mathbf{W} \cdot \boldsymbol{\Omega}]_{BC}$$

$$\tau_c^{-1} = \frac{6n}{5}[\mathbf{W\Omega} \vdots \mathbf{W\Omega}]_{BC}$$

$$v_{K1} = \tau_c\left(p - p_V + \frac{1}{2}\left(\frac{m}{I}\right)^{1/2}n^2[\mathbf{W} \cdot \boldsymbol{\Omega}, \mathbf{x}_{21} \cdot (\mathbf{N}_{12} - \mathbf{N}_{21})]_{BC}\right.$$

$$\left. - \frac{1}{2}(2kTm)^{1/2}n^2\{\mathbf{xW} \cdot \boldsymbol{\Omega}; \boldsymbol{\Omega}\}_{BC}\right)$$

$$v_{K2} = \tau_c^{-1}\left(p - p_V + \frac{3}{10}\left(\frac{m}{I}\right)^{1/2}n^2[\mathbf{W\Omega} \vdots \mathbf{x}_{21}(\mathbf{N}_{12} - \mathbf{N}_{21})]_{BC}\right.$$

$$\left. - \frac{3}{10}(2kTm)^{1/2}n^2\{\mathbf{x} \cdot \mathbf{W\Omega}; \boldsymbol{\Omega}\}_{BC}\right)$$

and may be compared with the analogous formulas of the Enskog and EP theories listed in Refs. 1 and 2. The integrals appearing in these expressions are defined by

$$[\alpha, \beta]_{BC} = n^{-2}\int dU\, f^0(1)f^0(2)\chi_2(\hat{\theta}_{12}\alpha)(\beta_1^* + \beta_2^*)$$

$$[\alpha, \beta]_{BC} = n^{-2}\int dU\, f^0(1)f^0(2)\chi_2(\hat{\theta}_{12}\alpha)\left(\int_0^\infty dt\, \beta(t)\right)$$

and

$$\{\mathbf{x}\alpha, \beta\}_{BC} = n^{-2}\int dU\, f^0(1)f^0(2)\mathbf{x}_{12}(\hat{\theta}_{12}\alpha)(\beta_1^* + \beta_2^*)$$

where

$$dU = d\mathbf{x}_{12}\left(\prod_{i=1,\,2} d\mathbf{c}_i\, d\mathbf{L}_i\, d\mathbf{e}_i\right).$$

The coupled equations satisfied by $\mathbf{C}_K$, $\mathbf{Q}_K{}^t$ and $\mathbf{Q}_K{}^r$ may be written in the matrix form $\partial_t a + Ma = Nd$, where $a$ and $d$ are the column matrices $(\mathbf{C}_K, \mathbf{Q}_K{}^t, \mathbf{Q}_K{}^r)$ and $(\mathbf{V} \times \boldsymbol{\ell}, \mathbf{V}T, \mathbf{V}n)$, respectively. $M$ and $N$ denote $3 \times 3$ matrices of which typical elements are given by

$$M_{11} = 2n[\mathbf{W} \times \boldsymbol{\Omega}; \mathbf{W} \times \boldsymbol{\Omega}]_{BC}$$

$$M_{13} = \frac{1}{3}\left(\frac{2I}{kT}\right)^{1/2} n[\mathbf{W} \times \boldsymbol{\Omega}; \mathbf{W}(\Omega^2 - 1)]_{BC}$$

$$\vdots$$

$$M_{33} = \frac{2n}{3}[\mathbf{W}\Omega^2; \mathbf{W}\Omega^2]_{BC}$$

$$N_{11} = -(p - p_V) + \frac{1}{2}\left(\frac{mkT}{2}\right)^{1/2} n^2 \underline{\boldsymbol{\varepsilon}} : \{\mathbf{x}\mathbf{W} \times \boldsymbol{\Omega}, \boldsymbol{\Omega}\}_{BC}$$

$$- \frac{1}{2}\left(\frac{m}{I}\right)^{1/2} n^2[\mathbf{W} \times \boldsymbol{\Omega}; \mathbf{x}_{21} \times (\mathbf{N}_{12} - \mathbf{N}_{21})]_{BC}$$

$$N_{12} = \frac{2}{3}\frac{n^2}{T}\left(\frac{I}{m}\right)^{1/2}\left[\mathbf{W} \times \boldsymbol{\Omega}; \mathbf{x}_{21}\left(\mathbf{G}_{12} \cdot \mathbf{F}_{12} + \frac{\mathbf{L}_1}{I} \cdot \mathbf{N}_{21}\right)\right]_{BC}$$

$$\vdots$$

$$N_{33} = \frac{1}{3}p\left(\frac{2kT}{m}\right)^{1/2}\left[\mathbf{W}\Omega^2; \left(\frac{\partial \ln \chi_2}{\partial \ln n}\right)_T \mathbf{G}_{12}\right]_{BC}$$

By solving (the Fourier transforms of) these equations, one can obtain the BC formulas for the coefficients $\lambda_K{}^t$, $\lambda_K{}^r$, $\pi_K{}^t$, $\pi_K{}^r$, $\mu_K$, and $\nu_{K3}$. The final step is the construction of the potential contributions to the fluxes. These are straightforward exercises which only require a bit of patience to work out. The difficult question and that of greatest interest is how the numerical predictions of the BC and EP theories compare with one another. Although Schrodt and Davis[51] already have conducted a numerical examination of the EP theory, no calculations based on the BC theory have been reported. Therefore we do not know the answer to this important question even for monatomic fluids. In order to test either of these theories for a Rice–Allnatt type of noncentral potential, it first will be necessary to device an accurate and reasonably economical procedure for evaluating the various collision integrals. The BC theory introduces a further difficulty, for in order to determine the "asterisk states" one must solve (at least approximately) the noncentral scattering problem. This is not an insuperable obstacle, but it does complicate matters greatly. Because of this and also because we are unaware of convincing evidence that the BC theory is more fundamentally sound that the EP theory, we are focusing our efforts on the latter. Although we and Condiff apparently agree on final results such as (3.26), he feels that our presentation severely distorts and misrepresents the BC theory. If this is so then it certainly happened by accident rather than design. Our intention was not to perform a critiùe but simply to indicate how this very promising theory could be extended to polyatomic fluids.

## IV. FINAL REMARKS

The message we have tried to convey in this article is that means now exist for conducting systematic computations of the nonequilibrium properties of dense polyatomic fluids. However, except for the few earlier studies of the rough-sphere and loaded-sphere models, the rigid and square-well ellipsoid calculations reported here are the only computations that actually have been performed. Therefore much hard work remains to be done before we shall be secure in our understanding of how the various phenomenological coefficients depend on molecular kinematic and interactional parameters. In this context it is as important to recognize the present limitations of our theory as it is to extol its virtues. Our ellipsoid computations only included the Enskog contributions, and in place of the requisite orientation-dependent equilibrium pair correlation functions we have substituted radial distribution functions for spherical molecules (with rigid sphere and square-well interactions). The task of correcting the second of these deficiencies is a nontrivial but tractable problem in equilibrium statistical mechanics. To proceed beyond the Enskog terms we must develop practical computational procedures for evaluating the soft-potential and additional-to-Enskog terms of the EP theory. Although this appears to present no fundamental difficulties, it is a problem of considerable complexity.

It already has been mentioned that we soon shall have evaluated the additional-to-Enskog terms for a monatomic square-well fluid. After this first step we intend to concentrate on models incorporating continuous long-range potentials. The simplest of these are the models discussed in Section III.D with short-ránge noncentral and long-range central interactions. Furthermore, it should be recognized that generalization to other (than ellipsoids) core shapes is not difficult to accomplish; smooth cores with tetrahedral symmetry (to similate real spherical top species such as methane and carbon tetrachloride) already have been the subject of dilute gas calculations.[52]

The only obstacle we foresee in extending the theory to fluid mixtures is the appalling amount of bookkeeping it entails. Thus the priority assigned to this task really depends only on whether the questions one can think to ask about mixtures are sufficiently interesting to justify this expenditure of effort. A job to which we currently attach a much higher priority is that of generalizing the theory to asymmetric interactions and internal distributions of molecular mass. This would provide a means for estimating the nonequilibrium properties of dense fluids composed of small polar molecules. Because we consider this to be the most important single generalization remaining to be done, it is the one we are pursuing with the greatest vigor. The problems we rank next in importance are those of determining how the theory can be modified to include quantal treatments of the rotational and vibrational degrees of freedom. Insofar as the moment method itself is concerned, the applications we presently are investigating include the scattering of light and of neutrons and the propagation and attenuation of ultrasonic disturbances.

Finally, we emphasize that, although the moment method can be used for any choice of the functional relationship between $f_1$ and $f_2$, we have considered here only a few specific examples. To be sure, these were the best we could find, but others certainly will be proposed. The task of comparing the numerical predictions of different, competing theories is one of critical importance. We believe that the versatility of the moment method, the wide variety of nonequilibrium phenomena to which it provides a unified approach, makes it an ideal tool for testing the quality of various statistical theories.

## Acknowledgment

We thank Dr. Kenneth Kulander for his help in performing the calculations reported here. Also, one of us (MT) wishes to express her appreciation to Professor I. Prigogine for financial support during a portion of the time she devoted to this research.

## References

1. M. Theodosopulu and J. S. Dahler, *J. Chem. Phys.*, **60**, 3567 (1974).
2. M. Theodosopulu and J. S. Dahler, *J. Chem. Phys.*, **60**, 4048 (1974).
3. (a) L. Waldmann, *Z. Naturforsch.*, **12a**, 660 (1957); (b) **13a**, 609 (1958); (c) R. F. Snider, *J. Chem. Phys.*, **32**, 1051 (1960); D. K. Hoffman and J. S. Dahler, *J. Stat. Phys.*, **1**, 521 (1970).
4. J. H. Irving and J. G. Kirkwood, *J. Chem. Phys.*, **18**, 817 (1950).
5. J. G. Kirkwood, F. P. Buff, and M. S. Green, *J. Chem. Phys.*, **17**, 988 (1949); R. W. Zwanzig, J. G. Kirkwood, K. F. Stripp, and I. Oppenheim, *ibid.*, **21**, 2050 (1953); R. W. Zwanzig, J. G. Kirkwood, I. Oppenheim and B. J. Alder, *ibid.*, **22**, 783 (1954).
6. J. S. Dahler, *J. Chem. Phys.*, **30**, 1447 (1959); *Phys. Rev.*, **129**, 1464 (1963).
7. J. S. Dahler in G. Temple and R. Seeger, Eds., *Research Frontiers in Fluid Dynamics*, Interscience, New York, 1965, Chap. 15.
8. J. S. Dahler and D. K. Hoffman in G. M. Burnett and A. M. North, Eds., *Transfer and Storage of Energy by Molecules*, Vol. 3, Wiley-Interscience, 1970, Chap. 1.
9. S. Chapman and T. G. Cowling, *The Mathematical Theory of Non-Uniform Gases*, Cambridge University Press, New York, 1953.
10. S. A. Rice and P. Gray, *The Statistical Mechanics of Simple Liquids*, Wiley, New York, 1966.
11. I. Prigogine, G. Nicolis, and J. Misguich, *J. Chem. Phys.*, **43**, 4516 (1965).
12. H. T. Davis, *Advan. Chem. Phys.*, **24**, 257 (1973).
13. H. Grad, *Phys. Fluids*, **6**, 147 (1963).

14. H. T. Davis, S. A. Rice, and J. V. Sengers, *J. Chem. Phys.*, **35**, 2210 (1961).

15. H. S. Green, *J. Math. Phys.*, **2**, 344 (1961).

16. J. C. McLennan, Jr., *Phys. Rev.*, **115**, 1405 (1959); *Phys. Fluids*, **3**, 493 (1960); *Phys. Fluids*, **4**, 1319 (1961).

17. J. G. Kirkwood and D. D. Fitts, *J. Chem. Phys.*, **33**, 1317 (1960).

18. C. S. Wang Chang, G. E. Uhlenbeck, and J. de Boer, in J. de Boer and G. E. Uhlenbeck, Eds., *Studies in Statistical Mechanics*, Vol. 2, p. 243, Interscience, New York, 1964.

19. B. J. McCoy, S. I. Sandler, and J. S. Dahler, *J. Chem. Phys.*, **45**, 3485 (1966). The thermal conductivity values reported in this paper are erroneous, and since have been corrected by W. M. Klein, D. K. Hoffman, and J. S. Dahler, *J. Chem. Phys.*, **49**, 2321 (1968).

20. J. H. Jeans, *Dynamical Theory of Gases*, Cambridge University Press, New York, 1904.

21. H. Grad, *Commun. Pure Appl. Math.*, **2**, 331 (1949).

22. L. Waldmann, *Z. Naturforsch.*, **15a**, 19 (1960); **18a**, 1033 (1963); S. Hess and L. Waldmann, *Z. Naturforsch.*, **21a**, 1529 (1966); S. Hess, *Z. Naturforsch.*, **23a**, 597, 898, 1095 (1968); **24a**, 1675, 1852 (1969); *Phys. Lett.*, **29A**, 108 (1969); **30A**, 239 (1969).

23. F. M. Chen and R. F. Snider, *J. Chem. Phys.*, **46**, 3937 (1967); **48**, 3185 (1968).

24. S. Yip and M. Nelkin, *Phys. Rev.*, **135**, A1241 (1964); S. Ranganathan and S. Yip, *Phys. Fluids*, **8**, 1956 (1965); **9**, 372 (1966); A. Sugawara and S. Yip, *ibid.*, **10**, 1911 (1967); A. Sugawara, S. Yip, and L. Sirovich, *ibid.*, **11**, 925 (1968).

25. J. O. Hirschfelder, C. F. Curtiss, and R. B. Bird, *Molecular Theory of Gases and Liquids*, Wiley, New York, 1965. Many other studies of convergence have been conducted within the context of the CE method. The relationship between this and the moment method, which is obvious on reflection, also has been addressed directly by I. L. McLaughlin and J. S. Dahler, *J. Chem. Phys.*, **44**, 4453 (1966).

26. W. Noll, *J, Ration. Mech. Anal.*, **4**, 627 (1955).

27. J. G. Kirkwood, *J. Chem. Phys.*, **14**, 180 (1946).

28. G. Severne, *Physica*, **31**, 877 (1965).

29. G. B. Brinser and D. W. Condiff, *J. Chem. Phys.*, **59**, 6599 (1973).

30. M. Theodosopulu and J. S. Dahler, *Phys. Fluids*, **15**, 1755 (1972).

31. D. W. Condiff, W.-K. Lu, and J. S. Dahler, *J. Chem. Phys.*, **42**, 3445 (1965).

32. T. G. Montgomery and D. K. Hoffman, *J. Chem. Phys.*, **53**, 848 (1970).

33. D. K. Hoffman, *J. Chem. Phys.*, **50**, 4823 (1969).

34. Y.-D. Chen and W. A. Steele, *J. Chem. Phys.*, **50**, 1428 (1969); **54**, 703 (1971); W. A. Steele and S. I. Sandler, *ibid.*, **61**, 1315 (1974); S. I. Sandler, *J. Mol. Phys.*, **28**, 1207 (1974); H. C. Andersen and D. Chandler, *J. Chem. Phys.*, **57**, 1918 (1972); **57**, 1930 (1972); L. J. Lowden and D. Chandler, *ibid.*, **59**, 6587 (1973); **61**, 5228 (1974).

35. A. Rotenberg, private communication; A. Rotenberg, *J. Chem. Phys.*, **43**, 1198 (1965).

36. E. R. Cooper and D. K. Hoffman, *J. Chem. Phys.*, **53**, 1100 (1970).

37. N. F. Sather and J. S. Dahler, *Phys. Fluids*, **5**, 754 (1962).

38. N. F. Sather and J. S. Dahler, *J. Chem. Phys.*, **37**, 1947 (1962).

39. S. I. Sandler and J. S. Dahler, *J. Chem. Phys.*, **46**, 3520 (1967).

40. J. S. Dahler, *J. Chem. Phys.*, **27**, 1428 (1957).

41. J. Misguich and G. Nicolis, *J. Mol. Phys.*, to be published.

42. P. M. Allen and G. H. A. Cole, *J. Mol. Phys.*, **14**, 413 (1968); **15**, 549, 557 (1968); P. M. Allen, *ibid.*, **18**, 349 (1970); M. J. Foster and G. H. A. Cole, *ibid.*, **20**, 417 (1971).

43. D. W. Condiff and J. S. Dahler, *J. Chem. Phys.*, **44**, 3988 (1966). Since this time many new studies of rotational Brownian motion have been reported. Several of these are described elsewhere in this volume.

44. B. J. McCoy and J. S. Dahler, *J. Chem. Phys.*, **50**, 2411 (1969).

45. J. Misguich, *J. Phys. Radium*, **30**, 221 (1969).

46. A. O. Cohen and R. A. Marcus, *J. Chem. Phys.*, **49**, 4509 (1968).

47. G. Dowling, Ph.D. Thesis, University of Minnesota, 1971, p. 33-I.
48. V. Silin, *Sov. Phys. JETP*, **11**, 1277 (1960).
49. W. D. Henline and D. W. Condiff, *J. Chem. Phys.*, **54**, 5346 (1971); G. B. Brinser and D. W. Condiff, *ibid.*, **59**, 2754 (1973).
50. D. T. Gillespie and J. V. Sengers, *Technical Report AEDC-TR-73-XX*, Arnold Engineering Development Center, Tenn.
51. I. B. Schrodt and H. T. Davis, *J. Chem. Phys.*, **61**, 323 (1974).
52. J. D. Verlin, E. R. Cooper and D. K. Hoffman, *J. Chem. Phys.*, **56**, 3740 (1972); E. R. Cooper, J. S. Dahler, J. D. Verlin, M. K. Matzen and D. K. Hoffman, *ibid.*, **59**, 403 (1973).

# THEORY OF LIQUID CRYSTALS*

## DIETER FORSTER†

*The James Franck Institute and The Department of Physics,
The University of Chicago, Chicago, Illinois*

## TABLE OF CONTENTS

## I. INTRODUCTION

Research on liquid crystalline phases is well into its ninth decade.[1,2] Attention paid to this subject has grown explosively in the last several years, as attested to by the more than 1000 publications devoted to liquid crystals that have appeared in the last 3 years. The coincidence of long-range orientational order in a system with the translational properties, and thus the flow characteristics, of a normal liquid is fascinating, and it gives liquid crystals a versatility that has led to research toward a wide range of applications, from optical and thermal devices to the function of mesophases in the living organism. Review articles,[3–5] books,[6,7] and conference proceedings[8–10] describe recent advances, and contain many literature references.

* I acknowledge the general support of the Materials Research Laboratory by the National Science Foundation.

† Present address: Department of Physics, Temple University, Philadelphia, Pa. 19122

In this article, I shall describe some aspects of the theory of liquid crystals. This is not meant to be an exhaustive review. Many topics of interest and importance go unmentioned because of limitations of space and time and because of the limited expertise of the author. I consider only thermotropic, that is, one-component, liquid crystals, and among these concentrate on the nematic mesophase which has been studied most extensively and is best understood. In keeping with my own taste and experience, I in particular focus on two aspects of the theory: (1) thermodynamic phenomena, namely, the phase transition from the isotropic liquid to the nematic phase, and related considerations including McMillan's recent theory of smectic A mesophases, and (2) dynamical collective phenomena, namely, the long-lived hydrodynamical fluctuations that are the cause of the intense light scattering observed from liquid crystals, and on whose proper treatment agreement seems to have been reached. These are topics that should be of interest to many-body physicists working in other fields, and they describe phenomena of a sufficiently general nature whose principal features can be understood without detailed consideration of the enormously complicated microscopic forces and microscopic structure of an anisotropic organic liquid. I have made an effort to present the theory from as unified a viewpoint as possible, and have attempted to use a language that facilitates ready comparison with any of the other favorite playgrounds of many-body physicists: ferromagnets, antiferromagnets, or superfluids.

Liquid crystals are formed by organic molecules of an elongated shape. The best-studied examples are $p$-azoxyanisole (PAA) and 4-methoxy-benzylidene-4'-$n$-butylaniline (MBBA), whose structures are given in Fig. 1. Both of these compounds have a nematic phase. The textbook picture of the isotropic and two liquid crystalline phases is given in Fig. 2. In the isotropic liquid the molecular axes are oriented randomly. In a nematic the molecular centers of mass are still randomly placed, except possibly for normal short-range order, but the long axes are aligned, on the average, along a common direction which is only prescribed by boundary effects or external fields. Consequently, nematics have essentially the flow properties of a normal liquid, but their optical and other properties are strongly anisotropic.

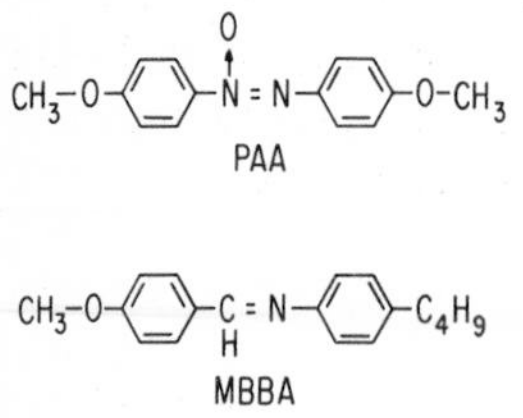

**Fig. 1** Structures of two nematogen molecules, $p$-azoxyanisole (PAA, $T_c = 136°C$) and 4-methoxybenzylidene-4'-$n$-butylaniline (MBBA, $T_c = 44°C$).

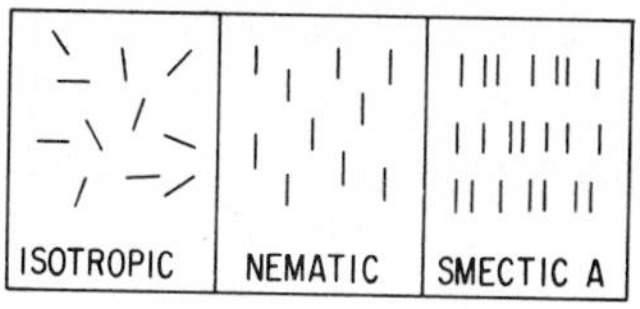

**Fig. 2** Textbook picture of the isotropic liquid, the nematic, and the smectic A phases, showing the orientational order in the nematic phase and the planar structure of the smectic A. The lines represent the long axes of the molecules.

Figure 2c gives the arrangement of molecules in the simplest smectic phase, smectic A. Here, in addition to orientational alignment, there is a density wave in the direction of the long molecular axis, resulting in a monolayered structure. The layers move easily with respect to each other, and within each layer the centers of mass are arranged chaotically as in a normal liquid. There are several variants of smectic liquid crystals (see Ref. 5).

## II. MACROSCOPIC EQUILIBRIUM PROPERTIES

### A. The Order Parameter

In any system in which long-range order is present, a convenient description is afforded by introducing an order parameter—a quantity that vanishes in the disordered (isotropic) phase but is different from zero in the ordered (nematic) phase. Many molecules that form nematics are essentially symmetric with respect to their center and, in any case, if $\mathbf{n}$ is a unit vector in the direction of alignment, $\mathbf{n}$ and $-\mathbf{n}$ are not physically distinct in thermal equilibrium—the molecules do not line up head to head. Consequently, as de Gennes[11] has pointed out, the simplest admissible description of liquid crystalline order is in terms of a symmetric and traceless tensor $\mathbf{Q}_{ij}$, which in a homogeneous system of uniaxial symmetry is of the canonical form

$$\mathbf{Q}_{ij} = S(n_i n_j - \tfrac{1}{3}\delta_{ij}) \tag{2.1}$$

where the unit vector $\mathbf{n}$ indicates the preferential direction, and the constant $S$, which depends on temperature, gives a measure for the degree of orientational order. A microscopic model for $\mathbf{Q}_{ij}$ has been discussed by Lubensky.[12] Considering the nematic as a gas of long, thin bars, with center-of-mass coordinates $\mathbf{r}^{\alpha}(t)$ and with the $\alpha$th molecule pointing in the direction $\mathbf{v}^{\alpha}(t)$ where $(\mathbf{v}^{\alpha})^2 = 1$, a local order parameter can be defined by

$$\hat{R}_{ij}(\mathbf{r}t) = \sum_{\alpha} [v_i^{\alpha}(t)v_j^{\alpha}(t) - \tfrac{1}{3}\delta_{ij}]\,\delta(\mathbf{r} - \mathbf{r}^{\alpha}(t)) \tag{2.2}$$

Averaged over any ensemble, this tensor field gives

$$\langle \hat{R}_{ij}(\mathbf{r}t) \rangle = \frac{\langle \rho(\mathbf{r}t) \rangle}{m}\,\mathbf{Q}_{ij}(\mathbf{r}t) \tag{2.3}$$

where $\langle \rho(\mathbf{r}t) \rangle$ is the mass density in the fluid, $m$ is the molecular mass, and the tensor $Q_{ij}(\mathbf{r}t)$ is of the form (2.1) except that, in a nonequilibrium situation, the "director" $\mathbf{n}(\mathbf{r}t)$ and the conventional dimensionless order parameter $S(\mathbf{r}t)$ may vary in space and time. In equilibrium $S$ is then identified as

$$S = \tfrac{1}{2}\langle 3 \cos^2 \theta - 1 \rangle \tag{2.4}$$

where $\theta$ is the angle between the molecular axis and the nematic symmetry axis. $S(T)$ vanishes in the isotropic liquid, and with decreasing temperature it is believed to increase from about 0.4, just below the clear point $T_c$, to values above 0.8, at the low-$T$ limit of the nematic range.

For microscopic calculations on real fluids (of which there are at present hardly any) a somewhat more useful definition is in terms of the quadrupolar term in the mass density[13]

$$R_{ij}(\mathbf{r}t) = \sum_{\alpha, k} m^{\alpha k}[(r^{\alpha k} - r^{\alpha})_i (r^{\alpha k} - r^{\alpha})_j - \tfrac{1}{3}\delta_{ij}(\mathbf{r}^{\alpha k} - \mathbf{r}^{\alpha})^2]\delta(\mathbf{r} - \mathbf{r}^{\alpha}) \tag{2.5}$$

where $\mathbf{r}^{\alpha k}$ and $m^{\alpha k}$ are the coordinate and mass of the $k$th particle in the $\alpha$th molecule. On the average,

$$\langle R_{ij}(\mathbf{r}t) \rangle = \alpha Q_{ij} \tag{2.6}$$

where

$$\alpha = (\langle \rho(\mathbf{r}) \rangle / m)(I_l - I_t)$$

in the uniform nematic, where $I_l = I_3$ and $2I_t = I_1 + I_2$ are given by

$$I_i = \sum_k m^{\alpha k}(r_i^{\alpha k} - r_i^{\alpha})^2 \tag{2.7}$$

$I_i$ is, essentially, the $i$th component of the molecular moment of inertia, evaluated in the system of principal molecular axes.

Equation (2.1) is valid for a uniaxial nematic and, with appropriate spatial dependence of $\mathbf{n}(\mathbf{r})$ and possibly $S(\mathbf{r})$, for any locally uniaxial mesophase. Lubensky[12] and Freiser[14] have considered the generalization of (2.1) necessary to describe biaxial liquid crystals. While no biaxial nematic has ever been found, there do seem to exist biaxial smectic systems,[15,16] designated smectic C. Alben,[17] on the basis of a model mean field calculation, has suggested that certain two-component nematic systems are the most likely to exhibit a biaxial nematic phase.

For macroscopic considerations the precise microscopic definition of $Q_{ij}$ does not matter very much, since any other symmetric, traceless tensor that defines a local property is proportional to $Q_{ij}$, at least for small $S^{11}$. Thus, for example, the pure tensor piece of the magnetic susceptibility $\chi_{ij}$ can be written as

$$\chi_{ij}(\mathbf{r}t) - \tfrac{1}{3}\delta_{ij} \, \text{tr} \, \underline{\chi}(\mathbf{r}t) \equiv \chi_{ij}^a(\mathbf{r}t) = \chi^a Q_{ij}(\mathbf{r}t) \qquad (2.8)$$

at least under circumstances when the local order varies only slowly in space and time. And, similarly, the dielectric tensor is given by (tr = trace)

$$\varepsilon_{ij}(\mathbf{r}t) - \tfrac{1}{3}\delta_{ij} \, \text{tr} \, \underline{\varepsilon}(\mathbf{r}t) \equiv \varepsilon_{ij}^a(\mathbf{r}t) = \varepsilon^a Q_{ij}(\mathbf{r}t) \qquad (2.9)$$

where $\chi^a$ and $\varepsilon^a$ are the anisotropies in the bulk magnetic and dielectric susceptibilities, respectively. Since $\chi^a$ is usually positive, a nematic aligns parallel to an external magnetic field.

## B. The Landau Free Energy

The tensorial order parameter $Q_{ij}$ can be used to construct an expression for the free-energy density for a homogeneous nematic constrained to exhibit a specified degree of order. In the presence of an external magnetic field $\mathbf{H}$, the Landau free-energy density[18] takes the form

$$F_L = F_0 + \tfrac{1}{2}A \, \text{tr} \, \mathbf{Q}^2 - \tfrac{1}{3}B \, \text{tr} \, \mathbf{Q}^3 + \tfrac{1}{4}C \, \text{tr} \, \mathbf{Q}^4 - \tfrac{1}{2}H_i \chi_{ij}^a H_j \qquad (2.10)$$

where an inconsequential second invariant of order $Q^4$ has been omitted. $F_0$, $A$, $B$, and $C$ are temperature- and pressure-dependent constants. Equations (2.1) and (2.8) give

$$F_L = F_0 + \tfrac{1}{3}AS^2 - \tfrac{2}{27}BS^3 + \tfrac{1}{18}CS^4 - \tfrac{1}{2}\chi^a[(\mathbf{H}\cdot\mathbf{n})^2 - \tfrac{1}{3}\mathbf{H}^2] \qquad (2.11)$$

In an isotropic ferromagnet in which the order parameter is a vector quantity, namely, the magnetization $\mathbf{M}$, no scalar contribution of third order can be constructed. The transition from the para- to the ferromagnetic phase is therefore second order, and occurs at a critical point in the absence of external fields. In liquid crystals the tensor character of the order parameter allows for terms odd in $S$, and consequently the transition from the isotropic to the nematic phase is *first order*. If we make the usual mean field theory ansatz,

$$A(T) = a(T - T^*) \qquad B \text{ and } C \text{ positive constants} \qquad (2.12)$$

and locate the minimum of $F_L$, in the absence of external fields, we find an isotropic-to-nematic transition at the clear point $T_c$, where

$$T_c - T^* = \frac{2}{27}\frac{B^2}{aC} \qquad S_c = \frac{2}{3}\frac{B}{C} \qquad (2.13)$$

$S_c$ is the jump in the degree of order as the transition occurs. A critical instability would occur at the temperature $T^* < T_c$. Experimentally, $T_c - T^*$ is found to be very small. For example,[19,20] in MBBA, $T_c = 315.5°\text{K}$ while

$T_c - T^* = 0.8°K$. Discontinuities in entropy and specific volume, which characterize a first-order transition, are thus expected, and found,[3,21] to be very small. One also expects pronounced pretransitional effects. These have been considered by Alben.[22,23] I will comment later on dynamical manifestations of the nearly second-order transition.

The magnetic term in (2.11) describes the alignment of nematics parallel to a magnetic field for systems of positive anisotropy $\chi^a > 0$. From (2.10) above $T_c$ a Curie–Weiss law for the static orientational response is obtained. Since by the principles of equilibrium statistical mechanics,

$$\alpha^2 \delta Q_{ij} = \chi_{ijkl} \delta(\tfrac{1}{2}\chi^a H_k H_l) \tag{2.14}$$

$$\chi_{ijkl} = (k_B T)^{-1} \int d^3(r - r') \langle R_{ij}(\mathbf{r}) R_{kl}(\mathbf{r}') \rangle$$

$$= \chi(T)(\delta_{ik}\delta_{jl} + \delta_{il}\delta_{jk} - \tfrac{2}{3}\delta_{ij}\delta_{kl}) \tag{2.15}$$

where $\alpha$ is the constant which relates $\langle R \rangle$ to $\mathbf{Q}$ [see (2.6)], we obtain from (2.10)

$$\chi(T) = \frac{\alpha^2}{a(T - T^*)} \qquad T > T_c > T^* \tag{2.16}$$

for the strength of pretransitional fluctuations of the local order in the isotropic phase. Equation (2.16) has been verified in MBBA to high accuracy[19,20] over a range of 20°C above $T_c$.

## C. The Frank Free Energy

One of the characteristic properties of liquid crystals is their ability to support certain static shear stresses.[22] By contrast, a simple liquid cannot support any static shear. Being a liquid, a liquid crystal does not support uniform shear as a solid does. However, the liquid crystal can be locally distorted, for example, by means of spatially varying external fields or boundary effects. Consequently, there is an elastic contribution to the free-energy density. By the standard phenomonological procedure, this contribution is obtained by considering the scalar terms which can be constructed from $\nabla_i Q_{jk}$. Only terms of order $\nabla$ and $\nabla^2$ need be considered if the spatial variation is slow on a molecular scale. Furthermore, nonuniformity in the degree of order $S$ can be disregarded; $S$ is determined by the microscopic interactions and can be changed only by extremely strong external fields. The direction of alignment, however, described by $\mathbf{n}$, is not determined by the molecular interactions and can be easily splayed, bent, and twisted by weak external agencies. On this basis, Lubensky[12] has rederived the Frank[23]–Oseen[24] expression for the elastic free-energy density:

$$F_F = \tfrac{1}{2}K_1(\nabla \cdot \mathbf{n})^2 + \tfrac{1}{2}K_2[\mathbf{n} \cdot (\nabla \times \mathbf{n}) + q_0]^2 + \tfrac{1}{2}K_3[\mathbf{n} \times (\nabla \times \mathbf{n})]^2$$
$$- \tfrac{1}{2}\chi^a[(\mathbf{H} \cdot \mathbf{n})^2 - \tfrac{1}{3}(\mathbf{H})^2] \tag{2.17}$$

where we have again included the contribution due to an inhomogeneous external field $\mathbf{H}$. The length $1/q_0$ is a pseudoscalar; it vanishes in nematics which are invariant under the parity operation. In cholesterics, $2\pi/q_0$ is the pitch of the helical structure; see, for example, Ref. 5. Stability requires that the three Frank coefficients $K_i$ be positive. They can be measured by a variety of techniques.[3] For MBBA, for example,[25] $K_1$ and $K_3$ are about $2 \times 10^{-7}$ dynes, at the clear point, increasing with decreasing temperature by a factor of 3 or 4.

A variety of interesting boundary and field effects[5] can be derived from the Frank free-energy expression. Moreover, minimizing $F_F$ leads to a variety of singular solutions called disclination lines which are the liquid-crystal equivalent of dislocations in crystalline solids and can be described in a similar fashion.[26,27] I cannot comment on this field.

### D. The Static Director Correlation Function

For many applications including light scattering, it is important to understand the Frank free energy in terms of equilibrium-averaged correlation functions. The procedure is orthodox. Given real local variables $A$ and $B$, we define their generalized susceptibilities by

$$\chi_{AB}(\mathbf{k}) = \frac{1}{k_B T} \int d^3(r - r') \, e^{-i\mathbf{k} \cdot (\mathbf{r} - \mathbf{r}')} \langle \delta A(\mathbf{r}) \delta B(\mathbf{r}') \rangle \tag{2.18}$$

where $\delta A \equiv A - \langle A \rangle$ and $\langle \ \rangle$ denotes an equilibrium average. It simplifies matters greatly to treat the macroscopic director $\mathbf{n}(\mathbf{r})$ as the nonequilibrium average of a local microscopic variable, also denoted by $\mathbf{n}(\mathbf{r})$ which is not likely to confuse the reader. It is then easy to see that for nematics (2.17) is equivalent to the statements[28]

$$\chi_{n_1 n_1}(\mathbf{k}) = (K_1 k_1^2 + K_3 k_3^2)^{-1} \quad \text{and} \quad \chi_{n_2 n_2}(\mathbf{k}) = (K_2 k_1^2 + K_3 k_3^2)^{-1} \tag{2.19}$$

Here, the coordinate system has been chosen so that the equilibrium director $\mathbf{n}^0$ is in the 3-direction, and the wave vector $\mathbf{k}$ is in the 1–3 plane. $\mathbf{n}(\mathbf{r})$ is microscopically defined in terms of the order parameter $R_{ij}(\mathbf{r})$ by

$$n_i(\mathbf{r}) \equiv \frac{\tfrac{2}{3}R_{i3}(\mathbf{r})}{\langle R_{33} \rangle} \qquad i = 1, 2 \tag{2.20}$$

or $n_i = \frac{2}{3}(\delta_{ik} - n_i^0 n_k^0)R_{kl}n_l^0/(n_r^0\langle R_{rs}\rangle n_s^0)$. The normalization is chosen so that $\mathbf{n}(\mathbf{r})$ transforms as a unit vector under rotations:

$$\langle[L_i, n_j(\mathbf{r})]\rangle = i\hbar\varepsilon_{ijk}n_k^0 \tag{2.21}$$

where $\mathbf{L}$ is the total angular momentum operator of the system, and we have chosen the more familiar quantum mechanical notation, [ , ] indicating a commutator.

According to (2.19), fluctuations in the local order are of long range, falling off in space as $1/r$. Equation (2.19) is the conventional fundamental statement of broken symmetry. Its equivalent is a superfluid, for example, are the long-ranged fluctuations in the quantum phase that accompany the broken gauge symmetry. The Frank constants, $K_i$, correspond to the superfluid density $\rho_s$ [see Eq. (4.24) of Ref. 29].

It would be nice if rigorous bounds could be put on the nature of the small-$k$ singularity of $\chi_{nn}$ (at least as $1/k^2$, say), as is possible for superfluids[30] and many other systems of broken symmetry.[31] These arguments start out from statements of broken symmetry analogous to (2.1) or (2.21), and use Bogoliubov inequalities. For nematics a corresponding more rigorous derivation of (2.19), from (2.21) rather than from the phenomenological (2.17), has not as yet been accomplished.[32] The difficulty is that the total angular momentum operator $\mathbf{L}$, which generates rotations, is not the integral of a local density. It is therefore not possible in nematics to derive from (2.21) the statements of the nonexistence of the ordered phase in two dimensions[30,31] that have fascinated us in other systems of broken symmetry. It is possible, however, to see that $\chi_{nn}(\mathbf{k})$ must be singular at $k = 0$; a continuity argument must accomplish the rest. For cholesterics, Lubensky[33,34] has pointed out that the Frank free energy would lead to a very weak instability, even in three dimensions, which is only removed by the boundaries.

The Frank coefficients depend on the temperature. A qualitative tendency can be inferred by noting that the microscopic "director" variables $n_1$ and $n_2$, as defined by (2.20), do not exist in the isotropic phase, but that the variables $R_{ij}$ do. It is therefore preferable to write (2.20) in terms of the latter variables, for example,

$$\chi_{R_{13}R_{13}}(\mathbf{k}) = \frac{\alpha^2}{(K_1/S^2)k_1^2 + (K_3/S^2)k_3^2} \qquad T < T_c \tag{2.22}$$

Since this expression is only modified above $T_c$ by addition of a constant term in the denominator [cf. (2.16)], one would infer that $K_i/S^2$ is relatively insensitive to changes in temperature, in other words, that

$$K_i(T) \approx \text{constant} \times S^2(T) \tag{2.23}$$

This prediction, motivated a little better by somewhat more microscopic considerations[35,36] (although it is easy to overestimate the power of inner field treatments as compared to a good old-fashioned Landau argument, which the one given here is), checks out remarkably well[37,25] except near the transition to a smectic phase.[38]

## III. MEAN FIELD THEORY

For a more detailed and microscopic understanding of liquid crystalline phases and of transitions between them, one has to turn to statistical mechanics. For the rigorous-minded, the difficulties involved are great; we do not know very much about the detailed interactions between molecules as complex as those forming liquid crystals, and if we did we would still be faced with the awesome problem of computing phase space integrals, and so on. It is very likely that both hard-core repulsion and softer anisotropic attractive interactions play a role in determining the structure. I concentrate on the latter and, following the ideas developed first by Maier and Saupe[39] for nematics, and more recently extended to cholesterics by Goossens[40] and to smectics by McMillan,[41] present a mean field theory approach to liquid crystals. An attempt has been made to make obvious the analogy to the familiar Weiss mean field theory for ferromagnets.

In contrast to the latter case, the isotropic-to-nematic transition is first order. As stated above, however, the transition takes place very close in temperature to what would be a critical instability at $T = T^*$. The dimensionless reduced temperature difference $\tau_c = (T_c - T^*)/T_c$ is very small, about $2.5 \times 10^{-3}$. An interesting parameter is the correlation length at $T_c$, $\xi_c$. Combining (2.16) and (2.22), $\xi$ is given by

$$\xi^2 = \frac{K/S^2}{a(T - T^*)} = \xi_0^2\left(1 - \frac{T}{T^*}\right)^{-1} \qquad T > T_c \qquad (3.1)$$

For MBBA, using measured values[25] at $T_c$ of $K \approx 2 \times 10^{-7}$ dynes, $T_c - T^* = 0.8°K$, $S = 0.4$, and[19] $a = 0.062 J/(cm^3)(°K)$, one would predict $\xi_c \approx 1600$ Å. This is clearly too large. A more realistic appraisal might be $\xi_c \approx 300$ Å, which follows by assuming the reasonable value $\xi_0 = 15$ Å in (3.1). In any case, at $T_c$ the correlations extend over many molecular lengths, and except in a narrow temperature range near $T_c$ one expects the pretransitional phenomena to occur that characterize a second-order phase transition. Fan and Stephen[42] have therefore examined the validity of a mean field description (which is known to fail for, say, ferromagnets in the "critical region" near $T_c$) and found that in nematics, critical, as contrasted to mean field, behavior may just be observable very close to $T_c$.

Suppose, then, that the correct microscopic system Hamiltonian $\mathcal{H}$ is known. The free energy is in principle given by

$$\beta F = -\log \operatorname{Tr} e^{-\beta \mathcal{H}} \tag{3.2}$$

where $\beta = 1/k_B T$, and Tr indicates the classical configurational integral. If $H$ is an approximation to $\mathcal{H}$ for which this integral can be computed, then for any $H$ the inequality holds:

$$\beta F \leq \beta F_H = -\log \operatorname{Tr} e^{-\beta H} + \beta \langle \mathcal{H} - H \rangle_H \tag{3.3}$$

Equation (3.3) is a restatement of the familiar inequality $\langle e^A \rangle \geq e^{\langle A \rangle}$. $\langle \cdots \rangle_H$ is a canonical average, computed with $H$ rather than $\mathcal{H}$. Standard mean field theory is the approximation obtained when an independent-particle Hamiltonian $H$ is used in (3.3). This Hamiltonian involves an "internal field" parameter with respect to which $F_H$ is minimized to locate the best possible approximation to $F$.

## A. The Isotropic-to-Nematic Transition

After these preliminaries we consider a system of rigid molecules, approximated by rods and interacting with a pair potential which depends on the relative angle as well as the distance between two molecules. We need the shape of this potential. The simplest assumption is that its angle dependence can be sufficiently characterized by the first term in its expansion in Legendre polynomials. For symmetric rods this is $P_2(\cos \theta^{\alpha\beta}) = (3 \cos \theta^{\alpha\beta} - 1)/2$ where $\theta^{\alpha\beta}$ is the relative angle between the $\alpha$th and $\beta$th molecules. Consequently, we write the pair Hamiltonian in the form

$$\mathcal{H} = \frac{1}{2} \sum_{\alpha \neq \beta} V^{\alpha\beta} - \frac{1}{2} \sum_{\alpha \neq \beta} J^{\alpha\beta} \hat{\mathbf{R}}^\alpha : \hat{\mathbf{R}}^\beta - \sum_\alpha \mathbf{h}^\alpha : \hat{\mathbf{R}}^\alpha \tag{3.4}$$

where

$$\hat{R}_{ij}^\alpha = v_i^\alpha v_j^\alpha - \tfrac{1}{3}\delta_{ij} \tag{3.5}$$

is the molecular order parameter introduced in (2.2), $h_{ij}$ is an external tensor field introduced here for calculational convenience, and the central and orientational potentials $V^{\alpha\beta}$ and $J^{\alpha\beta}$ are assumed to be functions of center-of-mass distance $r^{\alpha\beta} \equiv |\mathbf{r}^\alpha - \mathbf{r}^\beta|$ only. Their dependence on $r^{\alpha\beta}$ determines, one suspects, the density dependence of the free energy. Considering the dispersion forces between molecules, Maier and Saupe[39] have derived a Hamiltonian of the form (3.4), with $J^{\alpha\beta}$ attractive and falling off as $(r^{\alpha\beta})^{-6}$. Our presentation is similar to theirs which has more recently been improved by Chandrasekhar and Madhusudana,[43] who have taken account of the contributions of additional forces, that is, Legendre polynomials in (3.4). However, the density dependence that follows is either untested[43] or incorrect.[44] We shall therefore concentrate on the temperature dependence of the free energy predicted by (3.4). This allows us to simplify matters further by considering the molecules fixed on the sites of a three-dimensional simple cubic lattice. In this case only the orientational degrees of freedom remain to be considered, and the first term in (3.4) can be dropped. (We have also omitted the kinetic energy contribution to $\mathcal{H}$ which, for a classical system, can be separated out.)

In this case we have the classical Heisenberg Hamiltonian in (3.4), but with the important difference that the order parameter is a tensor, not a vector. For the ansatz Hamiltonian $H$ in (3.3), we use

$$H = -\sum_\alpha \mathbf{\eta}^\alpha : \hat{\mathbf{R}}^\alpha \qquad (3.6)$$

where $\eta_{ij}^\alpha$ is an "internal" field to be determined self-consistently. With this choice of $H$, we obtain, omitting the $V$ term in (3.4),

$$\langle \mathcal{H} \rangle_H = -\tfrac{1}{2} \sum_{\alpha \neq \beta} J^{\alpha\beta} \mathbf{Q}^\alpha : \mathbf{Q}^\beta - \sum_\alpha \mathbf{h}^\alpha : \mathbf{Q}^\alpha \qquad (3.7)$$

where

$$\mathbf{Q}^\alpha = \langle \hat{\mathbf{R}}^\alpha \rangle_H = \frac{\mathrm{tr}_\alpha \, \hat{\mathbf{R}}^\alpha \, e^{\beta \mathbf{\eta}^\alpha : \hat{\mathbf{R}}^\alpha}}{\mathrm{tr}_\alpha \, e^{\beta \mathbf{\eta}^\alpha : \hat{\mathbf{R}}^\alpha}} \qquad (3.8)$$

is only a function of $\mathbf{\eta}^\alpha$, and its calculation involves only the angle integral $\mathrm{tr}_\alpha \equiv \int d\Omega^\alpha$ for the $\alpha$th particle. The first and third term contributing to $F_H$ in (3.3) are also easily obtained. The internal field $\mathbf{\eta}$ is then determined by minimizing $F_H$. This results immediately in the equation $\delta F_H / \delta \eta^\alpha = \dot{0}$ or, equivalently,

$$\frac{\delta F_H}{\delta \mathbf{Q}^\alpha} = 0 = \mathbf{\eta}^\alpha - \mathbf{h}^\alpha - \sum_\beta J^{\alpha\beta} \mathbf{Q}^\beta \qquad (3.9)$$

This is the central equation of the mean field theory, determining the inner field $\mathbf{\eta}$ in terms of the externally applied field $\mathbf{h}$ and the anisotropy potential $J$. For vanishing external field $\mathbf{h} = 0$, (3.9) has a homogeneous, uniaxial solution:

$$Q_{ij}^\alpha = S(n_i^0 n_j^0 - \tfrac{1}{3}\delta_{ij}) \qquad (3.10)$$

where $\mathbf{n}^0$ is an arbitrary unit vector, and $S$ is determined from (3.8) and (3.9). The angle integrals can easily be done, leading to

$$S = \psi(\beta JS) \equiv \frac{3}{2} \frac{\int_0^1 dx \, (x^2 - \tfrac{1}{3}) e^{\beta JSx^2}}{\int_0^1 dx \, e^{\beta JSx^2}} \qquad (3.11)$$

where

$$J \equiv \sum_{\beta(\neq \alpha)} J^{\alpha\beta} \tag{3.12}$$

The function $\psi(y)$ is given in Fig. 3. Accordingly, there is always an isotropic solution $S = 0$. This solution is locally stable so long as $1 - \beta J \psi'(0)$ is positive, that is, for $T > T^*$, where

$$k_B T^* = J \psi'(0) = 0.1333J \tag{3.13}$$

A nematic solution, $S \neq 0$, exists only for $T < T_{\text{max}}$, where $T_{\text{max}}$ is determined so that $k_B T_{\text{max}} = J \psi'(\beta J S)$ holds together with (3.11). Numerically,

$$k_B T_{\text{max}} = 0.1486J \qquad S = 0.3235 \tag{3.14}$$

The isotropic $\leftrightarrow$ nematic transition temperature is determined by locating the global minimum of the free energy, which amounts to a Maxwell equal-area construction in Fig. 3. $T_c$ is given by

$$k_B T_C = 0.1468J \qquad S_C = 0.4292 \tag{3.15}$$

Mean field theory therefore predicts a universal value

$$\tau_c = \frac{T_c - T^*}{T_c} = 0.09 \tag{3.16}$$

This value is too large by more than an order of magnitude. However, the discontinuity $S_c$ and the dependence of the order parameter $S$ on reduced temperature, which follows from (3.11), are semiquantitatively correct.

The Maier–Saupe theory contains the additional assertion, obtained from a microscopic derivation of the intermolecular dispersion forces, that

$$J = \tilde{J} n^2 \tag{3.17}$$

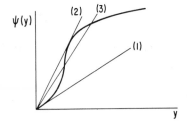

Fig. 3 $\psi(y)$ as a function of $y$ [see (3.11)]. The straight lines represent $S = (kT/J)y$ for three temperatures: (1) $T = T^*$, (2) $T = T_{\text{max}}$, and (3) $T = T_c$.

where $n$ is the density of molecules, and $\tilde{J}$ is a molecular constant. [The reader who is not too critical might note that in a continuum theory the lattice sum (3.12) should be replaced by

$$J = n \int d^3r \, g(r) J(r) \qquad (3.18)$$

where $g(r)$ functions as a probability distribution for the molecular centers of mass. The Maier–Saupe result (3.17) is inferred from the $1/r^6$ dependence of the dispersion interaction.] In this case expressions[45] for the volume jump, the transition entropy, and derivative quantities such as the compressibility and specific heat can be obtained as functions of $n^2/k_B T$. In particular, the theory predicts that the order parameter $S$ is a universal function of $n^2 T_c/n_c^2 T$ as is obvious from (3.11). Alben[44] has shown that this prediction is not experimentally valid; rather, for PAA a dependence of $S$ on $n^{4.3} T^{-1}$ is observed in the transition region. Chandrasekhar and Madhusudana[43] have improved the theory by including, within mean field theory, permanent dipole–dipole, dispersion, induction, and repulsion forces. A careful discussion of several model liquids of hard rods is given by Alben[46] who included, partially, the effects of repulsive packing, although still within a long-range-order molecular field description. As Alben has pointed out, the remaining disagreement between theory and experiment can only be remedied by someone who knows how to treat short-range order properly.

## B. The Frank Elasticity Coefficients

From the mean field theory just presented, one can calculate the order parameter correlation function, and therefore the Frank coefficients. Our aim here is mainly to expose the structure of the theory; only order-of-magnitude results are obtained. Similar, in principle though not in method, results have been obtained by Saupe;[47] see also Refs. 35 and 36.

Given the Hamiltonian (3.4), one can obtain the order parameter correlation function by varying the external field:

$$\chi_{ijkl}^{\alpha\beta} \equiv (k_B T)^{-1} (\langle \hat{R}_{Ij}^{\alpha} \hat{R}_{kl}^{\beta} \rangle - \langle \hat{R}_{ij}^{\alpha} \rangle \langle \hat{R}_{kl}^{\beta} \rangle) = \frac{\delta \langle \hat{R}_{ij}^{\alpha} \rangle}{\delta h_{kl}^{\beta}} \qquad (3.19)$$

Now $\langle \hat{R}^{\alpha} \rangle = Q^{\alpha}$ is given, in mean field approximation, by (3.8) and (3.9), as a function of the fields $h^{\alpha}$ which may still vary from site to site. The derivative (3.19) may therefore be calculated from (3.8) and (3.9). While it is easy to become entangled with indices, for present purposes it is sufficient to consider an external field whose only nonvanishing components are $\delta h_{13}^{\alpha} = \delta h_{31}^{\alpha}$, assuming the equilibrium director $\mathbf{n}_0$ to point in the 3-direction. For such a field, it is apparent from (3.8) and (3.9) that

$$\delta\eta^\alpha_{13} = \frac{\eta_{33}}{Q_{33}} \delta Q^\alpha_{13} = J\delta Q^\alpha_{13} \qquad (3.20)$$

with $J$ as defined by (3.12). Consequently, (3.9) and (3.19) yield

$$(J\delta^{\alpha\beta} - J^{\alpha\gamma})\chi^{\gamma\beta}_{1313} = \delta^{\alpha\beta} \qquad (3.21)$$

in the absence of external fields. A sum over the lattice site index $\gamma$ is implied. Equation (3.21) is easily solved by lattice Fourier transformation.

To make contact with (2.22), which defines the Frank coefficients, we have to obtain the continuous transformation for the correlation function of the field

$$\hat{R}_{ij}(\mathbf{r}) = \sum_\alpha \hat{R}^\alpha_{ij}\delta(\mathbf{r} - \mathbf{r}^\alpha) \qquad (3.22)$$

A few manipulations yield easily

$$\chi_{R_{13}R_{13}}(\mathbf{k}) = \frac{N/V}{J - J(k)} \qquad (3.23)$$

where

$$J(k) = \sum_\beta J^{\alpha\beta} e^{-i\mathbf{k}\cdot(\mathbf{r}-\mathbf{r}^\beta)} = J(1 - \tfrac{1}{2}k^2a^2 + \cdots) \qquad (3.24)$$

$N/V = a^{-3}$ is the density of molecules, $a$ is the unit length of our imaginary lattice, and in the second equation of (3.24) we have taken only nearest-neighbor interactions for simplicity; $a$ must then be chosen as an effective range of the potential $J(r)$. Note finally that, for the choice (3.22) of the order parameter, the coefficient $\alpha$ in (2.22) equåls $N/V = a^{-3}$. Apologizing for these little contortions, we find for the Frank coefficients

$$K = \frac{1}{2}\frac{J}{a}S^2 = \frac{15}{4}\frac{k_B T^*}{a}S^2 \qquad (3.25)$$

by comparison of (3.23) with (2.22), and replacing $J$ by $T^*$ as given in (3.13). Evidently, the theory here gives the same value for all three Frank constants, but this feature is improvable. In MBBA, taking $S = 0.45$, $T^* = 315°K$, and $a = 20$ Å, one obtains $K = 1.7 \times 10^{-7}$ dynes, in reasonable agreement with the measured[25] value. Note the temperature dependence of $K$ through $S(T)$, as surmised in (2.23).

## C. The Isotropic-to-Cholesteric Transition

The cholesteric state is a variant of the nematic state. As in the latter, the molecular centers of mass are moving chaotically. The molecular axes, however, are on the average arranged in a helical pattern, aligned in a plane as they are in nematics, but with the direction of alignment slowly rotating as one moves into the third direction. The order can thus be characterized by the tensor

$$Q_{ij}^0(\mathbf{r}) = S[n_i^0(\mathbf{r})n_j^0(\mathbf{r}) - \tfrac{1}{3}\delta_{ij}] \tag{3.26}$$

$$n_i^0(\mathbf{r}) = (\cos q_0 z, \sin q_0 z, 0) \tag{3.27}$$

The helical pitch $2\pi/q_0$ is generally very large, on the order of a few thousand ångstroms, comparable to the wavelength of visible light. Because of this feature, cholesterics are flexible materials for optical applications. Their optical properties have been studied by de Vries.[48] As the pitch goes to infinity, the nematic state is recovered.

Cholesteric phases are formed by optically active molecules, the standard showpiece being cholesteryl cinnamate whose chemical structure is given in Fig. 4. With the equilibrium state characterized by a screw axis, the lack of a

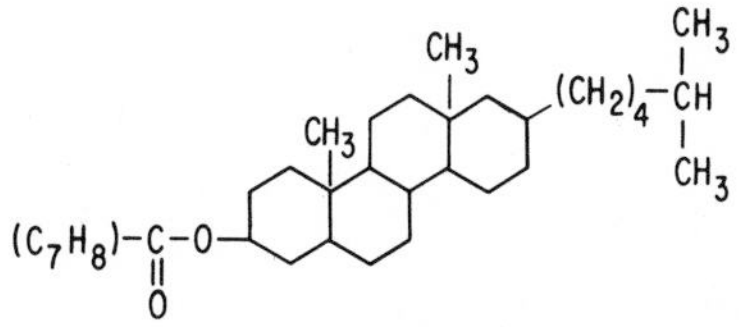

**Fig. 4**  Cholesteryl cinnamate forms a cholesteric phase from 156 to 197°C.

center of symmetry of the molecules must play an important role; the interaction energy between two molecules can now contain pseudoscalar contributions. In straightforward extension of (3.4), we therefore take the following pair Hamiltonian[49]

$$\mathscr{H} = -\tfrac{1}{2} \sum_{\alpha,\beta} \hat{\mathbf{R}}^\alpha : \mathbf{J}^{\alpha\beta} : \hat{\mathbf{R}}^\beta \tag{3.28}$$

where we have dispensed with the central part and the external field, and the pair energy is given by

$$J_{ij}^{\alpha\beta} = J(r^{\alpha\beta})\delta_{ij} - J'(r^{\alpha\beta})\varepsilon_{ijk}\frac{(\mathbf{r}^{\alpha\beta})_k}{r^{\alpha\beta}} \tag{3.29}$$

The first, scalar, term favors parallel alignment of two molecules, while the second, pseudoscalar, term favors orthogonal stacking. It is the competition between these two components that results in the helical order. It is obvious that optically active molecules will contribute terms of this form in a Legendre polynomial expansion of their interaction energy. Goossens[40] has distilled a pseudoscalar contribution out of the dipole-quadrupole van der Waals interaction. Equation (3.29) is of course the most general Hamiltonian that is bilinear in $\hat{R}^\alpha$ and $\hat{R}^\beta$.

The further procedure is the same, *mutatis mutandis*, as that employed for nematics. For simplicity we again consider the molecules to be placed on a cubic lattice, and we employ the inner field Hamiltonian (3.6). Corresponding to (3.9), we then obtain the equation

$$\eta^\alpha_{ij} = \sum_\beta J^{\alpha\beta}_{ik} Q^\beta_{kj} \tag{3.30}$$

where both sites ought to be symmetrized, and the trace subtracted. $\eta^\alpha$ is determined self-consistently by Eq. (3.8). We use as an ansatz

$$Q^\alpha_{ij} = S(n_i{}^\alpha n_j{}^\alpha - \tfrac{1}{3}\delta_{ij}) \tag{3.31}$$

$$n_i{}^\alpha = (\cos qz^\alpha, \sin qz^\alpha, 0) \tag{3.32}$$

Equation (3.30) then gives

$$\eta^\alpha_{ij} = S\tilde{J}(2q)(n_i{}^\alpha n_j{}^\alpha - \tfrac{1}{3}\delta_{ij}) + \delta\eta_{ij} \tag{3.33}$$

where

$$\tilde{J}(2q) = J(2q) - J'(2q)$$
$$J(2q) = \sum_\beta J(r^{\alpha\beta}) \cos (2qz^{\alpha\beta}) \tag{3.34}$$
$$J'(2q) = -\sum_\beta J'(r^{\alpha\beta}) \frac{z^{\alpha\beta}}{r^{\alpha\beta}} \sin (2qz^{\alpha\beta})$$

The last term in (3.33) is a constant of the form

$$\delta\eta_{ij} = \tfrac{1}{2}S[\tilde{J}(2q) - \tilde{J}(0)](q_i q_j - \tfrac{1}{3}\delta_{ij}) \tag{3.35}$$

where **q** is the unit vector in 3-direction. Anticipating that in equilibrium $q = q_0$ is extremely small, we omit this term. (It would describe nematic ordering not quite at right angles to the helical axis.) Then, using (3.8) the theory is self-consistent if

$$S = \psi[\beta\tilde{J}(2q)S] \tag{3.36}$$

where the function $\psi$ is defined in (3.11).

From the previous discussion of nematic ordering, it is evident that the system will order with the pitch $2\pi/q_0$ for which $\tilde{J}(2q)$ is at its maximum. The isotropic-to-cholesteric transition temperature is then given [cf. (3.15)] by

$$k_B T_c = 0.1468\tilde{J}(2q_0) \tag{3.37}$$

Note that $J(2q)$ is an even function of $q$, decreasing from $q = 0$, while $J'(2q)$ is an odd function of $q$. The maximum of $\tilde{J}(2q)$ will thus be at $q_0 \neq 0$, but $q_0$

will be very small since (if) the pseudoscalar contribution to the interaction energy is very small compared to the normal scalar one. Note that, as presented, the pitch is determined by properties of the interaction energy, and that it is therefore independent of temperature, in conflict with experiment. (A similar result is obtained in the theory of helical spin ordering in magnetic solids.[50]) One may surmise that since $J^{\alpha\beta}$ is an averaged force, in a fashion such as suggested in (3.18), the temperature dependence of $q$ is due to that of the short-range order and the fact that $J(r)$ and $J'(r)$ are not of equal range. If this is true, the temperature dependence of $q_0$ reflects that of the pair correlation function, and thus the compressibility. However, this suggestion is very tentative.

A very successful calculation of the pitch has been reported by Keating.[51] Taking the spontaneous nematiclike order for granted, Keating relates the pitch to cubic anharmonic terms in the forces resisting the relative twisting of neighboring molecules. Such terms can only exist for unsymmetric molecules. The calculation is similar to that used to obtain the expansion coefficient in solids. Keating finds that $q_0$ increases linearly with temperature, in good agreement with experimental data.

### D. The Smectic A Phase

Smectic A is the simplest one of the several smectic mesophases. The molecular arrangement in nematiclike monolayers is shown in Fig. 2. The long molecular axes point, on the average, at right angles to the layers whose spacing is on the order of 20 Å. As in nematics, the molecules are believed to rotate freely about the long axis. As a consequence, smectic A phases are optically uniaxial. Many materials exhibit an isotropic-to-nematic phase transition followed by, at lower temperature, a smectic phase before the substance finally crystallizes. The smectic-to-nematic phase transition is again found to be weakly first order, with transition entropies of the same order of magnitude as in the nematic-to-isotropic transition, about 2 cal/deg mole. One theoretical model should be able to give both mesophases. Our presentation of such a model is adapted from the work of McMillan.[41] Essentially the same model has also been proposed by Kobayashi.[52]

We begin with the anisotropic pair Hamiltonian already employed before, namely,

$$\mathscr{H} = -\tfrac{1}{2} \sum_{\alpha,\beta} J(r^{\alpha\beta}) \hat{\mathbf{R}}^{\alpha} : \hat{\mathbf{R}}^{\beta} \tag{3.38}$$

where we have dispensed with external forces and omitted the isotropic pair interaction. Since we want to discuss the emergence of a translationally layered structure, we abandon the artifice of a lattice. Otherwise, the calculation is again based on (3.3) and proceeds exactly as before. We allow for $H$

the general one-particle Hamiltonian

$$H = - \sum_\alpha U(\mathbf{r}^\alpha, \Omega^\alpha) \tag{3.39}$$

and choose $U$ so as to minimize the mean field free energy $F_H$. This results in the equation ($n = N/V$)

$$U(\mathbf{r}, \Omega) = n \int d^3r' \int d\Omega' \, J(\mathbf{r} - \mathbf{r}')[\hat{R}(\Omega) : \hat{R}(\Omega')]\rho(\mathbf{r}', \Omega') \tag{3.40}$$

where the averaged, angle-dependent, one-particle density $\rho$ is given by

$$\rho(\mathbf{r}, \Omega) = \frac{e^{\beta U(\mathbf{r}, \Omega)}}{\frac{1}{V} \int d^3r \int d\Omega \, e^{\beta U(\mathbf{r}, \Omega)}} \tag{3.41}$$

Equations (3.44) and (3.45) determine the self-consistent mean field. They correspond to our earlier (3.8) and (3.9). As they stand, they are too hard to solve.

Now in terms of $\rho$, the center-of-mass particle density at point $\mathbf{r}$ is given by

$$\langle n(\mathbf{r}) \rangle = n \int d\Omega \, \rho(\mathbf{r}, \Omega) \tag{3.42}$$

and the local tensor order parameter by

$$\langle \hat{R}_{ij}(\mathbf{r}) \rangle = \left\langle \sum_\alpha R_{ij}^\alpha \delta(\mathbf{r} - \mathbf{r}^\alpha) \right\rangle = n \int d\Omega \, \hat{R}_{ij}(\Omega)\rho(\mathbf{r}, \Omega) \tag{3.43}$$

To allow for both nematic and smectic order, we therefore us the ansatz

$$\rho(\mathbf{r}, \Omega) = \frac{\langle n(\mathbf{r}) \rangle}{4\pi n} + \frac{15}{8\pi} S(\mathbf{r})[n_i^0 \hat{R}_{ij}(\Omega)n_j^0]$$

$$= \frac{\langle n(\mathbf{r}) \rangle}{4\pi n} + \frac{15}{8\pi} S(\mathbf{r})(\cos^2 \vartheta - \tfrac{1}{3}) \tag{3.44}$$

where $\mathbf{n}^0$ is an arbitrary unit vector with respect to which the angle $\vartheta$ is measured. Equation (3.44) reproduces (3.42) and gives

$$\langle \hat{R}_{ij}(\mathbf{r}) \rangle = nS(\mathbf{r})(n_i^0 n_j^0 - \tfrac{1}{3}\delta_{ij}) \tag{3.45}$$

If we assume further that the degree of orientational order is sinusoidal in space, with periodicity $d$,

$$S(\mathbf{r}) = S + 2\sigma \cos \frac{2\pi z}{d} \tag{3.46}$$

we obtain, via (3.40), the self-consistent potential

$$U(\mathbf{r}, \Omega) = J\left[S + \sigma\alpha \cos\frac{2\pi z}{d}\right](\cos^2 \vartheta - \tfrac{1}{3}) \qquad (3.47)$$

where, to maintain similarity with the earlier nematic case, we have defined

$$J = n \int d^3r \, J(r)$$

$$\alpha J = 2n \int d^3r \, J(r) \cos\frac{2\pi z}{d} \qquad (3.48)$$

The two dimensionless order parameters $S$ and $\sigma$ are therefore determined by the equations

$$S = \langle \tfrac{3}{2}\cos^2 \vartheta - \tfrac{1}{2} \rangle_\rho \qquad (3.49)$$

and

$$\sigma = \left\langle \cos\frac{2\pi z}{d} (\tfrac{3}{2}\cos^2 \vartheta - \tfrac{1}{2}) \right\rangle_\rho \qquad (3.50)$$

where the average of a function of $z$ and $\cos \vartheta$ is defined by

$$\langle A(z, \cos \vartheta) \rangle_\rho \equiv \frac{\int_0^d dz \int_0^1 d \cos \vartheta \, A(z, \cos \vartheta) e^{\beta U}}{\int_0^d dz \int_0^1 d \cos \vartheta \, e^{\beta U}} \qquad (3.51)$$

with $U(z, \cos \vartheta)$ as given by (3.51). There is no independent self-consistency condition for the spatially varying density $\langle n(\mathbf{r}) \rangle$. This is reasonable, since it is only in conjunction with orientational order that a layered structure is established. Therefore, once $\sigma \neq 0$ is obtained, the density is periodic in space as well and is given by (3.42).

This theory is determined by two physical parameters $\alpha$ and $J$. $J$ simply sets the temperature scale, leaving the dimensionless interaction strength $\alpha$ which controls the competition between the nematic ($S \neq 0$, $\sigma = 0$) and the smectic ($S \neq 0$, $\sigma \neq 0$) phases. McMillan has solved (3.49) and (3.50) numerically for varying $\alpha$. We give some of his results in Fig. 5. Obviously, in the nematic phase the present theory is identical to the Maier–Saupe theory.

For $\alpha < 0.98$, a first-order transition takes place from the isotropic to the nematic state, followed at lower temperature by a transition to smectic A. This second transition is first order for $\alpha > 0.70$, but becomes second order for $\alpha < 0.70$. For $\alpha > 0.98$, the nematic phase is eliminated.

The theory does not determine the direction of the equilibrium director $\mathbf{n}^0$ relative to the smectic layers, and it does not predict the layer thickness $d$.

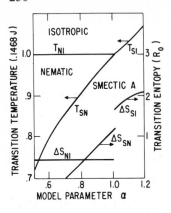

**Fig. 5a** Phase diagram for theoretical model parameter $\alpha$. Transition entropies are also shown in units of $R_0 = 1.986$ cal/deg mole. (According to McMillan.[41])

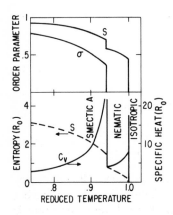

**Fig. 5b** Order parameters, entropy, and specific heat versus reduced temperature $T/T_{NI}$ for $\alpha = 0.85$ showing the first-order smectic-A–nematic transition. (According to McMillan.[41])

McMillan has argued that these parameters are determined by local packing effects, the competition between the long-range attractive force treated here, and short-range anisotropic repulsion. For a purely attractive Gaussian potential

$$J(r) = J_0 e^{-(r/r_0)^2} \tag{3.52}$$

the parameter $\alpha$ is given by

$$\alpha = 2e^{-(\pi r_0/d)^2} \qquad 0 < \alpha < 2 \tag{3.53}$$

If one therefore takes $d$ to be of order of $l$, the length of the molecule, then $\alpha$ increases with increasing length. On this basis, McMillan compared his results with experimental phase diagram data obtained by Arnold[53] on the homologous series 4-ethoxybenzal-4-amino-$n$-alkyl-$\alpha$-methylcinnamate and found good qualitative agreement. For many details the reader should consult McMillan's articles.[41]

As stated, mean field theory predicts that, for a range of parameters ($\alpha < 0.70$) the transition between the smectic A and nematic phases will be second order. Experimentally, the question is unsettled; it would be difficult to differentiate between a critical phase transition and one with a very small transition entropy. However, a subtle argument by Halperin, Lubensky, and Ma[54] is of interest to theoreticians. These investigators point out that the strong director fluctuations, left out of mean field theory, couple to the smectic order parameter and drive the system to a first-order transition.

## IV. HYDRODYNAMICAL FLUCTUATIONS

We now turn our attention to some interesting and remarkable dynamical phenomena which characterize an ordered liquid: anisotropic hydrodynamical flow and very strong light scattering. The latter, in particular, has puzzled physicists for decades. To explain the turbid appearance of nematics below their clear point $T_c$, Zocher[55] and others hypothesized the existence of large "swarms" of aligned molecules which scatter light, Tyndall-fashion as it were. As de Gennes[26] has pointed out, however, the mean linear size of such swarms would then be expected to be given by the range of the static director correlation function $\chi_{nn}(\mathbf{r} - \mathbf{r}')$ whose Fourier transform we have given in (2.19). This function is of infinite range.

Yet precisely this fact is the obvious and simple explanation[28] of the turbidity of nematics. It is by now well known (see, e.g., Ref. 28) that the intensity of inelastically scattered light is given by the equilibrium-averaged correlation function

$$I_{i \to j}(\mathbf{k}, \omega) = \int_{-\infty}^{\infty} d(t - t') \int d^3(r - r') \, e^{i\omega(t-t') - \mathbf{k}(\mathbf{r}-\mathbf{r}')} \langle \delta\varepsilon_{ij}(\mathbf{r}t)\delta\varepsilon_{ij}(\mathbf{r}'t') \rangle \quad (4.1)$$

where $\delta\varepsilon_{ij}(\mathbf{r}t)$ is a local fluctuation of the dielectric tensor, and an unimportant overall constant has been suppressed. $\omega = \Omega_{in} - \Omega_{out}$ and $\mathbf{k} = \mathbf{k}_{in} - \mathbf{k}_{out}$ are the shifts in photon frequency and wave vector, and $i$ and $j$ indicate the polarization of the incident and scattered photon, respectively. Now in local equilibrium, fluctuations of the dielectric constant are due to fluctuations of the mass density and of the local order [see also (2.9)]; that is,

$$\delta\varepsilon_{ij}(\mathbf{r}t) = \left(\frac{\partial \varepsilon}{\partial \rho}\right)_T \delta\rho(\mathbf{r}t)\delta_{ij} + \varepsilon^a \delta Q_{ij}(\mathbf{r}t) \quad (4.2)$$

Consequently, the total intensity for depolarized scattering is given, for example, by

$$I^{\text{tot}}_{1\to3}(\mathbf{k}) = \int d\omega\, I_{13}(\mathbf{k}\omega) \sim \chi_{R_{13}R_{13}}(k) \sim \frac{S^2}{Kk^2} \qquad (4.3)$$

where $K$ is a Frank coefficient, and we have not been careful with details of the angle dependence. The divergence of the director fluctuations at long wavelength is the explanation of the strong light scattering observed from liquid crystals. The line shape of the scattering spectrum follows from the hydrodynamical equations of motion. In this section I therefore consider the hydrodynamical fluctuations in a liquid crystal—those modes of collective motion whose lifetimes go to infinity rigorously as the wavelength becomes infinitely large. I will again restrict attention to nematics.

The development of a useful hydrodynamical theory for nematics[7] is due to Ericksen and Leslie. More recent derivations have been given by Stephen, Lubensky, and others. The Orsay group[28] and Lubensky[12] showed how the Leslie equations can be converted into statements about measurable correlation functions. These derivations justify the hydrodynamical theory, most of it, in orthodox fashion as a consequence of the microscopic conservation laws for mass, momentum, and energy. However, there is a fourth equation, namely, one for the director field, for which there is no separable conservation law, and whose validity was therefore in question.[22] (Reference 22 contains an error which was corrected in Refs. 13 and 56.) New derivations, which do in fact justify the Leslie theory as used by the Orsay group, have been given by Forster et al.[13] (see also Martin, Parodi, and Pershan[56]). I follow the presentation given in Ref. 57.

The point these rederivations make is that the director equation is a consequence not of a separate microscopic continuity equation for "internal angular momentum," but a consequence rather of broken rotational symmetry. The hydrodynamically slow director mode thus is a many-body equivalent of the Goldstone bosons in quantum field theory.[58] Its analogs are hydrodynamical spin waves in ferro- and antiferromagnets,[59] or second sound in superfluid helium.[60]

The thermodynamical statement of broken symmetry is the existence of a Frank free energy of macroscopic distortion, or the property of the static director correlation functions to diverge at long wavelengths [see (2.19)]. The divergence is due to the fact that it costs no energy to rotate the system uniformly through an angle $\delta\theta$, and thus that the energy necessary to create a sinusoidal alignment distortion of wavelength $\lambda$ goes to zero as $\lambda$ goes to infinity. To iron out the inhomogeneity, the system must transport "information about the local alignment direction" over a distance of order $\lambda$; if the distortion is allowed to relax, it does so with a hydrodynamical lifetime which approaches infinity as a power of $1/\lambda^2$—in fact, by a random-walk argument, $\tau \sim 1/\lambda^2$.

## A. Correlation Functions and Memory Functions

The hydrodynamical variables in a nematic are thus $A_\mu(\mathbf{r}t) = \{\rho; \mathbf{g}, \varepsilon; n_1, n_2\}$ namely, the conserved densities $\rho$, $\mathbf{g}$, and $\varepsilon$ of mass, momentum, and energy, and the "symmetry-breaking variables" $n_1$ and $n_2$ defined in (2.20). The equilibrium director $\mathbf{n}^0$ is again chosen in the three-direction. Only the two components $R_{13} \sim n_1$ and $R_{23} \sim n_2$ of the order parameter have divergent susceptibilities $\chi$ and are therefore hydrodynamical.

Defining the Fourier transforms

$$A_\mu(\mathbf{k}, t) = V^{-1/2} \int d^3r \; e^{i\mathbf{k}\cdot\mathbf{r}} A_\mu(\mathbf{r}t) \tag{4.4}$$

one studies the matrix of equilibrium-averaged time correlation functions

$$S_{\mu\nu}(\mathbf{k}z) = \int_0^\infty d(t - t') \, e^{iz(t-t')} \langle A_\mu^*(\mathbf{k}, t) A_\nu(\mathbf{k}, t') \rangle$$

$$= \langle A_\mu^*(\mathbf{k}) \frac{i}{z - \mathscr{L}} A_\nu(\mathbf{k}) \rangle \tag{4.5}$$

where $\mathscr{L}$ is the classical Liouville operator defined so that $\dot{A}(t) = i\mathscr{L}A(t)$, and $A_\mu(\mathbf{k}) \equiv A_\mu(\mathbf{k}, t = 0)$; also, Im $z > 0$. As Mori[61] has shown, $S(\mathbf{k}z)$ can always be represented by the dispersion relation

$$[z\delta_{\mu\lambda} - \Omega_{\mu\lambda}(\mathbf{k}) + i\Sigma_{\mu\lambda}(\mathbf{k}z)]S_{\lambda\nu}(\mathbf{k}z) = i\beta^{-1}\chi_{\mu\nu}(\mathbf{k}) \tag{4.6}$$

where

$$\chi_{\mu\nu}(\mathbf{k}) = \beta \langle A_\mu^*(\mathbf{k}) A_\nu(\mathbf{k}) \rangle \tag{4.6a}$$

$$\omega_{\mu\nu}(\mathbf{k}) \equiv \Omega_{\mu\lambda}(\mathbf{k})\chi_{\lambda\nu}(\mathbf{k}) = i\beta \langle \dot{A}_\mu^*(\mathbf{k}) A_\nu(\mathbf{k}) \rangle \tag{4.6b}$$

$$\sigma_{\mu\nu}(\mathbf{k}z) \equiv \Sigma_{\mu\lambda}(\mathbf{k}z)\chi_{\lambda\nu}(\mathbf{k}) = \beta \langle \dot{A}_\mu^*(\mathbf{k})Q \frac{i}{z - Q\mathscr{L}Q} Q\dot{A}_\nu(\mathbf{k}) \rangle \tag{4.6c}$$

$\chi_{\mu\nu}$ represents the strength of static fluctuations; its definition here agrees with that in (2.18). The frequency matrix $\omega_{\mu\nu}$ describes the instantaneous, reversible response to a disturbance. Because of the fluctuation-dissipation theorem,[62] $\omega_{\mu\nu}$ can also be obtained from the Poisson bracket

$$\omega_{\mu\nu}(\mathbf{k}) = i\langle [A_\mu^*(\mathbf{k}), A_\nu(\mathbf{k})]_{PB} \rangle \tag{4.7}$$

which can often be evaluated explicitly.

Equation (4.6) thus throws the complexity of the dynamics into the wave vector and frequency-dependent matrix of memory functions $\sigma_{\mu\nu}(kz)$. The simplification that is possible in the hydrodynamical regime of small $k$ and $|z|$ is due to the following features.

1. $Q$ in (4.6c) is a projector which rejects fluctuations of the variables $A_\mu$ themselves. If we have succeeded in assembling in the set $\{A_\mu\}$ all hydrodynamically slow modes, then the modified propagator $i(z - Q\mathscr{L}Q)^{-1}$ generates only microscopically fast processes. For small $|z|$ therefore, $\sigma_{\mu\nu}(kz)$ can be replaced by $\sigma_{\mu\nu}(k0)$.

2. If $A_\mu(\mathbf{k})$ is a microscopically conserved quantity, with the continuity equation $\dot{A}_\mu(k) = ikj_{A_\mu}(k)$, where vector indices are suppressed, then the corresponding $\sigma_{\mu\mu}(kz)$ is explicitly proportional to $k^2$. The eigenmodes are given by the poles of $S(kz)$. Conservation laws thus give rise to hydrodynamical poles, lifetimes at least of order $1/k^2$.

3. The symmetry-breaking variables $n_1$ and $n_2$ are not microscopically conserved so that, for example, $\sigma_{n_1 n_1}(kz)$ is finite even as $k \to 0$. However, the susceptibility $\chi_{n_1 n_1}(\mathbf{k})$ diverges as $1/k^2$, so that $\Sigma_{n_1 n_1}(kz)$ is again of order $k^2$ as $k \to 0$. Thus, nonconserved variables with divergent static susceptibilities also lead to slow hydrodynamical relaxation.

On this basis, and using the symmetry properties of the variables $A_\mu(\mathbf{r}t)$, the matrices $\Omega_{\mu\nu}(\mathbf{k})$ and $\Sigma_{\mu\nu}(\mathbf{k}, 0)$ can be calculated[57] to leading order in $k$, and the hydrodynamical motion represented by a set of thermodynamical (from $\chi$ and $\omega$) and transport (from $\sigma$) coefficients. For example,

$$\sigma_{g_2 g_2}(\mathbf{k}, 0) = v_2 k_1{}^2 + v_3 k_3{}^2 + (k^4) \tag{4.8}$$

where $g_2(\mathbf{k})$ is that component of the momentum density orthogonal to both the equilibrium director $\mathbf{n}^0$ and the wave vector $\mathbf{k}$, and where $v_2$ and $v_3$ are two positive shear viscosities. Equation (4.8) represents a Kubo relation for these coefficients. For another example,

$$\sigma_{n_i n_i}(\mathbf{k}, 0) = \xi + (k^2) \qquad i = 1, 2 \tag{4.9}$$

where $\xi$, conventionally called $\gamma_1^{-1}$, is a positive rotational relaxation coefficient.

Altogether, the full spectrum of hydrodynamical fluctuations in a compressible nematic is characterized by (1) five viscosities $v_1, \ldots, v_5$, (2) two heat conductivities $\kappa_\perp$ and $\kappa_\parallel$, (3) the director relaxation coefficient $\xi \equiv \gamma_1^{-1}$, and (4) one real coefficient $\lambda$ which gives the reversible coupling between director motion and symmetric stress. In addition, one has the usual thermodynamical derivatives, namely, the isotropic speed of sound $c = (\partial p/\partial\rho)_s^{1/2}$

and the specific heat ratio $c_p/c_v$, plus the three Frank constants $K_i$. Several of these 14 coefficients can be obtained from an analysis[13] of light-scattering experiments.[63]

## B. Macroscopic Equations of Motion

For a discussion of hydrodynamical flow experiments, it is more convenient to have equations of motion for the macroscopic near-equilibrium fluctuations $\delta \langle A_\mu(\mathbf{r}t) \rangle \equiv \langle A_\mu(\mathbf{r}t) \rangle_{\text{noneq}} - \langle A_\mu(\mathbf{r}t) \rangle_{\text{eq}}$. By means of the fluctuation-dissipation theorem, these can be easily obtained from the correlation function description. (The traditional procedure is the reverse; see Ref. 64 or 12.) The pertinent equation is that near equilibrium the undisturbed relaxation of $\delta \langle A_\mu(\mathbf{r}t) \rangle$ from its initial value $\delta \langle A_\mu(\mathbf{r}, 0) \rangle$ is given by

$$\delta \langle A_\mu(\mathbf{r}t) \rangle = S_{\mu\lambda}(\mathbf{r} - \mathbf{r}', t) \beta \chi_{\lambda\nu}^{-1}(\mathbf{r}' - \mathbf{r}'') \delta \langle A_\nu(\mathbf{r}'', 0) \rangle \qquad (4.10)$$

where two integrals over $d^3r'$ and $d^3r''$ have been omitted for clarity. An equation of motion [see (4.6)] for the correlation function is thus also an equation of motion for the macroscopic fluctuations $\delta \langle A_\mu \rangle$. On this basis the hydrodynamical equations are easily obtained.

The conserved densities obey the conservation laws:

$$\frac{\partial}{\partial t} \delta \langle \rho(\mathbf{r}t) \rangle + \mathbf{V} \cdot \delta \langle \mathbf{g}(\mathbf{r}t) \rangle = 0 \qquad \text{mass} \qquad (4.11a)$$

$$\frac{\partial}{\partial t} \delta \langle g_i(\mathbf{r}t) \rangle + \nabla_j \delta \langle \underline{\tau}_{ij}(\mathbf{r}t) \rangle = 0 \qquad \text{momentum} \qquad (4.11b)$$

$$\frac{\partial}{\partial t} \delta \langle \varepsilon(\mathbf{r}t) \rangle + \mathbf{V} \cdot \delta \langle \mathbf{j}^\varepsilon(\mathbf{r}t) \rangle = 0 \qquad \text{energy} \qquad (4.11c)$$

which identify the stress tensor $\underline{\tau}_{ij}$ and the energy current $\mathbf{j}^\varepsilon$. We add the statement

$$\frac{\partial}{\partial t} \delta \langle n_i(\mathbf{r}t) \rangle + \delta X_i(\mathbf{r}t) = 0 \qquad (4.12)$$

which simply defines the macroscopic quantity $\delta X_i(\mathbf{r}t)$. Equation (4.6) then gives rise to constitutive relations for the fluxes in (4.11) and (4.12). It is customary, and suggestive in view of the form of (4.6b) and (4.6c) to introduce the local forces $\delta a_\mu(\mathbf{r}t)$ conjugate to the variables $\delta \langle A_\mu(\mathbf{r}t) \rangle$, and we do so by the relation (in Fourier transforms)

$$\delta a_\mu(\mathbf{k}t) \equiv \chi_{\mu\nu}^{-1}(\mathbf{k}) \delta \langle A_\nu(\mathbf{k}t) \rangle \qquad (4.13)$$

where it is sufficient to use the susceptibility matrix to lowest order in $k$. The conjugate forces are then consistent with the designation

$$\delta a_\rho(\mathbf{r}t) = \frac{\delta p(\mathbf{r}t)}{\rho} \qquad \delta a_q = \frac{\partial T}{T} \qquad \delta a_{g_i} = v_i = \frac{\delta\langle g_i\rangle}{\rho} \qquad (4.14)$$

as fluctuations of the pressure, temperature, and average velocity. Again following custom,[64] we have introduced the entropy density $q(\mathbf{r}t) \equiv \varepsilon(\mathbf{r}t) - \rho(\mathbf{r}t)(\varepsilon + p)/\rho$ where $\varepsilon$, $p$, and $\rho$ are the equilibrium values of energy density, pressure, and mass density. The forces conjugate to the director fluctuations are

$$\delta a_{n_1}(\mathbf{r}t) \equiv \delta h_1(\mathbf{r}t) = [K_1\nabla_1{}^2 + K_3\nabla_3{}^2]\delta\langle n_1(\mathbf{r}t)\rangle \qquad (4.15a)$$

$$\delta a_{n_2}(\mathbf{r}t) \equiv \delta h_2(\mathbf{r}t) = [K_2\nabla_1{}^2 + K_3\nabla_3{}^2]\delta\langle n_2(\mathbf{r}t)\rangle \qquad (4.15b)$$

where, to avoid too many indices, we have again chosen the equilibrium alignment $n^0$ in the 3-direction, and spatial variations in the 1–3 plane. The forces $\delta h_i$ are the local stresses discusssed by the Orsay group[28] on the basis of the Frank free energy.

Given the conjugate forces, the derivation of hydrodynamical equations is standard. Expanding the fluxes in terms of the forces and their gradients yields the same results obtained on the basis of (4.6). It is difficult here to avoid complicated notation. The results are given explicitly in Ref. 13, and therefore are not repeated here. Reference 13 also gives the relations of transport coefficients, like the $v_i$ that enter (4.8), to those introduced by Leslie for the incompressible nematic. We want to point out that the stress tensor obtained in this fashion is symmetric, a point that has been discussed in detail by Martin, Parodi, and Pershan.[56]

The director equation obtained is of the form

$$\frac{\partial}{\partial t}\delta\langle n_i\rangle - (\boldsymbol{\omega} \times \mathbf{n}^0)_i = \lambda A_{ij}n_j{}^0 + \xi(K_i\nabla_1{}^2 + K_3\nabla_3{}^2)\delta\langle n_i\rangle \qquad (4.16)$$

for $i = 1$ and 2. Here $A_{ij} = \frac{1}{2}(\nabla_i v_j + \nabla_j v_i)$ and $\boldsymbol{\omega} = \frac{1}{2}(\nabla \times \mathbf{v})$. Because of the broken symmetry, $\delta\langle n_i(\mathbf{r}t)\rangle$ is macroscopically conserved even though there is no microscopic conservation law corresponding to this property. The term $\boldsymbol{\omega} \times \mathbf{n}^0$ guarantees solid-body rotation in the equilibrium rotating state. In addition, there is reversible coupling to symmetric shear flow, given by $\lambda$, and the last term describes the damping of director rotation.

The coefficient $\lambda$ can be extracted from light-scattering experiments.[13,63] More accurate values were obtained by flow-alignment measurements recently performed on several nematics and for a range of temperatures.[65,66] Namely, under a flow in the $x$-direction, with

$$v_x = ay \qquad v_y = v_z = 0 \qquad (4.17)$$

the director stabilizes in the $x$–$y$ plane, according to (4.16), at an angle $\theta$ to the direction of flow, which is given by

$$\cos 2\theta = \frac{1}{\lambda} \qquad (4.18)$$

Only if $\lambda > 1$ is stable flow alignment obtained. A microscopic calculation[67] of $\lambda$ based on a thermodynamical sum rule[57] results in

$$\lambda(T) = \frac{1}{3} + \frac{2}{3}\frac{I_l + 2I_t}{I_l - I_t} S^{-1}(T) \qquad (4.19)$$

where $I_l$ and $I_t$ are molecular moments of inertia, as defined in (2.7), and $S(T)$ is the conventional degree of order (2.4). Equation (4.19) predicts stable flow alignment at all temperatures in the nematic range, as reported in Ref. 66 but in contrast to the experiment of Ref. 65.

In cholesteric and in smectic phases, the hydrodynamical equations are more complicated. A unified treatment has been given by Martin, Parodi, and Pershan,[56] to which the reader is referred. From the fundamental point of view, cholesterics are particularly interesting. The range of validity of hydrodynamics is for low-frequency fluctuations whose wavelength is large compared to the "natural lengths" in the system. In cholesterics this would restrict hydrodynamics[34,68] to very small wave vectors $k \ll q_0$, where $2\pi/q_0$ is the very large helical pitch. It is not clear therefore whether the present hydrodynamical theory can be used to discuss light scattering in cholesteric systems.

## C. Pretransitional Fluctuations

In the isotropic phase the fluctuations in a nematic in the strict hydro-dynamical limit are those of any isotropic fluid.[64] However, since the iso-tropic-to-nematic transition is almost critical, there are strong pretransitional effects. These phenomena have been discussed by de Gennes.[69]

We pointed out above that, for a sizable region above $T_c$, the correlation length in a nematic is still much larger than it is in a normal liquid, or the static director correlation function at $k = 0$, $\chi(T)$, which diverges at $T = T^*$, remains large for $T$ not far above $T_c \approx T^*$ [see (2.16)]. The discussion following (4.6) suggests that the lifetime associated with orientational fluctuations above $T_c$ is given by[69]

$$\tau = \gamma^{-1}\chi(T) = \frac{\gamma^{-1}(T)}{a(T - T^*)} \qquad (4.20)$$

which near $T_c$ is therefore much larger than a normal microscopic decay time. $\gamma$ is a rotational diffusion coefficient (see below).

To make the discussion definite, we consider depolarized light scattering in the isotropic phase. The axes are chosen so that the incident and scattered wave vectors $\mathbf{k}_{in}$ and $\mathbf{k}_{out}$ are in the $yz$-plane and the momentum shift $\mathbf{k} = \mathbf{k}_{in} - \mathbf{k}_{out}$ points in the $z$-direction. Suppose that the incident light is polarized in the $x$-direction, and the scattered light selected is polarized in the scattering plane. With this geometry one observes what in scattering circles is known as the $I_V{}^H$ spectrum. In a simple liquid $I_V{}^H$ is zero. From (4.1), $I_V{}^H$ is given, except for an overall constant, by

$$I_V{}^H = \sin^2\left(\frac{\theta}{2}\right)\langle|\varepsilon_{12}^2|\rangle + \cos^2\left(\frac{\theta}{2}\right)\langle|\varepsilon_{13}^2|\rangle \tag{4.21}$$

where $\theta$ is the scattering angle, related to $\mathbf{k}$ by

$$k = 2k_{in}\sin\left(\frac{\theta}{2}\right) \tag{4.22}$$

and (4.21) is only meant to illustrate the components of $\varepsilon_{ij}$, and thus by (4.2) of the tensor $R_{ij}$, whose fluctuations determine the spectrum.

Within standard approximations this spectrum follows easily from (4.6). From symmetry the component $R_{12}$ is completely decoupled from the conserved densities as well as from the other tensor components $R_{ij}$. Under these circumstances, the frequency matrix element $\omega_{R_{12}R_{12}}$ in (4.6) vanishes from time-reversal symmetry, and we approximate the memory function by its value at $k = 0$ and $z = 0$, namely,

$$\alpha^{-2}\sigma_{R_{12}R_{12}}(0,0) \equiv \gamma = (k_B T\alpha^2)^{-1}\int_0^\infty dt \int d^3r \, \langle \dot{R}_{12}(\mathbf{r}t)\dot{R}_{12}(00)\rangle \tag{4.23}$$

Properly speaking, we should have inserted the projectors present in (4.6c), but we are not about to calculate $\gamma$. $\alpha$ is the constant contained in (2.16); it has been divided out for purposes of normalization. At $k = 0$ we then obtain from (4.6) the result

$$S_{R_{12}R_{12}}(k = 0, z) = \frac{i\beta^{-1}\chi(T)}{z + i\tau^{-1}} \tag{4.24}$$

where $\tau$ is as given in (4.20). At $k = 0$, $\langle|\varepsilon_{13}^2|\rangle$ and $\langle|\varepsilon_{12}^2|\rangle$ are identical, and thus the depolarized spectrum is given by the real part of (4.24), or

$$I_V{}^H = (k_B T)\chi(T) \cdot \frac{2\tau}{1 + \omega^2\tau^2} \tag{4.25}$$

Equation (4.25) is a good approximation as long as $\tau$ is much larger than the natural microscopic decay times in the system which will be true at least close to $T_c$. Stinson and Litster[19] have verified the Lorentzian line shape and measured $\tau$ as a function of temperature for MBBA for a range of 15°C above $T_c \cong 42°C$.

The constant $\gamma$ is, as (4.23) makes evident, a rotational diffusion coefficient. Rotational diffusion is a thermally activated process, so that one expects[70] that

$$\gamma(T) = \gamma_0 \, e^{-W/k_B T} \qquad (4.26)$$

($\gamma$ is a local property. It should be largely continuous across the isotropic-to-nematic phase transition. In the nematic range $\gamma$ is identical except for a proportionality constant to $\xi$ of (4.9), conventionally called $\gamma_1^{-1}$. $\gamma_1^{-1}$ has recently been measured as a function of temperature[66] and fits the Arrhenius law (4.26) very well. I apologize for the confused notation, but I believe that it characterizes the microscopic nature of this coefficient better.) The resulting equation

$$\tau^{-1} \equiv \Gamma = (\gamma_0 a) e^{-W/k_B T} (T - T^*) \qquad (4.27)$$

fits the temperature-dependent experimental line width very well,[19] and has also been verified by experiments on optical-field-induced ordering.[20] $W/k_B \approx 2800°K$ for MBBA, which is about half the value measured in the nematic phase.

Equation (4.25) omits the coupling of the orientational variable $R_{13}$ to the transverse momentum density $g_2$, that is, the coupling of collective molecular rotation and viscous flow, which is not strictly zero if $\mathbf{k} \neq 0$. Taking the coupling into account would result in a more complicated spectrum,[69] namely, a superposition of two Lorentzians. As a practical matter coupling to flow can be neglected as long as $\tau \gg \tau_\eta = (k^2 \eta/\rho)^{-1}$, where $\eta$ is the shear viscosity and $\rho$ is the mass density. With $\eta$ on the order of 2 cps, one obtains $\tau_\eta \approx 10^{-8}$ sec, while $\tau > 10^{-7}$ sec for $T - T_c < 15°C$ (see Refs. 19 and 20). However, $\tau$ decreases with increasing temperature, while $\tau_\eta$ increases. It may be worthwhile to look for the effects of the coupling (as already suggested by de Gennes[69]). What one would find in the range where $\tau \approx \tau_\eta$ is a doublet spectrum of exactly the type found by Stegeman and Stoicheff[71] in the depolarized scattering from many molecular liquids such as quinoline and aniline. While none of these liquids is a nematogen, the formal theory of the Stegeman–Stoicheff "shear waves" given by Andersen and Pecora[72] is in fact identical to the one suggested for nematics by de Gennes,[69] and follows immediately from (4.6) if the coupling of $R_{13}$ to $g_2$ is taken into account. Indeed, from available data on the temperature dependence[71,73] of the doublet splitting, the suggestion is strong that Stegeman–Stoicheff "shear waves" are in fact a pretransitional effect, heralding a transition to orientational order which is intercepted by crystallization.

# References

1. F. Reinitzer, *Wiener Monatsh. Chem.*, **9**, 421 (1888).
2. O. Lehman, *Z. Phys. Chem.* (Leipzig), **4**, 462 (1889); *Flüssige Kristalle*, Engelman, Leipzig, 1904.
3. I. B. Chistyakov, *Usp. Fiz. Nauk.*, **89**, 563 (1967).
4. G. H. Brown, J. W. Doane, and V. D. Neff, *Crit. Rev. Solid State Sci.*, **1**, 303 (1970).
5. A. Saupe, *Ann. Rev. Phys. Chem.*, **24**, 441 (1973).
6. G. W. Gray, *Molecular Structure and the Properties of Liquid Crystals*, Academic, New York, 1962.
7. P. G. de Gennes, *The Physics of Liquid Crystals*, Oxford University Press, London, 1974.
8. J. F. Johnson and R. S. Porter, Eds., *Liquid Crystals and Ordered Fluids*, Plenum, New York, 1970.
9. G. H. Brown and M. M. Labes, Eds., *Liquid Crystals 3*, Vols. I and II, Gordon and Breach, New York, 1972.
10. *Liquid Crystals, Symp. Faraday Soc.* No. 5 (1971).
11. P. G. de Gennes, *Phys. Lett.*, **30A**, 454 (1969).
12. T. C. Lubensky, *Phys. Rev.*, **A2**, 2497 (1970).
13. D. Forster, T. C. Lubensky, P. C. Martin, P. S. Pershan, and J. Swift, *Phys. Rev. Lett.*, **26**, 1016 (1971).
14. H. J. Freiser, *Mol. Cryst. Liq. Cryst.*, **14**, 165 (1971).
15. D. L. Uhrich, J. M. Wilson, and W. A. Resch, *Phys. Rev. Lett.*, **24**, 355 (1970).
16. T. R. Taylor, J. L. Fergason, and S. L. Arora, *Phys. Rev. Lett.*, **24**, 359 (1970).
17. R. Alben, *Phys. Rev. Lett.*, **30**, 778 (1973).
18. L. Landau and E. M. Lifschitz, *Statistical Physics*, Addison-Wesley, Reading, Mass., 1959.
19. T. W. Stinson, III, and J. D. Litster, *Phys. Rev. Lett.*, **25**, 503 (1970).
20. G. K. L. Wong and Y. R. Shen, *Phys. Rev. Lett.*, **30**, 895 (1973).
21. H. Arnold, *Z. Physik. Chem.* (Leipzig), **226**, 146 (1964).
22. P. C. Martin, P. S. Pershan, and J. Swift, *Phys. Rev. Lett.*, **25**, 844 (1970).
23. F. C. Frank, *Discuss. Faraday Soc.*, **25**, 19 (1958).
24. C. W. Oseen, *Trans. Faraday Soc.*, **29**, 883 (1933).
25. I. Haller, *J. Chem. Phys.*, **57**, 1400 (1972).
26. P. G. de Gennes, *Solid State Commun.* **10**, 753 (1972).
27. P. S. Pershan, *J. Appl. Phys.*, **45**, 1590 (1974).
28. Groupe d'Etude des Cristaux Liquides, *J. Chem. Phys.*, **51**, 816 (1968).
29. P. C. Hohenberg and P. C. Martin, *Ann. Phys.* (N.Y.), **34**, 291 (1965).
30. P. C. Hohenberg, *Phys. Rev.*, **158**, 383 (1967).
31. N. D. Mermin and H. Wagner, *Phys. Rev. Lett.*, **17**, 1133 (1966).
32. J. P. Straley, *Phys. Rev.*, **A4**, 675 (1972).
33. T. C. Lubensky, *Phys. Rev. Lett.*, **29**, 206 (1972).
34. T. C. Lubensky, *Phys. Rev.*, **A6**, 452 (1972).
35. J. Nehring and A. Saupe, *J. Chem. Phys.*, **56**, 5527 (1972).
36. R. G. Priest, *Mol. Cryst. Liquid Cryst.*, **17**, 129 (1972).
37. I. Haller and J. D. Litster, *Phys. Rev. Lett.*, **25**, 1550 (1970).
38. H. Gruler and G. Meier, *Z. Naturforsch.*, **A28**, 479 (1973).
39. W. Maier and A. Saupe, *Z. Naturforsch.*, **A14**, 882 (1959); **A15**, 287 (1960).
40. W. J. A. Goossens, *Mol. Cryst. Liquid Cryst.*, **12**, 237 (1971).
41. W. L. McMillan, *Phys. Rev.*, **A4**, 1238 (1971); **A6**, 936 (1972).
42. C. Fan and M. J. Stephen, *Phys. Rev. Lett.*, **25**, 500 (1970).
43. S. Chandrasekhar and N. V. Madhusudana, *Acta Crystallogr.*, **A27**, 303 (1971).

44. R. Alben, *Mol. Cryst. Liq. Cryst.*, **10**, 21 (1970).
45. A. Saupe, *Angew. Chem. Int. Ed.*, **7**, 97 (1968).
46. R. Alben, *Mol. Cryst. Liquid Cryst.*, **13**, 193 (1971).
47. A. Saupe, *Z. Naturforsch.*, **A15**, 815 (1960).
48. H. N. de Vries, *Acta Crystallogr.*, **4**, 219 (1951).
49. T. C. Lubensky, *Phys. Rev. Lett.*, **29**, 206 (1972).
50. T. Nagamiya, *Solid State Phys.*, **20**, 305 (1967).
51. P. N. Keating, *Mol. Cryst. Liq. Cryst.*, **8**, 315 (1969).
52. K. K. Kobayashi, *Phys. Lett.*, **31A**, 125 (1970); *J. Phys. Soc. Jap.*, **29**, 101 (1970).
53. H. Arnold, *Z. Phys. Chem.* (Leipzig), **239**, 283 (1968); **240**, 185 (1969).
54. B. I. Halperin, T. C. Lubensky, and Shang-keng Ma, *Phys. Rev. Lett.*, **32**, 292 (1974).
55. H. Zocher, *Ann. Phys.*, **31**, 570 (1938).
56. P. C. Martin, O. Parodi, and P. S. Pershan, *Phys. Rev.*, **A6**, 2401 (1972).
57. D. Forster, *Ann. Phys.* (N.Y.), **84**, 505 (1974).
58. A. Katz and Y. Frishman, *Nuovo Cimento*, **42A**, 1009 (1966).
59. B. I. Halperin and P. C. Hohenberg, *Phys. Rev.*, **188**, 898 (1969).
60. P. C. Hohenberg and P. C. Martin, *Ann. Phys.* (N.Y.), **34**, 291 (1965).
61. H. Mori, *Prog. Theor. Phys.* (Kyoto), **33**, 423 (1965).
62. P. C. Martin, in C. de Witt and R. Balian, Eds., *Many-Body Physics*, Gordon and Breach, New York, 1968.
63. Orsay Liquid Crystal Group, *Phys. Rev. Lett.*, **22**, 1361 (1969).
64. L. P. Kadanoff and P. C. Martin, *Ann. Phys.* (N.Y.), **24**, 419 (1963).
65. Ch. Gähwiller, *Phys. Rev. Lett.*, **28**, 1554 (1972).
66. S. Meiboom and R. C. Hewitt, *Phys. Rev. Lett.*, **30**, 261 (1973).
67. D. Forster, *Phys. Rev. Lett.*, **32**, 1161 (1974).
68. C. Fan, L. Kramer, and M. Stephen, *Phys. Rev.*, **A2**, 2482 (1970).
69. P. G. de Gennes, *Mol. Cryst. Liq. Cryst.*, **12**, 193 (1971).
70. See, for example, I. Z. Fisher, *Statistical Theory of Liquids*, University of Chicago Press, Chicago, 1964, p. 5.
71. G. I. A. Stegeman and B. P. Stoicheff, *Phys. Rev.*, **A7**, 1160 (1973).
72. H. C. Andersen and R. Pecora, *J. Chem. Phys.*, **54**, 2584 (1971).
73. G. R. Alms, D. R. Bauer, J. I. Brauman, and R. Pecora, *J. Chem. Phys.*, **59**, 5304 (1973).

# THEORY OF ELECTRON STATES IN LIQUID METALS

## L. E. BALLENTINE

*Simon Fraser University, Burnaby, British Columbia, Canada*

### TABLE OF CONTENTS

## I. INTRODUCTION AND BASIC CONCEPTS

The study of electron states in liquids presents greater theoretical difficulties than do the corresponding studies for gases and crystals. A single atom or small molecule contains only a modest number of particles, and it is feasible to solve the Schrödinger equation for it by well-tested approximation schemes. A crystal contains a practically infinite number of particles, but the periodicity of the lattice allows one to use Bloch's theorem to simplify the problem greatly. No corresponding simplification exists for a liquid.

Some of the ideas and techniques of solid-state theory can be taken over directly for the study of liquid metals. Because the ions are much heavier than the electrons, we may use the adiabatic approximation and consider the motion of electrons in the static field of the ions which are regarded as being at rest. The electrons will be regarded as independent (subject only to the Pauli exclusion principle), and the many-body nature of the problem will be taken into account only in the construction of self-consistently screened potentials.

The specifically liquid features of the system enter through the dependence of the potential energy of the electron, hence also of the wave functions, on the positions of the ions. Common sense suggests, and experiment confirms, that interesting physical quantities do not depend on the detailed instantaneous spatial arrangements of the ions, but only on certain statistical correlations. So it is necessary to average over the ensemble of all possible arrangements of the ions, thereby introducing pair, triplet, and so on, correlation functions. It is sometimes erroneously suggested that the introduction of this ensemble average depends on some ergodic hypothesis (replacement of a time average by an ensemble average), and therefore that it might be valid only for liquids but not for amorphous solids in which the disorder is "frozen." But in fact the ergodic or nonergodic nature of the system is irrelevant, for we may legitimately treat the liquid disorder as if it were static, neglect of the atomic motion being justified by the adiabatic approximation.* The justification in statistical physics for considering only the ensemble average of some quantity is that the average value should become overwhelmingly most probably in the limit as the system becomes arbitrarily large (the relative fluctuation from the average being typically of order $N^{-1/2}$, where $N$ is the number of atoms in the system). Since we have not proved such a theorem, our treatment lacks rigor to the same extent that most statistical mechanics textbooks lack rigor. In principle, all quantum mechanical predictions about the electrons are expressible as averages of products of wave functions. However, the wave functions possess so much irrelevant detail that it is seldom useful to work with them directly, and a more appropriate Green's function or density matrix formalism must be employed.

The Green function method was first introduced into the theory of liquid metals by Edwards.[28] The Green operator or resolvent is defined for arbitrary

---

* Greene and Kohn[32] showed that the effect of atomic motion on the electrical resistivity of sodium is less than 2%. Rice[65] also showed that atomic motion has a negligible effect on the electrical resistivity of liquid metals, and that its effect on the thermal conductivity is only a few percent.

complex $E$ as

$$G(E) = (E - H)^{-1} = \sum_n \frac{|\psi_n\rangle\langle\psi_n|}{E - E_n} \quad (1.1)$$

Here $H$ is the one-electron Hamiltonian, and $E_n$ and $|\psi_n\rangle$ are its eigenvalues and eigenvectors. The spectral operator is defined as

$$\rho(E) = \lim_{\eta \to 0} \left(\frac{-1}{2\pi i}\right)[G(E + i\eta) - G(E - i\eta)] = \sum_n |\psi_n\rangle\langle\psi_n|\delta(E - E_n) \quad (1.2)$$

The last member of (1.2) is only formal. To be precise and rigorous, one should pass to the limit of an infinite system while $\eta$ is small but nonzero and $\delta(E - E_n)$ is replaced by a peaked function of width $\eta$. If the portion of the energy spectrum within the range $E \pm \eta/2$ can be treated as a continuum, they we may formally let $\eta$ go to zero and obtain the Dirac distribution $\delta(E - E_n)$.

Although any representation could be used for these operators, it is convenient to use the momentum representation because of the translational invariance of the ensemble after averaging. We refer to the quantity

$$G(\mathbf{k}, E) = \langle\mathbf{k}|G(E)|\mathbf{k}\rangle$$

or its ensemble average

$$\mathscr{G}(k, E) = \langle G(\mathbf{k}, E)\rangle \quad (1.3)$$

as the Green function. Of fundamental importance is the spectral function

$$\rho(\mathbf{k}, E) = \langle\mathbf{k}|\rho(E)|\mathbf{k}\rangle = \frac{-1}{\pi} \text{Im } G(\mathbf{k}, E + i0) = \sum_n |\langle\mathbf{k}|\psi_n\rangle|^2\delta(E - E_n) \quad (1.4)$$

which gives us the momentum distribution of electrons with energy $E$. For a perfect crystal $\psi_n$ would be a Bloch wave, and $\rho(\mathbf{k}, E)$ would consist of an array of delta functions in $\mathbf{k}$ space, separated by reciprocal lattice vectors. The functional relationship between $E$ and the positions of these delta functions in $\mathbf{k}$ space defines the band structure $E(\mathbf{k})$. Neither momentum nor crystal momentum is a good quantum number for an electron in a disordered system such as a liquid. Therefore $\rho(\mathbf{k}, E)$ is generally a continuous function of $\mathbf{k}$ for a given $E$, and no precise $E(\mathbf{k})$ relation can be defined.

From the completeness of the set of states $\{\psi_n\}$, we obtain the sum rule

$$\int_{-\infty}^{\infty} \rho(\mathbf{k}, E)\, dE = 1 \quad (1.5)$$

The density of states per unit energy per unit volume (for one spin orientation) is given by

$$n(E) = \Omega^{-1}\,\mathrm{Tr}\,\rho(E) = (2\pi)^{-3}\int \rho(\mathbf{k}, E)\,d^3k \qquad (1.6)$$

where $\Omega$ is the volume of the system.

The ensemble average of the spectral operator $\rho(E)$ is translationally invariant, and its form in the position representation is

$$\bar{\rho}(R, E) = \langle\langle \mathbf{r} + \mathbf{R}|\rho(E)|\mathbf{r}\rangle\rangle \qquad (1.7a)$$

$$= (2\pi)^{-3}\int e^{i\mathbf{k}\cdot\mathbf{R}}\langle\rho(\mathbf{k}, E)\rangle\,d^3k \qquad (1.7b)$$

$$= \left\langle \sum_n \psi_n(\mathbf{r} + \mathbf{R})\psi_n{}^*(\mathbf{r})\delta(E - E_n)\right\rangle \qquad (1.7c)$$

This function, which may be called the wave function autocorrelation function, has received very little attention. Although it contains the same information as the spectral function, it presents it in a different way. Evaluating the trace in (1.6) in the position representation, we see that $\bar{\rho}(0, E)$ is just the ensemble average of the density of states $n(E)$. The spatial range of $\bar{\rho}(R, E)$ is clearly a measure of the range of phase coherence of the wave functions as seen from (1.7c).

One might think that the form of the autocorrelation function would distinguish between localized and nonlocalized states, but it is doubtful that this is true. If the states of energy $E$ are localized with a spatial extent $\lambda$, then clearly $\bar{\rho}(R, E)$ will go to zero rapidly for $R > \lambda$. But suppose that the states are extended over the entire volume of the system. Since (1.7c) is independent of $\mathbf{r}$, we may consider a typical term, $\psi(\mathbf{r} + \mathbf{R})\psi^*(\mathbf{r})$, and average it over $\mathbf{r}$. It is to be expected that this average will vanish for large enough $R$, because of the finite range of phase coherence, even though $|\psi(\mathbf{r})|$ may be of the same order of magnitude throughout the entire volume of the liquid metal. A satisfactory test for localization or nonlocalization of stationary states is the vanishing or nonvanishing of $\langle|\psi(\mathbf{r} + \mathbf{R})|^2|\psi(r)|^2\rangle$ for large $\mathbf{R}$. This information cannot be obtained from the ensemble average Green function, but it may be obtained from the average of the product of two Green functions. Such an average is indeed required to calculate the electrical conductivity, but we shall not consider such problems.

In this review we consider only the most fundamental aspects of the electronic states: their energy distribution, spectral composition, and closely related experiments. For other aspects of liquid metal physics, we refer the reader to the review articles by Cusack[24] and Wilson,[82] and the books by March[58] and Faber.[30]

## II. THE ELECTRON-ION POTENTIAL

We assume that a valence electron moves in the self-consistent potential due to the ion cores and the other valence electrons. This total potential is usually taken to be the sum of identical spherically symmetric potentials centered on each atom:

$$V(\mathbf{r}) = \sum_j v(\mathbf{r} - \mathbf{R}_j) \qquad (2.1)$$

Then, a matrix element of $V$ in momentum representation

$$\langle \mathbf{k} | V | \mathbf{k} + \mathbf{q} \rangle = \langle \mathbf{k} | v | \mathbf{k} + \mathbf{q} \rangle \sum_j e^{i\mathbf{q} \cdot \mathbf{R}_j} \qquad (2.2)$$

is a product of two factors: a form factor describing a single atom relative to its own center, and a structure factor depending only on the positions of the atoms. Thus any term of $n$th order in $V$ involves the $n$-particle distribution function which we discuss in subsequent sections.

Before continuing we should ask whether (2.1) is really justified. The potential due to the nuclei certainly can be written in this form, and the core electrons are so tightly bound that it is an excellent approximation to write the potential of the bare ion cores as

$$V_b(\mathbf{r}) = \sum_j v_b(\mathbf{r} - \mathbf{R}_j) \qquad (2.3)$$

But the valence (or conduction) electrons are not bound to any particular ion, so it is not obvious that the screening potential they produce can be written in this form. However, if the screening charge distribution is treated as a linear response to the self-consistent potential, then it follows that the sum of the bare ion potential and the screening potential does indeed have the form (2.1). If $V_b(\mathbf{r})$ is a local potential, as is implied by the notation (i.e., a function of one variable $\mathbf{r}$), then

$$\langle \mathbf{k} | V_b | \mathbf{k} + \mathbf{q} \rangle = V_b(\mathbf{q})$$

is independent of $\mathbf{k}$ and is just the Fourier transform of $V_b(\mathbf{r})$. Moreover, the transform of the self-consistent potential is given by

$$V(\mathbf{q}) = \frac{V_b(\mathbf{q})}{\varepsilon(q)} = \sum_j \frac{e^{i\mathbf{q} \cdot \mathbf{R}_j} V_b(q)}{\varepsilon(q)} \qquad (2.4)$$

where $\varepsilon(q)$ is a dielectric screening function. This is the Fourier transform of (2.1), provided we identify the effective potential centered on atom $j$, $v(\mathbf{r} - \mathbf{R}_j)$, as the inverse Fourier transform of $e^{i\mathbf{q} \cdot \mathbf{R}_j} v_b(q)/\varepsilon(q)$.

In this way we can regard a liquid metal as an assembly of identical *neutral pseudoatoms*.[83] Each pseudoatom is regarded as carrying its own screening charge cloud, and the superposition of these charge clouds auto-matically guarantees that $V(\mathbf{r})$ remains self-consistent as the pseudoatoms move about the liquid, provided only that the *linear* response model of screening is adequate. Of course the division of the valence electron density into a set of pseudoatom screening charge clouds is only a mathematical one, and an individual electron is not a permanent member of any one charge cloud.

The justification of (2.1) has now been reduced to the justification of linear screening theory. This will certainly be valid if the total potential $V$ is "small," but the appropriate criterion for smallness is not obvious. The usual criterion of smallness in applying perturbation theory, that the matrix elements of the perturbation be small compared with the energy-level spacing, is not applicable when the volume of the system becomes large and the energy levels form a quasi-continuum.

Because the matrix element (2.2) is a fluctuating quantity with zero mean, it is more appropriate to consider the ensemble average of its absolute square:

$$\langle|\langle\mathbf{k}|V|\mathbf{k}+\mathbf{q}\rangle|^2\rangle = N^{-1}|N\langle\mathbf{k}|v|\mathbf{k}+\mathbf{q}\rangle|^2 a(q) \tag{2.5}$$

where $a(q) = N^{-1}\langle|\sum_j e^{i\mathbf{q}\cdot\mathbf{R}_j}|^2\rangle$. If the number of atoms $N$ and the volume of the system $\Omega$ become very large (with $N/\Omega = $ constant), then $N\langle\mathbf{k}|v|\mathbf{k}+\mathbf{q}\rangle$ $=O(1)$ and $a(q) = O(1)$ for $q \neq 0$. But the mere fact that (2.5) is small, of order $N^{-1}$, does not justify perturbation theory because the number of terms that are second order in $V$ and which have nearly the same magnitude is of order $N$.

In subsequent developments it will become apparent that an appropriate criterion for smallness is that, for $q \neq 0$, $|N\langle\mathbf{k}|v|\mathbf{k}+\mathbf{q}\rangle|^2 a(q)$ be small compared with $\varepsilon_F{}^2$, where $\varepsilon_F$ is the kinetic energy of a free electron at the Fermi surface.* It is well known that (see, e.g., Ziman[83])

$$\lim_{q\to 0} N^{-1}\langle\mathbf{k}|v|\mathbf{k}+\mathbf{q}\rangle = -\tfrac{2}{3}\varepsilon_F \tag{2.6}$$

and that

$$\lim_{q\to 0} a(q) = nk_B T K_T \tag{2.7}$$

where $n = N/\Omega$ is the number density of atoms, $k_B$ is Boltzmann's constant, $T$ is the absolute temperature, and $K_T$ is the isothermal compressibility. Since (2.6) is not small, we see that the potential associated with a neutral pseudoatom can never be weak in a rigorous sense. But fortunately, (2.7) is typically about 0.03 for liquid metals near their melting point, so our weak-

---

* See (5.3) which shows that if this criterion is satisfied then the deviation of the electronic structure from the free-electron model will be small.

perturbation criterion is satisfied, at least for small $q$. But this example shows that the validity of our assumptions depends on the structure of the liquid, as well as on the nature of the atom. Near the critical point, where $K_T$ becomes infinite, our assumptions may not be valid.

It is clear that $\langle \mathbf{k}|V|\mathbf{k} + \mathbf{q}\rangle$ cannot be small for all larger values of $q$, because $V(\mathbf{r})$ is strong enough to bind the core electrons. But the core states are of little interest, being so tightly bound that they are essentially the same in a solid or liquid metal as in a free atom or ion. Therefore it is convenient to introduce a *pseudopotential* whose lowest eigenvalues correspond to the valence eigenvalues of the true potential. Instead of the usual Schrödinger equation

$$\left(\frac{p^2}{2m} + V\right)|\psi_n\rangle = E_n|\psi_n\rangle \tag{2.8}$$

we have a similar equation:

$$\left(\frac{p^2}{2m} + W\right)|\phi_n\rangle = E_n|\phi_n\rangle \tag{2.9}$$

containing the pseudopotential $W$ which is defined to be equal to the true potential $V$ in the interstitial region outside the ion cores, but which is much weaker than $V$ inside the ion cores. The valence pseudo-wave function $\phi_v(\mathbf{r})$ is equal to the true valence wave function $\psi_v(\mathbf{r})$ in the interstitial region, but $\phi_v(\mathbf{r})$ has no nodes within the core, whereas $\psi_v(\mathbf{r})$ must have several nodes in order to be orthogonal to the core wave functions.

There are many ways of constructing a pseudopotential. Austin, Heine, and Sham[6] showed that any operator of the form

$$W_{AHS} = V + \Sigma_c|\psi_c\rangle\langle F_c| \tag{2.10}$$

is a valid pseudopotential. Here the sum runs over all the core eigenfunction $\psi_c$, and the $F_c$ are completely arbitrary functions. For any pseudopotential of this form, the true valence wave function can be recovered from the pseudo-wave function by projecting out the core wave functions,

$$|\psi_v\rangle = |\phi_v\rangle - \Sigma_c|\psi_c\rangle\langle\psi_c|\phi_v\rangle \tag{2.11}$$

Alternatively, one can construct a model pseudopotential $W_M$ whose functional form within the ion core depends on some parameters adjusted to yield a smooth $\phi_v(\mathbf{r})$ which correctly matches $\psi_v(\mathbf{r})$ at the boundary of the ion core.[1,38,72,73] No equation such as (2.11) is valid in this case.

The pseudopotential is a nonlocal operator, that is, $\langle \mathbf{k}|W|\mathbf{k} + \mathbf{q}\rangle$ depends on $\mathbf{k}$ as well as on $\mathbf{q}$. In some forms $W$ also depends on the magnitude of the eigenvalue $E$. This leads to technical complications [e.g., a simple form like (2.4) for the screened pseudopotential is not valid], but the smallness of $W$, compared with $V$, serves to justify the treatment of screening as a linear response, and so finally we are able to justify writing the *total screened pseudopotential* as a superposition of contributions from identical neutral pseudoatoms:

$$\langle \mathbf{k} | W | \mathbf{k} + \mathbf{q} \rangle = \langle \mathbf{k} | w | \mathbf{k} + \mathbf{q} \rangle \sum_j e^{i\mathbf{q}\cdot\mathbf{R}_j} \qquad (2.12)$$

The many practical problems involved in a pseudopotential calculation are discussed in a book by Harrison.[37] By paying careful attention to self-consistency and many-body effects, one can calculate accurate model pseudopotentials from first principles, at least for simple metals.[11] However, it may be more convenient for some purposes to use an empirical pseudopotential which has been fitted to experimental data.[22] In most cases we may assume that an accurate electron–ion pseudopotential is known, although this assumption is reexamined in some of the cases to follow.

### III. THE CORRELATION FUNCTIONS

The structure of a liquid is characterized by a sequence of *distribution functions*. The $s$-particle distribution function is defined such that $n_s(\mathbf{R}_1 \dots \mathbf{R}_s)$ $d^3R_1 \cdots d^3R_s$ is the probability of finding a particle in the volume element $d^3R_1$ centered on $\mathbf{R}_1$, another particle in $d^3R_2$ centered on $\mathbf{R}_2$, and so on. It is normalized so that

$$\int \cdots \int n_s(\mathbf{R}_1 \cdots \mathbf{R}_s) \, d^3R_1 \cdots d^3R_s = N(N-1)\cdots(N-s+1) \quad (3.1)$$

where $N$ is the total number of atoms in the large but finite volume of the system.

In order to calculate the ensemble average of some power of the potential

$$V(\mathbf{r})V(\mathbf{r}') \cdots = \sum_{i,j,\dots} v(\mathbf{r} - \mathbf{R}_i)v(\mathbf{r} - \mathbf{R}_j) \cdots$$

we require other distribution functions which include the contributions of terms in which atoms $i$, $j$, and so on, are not all distinct. Therefore we introduce the *distribution functions including coincidences* $D_s(\mathbf{R}_1 \cdots \mathbf{R}_s)$ which are of the form:

$$D_1(\mathbf{R}_1) = n_1(\mathbf{R}_1) = n = N/\Omega$$
$$D_2(12) = n_2(12) + n\delta(12)$$
$$D_3(123) = n_3(123) + n_2(12)\delta(23) + n_2(23)\delta(31) + n_2(31)\delta(12) + n\delta(12)\delta(23)$$
$$D_4(1234) = n_4(1234) + [n_3(123)\delta(34) + \text{five similar terms}]$$
$$+ [n_2(12)\delta(13)\delta(24) + \text{two similar terms}]$$
$$+ [n_2(12)\delta(23)\delta(34) + \text{three similar terms}] + n\delta(12)\delta(23)\delta(34)$$

$$(3.2)$$

We use the abbreviated notations $n_2(12) \equiv n_2(\mathbf{R}_1, \mathbf{R}_2)$, $\delta(34) \equiv \delta(\mathbf{R}_3 - \mathbf{R}_4)$, and so on. The various contributions to $D_4(\mathbf{R}_1, \mathbf{R}_2, \mathbf{R}_3, \mathbf{R}_4)$ are for all four points distinct, one coincident pair, two coincident pairs, one coincident triplet, and all four points coincident. The normalization of these functions is

$$\int \cdots \int D_s(\mathbf{R}_1 \cdots \mathbf{R}_s) \, d^3 R_1 \cdots d^3 R_s = N^s \qquad (3.3)$$

Since all correlations are of finite range, the distribution functions must have the factorization property,

$$D_s(1 \cdots t \cdots s) \to D_t(1 \cdots t) D_{s-t}(t + 1 \cdots s) \qquad (3.4)$$

when the set of coordinates $\{\mathbf{R}_1 \cdots \mathbf{R}_t\}$ are far removed from the set $\{\mathbf{R}_{t+1} \cdots \mathbf{R}_s\}$. For any set of functions possessing this factorization property, one can introduce *cluster functions* $H_s(\mathbf{R}_1 \cdots \mathbf{R}_s)$ having the property

$$H_s(1 \cdots t \cdots s) \to 0 \qquad (3.5)$$

under the same conditions as for (3.4) The relation between distribution functions and cluster functions is[49,80]

$$D_1(\mathbf{R}_1) = H_1(\mathbf{R}_1) = n$$
$$D_2(12) = H_2(12) + H_1(1)H_1(2)$$
$$D_3(123) = H_3(123) + H_2(12)H_1(3) + H_2(23)H_1(1) + H_2(31)H_1(2)$$
$$\qquad + H_1(1)H_1(2)H_1(3)$$
$$D_4(1234) = H_4(1234) + H_3(123)H_1(4) + H_3(234)H_1(1)H_3(341)H_1(2)$$
$$\qquad + H_3(412)H_1(3) + H_2(12)H_2(34) + H_2(13)H_2(24)$$
$$\qquad + H_2(14)H_2(23) + H_2(12)H_1(3)H_1(4) + H_2(13)H_1(2)H_1(4)$$
$$\qquad + H_2(14)H_1(2)H_1(3) + H_2(23)H_1(1)H_1(4) + H_2(24)H_1(1)H_1(3)$$
$$\qquad + H_2(34)H_1(1)H_1(2) + H_1(1)H_1(2)H_1(3)H_1(4) \qquad (3.6)$$

The general rule is that $D_s(1 \cdots s)$ is equal to a sum of terms, each of which corresponds to a possible partitioning of the $s$ variables into clusters, the value of a term being the product of the corresponding cluster functions. It should be clear that there is no term-by-term equality between (3.2) and (3.6), even though the number of terms in the two sets of equations is the same.

It is convenient to introduce *dimensionless distribution functions*,

$$g_s(\mathbf{R}_1 \cdots \mathbf{R}_s) = \frac{n_s(\mathbf{R}_1 \cdots \mathbf{R}_s)}{n^s} \qquad (3.7)$$

which are normalized so that $g_s \equiv 1$ for uncorrelated particles. There is also a set of *dimensionless cluster functions*, $h_s(\mathbf{R}_1 \cdots \mathbf{R}_s)$, which are related to the

$g$ functions in the same way the $H$ functions are related to the $D$ functions:

$$g_1(\mathbf{R}_1) = h_1(\mathbf{R}_1) = 1$$
$$g_2(12) = h_2(12) + 1 \tag{3.8}$$
$$g_3(123) = h_3(123) + h_2(12) + h_2(23) + h_2(31) + 1$$

The relation between these $h$ functions and $H_s(1 \cdots s)$, the cluster functions including coincidences, is obtained by substituting (3.7) and (3.8) into (3.2) and (3.6):

$$H_1(1) = nh_1(1) = n$$
$$H_2(12) = n^2 h_2(12) + n\delta(12)$$
$$H_3(123) = n^3 h_3(123) + n^2[h_2(31)\delta(12) + h_2(12)\delta(23) \tag{3.9}$$
$$+ h_2(23)\delta(31)] + n\delta(12)\delta(23)$$

## IV. CLUSTER EXPANSIONS AND DIAGRAMMATIC REPRESENTATIONS

The primary calculational problem is to obtain the ensemble average of the resolvent operator,

$$\mathscr{G}(E) = \langle G(E) \rangle = \langle (E - H)^{-1} \rangle \tag{4.1}$$

from which the interesting physical quantities may be simply obtained, as discussed in Section I. If $G(E)$ is expanded as a series in powers of the potential or pseudopotential, then it is simple to perform the average term by term, an $n$th order term requiring the $n$-particle distribution function.

This expansion can be formally carried out either in terms of the potential $V$ or in terms of the pseudopotential $W$. In the former case the expansion will not be convergent, but it can be formally summed to infinite order by introducing certain invariant functions (e.g., the irreducible self-energy function, the $t$-matrix) whose existence does not depend on perturbation theory. In the latter case the series may converge (although it may be only asymptotic), but it will yield a spectral function [see (1.4)] which is related to the pseudo-wave functions rather than the true wave functions. Both approaches are useful. In this section we use the true potential, but a formally similar expansion clearly exists for the pseudopotential.

### A. The Irreducible Self-Energy Operator

Let us write the one-electron Hamiltonian in the form $H = H_0 + V$. The separation of the two terms is such that $V = \sum_j v_j$, where $v_j$ is an atomic potential centered at $\mathbf{R}_j$, and $H_0$ is independent of the positions of the atoms.

(Often $H_0$ is just the kinetic energy.) Then we can expand $G(E)$ as

$$G(E) = (E - H_0 - V)^{-1} = G_0 + G_0 V G_0 + G_0 V G_0 V G_0 + \cdots \quad (4.2)$$

where $G_0 = (E - H_0)^{-1}$.

The ensemble average of (4.2) can now be calculated term by term. For example,

$$\langle G_0 V G_0 V G_0 \rangle = G_0 \langle V G_0 V \rangle G_0 = G_0 \iint v_1 G_0 v_2 D_2(12)\, d^3 R_1\, d^3 R_2\, G_0$$

$$(4.3)$$

where $D_2(12) \equiv D_2(\mathbf{R}_1, \mathbf{R}_2)$ is given by (3.2). The simplicity of this equation is enhanced by our treating $v_1$, $v_2$, and $G_0$ as abstract operators. Written longhand in position representation, $v_1 G_0 v_2$ is

$$\langle \mathbf{r} | v_1 G_0 v_2 | \mathbf{r}''' \rangle = \int d^3 r' \int d^3 r'' \langle \mathbf{r} - \mathbf{R}_1 | v | \mathbf{r}' - \mathbf{R}_1 \rangle G_0(\mathbf{r}' - \mathbf{r}'')$$

$$\times \langle \mathbf{r}'' - \mathbf{R}_2 | v | \mathbf{r}''' - \mathbf{R}_2 \rangle$$

where $v$ is an atomic potential centered at the origin. If $v$ is a local operator (which is not generally the case), then the above expression simplifies to $v(\mathbf{r} - \mathbf{R}_1) G_0(\mathbf{r} - \mathbf{r}''') v(\mathbf{r}''' - \mathbf{R}_2)$.

Using the decomposition of the distribution function into a sum of products of cluster functions (3.6), we may substitute $D_2(12) = H_2(12) + H_1(1)H_1(2)$ into (4.3), obtaining

$$\langle G_0 V G_0 V G_0 \rangle = G_0 \iint v_1 G_0 v_2 H_2(12)\, d^3 R_1\, d^3 R_2\, G_0$$

$$+ G_0 \left[ \int v_1 H_1(1)\, d^3 R_1 \right] G_0 \left[ \int v_2 H_1(2)\, d^3 R_2 \right] G_0 \quad (4.4)$$

This result can be given a convenient symbolic representation:

$$(4.4')$$

in which a solid line represents $G_0$, a dashed line represents $v_j$, and the dot joining the dashed lines represents the cluster function $H_2(12)$. The first term of (4.4) is considered *irreducible*, because it cannot be factored (apart from the two trivial factors of $G_0$); and the second term is considered *reducible*, because it is equal to a product of irreducible factors. The generalization of this result to higher orders is obvious.

$$\langle G_0 V G_0 V G_0 V G_0 \rangle = G_0 \iiint v_1 G_0 v_2 G_0 v_3 D_3 (123) \, d^3R_1 \, d^3R_2 \, d^3R_3 \, G_0$$

$$(4.5)$$

which, using (3.6), we can represent as:

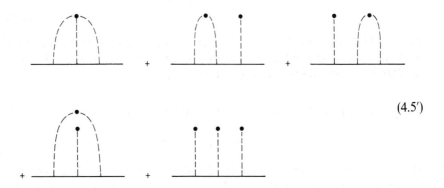

$$(4.5')$$

The first and fourth terms are irreducible (apart from trivial factors of $G_0$), and the others are reducible.

Because the set of all reducible diagrams in the expansion $\langle G \rangle$ is obtained by taking all possible products of irreducible diagrams in all possible orders, we may formally write the infinite series as

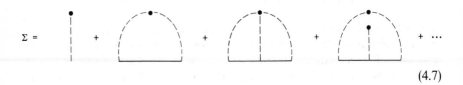

$$(4.6)$$

where the *irreducible self-energy operator* $\Sigma$ is defined as the sum of all irreducible diagrams (without $G_0$ factors on their extremities),

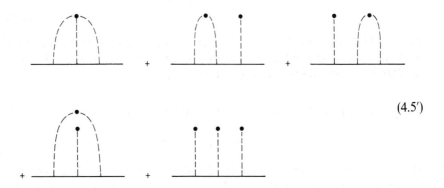

$$(4.7)$$

The sum of the operator series (4.6) is

$$\langle G \rangle = G_0(1 - \Sigma G_0)^{-1} = (G_0^{-1} - \Sigma)^{-1} = (E - H_0 - \Sigma)^{-1} \qquad (4.8)$$

If we now calculate the irreducible operator $\Sigma$ in some approximation, then (4.8) will automatically sum the infinite series of reducible terms that can be generated from it. This may appear to be only a computational device, but it will become apparent in the later sections that the summation of reducible terms, which is accomplished by (4.8), is an essential ingredient in any meaningful calculation.

The self-energy operator $\Sigma$ can also be defined without recourse to perturbation theory and diagrams:

$$\langle G \rangle \Sigma = \langle GV \rangle \qquad (4.9)$$

The infinite series (4.2) is equivalent to an implicit equation for $G$:

$$G = G_0 + G_0 VG \qquad (4.10)$$

Averaging (4.10) and using (4.9), we obtain (4.8).

Certain facts should be noted, since contrary statements exist in the literature:

1. The definition of $\Sigma$ and the use of (4.8) do not depend on perturbation theory.

2. No "geometric approximation" (replacing a nongeometric series by a geometric series) is involved in (4.8).

3. The use of (4.8) does not involve averaging $G^{-1}$ instead of $G$.

4. The essential property necessary for the definition of reducible and irreducible terms is the asymptotic factorization (3.4) of the distribution functions, which will be true provided all correlations are of finite range (i.e., range $\ll$ diameter of the system).

## B. Multiple Scattering and Cluster Expansions

The expansion described in the previous section was not a genuine cluster expansion, because the atoms in a cluster were not necessarily distinct. For example, the first term of (4.4) includes contributions in which $v_1$ and $v_2$ refer to distinct atoms and in which they refer to the same atom. Such physically different contributions can be systematically disentangled with the help of (3.9), and a cluster diagram representation can be introduced. Whereas in Section IV.A a dot connecting $s$ interaction lines represented a *cluster function including coincidences*, $H_s(1 \cdots s)$, we now separate the contributions from *clusters of distinct atoms*, as in (3.9). Thus the first term of (4.4) and (4.4′) is now written

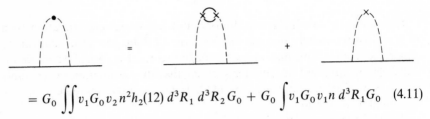

$$= G_0 \iint v_1 G_0 v_2 \, n^2 h_2(12) \, d^3R_1 \, d^3R_2 \, G_0 + G_0 \int v_1 G_0 v_1 n \, d^3R_1 G_0 \quad (4.11)$$

The general rules for evaluating a diagram are summarized in Table I.

## TABLE I

Cluster Diagram Rules and the Definition of the $t$ Matrix

| | |
|---|---|
| $G_0$ | |
| $v_j$ (centered at $R_j$) | |
| $nh(1) = n$ | |
| $n^2 h_2(12)$ | |
| $n^s h_s(1 \ldots s)$ | |
| Integrate: $\int \cdots \int (\cdots) \, d^3R_1 \cdots d^3R_s$ | |
| $t_j = v_j + v_j G_0 t_j$ | |

As an example of their use, the diagram

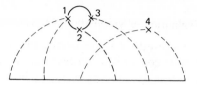

has the value

$$\iiiint v_1 G_0 v_1 G_0 v_4 G_0 v_2 G_0 v_3 G_0 v_4 n^3 h_3(123)\, d^3r_1\, d^3R_2\, d^3R_3\, n\, d^3R_4$$

Each cross corresponds to a distinct atom, and it is possible to develop a *multiple scattering* formulation of the theory. The first step is to define the *t matrix*, actually an operator whose matrix elements between free-particle states yield the scattering amplitude of a single atom (or rather a pseudo-atom):

$$t(E) = v + vG_0(E)t(E) = v + vG_0v + vG_0vG_0v + \cdots \qquad (4.12)$$

This equation may be given the symbolic representation

![diagram (4.12')]                                                    (4.12')

Now one may replace all the $v$-interaction lines with $t$-interaction lines in the expansions for $\langle G(E)\rangle$ and $\Sigma(E)$, provided one adds the *restriction* that successive interactions must refer to different atoms. The leading terms (up to third order in $t$) in the series for the irreducible self-energy operator are now

![diagram (4.13)]                                                      (4.13)

This formulation has the advantage that the $t$ matrix is a well-defined operator, even when a perturbation expansion in the potential $v$ is not valid.

It is also useful when one can obtain information about $t$ directly, without prior knowledge of $v$ and the use of (4.12).

### C. Evaluation in Momentum Representation

Because the previous results were developed in terms of abstract operators, one may use any convenient representation for their evaluation. Momentum representation is often convenient, because the translational invariance of the ensemble average implies that any averaged operator will be diagonal.

In momentum representation a matrix element of the potential of the atom at $\mathbf{R}_j$ differs from that of an atom at the origin by an exponential factor:

$$\langle \mathbf{k}|v_j|\mathbf{k} + \mathbf{q}\rangle = e^{i\mathbf{q}\cdot\mathbf{R}_j}\langle \mathbf{k}|v|\mathbf{k} + \mathbf{q}\rangle \tag{4.14}$$

Therefore the integrals over positions of the atoms in evaluation of cluster diagrams can be performed, yielding Fourier transforms of the cluster functions:

$$n^s \int \cdots \int e^{i(\mathbf{q}\cdot\mathbf{R}_1 + \cdots + \mathbf{q}_s\cdot\mathbf{R}_s)} h_s(\mathbf{R}_1 \cdots \mathbf{R}_s)\, d^3R_1 \cdots d^3R_s$$
$$= N\delta_{\mathbf{q}_1 + \cdots + \mathbf{q}_s, 0}\, b_s(\mathbf{q}_1 \cdots \mathbf{q}_s) \tag{4.15}$$

The Kronecker delta occurs because $h_s(\mathbf{R}_1 \cdots \mathbf{R}_s)$ is unchanged by a simultaneous displacement of all of its arguments, and so the function $b_s(\mathbf{q}_1 \cdots \mathbf{q}_s)$ is defined only when $\mathbf{q}_1 + \mathbf{q}_2 + \cdots + \mathbf{q}_s = 0$. In the most common case of $s = 2$, we use this fact and rotational invariance to simplify the notation, writing

$$b_2(\mathbf{q}, -\mathbf{q}) = b(q) = n \int e^{i\mathbf{q}\cdot\mathbf{R}} h(R)\, d^3R \tag{4.16}$$

where $h(R) = h_2(\mathbf{R}_1, \mathbf{R}_2)$, and $R = |\mathbf{R}_1 - \mathbf{R}_2|$.

The evaluation of diagrams in momentum representation is described in Table II. If two or more of the scattering operators ($v$ or $t$) refer to the same atom, there will be a factor like $e^{i\mathbf{q}\cdot\mathbf{R}_j}$ from each, and the sum of the corresponding momentum variables will occur in the Fourier transform of the cluster function.

As an example, the diagram

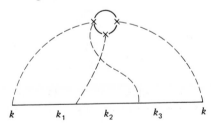

## TABLE II

Evaluation of Diagrams in Momentum Representation, and the Relation of
Structure Functions to Cluster Functions

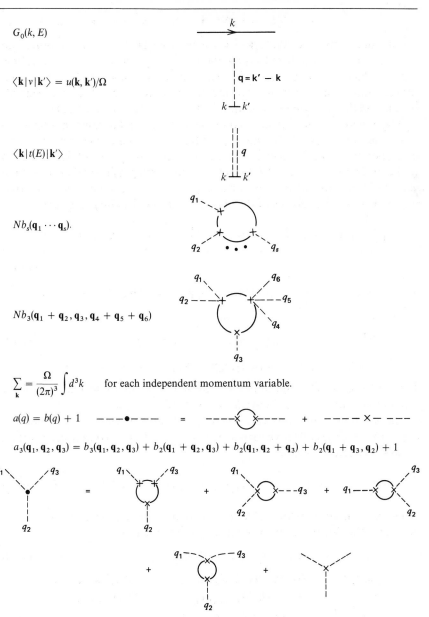

$G_0(k, E)$

$\langle \mathbf{k} | v | \mathbf{k}' \rangle = u(\mathbf{k}, \mathbf{k}')/\Omega$     $\mathbf{q} = \mathbf{k}' - \mathbf{k}$

$\langle \mathbf{k} | t(E) | \mathbf{k}' \rangle$

$N b_s(\mathbf{q}_1 \cdots \mathbf{q}_s)$.

$N b_3(\mathbf{q}_1 + \mathbf{q}_2, \mathbf{q}_3, \mathbf{q}_4 + \mathbf{q}_5 + \mathbf{q}_6)$

$$\sum_{\mathbf{k}} = \frac{\Omega}{(2\pi)^3} \int d^3k \qquad \text{for each independent momentum variable.}$$

$a(q) = b(q) + 1$

$a_3(\mathbf{q}_1, \mathbf{q}_2, \mathbf{q}_3) = b_3(\mathbf{q}_1, \mathbf{q}_2, \mathbf{q}_3) + b_2(\mathbf{q}_1 + \mathbf{q}_2, \mathbf{q}_3) + b_2(\mathbf{q}_1, \mathbf{q}_2 + \mathbf{q}_3) + b_2(\mathbf{q}_1 + \mathbf{q}_3, \mathbf{q}_2) + 1$

279

has the value

$$\sum_{\mathbf{k}_1} \sum_{\mathbf{k}_2} \sum_{\mathbf{k}_3} \langle \mathbf{k}|v|\mathbf{k}_1\rangle G_0(k_1, E)\langle \mathbf{k}_1|v|\mathbf{k}_2\rangle G_0(k_2, E)\langle \mathbf{k}_2|v|\mathbf{k}_3\rangle$$

$$\times\, G_0(k_3, E)\langle \mathbf{k}_3|v|\mathbf{k}\rangle N b_3(\mathbf{k} - \mathbf{k}_1 + \mathbf{k}_2 - \mathbf{k}_3, \mathbf{k}_1 - \mathbf{k}_2, \mathbf{k}_3 - \mathbf{k})$$

One must remember that $\langle \mathbf{k}|v|\mathbf{k}'\rangle$ is of order $\Omega^{-1}$, since the momentum eigenvectors are normalized over a volume $\Omega$. These factors always combine with the factor $N$ associated with each function $b_s$ and the factor $\Omega$ from converting a sum over $\mathbf{k}$ into an integral, so that the result is independent of $\Omega$ in the thermodynamical limit ($N, \Omega \to \infty$, $N/\Omega = n$, constant).

It is sometimes convenient to combine into one all diagrams that have the same factors of $\langle \mathbf{k}|v|\mathbf{k}'\rangle$ and $G_0(k, E)$ and the same connectivity. This corresponds to going backward from the cluster diagrams of Section IV.B to the diagrams of Section IV.A which do not distinguish whether or not the various atoms are distinct. Accordingly, we define *structure functions* that are the Fourier transforms of the *cluster functions including coincidences*,

$$a_s(\mathbf{q}_1 \cdots \mathbf{q}_s) = N^{-1} \int \cdots \int e^{i(\mathbf{q}_1 \cdot \mathbf{R}_1 + \cdots + \mathbf{q}_s \cdot \mathbf{R}_s)} H_s(\mathbf{R}_1 \cdots \mathbf{R}_s)\, d^3R_1 \cdots d^3R_s$$

$$(4.17)$$

with $\mathbf{q}_1 + \mathbf{q}_2 + \cdots + \mathbf{q}_s = 0$. From (3.9) and (4.15) we obtain

$$a_0 = 1$$

$$a_2(\mathbf{q}, -\mathbf{q}) = b(q) + 1 \equiv a(q) \qquad (4.18)$$

$$a_3(\mathbf{q}_1, \mathbf{q}_2, \mathbf{q}_3) = b_3(\mathbf{q}_1, \mathbf{q}_2, \mathbf{q}_3) + b(q_1) + b(q_2) + b(q_3) + 1$$

Table II shows the diagrammatic representation of these relations. In a random system one would have $b_s = 0$ ($s > 0$) and $a_s = 1$.

The structure functions can be obtained more directly by considering correlations of the collective variables $\rho_\mathbf{q} = \sum_j e^{i\mathbf{q} \cdot \mathbf{R}_j}$. The average of a product of these collective variables,

$$\langle \rho_{\mathbf{q}_1} \rho_{\mathbf{q}_2} \cdots \rho_{\mathbf{q}_s} \rangle = \int \cdots \int \rho_{\mathbf{q}_1} \rho_{\mathbf{q}_2} \cdots \rho_{\mathbf{q}_s} D_s(1 \cdots s)\, d^3R_1 \cdots d^3R_s \quad (4.19)$$

is just the Fourier transform of $D_s(1 \cdots s)$. If one substitutes (3.6) for $D_s(1 \cdots s)$ and takes the Fourier transform term by term, one obtains

$$\langle \rho_{\mathbf{q}_1} \rho_{\mathbf{q}_2} \rangle = N a(q_1) \delta_{\mathbf{q}_1 + \mathbf{q}_2, 0} + N^2\, \delta_{\mathbf{q}_1, 0}\, \delta_{\mathbf{q}_2, 0} \qquad (4.20)$$

$$\langle \rho_{\mathbf{q}_1} \rho_{\mathbf{q}_2} \rho_{\mathbf{q}_3} \rangle = N a_3(\mathbf{q}_1, \mathbf{q}_2, \mathbf{q}_3) \delta_{\mathbf{q}_1 + \mathbf{q}_2 + \mathbf{q}_3, 0} + N^2 a(q_1) \delta_{\mathbf{q}_1 + \mathbf{q}_2, 0}\, \delta_{\mathbf{q}_3, 0}$$

$$+\, N^2 a(q_2) \delta_{\mathbf{q}_2 + \mathbf{q}_3, 0}\, \delta_{\mathbf{q}_1, 0} + N^2 a(q_3) \delta_{\mathbf{q}_3 + \mathbf{q}_1, 0}\, \delta_{\mathbf{q}_2, 0}$$

$$+\, N^3 \delta_{\mathbf{q}_1, 0}\, \delta_{\mathbf{q}_2, 0}\, \delta_{\mathbf{q}_3, 0}$$

In general $\langle \rho_{\mathbf{q}_1} \cdots \rho_{\mathbf{q}_s} \rangle$ is equal to $N a_s(\mathbf{q}_1 \cdots \mathbf{q}_s)$ if $\sum_{i=1}^s \mathbf{q}_i = 0$ but no smaller subset of the $\mathbf{q}$ vectors sums to zero. But if, for example, there were three disjoint subsets of $\mathbf{q}$ vectors that separately summed to zero, $\mathbf{q}_1 + \cdots + \mathbf{q}_l = \mathbf{q}_{l+1} + \cdots + \mathbf{q}_r = \mathbf{q}_{r+1} + \cdots + \mathbf{q}_s = 0$, then there would be an additional contribution, $N^3 a_l(\mathbf{q}_1 \cdots \mathbf{q}_l) a_{r-l}(\mathbf{q}_{l+1} \cdots \mathbf{q}_r) a_{s-r}(\mathbf{q}_{r+1} \cdots \mathbf{q}_s)$, which would be dominant in the limit $N \to \infty$.

The Green function and the self-energy operator are diagonal in momentum representation. Thus (4.8) can be written as an algebraic equation instead of an operator equation:

$$\mathscr{G}(k, E) = [G_0^{-1}(k, E) - \Sigma(k, E)]^{-1} \tag{4.21}$$

This function, whose physical content was discussed in Section I, is the object of most calculations.

## V. A SIMPLE CALCULATION FOR ALUMINUM

### A. Second-Order Perturbation Theory

The simplest calculation one can make is to evaluate (4.7) to the lowest nontrivial order of perturbation theory. The Green function is then given by

$$\mathscr{G}(k, E) = [E - k^2 - \Sigma(k, E)]^{-1} \tag{5.1}$$

(choosing units $\hbar = 2m = 1$), the self-energy function $\Sigma(k, E)$ being approximated by the first two terms of (4.7). If we assume the effective electron–ion interaction to be a *local potential* $v(r)$, then its momentum representation,

$$\langle \mathbf{k} | v | \mathbf{k} + \mathbf{q} \rangle = \Omega^{-1} u(q) = \Omega^{-1} \int e^{i\mathbf{q}\cdot\mathbf{r}} v(r) \, d^3 r \tag{5.2}$$

is independent of $\mathbf{k}$. The first-order contribution to $\Sigma(k, E)$, $N \langle \mathbf{k} | v | \mathbf{k} \rangle$, is just average potential and, being independent of $\mathbf{k}$, it may be eliminated by redefining the zero of the energy scale. Thus our approximation is

$$\Sigma(k, E) = \sum_{\mathbf{q} \neq 0} \frac{|\langle \mathbf{k} | v | \mathbf{k} + \mathbf{q} \rangle|^2 N a(q)}{E - (\mathbf{k} + \mathbf{q})^2} \tag{5.3a}$$

$$= \frac{n}{(2\pi)^3} \int d^3 q \, \frac{|u(q)|^2 a(q)}{E - (\mathbf{k} + \mathbf{q})^2} \tag{5.3b}$$

where $n = N/\Omega$ is the atomic number density.

The density of states (per unit energy range, unit volume, and one-spin orientation) is determined from (1.6) to be

$$n(E) = (2\pi^2)^{-1} \int_0^\infty k^2 \rho(k, E) \, dk \tag{5.4}$$

where $\rho(k, E) = -(\pi)^{-1} \operatorname{Im} \mathscr{G}(k, E + i0)$.

In Fig. 1 the structure factor $a(q)$ and the pseudopotential form factor $u(q)$ are shown schematically. It happens that the zero of $u(q)$ approximately coincides with the peak of $a(q)$. Thus the product $|u(q)|^2 a(q)$, which occurs in (5.3), is small for all $q$ because one factor is small whenever the other is large. So we expect the density of states $n(E)$ to deviate by only a small amount from the free-electron value.

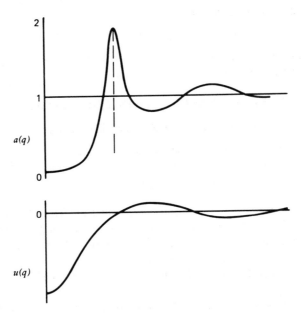

**Fig. 1**   Structure factor $a(q)$ and pseudopotential form factor $u(q)$ for a weak-scattering liquid metal.

As may be seen from the dot-dash curve in Fig. 2 (from Ballentine[8]), this is indeed the case except near $E = 0$. But at $E = 0$ the result is very strange. According to (5.3), the imaginary part of $\Sigma(k, E)$ is generated only by a contour integral around the pole in the integrand (one must replace $E$ by $E + i\delta$, and the let $\delta \to 0$). When $E$ is negative, there is no pole for real values of $q$, and we obtain Im $\Sigma(k, E) = 0$ for $E < 0$. In fact Im $\Sigma(k, E)$ goes continuously to zero as $E$ goes to zero. Since $n(E)$ is obtained from Im $\mathscr{G}(k, E)$, which is proportional to Im $\Sigma(k, E)$, we might expect $n(E)$ to go continuously to zero also, but this does not happen. The fault is with our overly simple approximation which must be improved in the next section.

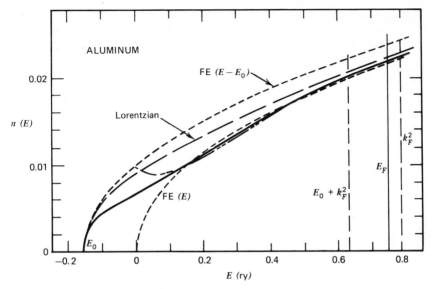

**Fig. 2** Density of states for aluminum, calculated with self-consistent internal propagator (solid curve) and without it (dot-dash curve), compared with free-electron curves FE $(E)$ and FE $(E - E_0)$, and Lorentzian approximation.

## B. Self-Consistent Internal Propagator

According to the previous approximate calculation, Im $\Sigma(k, E) = 0$ for $E < 0$. But Re $\Sigma(k, E) < 0$ in this regime, therefore (5.1) shows that the Green function $\mathscr{G}(k, E)$ has a pole on the real axis (at $k = k_0(E)$, say) for a certain range of negative $E$ values. This in turn implies, through the defining equations (1.1) to (1.4), that some plane wave $|\mathbf{k}_0\rangle$ is a stationary state of the system, a conclusion that is certainly false.

This inconsistency produced by the approximation (5.3) can be corrected by replacing the free-particle Green function in the integrand by the full Green function:

$$\Sigma(k, E) = \frac{n}{(2\pi)^3} \int d^3k' \frac{|u(\mathbf{k}' - \mathbf{k})|^2 a(\mathbf{k}' - \mathbf{k})}{E - (k')^2 - \Sigma(k', E)} \qquad (5.5)$$

and evaluating $\Sigma(k, E)$ self-consistently from the resulting integral equation. The Green function now cannot have a pole on the positive $k$ axis for, if we assume such a pole exists, then the contour integral around it on the rhs of (5.5) will produce a nonzero value of Im $\Sigma$ on the left-hand side, contradicting the assumption.

Of course (5.5) is still an approximation, the nature of which can be seen by iteration. Beginning with $\Sigma = 0$ in the denominator we obtain a first approximation to $\Sigma$, which may be put back into the denominator to obtain the second approximation, and so on. So the approximation (5.5) consists of summing the infinite set of terms illustrated in Fig. 3.

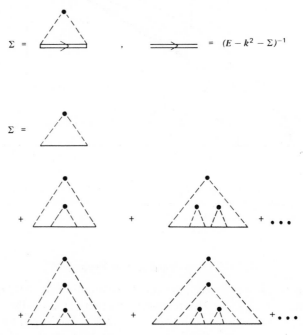

**Fig. 3**  Diagrammatic representation of (5.5) and its iterative solution.

This principle for avoiding the sort of disaster at low energies that occurred in the previous section may be generalized beyond the approximation (5.5). It is simply that all terms in a diagrammatic expansion that involve the modification of an internal Green function line should be implicitly summed by replacing all free-particle Green functions with full Green functions. The remaining diagrams, which may be called *skeleton diagrams*, appear in Fig. 4. A specific approximation is obtained by retaining only some tractable subset of the skeleton diagrams, and evaluating the Green function by means of the resulting pair of self-consistent equations (Fig. 4a and b). Equation (5.5) is the lowest nontrivial member of this sequence of approximations.

The importance of using a self-consistent internal Green function is illustrated by a calculation,[8] the results of which are shown in Fig. 2. Equation

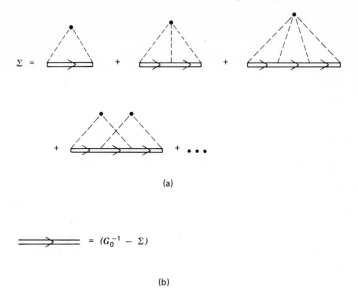

(a)

$$\Longrightarrow\!\!\!=\!\!\!= \;=\; (G_0^{-1} - \Sigma)$$

(b)

**Fig. 4**  Skeleton self-energy diagrams with the self-consistent internal propagator.

(5.5) was solved with the same pseudopotential $u(q)$, and structure factor $a(q)$ as was employed in the simpler calculation of Section V.A. To simplify the numerical integration, the $k'$ dependence of $\Sigma(k', E)$ was neglected in the denominator of (5.5). This was accomplished by replacing the denominator by $F - (k')^2$, where $F = E - \Sigma(k_0, E)$ with $k_0$ was determined by the equation

$$E - k_0^2 - \text{Re}\, \Sigma(k_0, E) = 0 \tag{5.6}$$

Now (5.5) has the same form as (5.3), except for the *complex energy approximation* of replacing $E$ by $F$. This approximation is accurate, provided $\Sigma(k, E)$ is not too rapidly varying over the range of $k$ within which $\mathscr{G}(k, E)$ is large.

We see from the results in Fig. 2 that the use of a self-consistent internal propagator has a qualitatively important effect on the density of states for $E < 0.1$ ry, but it makes very little difference near the Fermi surface. Paradoxically, this does not mean that one can use the simpler theory of Section V.A to calculate $n(E_F)$, because one must know the total area under the $n(E)$ curve for all $E < E_F$ in order to locate the Fermi level.

Also shown is a "Lorentzian" approximation obtained by replacing $\Sigma(k, E)$ by $\Sigma(k_0, E)$ (independent of $k$) in (5.4), thereby approximating $\rho(k, E)$ by a Lorentzian function of $k^2$. We see that the density of states is seriously altered by this further approximation, which demonstrates that the spectral function has significantly more structure than a Lorentzian function, although this would not be evident from a graph of $\rho(k, E)$. Since the Lorentzian approximation causes a 20% error in $n(E)$, we may expect the "complex-energy" approximation to commit a similar error in $\Sigma(k, E)$, but this should only cause $(0.20)^2 = 4\%$ error in $n(E)$.

This calculation is the simplest nontrivial approximation that may be expected to yield reasonable results for $n(E)$ in a liquid metal.

## VI. CAN ONE USEFULLY DEFINE AN $E$-VERSUS-$k$ RELATION?

The reader may be skeptical about the final sentence of the previous section. Can one not, at least as an approximation, obtain an $E$-versus-$k$ relation by perturbation theory and use it to evaluate $n(E)$? Since there is no preferred direction in a liquid metal, any relation $E = E_L(k)$ would necessarily be spherically symmetric and, by assuming a uniform distribution of states in $k$ space, one would obtain

$$n(E) = \frac{k^2}{2\pi^2} \left[ \frac{\partial E_L(k)}{\partial k} \right]^{-1} \tag{6.1}$$

This would certainly be a simpler calculation than that of Section V.B, if indeed it were valid.

Since the stationary states of free electrons can be labeled by a unique $\mathbf{k}$ vector, common sense tells us that this must be approximately so for an electron in a liquid metal if the scattering is sufficiently weak. Indeed, the peak of the spectral function $\rho(k, E)$ or the quantity $k_0 = k_0(E)$ from (5.6) provides possible definitions for the characteristic $k$ to be associated with $E$. But the key word in the title is *usefully*—it is not assured that such a definition of $k(E)$ or $E(k)$ would possess any of the useful properties of a genuine band structure relation. To define a wave vector for a stationary state of an electron in a liquid is useless, unless it permits us to use some of the intuition we have developed from the study of crystal band structures.

Intuition suggests, and perturbation calculations confirm,[74] that in a simple liquid metal one would have a band structure $E_L(k)$ that is more or less the spherical average of that in the solid $E_s(\mathbf{k})$, but with the sharp features of the solid band structure smoothed out by the disorder. This is illustrated in Fig. 5 for the lowest energy band of a hypothetical metal. One of the important properties of $E_s(\mathbf{k})$ is that its normal gradient vanishes on a zone boundary. Thus it seems very likely that $\partial E_L(k)/\partial k$ will vanish somewhere near the average zone boundary. But then (6.1) implies that $n_L(E)$ will be infinite at the band edge in the liquid. This contradicts both our intuition and the experimental evidence of optical and soft x-ray spectroscopy which

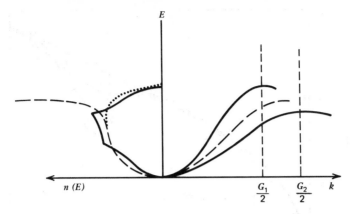

**Fig. 5** Hypothetical $E(k)$ and $n(E)$ for a solid and a liquid [assuming (6.1), dashed curve; actual, dotted curve].

suggest that $n_L(E)$ should be at least qualitatively like a smoothed version of $n_s(E)$. Evidently, any $E_L(k)$ that satisfies (6.1) must possess an infinite slope at the band edge, rather than zero slope, and so it could not have the dynamical significance an $E$-$k$ relation possesses in a crystal.

It may be objected that we have demanded too much of $E_L(k)$. Perhaps it can be usefully defined only for simple metals which do not have any absolute gaps in $n(E)$. This possibility was tested[10] by considering a model Green function:

$$\Sigma(k, E) = E_L(k) - \frac{k^2}{2m} - ibE^{1/2}$$

$$\mathscr{G}(k, E) = [E - E_L(k) + ibE^{1/2}]^{-1} \qquad (6.2)$$

This is the simplest model that automatically satisfies the sum rule [see (1.5)] by virtue of its analytic properties. For some plausible choice of $E_L(k)$, we then vary the magnitude of Im $\Sigma$ by varying the parameter $b$, and compare the density of states yielded by (5.4) and by (6.1).

Figure 6 shows this comparison for the model

$$E_L(k) = k^2 + 0.1(\{1 + \exp [(0.975 - k)/0.01]\}^{-1} - [1 + \exp (97.5)]^{-1}),$$

with units $\hbar = 2m = k_F = 1$. [The parameters were chosen so as to approximately represent $n(E)$ for mercury.] As expected, the prediction of (6.1) becomes exact in the limit Im $\Sigma \to 0$ (i.e., $b \to 0$), but it is seriously in error when Im $\Sigma$ is only 5% of $E_F$($b = 0.05$).

We conclude that the use of a relation $E = E_L(k)$ and (6.1) may be valid when it predicts a negligible deviation of $n(E)$ from the free-electron parabola,

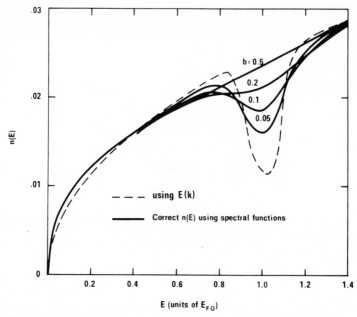

**Fig. 6**   Density of states for the model (6.2). (From Ballentine and Chan.[11])

but that it cannot accurately predict the magnitude of the deviations from the free-electron parabola even when these are quite small.

For any dispersion relation one can define an effective mass

$$m^* = k\left(\frac{\partial E}{\partial k}\,k\right)^{-1} \tag{6.3}$$

If (6.1) holds, then one has the relation

$$\frac{m^*}{m} = \frac{n(E_k)}{n_0(E_k^{(0)})} \tag{6.4}$$

where $E_k^{(0)} = k^2/2m$ and $n_0(E)$ is the free-electron density of states. It should be noted that $n(E)$ and $n_0(E)$ are to be compared for the same $k$, and not for the same $E$ as might be suggested by imprecise notation [e.g., equation (2.9) of Shaw and Smith[74]]. However, the relation (6.4) between the effective mass and the density of states does not hold in a liquid. The effective mass ratio $m^*/m$ in Fig. 7 was obtained from the data of Ballentine[8] by using (5.6) to define a dispersion relation. Of course the notion of effective mass cannot be made precise for a liquid metal, but this seems to be as reasonable a definition of it as is possible. Because it omits the first-order contribution to $m^*$ from the nonlocal pseudopotential, this data should not be compared with experiment

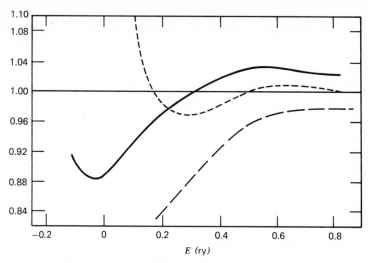

**Fig. 7** $m^*/m$ (solid line), $n(E_k)/n_0(k^2)$ (long dashes), and $n(E)/n_0(E)$ (short dashes) for aluminum using local pseudopotential.

but should only be regarded as a test of (6.4). Although the curves of $m^*/m$ and $n(E_k)/n_0(E_k^{(0)})$ have the same general shape, there is a significant systematic difference between them. Also shown in Fig. 7 is the ratio $n(E)/n_0(E)$, which happens to be closer to $m^*/m$ at large energies but which becomes infinite at $E = 0$. The nonvalidity of (6.4) for liquid metals is significant, because most of the calculations that report $n(E)$ determined from an $E(k)$ are actually calculations of $m^*/m$.

## VII. PSEUDOPOTENTIAL CALCULATIONS

In Section V we assumed that the effective electron–ion interaction could be represented by a local potential $v(\mathbf{r})$. This simplest possible assumption was made in order to illustrate the essential features of a Green function calculation with a minimum of complication. But in fact the pseudopotential is in general a nonlocal, energy-dependent operator, and in many cases this is quantitatively important.

Nonlocality means that the momentum representation matrix element $\langle \mathbf{k}|W|\mathbf{k} + \mathbf{q}\rangle$ depends on $\mathbf{k}$ as well as on $\mathbf{q}$. This somewhat complicates the numerical evaluation of sums like (5.3a), but it causes no difficulty in principle. If the pseudopotential $W = W(E)$ is energy-dependent, then the eigenvectors and eigenvalues

$$\left[\frac{p^2}{2m} + W(E)\right]|\phi_n(E)\rangle = \varepsilon_n(E)|\phi_n(E)\rangle \tag{7.1}$$

will also be energy-dependent. The true energy levels are obtained by setting the parameter $E$ equal to the current eigenvalue, that is,

$$E_n = \varepsilon_n(E_n) \tag{7.2}$$

This is a more serious complication than is nonlocality, because the density of eigenvalues $\varepsilon_n(E)$ for fixed $E$ may be different from the true density of states, the density of $E_n$. Several different methods have been employed to deal with these complications.

## A. Green's Function Calculations—Bismuth

For the liquid metal bismuth a series of calculations has been made in which the pseudopotential was treated in different ways, thereby illustrating the quantitative significance of various refinements. All these calculations are based on the Heine–Abarenkov (HA) model potential,[1,3,38] or on some refinement of it. The HA model potential is constructed by replacing the strong potential inside the ion core by a constant $A_l(E)$. The parameter $A_l(E)$ is usually different for different partial waves $l$. Its $E$ dependence is obtained by adjusting it to fit the spectroscopic energies of the excited states of the ion.

To a first approximation the effect of the $\mathbf{k}$ and $E$ dependences of

$$\langle \mathbf{k} | W(E) | \mathbf{k} + \mathbf{q} \rangle$$

can be absorbed into an effective mass.[20] The Hamiltonian

$$H = \frac{p^2}{2m} + W(E) \tag{7.3}$$

is separated into two terms, $H_0$ and $H_1$, but instead of taking $H_0$ to be $p^2/2m$, as was done in Section V, we now choose

$$H_0 = \frac{p^2}{2m} + \sum_{\mathbf{k}} |\mathbf{k}\rangle\langle\mathbf{k}| W(\varepsilon_k^0)|\mathbf{k}\rangle\langle\mathbf{k}| \tag{7.4}$$

$$H_1 = H_1{}^d + H_1^{od} \tag{7.5}$$

where $H_1{}^d$ and $H_1^{od}$ are the diagonal and off-diagonal parts of $H_1$ in momentum representation.

$$H_1{}^d = \sum_{k} |\mathbf{k}\rangle\langle\mathbf{k}| \{\langle\mathbf{k}| W(E)|\mathbf{k}\rangle - \langle\mathbf{k}| W(\varepsilon_k^0)|\mathbf{k}\rangle\} \tag{7.6}$$

$$H_1^{od} = \sum_{\mathbf{k}\neq\mathbf{k}'}\sum |\mathbf{k}\rangle\langle\mathbf{k}| W(E)|\mathbf{k}'\rangle\langle\mathbf{k}'| \tag{7.7}$$

The parameter $\varepsilon_k{}^0$ is chosen to minimize the residual $H_1{}^d$ in the most important region of $k$-$E$ space. There is no unique way to do this, but a reasonable choice is to make $\varepsilon_k{}^0$ an eigenvalue of $H_0$, that is, a solution of

$$\varepsilon_k{}^0 = \frac{k^2}{2m} + \langle \mathbf{k} | W(\varepsilon_k{}^0) | \mathbf{k} \rangle \tag{7.8}$$

It was found[17] that $\varepsilon_k$ is practically a linear function of $k^2$, so that an effective-mass formula

$$\varepsilon_k{}^0 = \frac{k^2}{2m^*} + E_0 \tag{7.9}$$

is an excellent approximation to (7.8). Since $E_0$ merely shifts the zero of energy, it may be ignored.

The essential approximation of the method is to neglect the residual $H_1{}^d$ and to neglect the $E$ dependence of $H_1^{od}$ by evaluating it at $E = E_F$. We may now apply the perturbation theory of Section IV with $G_0(k, E) = (E - k^2/2m^*)^{-1}$. The ensemble average Green function is

$$\mathscr{G}(k, E) = \left[ \frac{E - k^2}{2m^*} - \Sigma(k, E) \right]^{-1} \tag{7.10}$$

There is no first-order contribution of $\Sigma(k, E)$, because $H_1{}^d$ has been absorbed into the effective mass, and the self-consistent second-order approximation of Section V.B becomes

$$\Sigma(k, E) = \frac{n}{(2\pi)^3} \int \frac{|u(\mathbf{k}, \mathbf{k}'; E_F)|^2 a(q)}{E - (k')^2/2m^* - \Sigma(k', E)} d^3k' \tag{7.11}$$

where $n$ is the atomic number density, $q = |\mathbf{k}' - \mathbf{k}|$, and $u(\mathbf{k}, \mathbf{k}'; E_F) = \Omega \langle \mathbf{k} | w(E_F) | \mathbf{k}' \rangle$.

The first calculation in the series[8] employed a local pseudopotential $u(q)$, obtained from the nonlocal $u(\mathbf{k}, \mathbf{k}'; E_F)$ by setting $|\mathbf{k}| = k_F$, and $|\mathbf{k}'| = k_F$ for $q < 2k_F$, $\mathbf{k}'$ antiparallel to $\mathbf{k}$ for $q > 2k_F$. It also employed the *complex energy approximation* described in Section V.B and set $m^*/m = 1$. The resulting density of states $n(E)$, shown as case 1 in Fig. 8, has a small peak followed by a dip. The dip occurs when $k_0(E)$ [defined by (5.6)] is near $k_p/2$, $k_p$ being the position of the peak in $a(q)$. No such dip in $n(E)$ was obtained for aluminum (see Fig. 2), because in that case $u(q)$ is zero very near to $k_p$, and so $|u|^2 a(q)$ has no sharp peak for aluminum.

In subsequent calculations[20] these approximations were removed one by one. The correct evaluation of the integral in (7.11), without the complex energy approximation, makes only a small change in $n(E)$ (Fig. 8, case 2) and not more than 0.01 ry difference in $\Sigma(k, E)$.

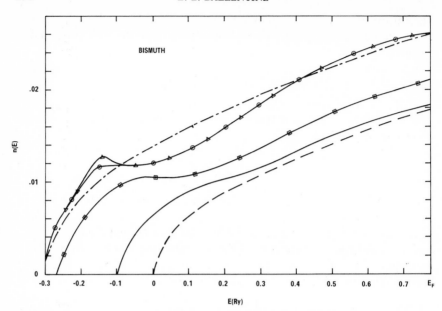

**Fig. 8** Density of states per unit volume per unit energy range $n(E)(a_0^{-3}/\text{ry})$ for bismuth. △, Case 1, local energy-independent model potential, complex energy approximation; ⊘, case 2, local energy-independent model potential; ⊞, case 3, local model potential, effective-mass correction; —, case 4, nonlocal model potential effective-mass correction; — — —, free-electron parabola (origin shifted); — —, free-electron parabola corresponding to $m^*/m = 0.87$. (From Chan and Ballentine.[20])

    In case 3 (Fig. 8) effective mass $m^*/m = 0.87$, determined by (7.8) and (7.9), is included but the local $u(q)$ is retained in (7.11). The only apparent effect is to reduce $n(E)$ by the factor $m^*/m$ and to extend the energy scale by the reciprocal of that factor. (The position of the bottom of the band has no significance.) Finally, in case 4 the nonlocal pseudopotential is employed in (7.11), resulting in an almost complete smoothing out of the dip in $n(E)$. The self-energy function $\Sigma(k, E)$ for this final case (Figs. 9 and 10) looks very different in magnitude from that of the first case,[8] however, we should not compare them for the same $E$ values on arbitrary scales but rather for the same positions within the band relative to $E_F$ or to the bottom of the band. Figure 11 shows Im $\Sigma(k, E)$ at $E = E_F$ for cases 1 and 4, giving us an indication of the difference between the effect of the nonlocal pseudopotential and of the local approximation to it.

    The spectral function $\rho(k, E)$ [see (1.4)] is shown in Fig. 12. The density of states is obtained from it by integrating over **k** space [see (5.4)]. By integrating

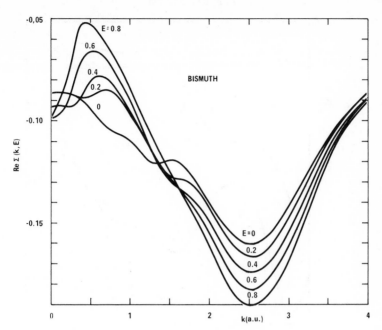

**Fig. 9** Real part of the self-energy function for bismuth, case 4. (From Chan.[17])

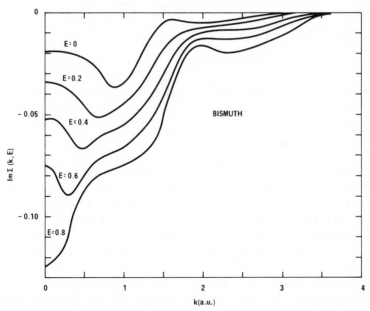

**Fig. 10** Imaginary part of the self-energy function for bismuth, case 4. (From Chan.[17])

293

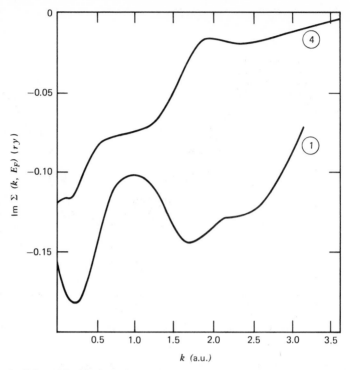

**Fig. 11**  Im $\Sigma(k, E_F)$ for bismuth. Case 1, local pseudopotential, $E_F = 0.457$ ry (Ballentine.[8])
Case 4, nonlocal pseudopotential, $E_F = 0.775$ ry. (Chan and Ballentine.[20])

over $E$ we obtain the momentum distribution of all the conduction electrons:

$$P(\mathbf{k}) = \int_{-\infty}^{E_F} \rho(\mathbf{k}, E) \, dE \qquad (7.12)$$

which is shown in Fig. 13 for cases 1 and 4. Although it resembles a Fermi distribution, it is entirely due to scattering of the electrons by the disordered potential, the much smaller effect of finite temperature having been omitted from the calculation. A check on the computational accuracy is provided by the sum rule [see (1.5)], which was satisfied to within 1 % for aluminum but which seemed to fall short by about 4 % for bismuth in the work of Ballentine.[8]

In the original articles the "thickness of the Fermi surface" was characterized by a parameter $\Delta k_F$ equal to the value of $(dP(k)/dk)^{-1}$, evaluated where $P = 0.5$. While this is not unreasonable in principle, it is perhaps unwise to

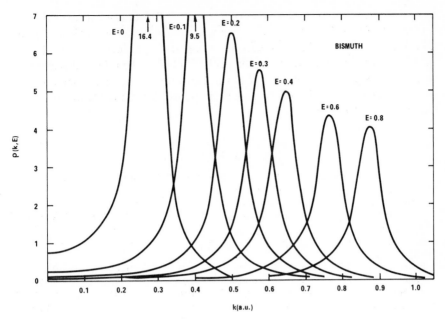

**Fig. 12**   Spectral function for bismuth, case 4. (From Chan.[17])

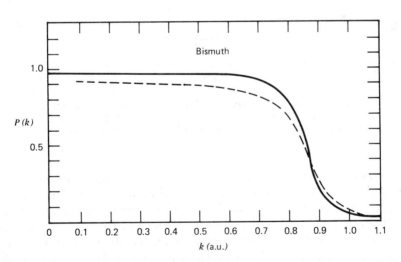

**Fig. 13**   Momentum distribution of occupied states for bismuth. Dashed line, case 1; solid line, case 4.

give such emphasis to a parameter obtained by differentiating a curve that is the result of so much numerical computation. The width at half-maximum of $\rho(k, E_F)$ is likely to be a numerically more stable parameter. If $\Sigma(k, E_F)$ is slowly varying over the width of the peak of $\rho(k, E_F)$, then the width at half-maximum is given by $\text{Im } \Sigma(k_F, E_F)/k_F$. The ratio of this parameter to $k_F$ is tabulated in Table III, so that one can compare different metals and different methods of calculation.

TABLE III

Bandwidth and Spectral Function Width Parameters

| Metal and method of calculation[a] | $k_F$ (a.u.) | $m^*/m$ | $E_F - E_0$ (ry) | $\dfrac{E_F - E_0}{k_F^2}$ | $\text{Im } \Sigma(k_F, E_F)$ (ry) | $\dfrac{\text{Im } \Sigma(k_F, E_F)}{k_F^2}$ |
|---|---|---|---|---|---|---|
| Bismuth | | | | | | |
| Method 1 | 0.86 | 1 | 0.763 | 1.03 | 0.104 | 0.14 |
| Method 2 | | 1 | 0.768 | 1.04 | 0.110 | 0.15 |
| Method 3 | | 0.87 | 0.861 | 1.16 | 0.093 | 0.125 |
| Method 4 | | 0.87 | 0.861 | 1.16 | 0.076 | 0.10 |
| Method 5 | | | 1.18 | 1.59 | | |
| | | | $m/m^* = 1.15$ | | | |
| Aluminum | | | | | | |
| Method 1 | 0.886 | 1 | 0.898 | 1.14 | 0.075 | 0.095 |
| Indium | | | | | | |
| Method 4 | 0.785 | 0.90 | 0.675 | 1.10 | 0.0223 | 0.0362 |
| | | | $m/m^* = 1.11$ | | | |
| Mercury | | | | | | |
| Method 4 | 0.709 | 1.12 | 0.452 | 0.90 | 0.022 | 0.044 |
| | | | $m/m^* = 0.89$ | | | |

[a] 1, Local energy-independent pseudopotential, complex energy approximation;[8] 2, same but without complex energy approximation;[17] 3, as 2 but with effective-mass correction;[17] 4, non-local pseudopotential and effective-mass correction;[17,20] 5, integrated density of states.[17]

The momentum distribution $P(\mathbf{k})$ is measurable in principle. By measuring the angle between the pair of x-rays emitted during the annihilation of a positron with an electron, one determines the total momentum of the electron-positron pair. If the momentum of the positron could be neglected we could determine the electron momentum distribution directly. West et al.[81] found agreement between their measurements for bismuth and the calculations of Ballentine.[8] This agreement was fortuitous, because it is now established that the positron momentum distribution is appreciable, and that it is affected by vacancies in the solid and so presumably also by disorder in the liquid. Thus the "thickness of the Fermi surface," the unsharpness of the cutoff of $P(\mathbf{k})$ at $k_F$, must actually be less than observed broadening of the Fermi cutoff in the experimental results. It is interesting that the use of the nonlocal pseudopotential yields a sharper cutoff in $P(\mathbf{k})$ (see Fig. 13), but no quantitative comparison between theory and experiment is as yet possible.

For a recent review of the experimental situation, one should consult the *Proceedings of the Tokyo Conference*, particularly the article by Cusack.[25] Lest the reader despair that fundamental quantities like $\rho(k, E)$, $n(E)$, and $P(\mathbf{k})$ will never be measured, we note that Compton scattering can give a quite direct measurement of the electron momentum distribution,[23] and this technique does not suffer from the difficulties of interpretation that affect positron annihilation. However, this technique has not yet been applied to liquid metals.

In Table III we list the bandwidth and spectral function width parameters for this sequence of calculations. The bandwidth is determined mainly by the first-order effective-mass correction from the nonlocal pseudopotential, the other refinements being less important. However, the various refinements in the sequence are of considerable importance in determining the width of the spectral function, or thickness of the Fermi surface, as it might be called.

## B. Integrated Density of States

Although the calculation described above fully accounts for the nonlocal nature of the pseudopotential, it can include energy dependence only in the diagonal part of the pseudopotential, and only approximately, since no choice of $\varepsilon_k{}^0$ in (7.6) can make $H_1{}^d = 0$ for all $E$ and $k$. That some such limitation is unavoidable within the present form of the Green function method can be seen as follows. If the pseudopotential is energy-dependent, as in (7.1), then instead of (1.2) the spectral operator will have the form

$$\rho(E) = \sum_n |\phi_n(E)\rangle\langle\phi_n(E)| \delta[E - \varepsilon_n(E)] \tag{7.13}$$

Its trace will be

$$\text{Tr } \rho(E) = \sum_n \delta[E - \varepsilon_n(E)] = \sum_n \frac{\delta(E - E_n)}{|1 - d\varepsilon_n/dE|} \tag{7.14}$$

where $E_n$ is the true energy level given by (7.2). This differs from the density of states per unit energy, because of the factor $|1 - d\varepsilon_n/dE|$ in the denomination, and thereby demonstrates that this method cannot yield the true density of states for an arbitrarily energy-dependent potential.

Following an idea due to Lloyd,[57] Chan and Ballentine[18] developed a method that is valid for a general energy-dependent potential. Let $H(E)$

denote the energy-dependent Hamiltonian of (7.1). Then the total number of energy levels below $E$ is given by

$$N(E) = (2\pi i)^{-1} \lim_{\eta \to 0} \text{Tr} \left[ \ln (H - E + i\eta) - \ln (H - E - i\eta) \right]$$

$$= \pi^{-1} \sum_n \text{Im} \ln \left[ \varepsilon_n(E) - (E + i0) \right]$$

$$= \sum_n \theta[E - \varepsilon_n(E)] \tag{7.15}$$

where $\theta(x) = 1$ for $x > 0$, and $\theta(x) = 0$ for $x < 0$. According to the last line, $N(E)$ is equal to the number of eigenvalues $\varepsilon_n(E)$ of $H(E)$ that are less than $E$. But this is the same as the number of true energy levels $E_n$ [see (7.2)] provided $d\varepsilon_n/dE < 1$ so that, as $E$ increases, an $\varepsilon_n(E)$ just below $E$ does not rise above $E$.

In order to use perturbation theory, we write

$$\ln (H - E) = \ln \left[ (H_0 - E)(1 - G_0 W) \right] \tag{7.16}$$

where $H = H_0 + W$, and $G_0 = (E - H_0)^{-1}$. Now $\ln (AB) \neq \ln A + \ln B$ if the operators $A$ and $B$ do not commute, but it is still true that

$$\text{Tr} \ln (AB) = \ln (\det |AB|)$$

$$= \ln (\det A) + \ln (\det B)$$

$$= \text{Tr} \ln A + \text{Tr} \ln B \tag{7.17}$$

Thus we have

$$N(E) = -\pi^{-1} \text{Tr Im} \left[ \ln (H_0 - E) + \ln (1 - G_0 W) \right] \tag{7.18}$$

The first term is just $N_0(E)$, the number of energy levels below $E$ for the unperturbed Hamiltonian $H_0$, and the second term can be evaluated by a form of perturbation theory very similar to that employed in the usual Green function approach.

This theory has been evaluated for bismuth in a self-consistent second-order approximation, retaining the full-energy dependence of the nonlocal pseudopotential. The computation is very lengthy, because it is necessary to solve an integral equation of the same complexity as (7.11) not only with the actual pseudopotential $W$, but with $\lambda W, 0 \le \lambda \le 1$. The calculation (for details of which we refer to Chan and Ballentine[18] essentially evaluates $N(E)$ by counting the energy levels pulled below $E$ as the coupling parameter $\lambda$ is increased. Thus $N(E)$ is determined not only for $H = H_0 + W$, but also for $H = H_0 + \lambda W$, $0 \leq \lambda \leq 1$. Unfortunately, there is little use for all this extra information.

The resulting $N(E)$ is shown in Fig. 14, where it is compared with the integral of $n(E)$ from Fig. 8, case 4, which was the most sophisticated Green function approximation including an effective mass. Because of numerical difficulties at small $E$, the bandwidth was determined by plotting $[N(E)]^{2/3}$ versus $E$ (which would be linear for free electrons) and extrapolating to the bottom of the band. The slope of this line at small energy corresponds to $m^*/m = 0.63$. Although the bandwidth so obtained may be inaccurate, there is no doubt that this method (case 5 in Table III) predicts a much wider band than does the calculation of Section VII.A.

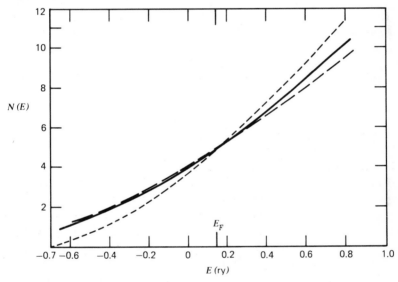

**Fig. 14** Number of electron states below energy $E$ per atom of liquid bismuth. Solid curve, result according to the theory of this section; dashed curve, best fitting free-electron curve ($m^*/m = 0.63$); dotted curve, result of integrating $n(E)$, case 4, Fig. 8. (From Chan.[17])

There are two possible explanations for this difference. It may be genuine effect of the energy dependence of the nondiagonal matrix elements of $W(E)$, which was neglected in Section VII.A. However, while both methods must agree when evaluated exactly with an energy-independent potential, it has not been possible to determine whether they agree when both are evaluated to the same order of perturbation theory. So we cannot be sure that we are comparing equivalent approximations.

## C. Calculation of $E_L(k)$

Several investigators have attempted to determine the electronic structure of liquid metals from the second-order perturbation correction for the energy of a free-electron state of momentum $\mathbf{k}$:

$$E(\mathbf{k}) = \frac{k^2}{2m} + \langle \mathbf{k}|W(E)|\mathbf{k}\rangle + 2m \sum_{q \neq 0} \frac{\langle \mathbf{k}|W(E)|\mathbf{k}+\mathbf{q}\rangle\langle \mathbf{k}+\mathbf{q}|W(E)|\mathbf{k}\rangle}{k^2 - (\mathbf{k}+\mathbf{q})}$$

(7.19)

Its ensemble average $E_L(k) = \langle E(\mathbf{k})\rangle$ is regarded as the dispersion relation for an electron in the liquid, and the density of states is then obtained from (6.1). Although we showed in Section VI that this procedure is liable to overestimate any structure in $n(E)$, and can even be nonsensical in extreme cases, there have been so few calculations of $n(E)$ for liquid metals that examining these results is still worthwhile. Although they may lack quantitative significance, they can still provide qualitative comparisons between metals or between pseudopotentials.

In a notation similar to that employed in (5.3) or (7.11), we obtain the expression

$$E_L(k) = \frac{k^2}{2m} + nu(\mathbf{k}, \mathbf{k}; E_L) + 2m \frac{n}{(2\pi)^3} \int d^3q \, \frac{a(q)|u(k, k+q; E_L)|^2}{k^2 - (\mathbf{k}+\mathbf{q})^2}$$

(7.20)

The nonlocality and energy dependence of the pseudopotential are only minor complications in the evaluation of (7.20), making possible a straightforward assessment of their significance in the final results.

Shaw and Smith[74] evaluated $E_L(k)$, and from it $m^*/m$ (6.3), and compared an exact treatment of the nonlocal pseudopotential with certain approximate treatments. Figure 15 shows their results for lithium. A striking feature is that an evaluation of (7.20) to second order yields a variation of $m^*/m$ by 26%, but an evaluation to only first order would predict $m^*/m$ to be almost independent of $k$. This might lead one to doubt the convergence of perturbation theory, but in fact the first-order contribution to $E_L(k)$ is about 10 times larger than the second-order contribution.

Also shown in Fig. 15 are the results of two local approximations to the pseudopotential. One, called the *Fermi-surface approximation*, is to make $\langle \mathbf{k}|W(E)|\mathbf{k}+\mathbf{q}\rangle$ into a function of $q$ only by setting $E = E_F$, $|\mathbf{k}| = k_F$, and $|\mathbf{k}+\mathbf{q}| = k_F$ for $q < 2k_F$, $\mathbf{k}$ and $\mathbf{q}$ antiparallel for $q > 2k_F$. This is the same approximation used by Ballentine,[8] and we see that it causes very little error. The other, called the *energy-shell* approximation, attempts to allow partially for the $E$ and $k$ dependence by setting $E = k^2/2m$, and $|\mathbf{k}+\mathbf{q}| = k$ for $q < 2k$, $\mathbf{k}$ and $\mathbf{q}$ antiparallel for $q > 2k$. Surprisingly, this

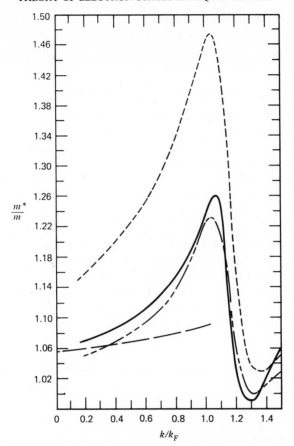

**Fig. 15** Effective mass for lithium, from the data of Shaw and Smith.[74] ————, nonlocal pseudopotential, to second order; ——— ———, nonlocal, to first order; ——— -- ———, local (Fermi surface approximation) to second order, — — —, local (energy shell approximation) to second order.

leads to considerably less accurate results than the Fermi-surface approximation. A similar conclusion was obtained for cadmium and indium.

Although the authors reported their results as $n(E)/n_0(E)$, we emphasize that if (6.4) is not valid the proper interpretation of the results is as $m^*/m$. However, the error committed by (6.4) tends to cancel out of the ratio of the densities of state for the liquid and solid at the melting point. (Since the change in resistivity at melting is usually no more than a factor of 2, the widths of $\rho(\mathbf{k}, E)$ are also comparable for the liquid and solid at melting, and so the error committed by (6.4) should be similar in both cases.) There-fore the agreement obtained for lithium between the predicted 2.5% decrease in $n(E_F)$ on melting and the observed decrease (2%, deduced from electron spin resonance experiments by Hahn and Enderby[36]) remains significant.

## D. A Survey of Results for Various Metals

In the appendix we have tabulated calculations of the density of electronic states and related quantities for different metals throughout the periodic table. The majority of these calculations employed pseudopotentials, and we now describe the most interesting results.

Several investigators agree that the densities of states in the *alkali* metals (except lithium) are very much like the free-electron (FE) model. For the exceptional case of *lithium*, Shaw and Smith[74] predict that $n(E)$ should have a broad peak near $E_F$, which rises about 25% above the FE value. Ichikawa,[42] using the same method, finds a similar broad peak but at a rather lower energy. However, he finds that the peak is suppressed when he employs the Animalu–Heine pseudopotential instead of Shaw's pseudopotential.

Two calculations of $m^*/m$ for *cadmium*, both of them by the method of Section VII.C, are in striking disagreement (Fig. 16). The result of Shaw and Smith[74] is based on Shaw's[73] model potential which is an optimized version

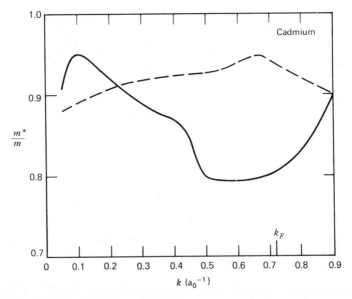

**Fig. 16**  Effective mass for cadmium. Solid line, Jena and Halder[46]; broken line, Shaw and Smith.[74]

of the Animalu–Heine[3] model potential, whereas Jena and Halder[45] used an empirical nonlocal pseudopotential fitted to Fermi surface data.[77] We favor the latter result, because the empirical pseudopotential can account for several other experimental results, including the large temperature dependence of the Knight shift in solid cadmium and its remarkable increase on melting by a factor of 1.34. (It should be pointed out that Jena and Halder seem to have used the value of $k_F$ appropriate for solid cadmium. If one uses the value for liquid cadmium, then their Fig. 1 yields $m^*/m = 0.805$ rather than their reported value of 0.827.) A note of disonance is introduced by the fact that the older model potential predicted the resistivity within 40% of its observed value, whereas the empirical pseudopotential yields a theoretical value five times too large. Greenfield and Wiser[33] suggest that the Born approximation in the resistivity calculation is at fault, but no estimate of the magnitude of the correction has been attempted.

Two independent calculations have concluded that the density of states for liquid *indium* is very nearly free-electron-like, except for a rescaling of the curve by an average effective-mass factor, $m^*/m = 0.90$. The first calculation[74] employed the $E_L(k)$ method of Section VII.C with Shaw's[73] optimized model potential. The second[20] used the Green function method of Section VII.A with the Heine–Animalu form of the model potential. However, some of the parameters were adjusted to agree with the analysis of the Fermi surface of solid indium by Ashcroft and Lawrence.[4] The agreement between such different calculations gives us sufficient confidence in the correctness of the results that we are inclined to discount the interpretation of photoemission data[54,61] in terms of structure in the density of states. This attitude is not so radical, for there are both surface and bulk contributions to the photoemission spectrum, and even the bulk component is not merely proportional to the density of states. The only obvious weakness in the theoretical calculations lies in the fact that the fit of a pseudopotential to a Fermi surface is not unique. Ashcroft and Lawrence found two possible models, one of which was recommended as providing the best fit to cyclotron resonance and specific-heat data. In their calculations Chan and Ballentine used only the preferred model.

The electronic properties of *mercury* seem unusual by comparison with other liquid metals. In particular the dc resistivity is rather large, and it decreases on addition of most metallic impurities. These and other empirical facts were reviewed by Mott,[59] who proposed that they might be explained by the existence of a "pseudogap," that is, a reduced density of states at the Fermi energy. He conjectured that the ratio of the density of states at the Fermi energy to its free-electron value, $g \equiv n(E_F)/n_0(E_F)$, should be approximately $g = 0.7$. Although the basis of Mott's conjecture was empirical, he suggested as a physical origin of the pseudogap that there might be a kink in $E_L(k)$ [see (7.20)] due to the main peak of the structure factor $a(q)$ which, according to (6.1), would reduce $n(E)$. However, the Animalu–Heine[3] model potential form factor, which was then accepted, vanishes very near the position of the main peak, and so no such structure in $E_L(k)$ would be expected to exist.

Subsequently, Evans[29] recalculated the model potential, giving particular attention to the influence of the $d$ states near the bottom of the conduction band. As can be seen in Fig. 17, this leads to a qualitatively different form

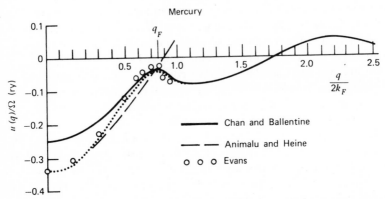

**Fig. 17** Pseudopotential Fermi-energy form factor for mercury, according to Animalu and Heine,[3] and Evans[29] who considered the effect of $d$ states, and Chan and Ballentine[20] who included certain effective-mass modifications in the screening.

factor with which Evans was able to explain the unusual transport properties of mercury and its alloys without invoking Mott's pseudogap hypothesis. Jones and Datars[47,48] empirically fitted a nonlocal pseudopotential to measurements of the Fermi surface of solid mercury. They found three qualitatively different pseudopotentials which provide a reasonable fit to the data, one of them being similar to the Animalu–Heine pseudopotential, and another being very close to Chan and Ballentine's[20] modification of Evans' pseudopotential.

The density of states in mercury was calculated by Chan and Ballentine,[19,20] using the Green function method of Section VII.A and a nonlocal pseudopotential, the Fermi level form factor shown in Fig. 17. Their result (Fig. 18) predicts only a small dip of $n(E)$ below the free-electron curve, and so does not support Mott's pseudogap conjecture. Also shown in Fig. 18 is a calculation by Evans and Mount (unpublished) based on the same nonlocal pseudopotential (except for effective mass corrections) but using the method of

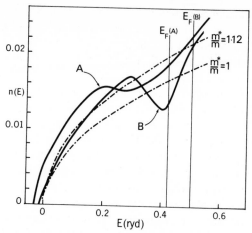

**Fig. 18** Density of states for mercury. *A*, Chan and Ballentine;[19,20] *B*, Evans and Mount (unpublished); and free electron curves for the corresponding effective masses.

Section VII.C. This method apparently overestimates the depth of the pseudogap below the free-electron curve by a factor of 2. Since Im $\Sigma(k_F, E_F)$ is about 5% of the bandwidth, this exaggeration of the pseudogap by a factor of 2 is exactly what was predicted by Fig. 6 (compare $b = 0.05$ with $b = 0$). Ichikawa[42] performed a calculation similar to that of Evans and Mount, and his result is very similar. Itami and Shimoji[44] performed a Green function calculation using the methods and approximations of Ballentine[8] (see Section V), and they obtain only a slight dip in $n(E)$. All these investigators agree that any pseudogap lies below $E_F$, rather than at $E_F$ as Mott conjectured.

In a later paper Mott[60] withdrew the pseudogap hypothesis as being relevant to the transport properties of mercury at room temperature and pressure (although presumably it plays an important role at lower densities near the critical point, a subject that lies outside the scope of this article). However, he still conjectures that it may exist, because the magnetic susceptibility of mercury increases as one raises the temperature or adds indium.[34] This is interpreted as being due to an increase in $n(E_F)$ as the pseudogap is filled up. Such an interpretation of the temperature dependence is problematical, because the structure factor $a(q)$ is nearly independent of temperature. Thus, even if the conclusions of the pseudopotential calculations were wrong and a pseudogap did exit at $E = E_F$, it should not be significantly modified by a 100 degree increase in temperature.

Devillers, Young, and Vroomen[27] recently proposed that the pseudopotential of mercury should be quite different from that reported by Evans,[29] giving empirical evidence in favor of their proposal. Norris et al.[62] have reported a measurement of the photoemission spectrum of mercury. If interpreted as a measurement of the bulk density of states, their data would support a model such as Mott's with a reduced density of states at the Fermi level. But as previously stated, we must caution against such a naive interpretation of photoemission data. Evidently, however, there remain open questions about the electronic structure of liquid mercury.

Itami and Shimoji[44] calculated the density of states in several *liquid alloys* (Na–K, Ag–Au, Pb–Sn, Cu–Sn, Hg–Bi, and Hg–In) for various compositions. They employed the method and approximations of Ballentine,[8] which are described in Section V.B. These results are of interest, being the first such calculations for alloys. But because they used local energy-independent pseudopotentials, the results can only be regarded as tentative.

One conclusion that clearly emerges from these calculations is that the reliability of results depends strongly on the reliability of the pseudopotential used. The nonlocal nature of the pseudopotential is often quantitatively important. A calculation based on a hastily chosen pseudopotential, such as an exponentially screened Coulomb potential, has no value at all.

Calculations based on the assumed existence of a dispersion relaxation $E_L(k)$ in the liquid state may be quantitatively and, in extreme cases, qualitatively misleading. But if their limitations are understood, they are not entirely useless.

## VIII. BEYOND THE WEAK-SCATTERING LIMIT—TWO PROBLEMS

All the calculations to this point are restricted to the weak-scattering limit, in which the effect of the ions on the electrons may be accounted for by treating a weak pseudopotential as a perturbation. We did not merely use finite-order perturbation theory, the summation of certain infinite classes of terms by introducing the irreducible self-energy operator and the self-consistent internal propagator being necessary in order to avoid pathological results. Nevertheless, the best we have done is really only a self-consistent second-order perturbation theory. Two problems arise in any attempt to go beyond the weak-scattering limit. First, the scattering of a single ion is not weak enough to be treated by the Born approximation. Second, stronger single-ion scattering implies that multiple scattering should become significant, so we need to know the triplet and higher-order atomic correlation functions.

### A. The Single-Atom Problem

The first problem one encounters is the construction of a self-consistently screened potential for, if the weak-scattering approximation is not valid, then the linear screening theory of Section II is not valid. We have nothing to say about this problem, because practically no research has been done on non-linear screening in a disordered system. It is not even clear that we should write the total potential as the sum of identical potentials centered on each atom, as in (2.1), but we shall continue to do so because that form is known to be correct for the extremes of weak and tight binding.

The lack of a potential may not be fatal, because one can in principle determine the distribution of energy levels without detailed knowledge of ion cores; one needs only the scattering amplitudes outside some nonoverlapping ionic spheres. A formal solution to this problem is provided by the $t$ matrix introduced in Section IV.B. However, the $t$ matrix defined by (4.12) describes the scattering by a single ion in free space, and it is not well suited to our problem because of its qualitatively different behavior for $E > 0$ and $E < 0$. If $v$ is spherically symmetric, then $\langle \mathbf{k} | t(E) | \mathbf{k}' \rangle$ is complex for $E > 0$ and real for $E < 0$, the difference reflecting the complex and real exponential forms of the corresponding wave functions. In an approximate calculation this will likely lead to pathological behavior for $E \leq 0$, similar to that found in Section V.A. This occurs, for example, in the theory of Beeby[12] as shown in detail by Ballentine.[7] *

The cure for this pathology was introduced in Section V.B: the *self-consistent internal propagator*. We define a self-consistent or medium-adapted $t$ matrix:

$$\hat{t}(E) = v + v\mathcal{G}(E)\hat{t}(E) \tag{8.1}$$

or in a graphical notation:

$$\tag{8.1'}$$

This differs from the ordinary $t$ matrix (4.12) by the replacement of the free-particle propagator $G_0$ with the full-ensemble average propagator

$$\mathcal{G}(E) = [G_0^{-1} - \Sigma(E)]^{-1} \tag{8.2}$$

which we represent by a double horizontal line. Because $\Sigma(E)$ also depends on $\hat{t}$, we have a set of equations that must be solved self-consistently. The leading

---

* The use of the reaction matrix in the form proposed by DeDycker and Phariseau[26] is unsatisfactory for the same reason. The imaginary part of their expansion to second order vanishes for $E < 0$, but $E = 0$ is not the bottom of the band.

terms up to third order in $\hat{t}$ are [compare (4.13)]

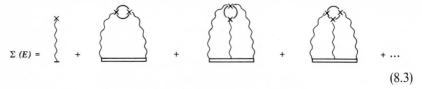

$$\Sigma\,(E) = \qquad\qquad\qquad\qquad\qquad\qquad\qquad\qquad\qquad + \cdots \tag{8.3}$$

Tables I and II give the key to interpreting these diagrams.

Because the momentum representation matrix element is inversely proportional to $\Omega$ (the volume of the system), we introduce

$$\hat{t}_{\mathbf{k},\mathbf{k}'} = \Omega\langle\mathbf{k}|\hat{t}|\mathbf{k}'\rangle \tag{8.4}$$

and

$$v_{\mathbf{k},\mathbf{k}'} = \Omega\langle\mathbf{k}|v|\mathbf{k}'\rangle$$

which are finite as $\Omega \to \infty$. The medium-adapted $t$ matrix is now the solution of the equation

$$\hat{t}_{\mathbf{k},\mathbf{k}'} = v_{\mathbf{k},\mathbf{k}'} + \frac{1}{(2\pi)^3}\int \frac{v_{\mathbf{k},\mathbf{k}''}\,\hat{t}_{\mathbf{k}'',\mathbf{k}'}}{E - (k'')^2 - \Sigma(k'', E)}\,d^3k'' \tag{8.5}$$

Because the denominator contains $E - \Sigma$, the medium-adapted $t$ matrix is unaffected by the shift of energy scale when any constant is added to the potential $V(\mathbf{r})$, whereas the choice of the zero of $E$ is a problem in any calculation employing the ordinary $t$ matrix [see (4.12)]. Of course the self-consistent internal propagator accomplishes much more than merely eliminating the energy origin problem.

Keeping only the first term in (8.3) yields a self-consistent single-site approximation:

$$\Sigma(k, E) = n\hat{t}_{\mathbf{k},\mathbf{k}}(E) \tag{8.6}$$

No realistic applications of this approximation have been made for liquid metals* but, since the single-site approximation is the leading term in a density expansion, we would expect it to be suitable only for a low-density gas. This opinion is supported by a calculation of Cloizeaux[21] for a one-dimensional random chain of delta-function atoms, which can be compared

---

* Anderson and McMillan[2] performed a type of single-site calculation for liquid iron. Their work is analyzed and criticised by Schwartz and Ehrenreich,[70] who performed a modified single-site calculation for liquid copper. They obtained a very deep minimum in the density of states near the center of the band, which is almost certainly an artefact of replacing the actual neighborhood of an atom with a spherically symmetric continuum.

with some quite accurate numerical results. He found that a non-self-consistent approximation including multiple scattering between two atoms (his Fig. 10) gave a better representation of the general shape of the numerically computed density of states (his Fig. 1) than did the self-consistent single-site approximation (his Fig. 6), although the approximation that lacked the self-consistent propagator inevitably possessed unphysical singularities. This result suggests that the approximation of keeping two terms in (8.3) might be of interest. Methods of summing some infinite subsets of these diagrams are discussed in Section IX.

## B. Higher-Order Correlations

The evaluation of higher-order terms in expansions like (4.7) and (8.3) requires not only the two-particle correlation function, which is known experimentally, but also higher-order correlation functions which are not directly measurable. To be suitable an approximate correlation function must be tractable, as well as accurate. The *superposition approximation*:

$$g_3^s(\mathbf{R}_1, \mathbf{R}_2, \mathbf{R}_3) = g_2(\mathbf{R}_1 - \mathbf{R}_2)g_2(\mathbf{R}_2 - \mathbf{R}_3)g_2(\mathbf{R}_3 - \mathbf{R}_1) \qquad (8.7)$$

is physically plausible but has seldom been applied to this problem because the presence of a factor $g_2$ for every pair of particles does not permit any factoring of the multiple integrals that arise—a difficulty that becomes more formidable as one considers higher orders.

A more tractable but less plausible form for the higher-order correlation functions we shall call the *chain approximation*. It has appeared, under different names, in many different contexts. Suppose we wish to evaluate $\langle v_\alpha v_\beta v_\gamma \rangle$, which requires the three-particle distribution function $g_3(\mathbf{R}_\alpha, \mathbf{R}_\beta, \mathbf{R}_\gamma)$. The averaging can be done in a three-step process, $\langle v_\alpha \langle v_\beta \langle v_\gamma \rangle_{\alpha\beta} \rangle_\alpha \rangle$, averaging first over all positions of $\gamma$ with $\alpha$ and $\beta$ fixed, then over all positions of $\beta$ with $\alpha$ fixed, and finally over all positions of $\alpha$. This corresponds to writing the three-particle distribution function in the form

$$g_1(\mathbf{R}_\alpha) \frac{g_2(\mathbf{R}_\alpha - \mathbf{R}_\beta)}{g_1(\mathbf{R}_\alpha)} \frac{g_3(\mathbf{R}_\alpha, \mathbf{R}_\beta, \mathbf{R}_\gamma)}{g_2(\mathbf{R}_\alpha, \mathbf{R}_\beta)} \qquad (8.8)$$

which may be read from right to left as the probability that a third particle is at $\mathbf{R}_\gamma$ given that there are particles at $\mathbf{R}_\beta$ and $\mathbf{R}_\alpha$, multiplied by the probability that a particle is at $\mathbf{R}_\beta$ given that one is at $\mathbf{R}_\alpha$, multiplied by the probability that a particle is at $\mathbf{R}_\alpha$. The chain approximation is made by assuming that the probability of a particle being at $\mathbf{R}_\gamma$, when there are particles fixed at both $\mathbf{R}_\beta$ and $\mathbf{R}_\alpha$ (the rightmost factor in 8.8), is the same as if there were only one particle fixed at $\mathbf{R}_\beta$. This is equivalent to replacing $g_3(\mathbf{R}_\alpha, \mathbf{R}_\beta, \mathbf{R}_\gamma)$ by $g_2(\mathbf{R}_\alpha - \mathbf{R}_\beta)g_2(\mathbf{R}_\beta - \mathbf{R}_\gamma)$. Its generalization:

$$g_s^c(\mathbf{R}_1 \cdots \mathbf{R}_s) = g_2(\mathbf{R}_1 - \mathbf{R}_2)g_2(\mathbf{R}_2 - \mathbf{R}_3) \cdots g_2(\mathbf{R}_{s-1} - \mathbf{R}_s) \qquad (8.9)$$

suggests the name *chain approximation*.

Because (8.9) lacks the full permutation symmetry of the true distribution function, the definition of the corresponding cluster function $h_s(\mathbf{R}_1 \cdots \mathbf{R}_s)$ is ambiguous. If one uses (3.8) to define it from (8.9), it fails to vanish in all the limits in which it should vanish, in other words it is *not* a cluster function. It seems appropriate to define the chain approximation of the cluster function to be

$$h_s^c(\mathbf{R}_1 \cdots \mathbf{R}_s) = h_2(\mathbf{R}_1 - \mathbf{R}_2)h_2(\mathbf{R}_2 - \mathbf{R}_3) \cdots h_2(\mathbf{R}_{s-1} - \mathbf{R}_s) \qquad (8.10)$$

even though this requires us to omit some of the lower-order cluster functions from (3.8) when we employ $g_s^c$ and $h_s^c$. The lack of permutation symmetry in the chain approximation is clearly responsible for this inconsistency.

This approximation seems to have been first introduced by Lax[55,56] in the problem of multiple scattering of waves, where it was called the quasi-crystalline approximation. More recently it was employed in the coherent potential method, which is discussed in the next section.

To determine the form of the chain approximation in momentum representation, let us consider a four-particle example, $\langle v_\alpha v_\beta v_\gamma v_\delta \rangle$. The averaging of its matrix elements in momentum representation involves only certain exponential factors,

$$\langle \exp\left[-i(\mathbf{k} - \mathbf{k}') \cdot \mathbf{R}_\alpha - i(\mathbf{k}' - \mathbf{k}'') \cdot \mathbf{R}_\beta - i(\mathbf{k}'' - \mathbf{k}''') \cdot R_\gamma - i(\mathbf{k}''' - \mathbf{k}) \cdot R_\delta\right]\rangle$$
$$= \langle \exp\left[i(\mathbf{k}' - \mathbf{k}) \cdot (\mathbf{R}_\alpha - \mathbf{R}_\beta)\right] \exp\left[i(\mathbf{k}'' - \mathbf{k}) \cdot (\mathbf{R}_\beta - \mathbf{R}_\gamma)\right]$$
$$\times \exp\left[i(\mathbf{k}''' - \mathbf{k}) \cdot (\mathbf{R}_\gamma - \mathbf{R}_\delta)\right]\rangle$$

Introducing the chain approximation here is clearly equivalent to replacing the average of the product of exponentials with the product of averages. This implies the approximation for the structure function

$$a_4^c(\mathbf{q}_1, \mathbf{q}_2, \mathbf{q}_3, \mathbf{q}_4) = a(\mathbf{q}_1)a(\mathbf{q}_1 + \mathbf{q}_2)a(\mathbf{q}_1 + \mathbf{q}_2 + \mathbf{q}_3) \qquad (8.11)$$

where we have written $\mathbf{q}_1 = \mathbf{k}' - \mathbf{k}$, $\mathbf{q}_2 = \mathbf{k}'' - \mathbf{k}'$, $\mathbf{q}_3 = \mathbf{k}''' - \mathbf{k}''$, and $\mathbf{q}_4 = \mathbf{k} - \mathbf{k}'''$. A similar approximation form holds for the cluster function $b_4(\mathbf{q}_1, \mathbf{q}_2, \mathbf{q}_3, \mathbf{q}_4)$, with obvious generalization to arbitrary order $s$. However, the inconsistent nature of the chain approximation reveals itself by the fact that some of the lower-order terms are absent from (4.18) when $a_s^c$ and $b_s^c$ are employed, just as occurs for (3.8).

The chain approximation in momentum space was used by Beeby[12] in a sophisticated Green function theory in which it was called the "geometric approximation." As noted in Section VIII.A, this theory failed because it lacked the self-consistent internal propagator, and so its failure need not reflect unfavorably on the chain approximation. Ashcroft and Schaich[5] used the chain approximation to calculate higher-order corrections for the electrical resistivity of liquid metals. The higher-order terms turned out to be surprisingly (one is inclined to say unbelievably) large, and in this case the trouble can be linked to the chain approximation.

In all its applications the chain approximation has been introduced primarily for computational convenience. No physical arguments in its favor have been given, except for the trivial observations that it is correct in the extreme cases of an ideal gas (for which $g_s \equiv 1$) and a perfect crystal (for which all distribution functions become lattices of delta functions, the statistical aspect of the problem, which gave rise to the correlation functions, having disappeared). However, criticism of it has seldom gone beyond the trivial observation that (8.9) lacks full permutation symmetry. However, the following argument, originally introduced in another context,[9] seems to be relevant.

If there is such a thing as a weak-scattering liquid metal, then its electron-scattering amplitude should be small in all orders of perturbation theory. The term of order $s$ in the scattering potential involves $v(q_1)v(q_2) \cdots v(q_s)a_s(\mathbf{q}_1, \mathbf{q}_2, \ldots, \mathbf{q}_s)$, as can be seen from the form of the potential (2.2) and the definition of the structure functions (4.20). But (2.6) implies that $v(q) = \langle \mathbf{k}|v|\mathbf{k} + \mathbf{q} \rangle$ is never small in the limit $q \to 0$. Therefore the $s$-order term in the scattering amplitude can be small only if $a_s(\mathbf{q}_1, \mathbf{q}_2, \ldots, \mathbf{q}_s)$ becomes small when *any one* of its arguments approaches zero. This is well known to occur for $s = 2$, according to (2.7), but no similar results for $s > 2$ have been previously reported.

The above argument is only suggestive, but the conditions for the validity of its conclusion can be rigorously established. According to (4.19) and (4.20), the structure functions represent correlations of density fluctuations. A long-wavelength ($q \to 0$) fluctuation may be treated thermodynamically. Callen (Ref. 16, Chap. 15) has set up a general thermodynamical fluctuation theory applicable to this problem, from which we can obtain the result[11b].

$$a_s(\mathbf{q}_1, \mathbf{q}_2, \ldots, \mathbf{q}_{s-1}, 0) = a(0)a_{s-1}(\mathbf{q}_1, \mathbf{q}_2, \ldots, \mathbf{q}_{s-1})$$

$$+ \, nk_B T \, \frac{\partial}{\partial P} \, a_{s-1}(\mathbf{q}_1, \mathbf{q}_2, \ldots, \mathbf{q}_{s-1}) \quad (8.12)$$

The pressure derivative is to be taken at constant temperature $T$, and $a(0)$ is given by (2.7). In (8.12) it is assumed that the limit $q_s \to 0$ has been taken so that no subset of the remaining $\mathbf{q}_1 \cdots \mathbf{q}_{s-1}$ sums to zero, hence the terms in (4.20) containing more than one delta function are always zero.

This result, which may be regarded as a generalization of (2.7), shows that if the liquid structure is sufficiently resistant to compression any partial long-wavelength limit of any order structure function will be small, and so will cancel the large value of the screened potential in that limit.

From the measurements by Egelstaff et al.[28b] of $[\partial a(q)/\partial P]_T$ for rubidium, we have verified that $a_3(\mathbf{q}, -\mathbf{q}, 0)$ is small compared with unity for all $q$ values, having a maximum value of 0.2. They also calculated $[\partial a(q)/\partial P]_T$ by the superposition approximation, obtaining a result much larger than the experimental value. The worst disagreement occurs for $a_3(0, 0, 0)$, which should be about $-10^{-3}$, whereas the superposition approximation yields $+4.5$. It is therefore very likely that the superposition approximation seriously overestimates third-order contributions. The chain approximation is also unsatisfactory, since $a_3{}^c(\mathbf{q}_1, \mathbf{q}_2, \mathbf{q}_3) = a(\mathbf{q}_1)a(\mathbf{q}_1 + \mathbf{q}_2)$ does not become small in the limit $q_2 \to 0$, $a_4{}^c$ [see (8.11)] does not become small in the limits $q_2 \to 0$ or $q_3 \to 0$, and so on.

This subtle cancelation in the long-wavelength limit between the large potential of a screened pseudoatom and the small structure functions illustrates the difficulties in calculating higher-order terms in perturbation or multiple-scattering series.

## IX. THE COHERENT POTENTIAL METHOD

The coherent potential method (CPM) was originally formulated by Soven[75] for random substitutional alloys. Although both restrictions appeared to be essential in the original formulation, the method has been generalized by Faulkner,[31] Gyorffy,[35] and Thornton,[78] who eliminated the requirements of randomness and lattice structure and made the CPM applicable to liquid metals. The starting point is quite different from that of diagrammatic perturbation theory, and therefore it is rather surprising that it leads to much the same result.

The idea underlying the CPM is to divide the Hamiltonian into two parts, not kinetic plus potential energy, $H = H_0 + V$, as is done in perturbation theory, but instead

$$H = (H_0 + \Sigma) + (V - \Sigma) = K + \tilde{V} \qquad (9.1)$$

The "coherent potential" $\Sigma$ is to be chosen so that the scattering from the residual potential $\tilde{V}$ is zero on average.

Let $\mathscr{G}$ be the Green operator for the coherent medium:

$$\mathscr{G} = (E - K)^{-1} = (E - H_0 - \Sigma)^{-1} \qquad (9.2)$$

Then the Green operator of the entire system, $G = (E - H)^{-1}$, can be written

$$G = \mathscr{G} + \mathscr{G}\tilde{V}G \tag{9.3a}$$

$$= \mathscr{G} + \mathscr{G}\tilde{T}\mathscr{G} \tag{9.3b}$$

Equation (9.3b) defines $\tilde{T}$, the transition operator that describes the scattering by the residual potential $\tilde{V}$. The coherent potential operator $\Sigma$, which is itself independent of the specific positions of the atoms, is to be chosen so that the scattering is zero on average:

$$\langle \tilde{T} \rangle = 0 \tag{9.4}$$

When this condition is satisfied, (9.3b) yields

$$\langle G \rangle = \mathscr{G} \tag{9.5}$$

hence $\mathscr{G}$ and $\Sigma$ turn out to be identical to the ensemble average Green operator and irreducible self-energy operator denoted by the same symbols in previous sections.

Of course (9.4) is no easier to solve than any other exact equation, and the inevitable approximations are yet to be introduced. Let us write the potential operators as sums of the contributions from the individual atoms: $\tilde{V} = \sum_{j=1}^{N} \tilde{v}_j$, $\tilde{v}_j = v_j - \sigma_j$, $\Sigma = \sum_{j=1}^{N} \sigma_j$. Now $\Sigma$ must possess the full symmetry of the ensemble average. This is possible in a liquid only if $\sigma_j = \sigma$ is independent of position (although it is normally a function of momentum), hence we have $\Sigma = N\sigma$. (In a substitutional alloy $\Sigma$ need only have the symmetry of the lattice. Thus $\sigma_j$ must have the same form for all equivalent sites, but it may be a function of position relative to $\mathbf{R}_j$. Thornton and Sampanthar[79] have discussed the differences between the CPM for a liquid and for a lattice.) The transition operator $\tilde{T}$ has a standard multiple-scattering expansion

$$\tilde{T} = \sum_j \tilde{t}_j + \sum_{j,k}{}' \tilde{t}_j \mathscr{G} \tilde{t}_k + \sum_{j,k,l}{}' \tilde{t}_j \mathscr{G} \tilde{t}_k \mathscr{G} \tilde{t}_l + \cdots \tag{9.6}$$

$$\tilde{t}_j = \tilde{v}_j + \tilde{v}_j \mathscr{G} \tilde{t}_j \tag{9.7}$$

where the primes on the summations imply that consecutive $T$ matrices must belong to different sites ($j \neq k$, $k \neq l$, etc.).

The *single-site approximation* is obtained by replacing the exact condition (9.4) with $\langle \tilde{t}_j \rangle = 0$. This is not exact even for a random system, since it does not correctly average the third-order term of (9.6) in which $j = l$. To determine the consequences of this approximation, let us calculate the Green operator for a system consisting of the hypothetical coherent medium plus the additional potential $\tilde{v}_j = v_j - \sigma_j$:

$$[E - H_0 - \Sigma - (v_j - \sigma_j)]^{-1} = \mathscr{G} + \mathscr{G}\tilde{t}_j\mathscr{G} \qquad (9.8)$$

This equation may be regarded as defining $\tilde{t}_j$, it being equivalent to (9.7) in this respect. This system can also be formed by adding the potential $v_j$ to the coherent medium from which $\sigma_j$ has been removed,

$$[E - H_0 - (\Sigma - \sigma_j) - v_j]^{-1} = \mathscr{G}'_j + \mathscr{G}'_j t'_j \mathscr{G}'_j \qquad (9.9)$$

where $\mathscr{G}'_j = [E - H_0 - (\Sigma - \sigma_j)]^{-1}$, and $t'_j$ is defined by this equation. Combining (9.8) and (9.9) and using the condition $\langle \tilde{t}_j \rangle = 0$, we can solve for the coherent potential

$$\sigma_j = \langle t'_j \rangle \mathscr{G}'_j \mathscr{G}^{-1} \qquad (9.10)$$

Now in a liquid metal, where $\sigma_j = \sigma$ is a constant throughout space, the difference between $\mathscr{G} = [E - H_0 - N\sigma]^{-1}$ and $\mathscr{G}'_j = [E - H_0 - (N - 1)\sigma]^{-1}$ becomes insignificant as $N \to \infty$. Therefore $t'_j$, which satisfies

$$t'_j = v_j + v_j \mathscr{G}'_j t'_j \qquad (9.11)$$

becomes indistinguishable from $\tilde{t}_j$, the medium-adapted $t$ matrix defined by (8.1). Thus the single-site coherent potential approximation for a liquid metal is identical to (8.6).

We see now that the essential principle of the CPM is equivalent to the use of the *self-consistent internal propagator*. So in spite of its apparently different starting point, the CPM does not lead to a set of approximations different to those of the diagrammatic expansion method, but only to a new view of the same approximations.

Of course the single-site approximation is not adequate for a liquid, and some account must be taken of the correlation among atoms in the higher-order terms of (9.6). Gyorffy[35] suggested using the chain approximation. It leads to an integral equation which may be written

$$\sigma_i = \tilde{t}_i + \int \tilde{t}_i \mathscr{G}\sigma_j nh_2(\mathbf{R}_i - \mathbf{R}_j)\, d^3R_j \qquad (9.12)$$

which, upon iteration, generates an infinite series

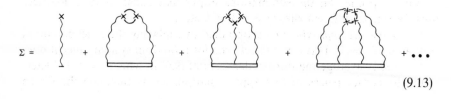

$$(9.13)$$

in which the cluster functions for three or more atoms are replaced by the chain approximation (8.10). Furthermore, many diagrams such as the third in (8.3) are entirely omitted.

This application of the chain approximation in the CPM has unfortunately come to be called the "coherent potential approximation," thereby obscuring the essential features of the CPM, the self-consistent internal propagator and the medium-adapted $t$ matrix, in favor of an inessential (and not particularly reliable) approximation for higher-order correlation functions. Equally unfortunate is the use of the term "quasi-crystalline approximation" (which was Lax's original name for the chain approximation) for a non-consistent version of (9.12) in which $\hat{t}$ and $\mathcal{G}$ are replaced by $t$ and $G_0$. The chain approximation actually refers only to correlation functions, and it can be employed in self-consistent or in nonconsistent theories.

Schwartz and Ehrenreich[70] proposed a more symmetric alternative to the chain approximation. They replaced the probability of finding a particle at $\mathbf{R}_3$ when there are particles fixed at $\mathbf{R}_2$ and $\mathbf{R}_1$, $g_3(123)/g_2(12)$, by the approximation $g_2(13) + g_2(23) - 1$. This gives $g_3(123)$ its correct form (3.8), but with the three-particle cluster function $h_3(123)$ replaced by $h_2(12)h_2(13) + h_2(12)h_2(23)$, which is more symmetric than the chain approximation but which still lacks full symmetry. This ansatz and its higher-order generalizations lead to the integral equation

$$\sigma_i = \hat{t}_i + \int \sigma_i \mathcal{G} n h_2(\mathbf{R}_i - \mathbf{R}_j)\sigma_j(1 + \mathcal{G}\hat{t}_i) \, d^3 R_j \qquad (9.14)$$

This equation reproduces all the third-order terms of (8.3), except for the aforementioned approximation of $h_3(123)$ in the last term. Schwartz and Ehrenreich[70] and Schwartz[71] carefully analyzed the differences between (9.12) and (9.14), but no actual calculations have been made for either equation. [In particular it should be noted that the calculation for copper reported by Schwartz and Ehrenreich is not based on (9.14), but rather on a cruder approximation.]

## X. TIGHT-BINDING METHODS

In a tight-binding calculation one takes as basis functions a set of localized orbitals centered on the various atoms. Two different kinds of localized basis functions can be used for a crystal: atomic orbitals or Wannier functions. Atomic orbitals on neighboring atoms overlap and are not orthogonal. Wannier functions (see Ref. 13 for a general discussion) are less tightly localized than atomic orbitals, but they are often more convenient because they are orthogonal.

Wannier functions (orthogonal localized orbitals) have never been defined
for a liquid. Presumably, it would not be useful to do so, because the form of
such a function localized about a particular atom depends also on the environment of the atom. Thus no two atoms would have identical orbitals.
The only practical possibility is to use identical atomic orbitals on each
atom, the nonorthogonality parameters of which depend on interatomic
separation:

$$\int \phi^*(\mathbf{r} - \mathbf{R}_i)\phi(\mathbf{r} - \mathbf{R}_j)\, d^3r = S(\mathbf{R}_i - \mathbf{R}_j) \tag{10.1}$$

If one introduces creation and annihilation operators based on these atomic
orbitals, they will not obey the standard anticommutation relations for
fermions, but rather

$$a_i^\dagger a_j + a_j a_i^\dagger = S(\mathbf{R}_i - \mathbf{R}_j) \tag{10.2}$$

This fact has not been taken into account in many calculations.[43,50]

A general tight-binding formalism based on nonorthogonal orbitals was
set up by Roth[66] and applied to some simple models. To illustrate the importance of the overlap of nonorthogonal orbitals, we present results (L. Roth,
unpublished) for a simplified hard-sphere liquid in which the pair distribution function $g(R)$ is constant for $R$ greater than the hard-sphere diameter $\sigma$

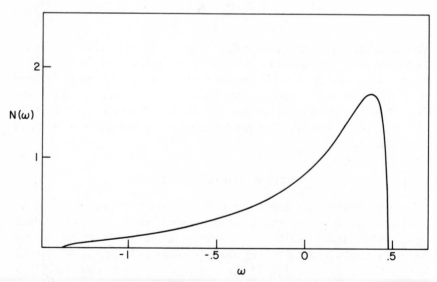

**Fig. 19** Density of states for a hard-sphere liquid, assuming orthogonal orbitals. (Roth,
private communication)

and vanishes for $R$ less than $\sigma$. The calculation employed an effective medium approximation[66b] similar in spirit to the CPM, and treated triplet correlations by the superposition approximation. Figure 19 shows the density of states for the model, assuming orthogonal orbitals. Figure 20 shows the density of states obtained when the overlap of exponential orbitals is taken into account. The results emphasize the importance of the overlap function $S(\mathbf{R}_i - \mathbf{R}_j)$.

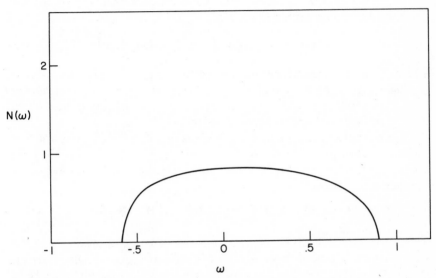

**Fig. 20** Density of states for a hard-sphere liquid with nonorthogonal orbitals. (Roth, private communication)

The calculations in the above-mentioned papers are for artificial models. Realistic calculations encounter the difficulty that the matrix elements of the Hamiltonian and the overlap function decrease rapidly with increasing interatomic distance $R_{ij}$ in the same region where the pair distribution $g(R_{ij})$ increases rapidly to its peak. Thus the results are very sensitive to errors in any of these quantities.

An entirely different kind of tight-binding theory was formulated by Rousseau, Stoddart, and March[67] (RSM), based in the canonical density matrix

$$C(\mathbf{r}, \mathbf{r}', \beta) = \sum_n \psi_n^*(\mathbf{r})\psi_n(\mathbf{r}')e^{-\beta\varepsilon_n} \tag{10.3}$$

where $\psi_n(\mathbf{r})$ and $\varepsilon_n$ are eigenfunctions and eigenvalues of the one-electron Hamiltonian. Its trace, the partition function for one electron, is the Laplace transform of the density of states:

$$Z(\beta) = \int C(\mathbf{r}, \mathbf{r}, \beta)\, d^3r$$

$$= \sum_n e^{-\beta\varepsilon_n} = \int n(\varepsilon)e^{-\beta\varepsilon}\, d\varepsilon \tag{10.4}$$

The density matrix for an electron in the presence of a single-scattering potential $v(\mathbf{r})$ can be written in the form

$$C_1(\mathbf{r}, \mathbf{r}', \beta) = C_0(\mathbf{r}, \mathbf{r}', \beta) \exp\left[-\beta U(\mathbf{r}, \mathbf{r}', \beta)\right] \tag{10.5}$$

where $C_0$ is the free-electron density matrix, and $U$ is called an effective potential matrix. As $\beta$ approaches zero, $U(\mathbf{r}, \beta) \equiv U(\mathbf{r}, \mathbf{r}, \beta)$ approaches the actual potential. Its calculation for arbitrary $\beta$ is discussed by Hilton, March, and Curtis[39] and by Sah and Rajagopal.[68]

The essential assumption made by RSM is that if the potential is a sum of identical potentials centered on each atom, $V(\mathbf{r}) = \sum_i v(\mathbf{r} - \mathbf{R}_i)$, then the effective potential may also be written as a sum of contributions from each "pseudoatom," hence

$$C(r, r, \beta) = C_0(\mathbf{r}, \mathbf{r}, \beta) \prod_i \exp\left[-\beta U(\mathbf{r} - \mathbf{R}_i, \beta)\right] \tag{10.6}$$

This is the most difficult step in their theory to analyze or generalize. RSM claim that it works well under any one of the following conditions: (1) $v(\mathbf{r})$ is small, (2) $\beta$ is small, (3) $U(\mathbf{r}, \beta)$ is slowly varying in space, and (4) the overlaps between $\nabla U(\mathbf{r} - \mathbf{R}_i, \beta)$ for different atoms are small. It is the last of these conditions that led us to classify the theory as a tight-binding method. However, its scope is apparently greater than that of the conventional tight-binding approximation.

In order to introduce the ensemble average, RSM expand (10.6) as a power series in $f(r, \beta) = \exp\left[-\beta U(r, \beta)\right] - 1$, the form of which is analogous to Mayer's cluster expansion in classical statistical mechanics. They approximate the higher-order distribution functions by the expressions

$$g_3(\mathbf{R}_1, \mathbf{R}_2, \mathbf{R}_3) = \tfrac{1}{3}\{g(R_{12})g(R_{13}) + g(R_{23})g(R_{21}) + g(R_{31})g(R_{32})\} \tag{10.7}$$

$$g_s(\mathbf{R}_1 \cdots \mathbf{R}_s) = \frac{1}{s} \sum_{i=i}^{s} \prod_{j\neq i} g(R_{ij}) \tag{10.8}$$

where $R_{ij} = |\mathbf{R}_i - \mathbf{R}_j|$. This approximation is fully symmetric and has the correct asymptotic behavior, but it fails to vanish when two particles overlap.

Whatever its limitations, this ansatz enables one to sum the infinite series, obtaining the final result

$$Z(\beta) = (2\pi\beta)^{-3/2} \left\{ 1 + \int \frac{f(\mathbf{r}_1, \beta)[\exp(n\gamma(\mathbf{r}_1)) - 1]}{\gamma(\mathbf{r}_1)} d^3r_1 \right\} \qquad (10.9)$$

where

$$\gamma(\mathbf{r}_1) = \int f(\mathbf{r}_2, \beta)g(\mathbf{r}_1 - \mathbf{r}_2)\, d^3r_2 \qquad (10.10)$$

and $n$ is the number of atoms per unit volume.

Inversion of the Laplace transform (10.4) to obtain the density of states would require us to compute $Z(\beta)$ for complex $\beta$. Therefore RSM chose models for $n(E)$ and compared $Z(\beta)$ for the models with the numerical results from (10.9). This is an inconvenient, indirect method to obtain $n(E)$. Pant, Das, and Joshi[63,64] inverted the Laplace transform by means of a first-order steepest-decent approximation. Hoare and Ruijgrok[40] showed that the error of that approximation is comparable to that of the Stirling approximation for $N!$, $N$ being the number of degrees of freedom of the system. We are dealing here with the one-electron partition function, hence $N = 3$, and therefore the accuracy of the results of Pant, Das, and Joshi is seriously in doubt.

Very few applications of this theory have been made, which is unfortunate if indeed it is correct in the diverse conditions listed by RSM. RSM compared solid, liquid, and random $[g(R_{12}) = 1]$ beryllium. Their results indicate that $n(E)$ for the liquid is not greatly different from that for the solid, but that it is very different from the random case [in which $n(E)$ has a long tail connecting the $2s$ and $1s$ bands). These results by RSM do not agree with those of Pant, Das, and Joshi.[64] As mentioned above, the latter are in doubt, however, the uniqueness of the model $n(E)$ used by RSM to fit $Z(\beta)$ has not been demonstrated. The results of Pant, Das, and Joshi[63] for aluminum are in reasonable agreement with the pseudopotential calculations of Ballentine.[8]

## XI. CLUSTER SCATTERING

The traditional theory of the solid state attributes a dominant role to the lattice structure, Brillouin zone boundaries, Bloch's theorem, and other such manifestations of long-range order. But it is an empirical fact that liquid metals and amorphous semiconductors also possess many of the electric properties of their crystalline counterparts. This suggests that short-range order may be more important than long-range order. To test this idea we need a method of calculation in which the range of order can be varied. In principle this could be done within the theoretical formalisms already discussed by going to higher and higher orders, thereby bringing in correlation functions of larger numbers of particles. But in practice such calculations would soon become prohibitively lengthy, even if the necessary correlation functions were available. However, another method has been developed and applied with considerable success.

This method is based on a theory due to Lloyd.[57] The development of the basic theory is rather lengthy, so we only describe the main points. Lloyd begins with (7.15) for the integrated density of states but, instead of using a perturbation expansion in momentum representation, he transforms to a representation involving angular momentum relative to each of the atomic centers. His expression for the integrated density of states (of one-spin orientation) below energy $E$ can be written

$$N(E) = N_0(E) - (\pi\Omega)^{-1} \operatorname{Im} \ln \left[ \det \| \delta_{LL'} \delta_{ij} - \sum_{L''} G_{LL''}(\mathbf{R}_i - \mathbf{R}_j) K_{L''L'}(\mathbf{R}_i) \| \right]$$

$$(11.1)$$

Here $N_0(E)$ is the integrated density of states for free electrons. The determinant is of a matrix whose rows are labeled by a composite index consisting of angular momentum $L = (l, m)$ and $i$, and whose columns are labeled $L'$ and $j$. $K_{L''L}(\mathbf{R}_i)$ is the reaction matrix for the atom at $\mathbf{R}_i$. For a spherically symmetric atom it is a diagonal matrix having elements $(-\tan \eta_l)/\kappa$, where $\eta_l$ is a phase shift and $\kappa = \sqrt{E}$. Finally

$$G_{LL'}(\mathbf{R}) = -i\kappa \sum_{L''} C_{L'L''}^L 4\pi i^{l''} h_{l''}^+(\kappa R) Y_{L''}(\hat{R}) \qquad |\mathbf{R}| \neq 0$$

$$= -i\kappa \delta_{LL'} \delta_{ij} \qquad |R| = 0$$

where $Y_L(\hat{R})$ is a spherical harmonic, $h_l^+(\kappa R)$ is a spherical Hankel function, and

$$C_{L'L''}^L = \int Y_L(\hat{R}) Y_{L'}^*(\hat{R}) Y_{L''}^*(\hat{R}) \, d\Omega_{\hat{R}}$$

The expression is valid for nonoverlapping atomic potentials, and $E$ is measured relative to the uniform potential in the interstitial region. Although (11.1) seems formidable at first sight, its ingredients are quite intelligible: a matrix $K_{L''L}$ which describes the scattering from a single scatterer, and a propagator between scatterers $G_{LL''}(\mathbf{R}_i - \mathbf{R}_j)$.

Lloyd shows how his formal expression can be given a diagrammatic expansion, which can in turn be resummed in terms of a "medium $t$ matrix" and a "medium propagator," the steps being formally analogous to our introduction of the self-consistent $t$ matrix and propagator. As an example,

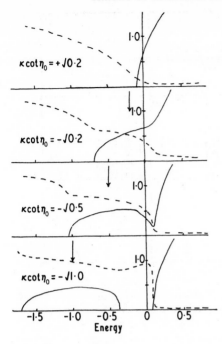

Fig. 21   Density of states for a model of uncorrelated point atoms. (From Lloyd.[57]).

he solves his equations for a simple model of uncorrelated point atoms whose phase shifts are nonzero only for $s$-wave scattering. The result (Fig. 21) is satisfying in many respects. The density of states continues smoothly through $E = 0$, in spite of the formal dependence of (11.1) on the artificially chosen origin for $E$. This was made possible by the self-consistent resummation of the diagrammatic expansion. The bound state of the single atom, which exists when the potential is attractive, is broadened in the liquid. When the potential becomes strong enough a bound band containing one state per atom separates from the free-electron-like band.

The next step, taken by Klima, McGill, and Ziman[53] is to regard the system as being made up of clusters of $N_c$ of atoms. Equation (11.1) still applies if we reinterpret $K_{L''L}(\mathbf{R}_i)$ as the reaction matrix for a cluster centered at $\mathbf{R}_i$, the labels $i$ and $j$ now referring to clusters. Although there is no difficulty in formally introducing a medium $t$ matrix and a medium propagator, as Lloyd did, the computational difficulty in evaluating the resultant expression would be great. Instead, Klima, McGill, and Ziman entirely neglect the effects of multiple scattering between clusters. This is equivalent to evaluating the second term of (11.1) for a finite cluster and averaging over the various possible clusters of $N_c$ atoms.

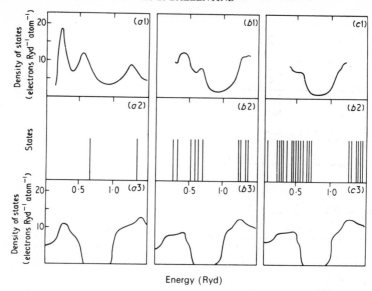

**Fig. 22**  Density of states from finite clusters of carbon atoms with various boundary conditions. Rows 1, 2 and 3 correspond to the three boundary conditions shown in Fig. 23. Columns $a$, $b$, and $c$ correspond to clusters of 2, 18, and 30 atoms. (From Keller.[51])

The objective is now to study the dependence of the density of states on the size of the cluster $N_c$ and on the short-range order within the cluster. The first row of Fig. 22 shows such a calculation[51] for a model of amorphous carbon. There is a region of low density of states, which becomes sharper and deeper as $N_c$ increases. Because the calculation lacks the medium propagator of Lloyd's full theory, it cannot produce an actual gap in the spectrum. It can only yield a strictly continuous state density for all energies above the constant interstitial potential, and an unbroadened discrete set of states below the interstitial level. The use of a self-consistent medium propagator can be looked on as taking into account the correct boundary conditions for a cluster imbedded in a disordered medium. Keller estimates the effect of improper boundary conditions by comparing with two extreme cases which are schematically illustrated in Fig. 23. The second row of Fig. 22 shows the discrete bound states of an isolated cluster in vacuum, and the third row shows the density of states for a superlattice of clusters. Although the density of states depends quantitatively on the boundary conditions, the gross distribution of the states is apparently determined by short-range order within the cluster.

Of course the cluster scattering method does not circumvent the problem

Energy

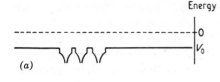

(a)

(b)

(c)

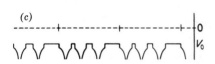

**Fig. 23**  Schematic illustration of the potentials and boundary conditions used in the calculation of Fig. 22. (a) Potential outside of the cluster equal to the interstitial potential. (b) Outside potential equal to the vacuum potential. (c) Periodic boundary conditions. (From Keller.[51])

of determining higher-order correlation functions which describe the arrangement of atoms within a cluster. However, it allows us to determine the consequences of any structural hypothesis, and so it may indirectly shed some light on the atomic correlation. There is good reason to believe that amorphous covalent semiconductors such as silicon and germanium are tetrahedrally coordinated, having bond lengths and bond angles differing by only a few percent from their values in the crystalline state. The calculations of Klima, McGill, and Ziman[53] and Keller[51] are in agreement with experimental evidence that such amorphous semiconductors possess a band gap similar to that of the corresponding crystals. Keller's results suggest that states would appear in the center of the gap if the bond angles were to vary by as much as 15%.

Unfortunately, there is no reason to suppose that any particular characteristic cluster dominates the short-range order of a liquid metal, and so we seem to be forced to average over all possible clusters with unknown weighting factors. Some simplification may be possible if the atomic scattering amplitude is dominated by a sharp resonance in one partial wave, such as the $d$ resonance in transition and noble metals. Keller and Jones[52] obtained reasonable results for iron and copper by considering only the $l = l' = 2$ block of Lloyd's determinant [see (11.1)]. They suggest that, in the limit of a sharp resonance, the only structural features that are of importance for the density of states are the coordination number and nearest-neighbor distance. This can be understood by thinking of a resonance as a virtual bound state; the sharper the resonance, the more nearly localized is the state, and the situation is like tight binding. House and Smith[41] showed that, although such an approximation is valid for copper, it is not adequate for iron.

Taken as a whole, the cluster calculations give support to the idea that the local short-range order, and of course the single-atom potential, determine the electronic structure of a substance, be it crystalline, amorphous, or liquid. However, full quantitative accuracy has not yet been achieved. Two kinds of possible improvements are: treating correlations (particularly angular correlations) between clusters, and combining the best features of the coherent potential method and the cluster method by introducing the self-consistent propagator into the cluster method. The latter suggestion has been implemented with very good result[14,15] for a one-dimensional random alloy model.

## References

1. I. V. Abarenkov and V. Heine, *Phil. Mag.*, **12**, 529 (1965).
2. P. W. Anderson and W. L. McMillan, *Proc. Int. Sch. Phys.*, **37**, 50 (1967).
3. A. O. E. Animalu and V. Heine, *Phil. Mag.*, **12**, 1249 (1965).
4. N. W. Ashcroft and W. E. Lawrence, *Phys. Rev.*, **175**, 938 (1968).
5. N. W. Ashcroft and W. Schaich, *Phys. Rev.*, **B1**, 1370 (1970); Erratum, **B3**, 1511 (1971).
6. B. J. Austin, V. Heine, and L. J. Sham, *Phys. Rev.*, **127**, 276 (1962).
7. L. E. Ballentine, Ph.D. Thesis, University of Cambridge, England, 1965.
8. L. E. Ballentine, *Can. J. Phys.*, **44**, 2533 (1966).
9. L. E. Ballentine, *Proc. Phys. Soc.*, **89**, 689 (1966).
10. L. E. Ballentine and T. Chan, in *The Properties of Liquid Metals, Proceedings of the Second International Conference, Tokyo, 1972*, (ed. S. Takeuchi), Taylor and Francis, London, 1973, p. 197.
11. L. E. Ballentine and O. P. Gupta, *Can. J. Phys.*, **49**, 1549 (1971).
11b. L. E. Ballentine and A. Lakshmi, *Can. J. Phys.*, **53**, 372 (1975).
12. J. L. Beeby, *Proc. Roy. Soc.*, **A279**, 82 (1964).
13. E. I. Blount, *Solid State Phys.*, **13**, 306 (1962).
14. W. H. Butler, *Phys. Lett.*, **39A**, 203 (1972).
15. W. H. Butler, *Phys. Rev.*, **B8**, 4499 (1973).
16. H. B. Callen, *Thermodynamics*, Wiley, New York, 1960.
17. T. Chan, Ph.D. Thesis, Simon Fraser University, Burnaby, B.C., Canada, 1971.
18. T. Chan and L. E. Ballentine, *Phys Chem. Liq.*, **2**, 165 (1971).
19. T. Chan and L. E. Ballentine, *Phys. Lett.*, **35A**, 385 (1971).
20. T. Chan and L. E. Ballentine, *Can. J. Phys.*, **50**, 813 (1972).
21. J. des Cloizeaux, *Phys. Rev.*, **139**, A1531 (1965).
22. M. L. Cohen and V. Heine, *Solid State Phys.*, **24**, 37 (1970).
23. M. Cooper, *Advan. Phys.*, **20**, 453 (1971).
24. N. E. Cusack, *Rep. Prog. Phys.*, **XXVI**, 361 (1963).
25. N. E. Cusack, in *The Properties of Liquid Metals, Proceedings of the Second International Conference, Tokyo, 1972*, (ed. S. Takeuchi), Taylor and Francis, London, 1973, p. 157.
26. E. DeDycker and P. Phariseau, *Advan. Phys.*, **16**, 401 (1967).
27. M. A. C. Devillers, W. H. Young, and A. R. de Vroomen, *J. Phys. F: Met. Phys.*, **3**, L220 (1973).
28a. S. F. Edwards, *Proc. Roy. Soc.*, **A267**, 518 (1962).
28b. P. A. Egelstaff, D. I. Page, and C. R. T. Heard, *J. Phys. C: Solid State Phys.*, **4**, 1453 (1971).
29. R. Evans, *J. Phys. C.: Met. Phys.*, *Suppl.* S137 (1970).

30. T. E. Faber, *Introduction to the Theory of Liquid Metals*, Cambridge University Press, New York, 1972.
31. J. S. Faulkner, *Phys. Rev.*, **B1**, 934 (1970).
32. M. P. Greene and W. Kohn, *Phys. Rev.*, **137**, A513 (1965).
33. A. J. Greenfield and N. Wiser, *J. Phys. F: Met. Phys.*, **3**, 1397 (1973).
34. H. J. Güntherhodt, A. Menth, and Y. Tièche, *Phys. Kondens. Mat.*, **5**, 392 (1966).
35. B. L. Gyorffy, *Phys. Rev.*, **B1**, 3290 (1970).
36. C. E. W. Hahn and J. E. Enderby, *Proc. Phys. Soc.*, **92**, 418 (1967).
37. W. A. Harrison, *Pseudopotentials in the Theory of Metals*, Benjamin, New York, 1966.
38. V. Heine and I. Abarenkov, *Phil. Mag.*, **9**, 451 (1964).
39. D. Hilton, N. H. March, and A. R. Curtis, *Proc. Roy. Soc.*, **A300**, 391 (1967).
40. M. R. Hoare and Th. W. Ruijgrok, *J. Chem. Phys.*, **52**, 113 (1970).
41. D. House and P. V. Smith, *J. Phys. F: Met. Phys.*, **3**, 753 (1973).
42. K. Ichikawa, *Phil. Mag.*, **27**, 177 (1973).
43. Y. Ishida and F. Yonezawa, *Prog. Theor. Phys.*, **49**, 731 (1973).
44. T. Itami and M. Shimoji, *Phil. Mag.*, **25**, 229 (1972).
45. P. Jena and N. C. Halder, *Phys. Rev. Lett.*, **26**, 1024 (1971).
46. P. Jena and N. C. Halder, *Phys. Rev.*, **B6**, 2131 (1972).
47. J. C. Jones and W. R. Datars, *J. Phys. F: Met. Phys.*, **1**, L56 (1971).
48. J. C. Jones and W. R. Datars, *Can. J. Phys.*, **50**, 1659 (1972).
49. B. Kahn and G. E. Uhlenbeck, *Physica*, **5**, 399 (1938).
50. I. Katz and S. A. Rice, *J. Phys. C: Solid State Phys.*, **5**, 1165 (1972).
51. J. Keller, *J. Phys. C: Solid State Phys.*, **4**, 3143 (1971).
52. J. Keller and R. Jones, *J. Phys. F: Met. Phys.*, **1**, L33 (1971).
53. J. Klima, T. C. McGill, and J. M. Ziman, *Discuss. Faraday Soc.*, **50**, 20 (1970).
54. R. Y. Koyama and W. E. Spicer, *Phys. Rev.*, **4**, 4318 (1971).
55. M. Lax, *Rev. Mod. Phys.*, **23**, 287 (1951).
56. M. Lax, *Phys. Rev.*, **85**, 621 (1952).
57. P. Lloyd, *Proc. Phys. Soc.*, **90**, 207 and 217 (1967).
58. N. H. March, *Liquid Metals*, Pergamon, Oxford, 1968.
59. N. F. Mott, *Phil. Mag.*, **13**, 989 (1966).
60. N. F. Mott, *Phil. Mag.*, **26**, 505 (1972).
61. C. Norris, D. C. Rodway, and G. P. Williams, in *The Properties of Liquid Metals, Proceedings of the Second International Conference, Tokyo, 1972*, (ed. S. Takeuchi), Taylor and Francis, London, 1973, p. 181.
62. C. Norris, D. C. Rodway, G. P. Williams, and J. E. Enderby, *J. Phys. F: Met. Phys.*, **3**, L182 (1973).
63. M. M. Pant, M. P. Das, and S. K. Joshi, *Phys. Rev.*, **B4**, 4379 (1971).
64. M. M. Pant, M. P. Das, and S. K. Joshi, *Phys. Rev.*, **B7**, 4741 (1973).
65. M. J. Rice, *Phys. Rev.*, **B2**, 4800 (1970).
66a. L. M. Roth, *Phys. Rev.*, **B7**, 4321 (1973).
66b. L. M. Roth, *Phys. Rev.*, **B9**, 2476 (1974).
67. J. Rousseau, J. C. Stoddart, and N. H. March, *Proc. Roy. Soc.*, **A317**, 211 (1970).
68. P. Sah and A. K. Rajagopal, *J. Phys. C: Solid State Phys.*, **5**, 1207 (1972).
69. T. Schneider and E. Stoll, *Advan. Phys.*, **16**, 731 (1967).
70. L. Schwartz and H. Ehrenreich, *Ann. Phys.*, **64**, 100 (1971).
71. L. M. Schwartz, *Phys. Rev.*, **B7**, 4425 (1973).
72. R. W. Shaw and A. W. Harrison, *Phys. Rev.*, **163**, 604 (1967).
73. R. W. Shaw, Jr., *Phys. Rev.*, **174**, 769 (1968).
74. R. W. Shaw and N. V. Smith, *Phys. Rev.*, **178**, 985 (1969).
75. P. Soven, *Phys. Rev.*, **156**, 809 (1967).
76. S. K. Srivastava and P. K. Sharma, *Ind. J. Pure Appl. Phys.*, **7**, 644 (1969).

77. R. W. Stark and L. M. Falicov, *Phys. Rev. Lett.*, **19**, 795 (1967).
78. D. E. Thornton, *Phys. Rev.*, **B4**, 3371 (1971).
79. D. E. Thornton and S. Sampanthar, *J. Phys. C.*, **4**, L271 (1971).
80. G. E. Uhlenbeck and G. W. Ford, in *Studies in Statistical Mechanics*, (ed. J. de Boer and G. E. Uhlenbeck), Vol. 1, North-Holland, Amsterdam, 1962.
81. R. N. West, R. E. Bortand, J. R. A. Cooper, and N. E. Cusack, *Proc. Phys. Soc.*, **92**, 195 (1967).
82. J. R. Wilson, *Metal. Rev.*, **10** (40), 381 (1965).
83. J. M. Ziman, *Advan. Phys.*, **13**, 89 (1964).

## APPENDIX: CALCULATIONS OF THE DENSITY OF ELECTRONIC STATES IN LIQUID METALS

Beside each reference number is a symbol (G, E, N, PA, SS, C) representing the method of calculation. If a pseudopotential was used, another symbol indicates whether it was local (l) or nonlocal (nl). For a given element the order of listing is intended to correlate with the estimated reliability of the result.

| Li | Na | K | Rb | Cs | Be |
|---|---|---|---|---|---|
| 9. E, l, and nl | 4. G, l | 4. G, l | 5. E, l | 5. E, l | 12. PA |
| 5. E, l | 5. E, l | 5. E, l | 7. E, l | 7. E, l | 14. PA |
| 10. E, l | 7. E, l | 7. E, l | 10. E, l | 10. E, l | |
| | 8. E, l | 8. E, l | | | |
| | 10. E, l | 10. E, l | | | |

| Ca | Ba | Zn | Cd | Hg | Al |
|---|---|---|---|---|---|
| 10. E, l | 10. E, l | 1. G, l | 6, 7. E, nl | 2, 3. G, nl | 1. G, l |
| | | | 9. E, l, and nl | 4. G, l | 8. E, l |
| | | | | 5. E, l | 13. PA |
| | | | | | 10. E, l |

| Ga | In | Tl | Sn | Pb | Bi |
|---|---|---|---|---|---|
| 7. E, l | 3. G, nl | 7. E, l | 4. G, l | 4. G, l | 3. G, nl |
| | 9. E, l, and nl | | 7. E, l | 8. E, l | 11. N, nl |
| | 7. E, l | | | 10. E, l | 1. G, l |
| | | | | | 4. G, l |

| Cu | Ag | Au | Fe | Ni | Alloys |
|---|---|---|---|---|---|
| 17. C | 4. G, l | 4. G, l | 17. C | 17. C | 4. G, l |
| 18. C | | | 18. C | | |
| 4. G, l | | | 15. SS | | |
| 16. SS | | | | | |

## References for Appendix

G: Green Function method:
1. L. E. Ballentine, *Can. J. Phys.*, **44**, 2533 (1966).
2. T. Chan and L. E. Ballentine, *Phys. Lett.*, **35A**, 385 (1971).
3. T. Chan and L. E. Ballentine, *Can. J. Phys.*, **50**, 813 (1972).
4. T. Itami and M. Shimoji, *Phil. Mag.*, **25**, 229 (1972).

E: $E_L(k)$ method:
5. K. Ichikawa, *Phil. Mag.*, **27**, 177 (1973).
6. P. Jena and N. C. Halder, *Phys. Rev. Lett.*, **26**, 1024 (1971).
7. P. Jena and N. C. Halder, *Phys. Rev.*, **B6**, 2131 (1972).
8. T. Schneider and E. Stoll, *Advan. Phys.*, **16**, 731 (1967).
9. R. W. Shaw and N. V. Smith, *Phys. Rev.*, **178**, 985 (1969).
10. S. K. Srivastava and P. K. Sharma, *Ind. J. Pure Appl. Phys.*, **7**, 644 (1969). Reports $n(E)$ at $E = E_F$ only.

N: Integrated density of states:
11. T. Chan and L. E. Ballentine, *Phys. Chem. Liq.*, **2**, 165 (1971).

PA: Pseudoatom method:
12. J. Rousseau, J. C. Stoddart, and N. H. March, *Proc. Roy. Soc.*, **A317**, 211 (1970).
13. M. M. Pant, M. P. Das, and S. K. Joshi, *Phys. Rev.*, **B4**, 4379 (1971).
14. M. M. Pant, M. P. Das, and S. K. Joshi, *Phys. Rev.*, **B7**, 4741 (1973).

SS: Single-Site theory:
15. P. W. Anderson and W. L. McMillan, *Proc. Int. Sch. Phys.*, **37**, 50 (1967).
16. L. Schwartz and H. Ehrenreich, *Ann. Phys.*, **64**, 100 (1971).

C: Finite cluster theory:
17. D. House and P. V. Smith, *J. Phys. F: Met. Phys.*, **3**, 753 (1973). The calculation is for a crystal, but it is relevant also to a liquid if the coordination number and nearest-neighbor distance are the same.
18. J. Keller and R. Jones, *J. Phys. F: Met. Phys.*, **1**, L33 (1971).

*Note added in proof:*

J. Keller, J. Fritz, and A. Garritz, *J. Physique* **35**, C4-379 (1974) report calculations by method C for Fe, Co, Ni, Cu, Sr, Ba, Ce, and Cu-Ni, Ce-Co alloys.

R. Srivastava and A. Jain, *Phys. Stat. Sol.* (b) **61**, K81 (1974) report calculations by method E for Ca, Sr and Ba.

# LOW-ENERGY ELECTRONS IN NONPOLAR FLUIDS

## H. TED DAVIS AND ROGER G. BROWN

*Departments of Chemistry and Chemical Engineering*
*University of Minnesota, Minneapolis, Minnesota*

## TABLE OF CONTENTS

## I. INTRODUCTION

The nature of electronic states and transport in disordered systems has been under intense investigation for the past 15 years. For amorphous semiconductors, liquid metals, polar liquids, metal–polar liquid solutions and nonpolar liquids, much progress has been made toward understanding the relationship between electronic states and transport and the statics and dynamics of the molecular geometry of the media. It is generally accepted

now that the electronic states in amorphous systems include (a) quasi-free
or delocalized states in which the electrons are highly mobile, (b) a continuum
of low-mobility localized states (localized in the Anderson sense) resulting
from fluctuations in local potential, and (c) low-mobility states in which the
electron is self-trapped in "bubbles" or "cavities." Of course, impurities
may provide electron trap states by ion formation, but these states are less
dependent on the molecular structure and dynamics of the medium than
are (a) to (c).

Attention in this article is focused on the behavior of excess electrons in
nonpolar fluids (dilute gases are not dealt with because our interest is in the
density regime where collective effects occur). Studies of electron drift
velocities and mobilities (defined as the drift velocity divided by the electric
field causing the drift), work functions, ranges in radiation pulses, absorption
spectroscopy, and reaction rates with scavengers provide the major source
of experimental information on excess electrons in nonpolar fluids. In
Sections II to V experimental data are presented with a qualitative discussion
of their implications. In later sections detailed theoretical models are treated.

## II. MOBILITY AND WORK FUNCTION STUDIES

Aside from water, liquid helium has probably fascinated scientists more
than any other fluid because of the very low solidification temperatures of
helium and the lambda-point superfluid transition of $^4$He. In fact, in the
early mobility studies of Williams[1] and of Reif and Meyer[2] on charge transport in liquid helium, the temperature effects associated with superfluid
behavior received more attention than did effects hinting at the nature of the
negative-charge carriers. In their rather extensive study of the temperature
dependence of positive and negative mobilities $\mu$ (where

$$\mu \equiv \frac{v_d}{E} \qquad (2.1)$$

and $v_d$ is the drift velocity caused by the electric field $E$) in the superfluid
region of $He^4$, Meyer and Reif's[2] primary interest was in using charge
mobility to study the superfluid properties of $^4$He. However, they observed
that the negative species (produced by alpha-particle irradiation from a
$^{210}$Po source) behaved in such a way as to indicate that it is not simply a
negative ion. Some of their low-field mobility results are shown in Fig. 1.
From their analysis of charged particle–roton scattering, Meyer and Reif
concluded that the negative species had an effective mass on the order of a
molecular mass. However, as opposed to the positive ion behavior (and to
what would be expected of an ion), the negative species mobility increases
with density at low pressures. This behavior is shown in Fig. 1.

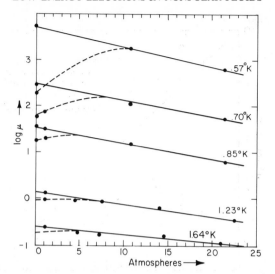

**Fig. 1**  Pressure dependence of the mobilities at various temperatures (Ref. 2b). The mobility $\mu$ is measured in cm$^2$/(V)(sec), the pressure is in atmospheres. The solid curves refer to the positive ion, the dashed curves to the negative ion. The point at 0.57°K and zero pressure was obtained by extrapolating the curve of ln $\mu$ versus $T^{-1}$ to this temperature, neglecting its bending over due to phonon scattering.

To determine whether or not the peculiarities of the negative species were related to the quantum behavior of $^4$He, Meyer et al.,[3] and later Magistris, Modena, and Scaramuzzi,[4] studied the density dependence of the mobility of the negative species in liquid He$^3$ and normal liquid $^4$He. Some of their results are shown in Figs. 2 and 3. A density region in which the mobility of the negative species increases with density is observed for both helium isotopes.

Meyer's[5] magnetic deflection experiments in superfluid helium and Donnelly's[6] studies of the interaction of charged species and quantized vortices in rotating superfluid helium also implied large electronic effective masses (of the order of 50 to 100 helium atom masses).

To explain what might be happening in liquid helium, Ferrel[7] and later Feynman suggested what eventually turned out to be the accepted model, namely, one in which the electron is trapped in a cavity or bubble. Such an event could occur if the electron-bubble free energy is less than the free energy of a quasi-free electron (i.e., a delocalized electron). From a simple model, Kuper[7] estimated a bubble radius $R$ of about 12 Å. With this value Stokes' law $\mu = e/6\pi\eta R$, where $\eta$ is the fluid viscosity and $e$ is the electronic charge, gives a fair estimate of the observed mobility of the negative species. If it is

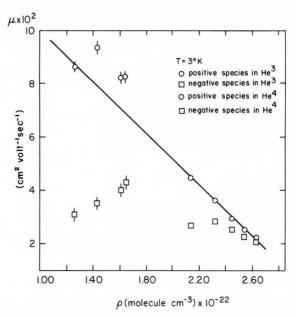

Fig. 2 Mobilities of the positive and negative species in $^4$He and $^3$He as a function of density at approximately 3°K (Ref. 3). Note that the mobility of the positive species seems to be a linear function of the density independent of the isotope.

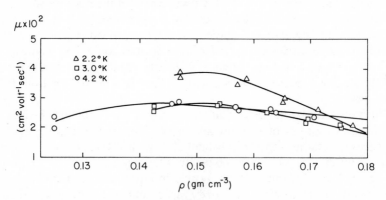

Fig. 3 Mobility of the negative species in HeI as a function of density at 2.2°, 3.0°, and 4.2°K. (Ref. 3).

assumed that the bubble radius decreases with increasing pressure (density) faster than the viscosity increases, then the observed trend of mobility with density may be explained. The decrease in $R$ predicted by Kuper's model was not fast enough to explain experimental results. However, more refined models, to be discussed in the theoretical sections to follow, give the proper variation of $R$.

Since the free energy of an electron in a helium bubble is positive (the polarization energy of an electron in helium being exceedingly small), there must be a barrier to injection of an electron into liquid helium if the bubble model is correct. The existence of such a barrier was demonstrated by Sommer,[8] who found a 1.3-eV work function (see Table I) for injection of electrons into liquid helium (he presumably measured the free energy of quasi-free or extended electrons, since the electrons enter the liquid faster than the liquid structure can relax to form a cavity).

At about the same time mobility studies were being done in liquid helium, Levine and Sanders[9,10] found that the mobility of electrons photoinjected into helium gas underwent a dramatic drop of some four orders of magnitude over a pressure range of about 20% of the vapor pressure. Some of their low-field results are shown in Figs. 4 and 5. The magnitude and field dependence of the mobility in a low-pressure region (see Fig. 5a) is typical of quasi-free electron mobility, whereas at about saturated vapor pressure the magnitude and field independence (see Fig. 5c) of the mobility are characteristic of heavy-ion mobility. Thus the electrons tend to undergo a sharp transition from a quasi-free state to a localized state as the density increases. Levine and Sanders found that the theoretical formula

$$\mu = \frac{2}{3}\left(\frac{2}{\pi m k T}\right)^{1/2}\frac{e}{n\sigma} \tag{2.2}$$

and the semihydrodynamical formula[11] (an interpolation formula giving the hydrodynamical limit as $R \to \infty$ and the kinetic theoretical value for heavy ion–atom hard spheres in the limit $R \to 0$)

$$\mu = \frac{e}{6\pi\eta R}\left[1 + \frac{9\pi\eta}{4nR(2\pi M_{He} kT)^{1/2}}\right] \tag{2.3}$$

accounted for the low field mobilities in low- and high-pressure regions, respectively. The quantity $m$ is the electron mass, $\sigma$ is the electron–helium momentum cross section, $k$ is Boltzmann's constant, $T$ is the temperature, $n$ is the fluid density, $R$ is the bubble radius (estimated from an optical potential model), $\eta$ is the fluid viscosity, and $M_{He}$ is the mass of a helium atom.

## TABLE I

## Mobilities and Work Functions Observed for Electrons in Nonpolar Hydrocarbon and Inert Gas Liquids

| Liquid | Arrhenius fit of mobility data, $\mu = \mu_0 \exp(-E_A/RT)$ | | $T$ (°K) | Mobility $\mu$ at $T$ [cm²/(V)(sec)] | Largest field $E_0$ below which $\mu$ is known to be field-independent. Entries marked with * correspond to fields above which $\partial\mu/\partial E \neq 0$ (kV/cm) | Work function $V_0$ at $T$ (eV) |
|---|---|---|---|---|---|---|
| | $\mu_0$ [cm²/(V)(sec)] | $E_A$ (kcal/mole) | | | | |
| Methane | 170 | ~0 | 111 | 400[a,45] | 1.5* | 0[f] |
| Ethane | 210 | 2.1 | 200 | 0.84[49,48] | 80* | |
| Propane | | 2.6 | 208 | 0.4[47] | 50 | |
| Butane | | | 296 | 0.4[39,40,59] | | |
| n-Pentane | | | 300 | 0.075[38] | | |
| | | | 296 | 0.16[39,40,59] | | |
| n-Hexane | 67 | 4.06 | 296 | 0.093[8,39,40,59] | 80 | 0[51] |
| | | | | | | −0.09[52] |
| | | | | | | 0.04[50] |
| | | | | | | 0.16[53] |
| Neopentane | | | 296 | 67[46] | 5* | −0.43[52] |
| | | | 298 | 70[47] | | −0.43[50] |
| | | | 295 | 70[49] | | −0.35[51] |
| | | | 296 | 59[b] | | |
| | | | 258 | 44[b] | | |
| TMS | 194 | 0.412 | 296 | 90[39,40,59] | | −0.61[52] |
| | | | | | | −0.62[50] |
| | | | | | | −0.55[51] |
| Cyclopentane | | 3.5[e] | 296 | 1.1[39,40,59] | 62 | 0.17[51] |
| | | | | | | 0.18[52] |
| | | | | | | 0.28[50] |
| Cyclohexane | 288.9 | 3.979 | 306 | 0.45[43] | | |
| Cyclohexene | 386.1 | 3.462 | 293 | 1.0[c] | | |
| Methylcyclohexane | 169.1 | 4.612 | 296 | 0.068[b] | | 0.14[52] |
| Toluene | | | 298 | 0.54[46] | 3 | −0.22[53] |
| trans-Butene-2 | 350 | 5.4 | 312 | 0.051[c] | 14 | |
| cis-Butene-2 | 820 | 3.7 | 293 | 2.2[c] | 14 | |

| Compound | | | T (K) | | | |
|---|---|---|---|---|---|---|
| Isobutene | | | 293 | 1.44 | | |
| Butene-1 | | | 293 | 0.064[c] | | |
| (CH$_3$)$_3$CH | | | 294 | 5.1[41] | | |
| 2-Methylbutene-2 | 90.1 | 2.0 | 281 | 2.4[37,38] | 3 | −0.24[52] |
| 2,2-Dimethylbutane | 84.1 | 1.2 | 296 | 10.9[39,40,59] | 35 | −0.15[53] |
| 2,3-Dimethylbutene-2 | 131.9 | 1.84 | 293 | 5.8[c] | 14 | −0.15[52] |
| 2,2,4-Trimethylpentane | | | 296 | 7[39,40,59] | 35 | −0.18[50] |
| 2,2,4,4-Tetramethylpentane | 308 | 1.48 | 295 | 24[3] | | −0.14[51] |
| 2,2,3,3-Tetramethylpentane | 61.12 | 1.452 | 295 | 5.2[43] | | |
| 2,2,5,5-Tetramethylhexane | 79.34 | 1.103 | 293 | 12[43] | | |
| Toluene | 10.17 | 2.95 | 298 | 0.54[46] | 3 | −0.22 |
| | | | 304 | 0.077[d] | | |
| Benzene | 30190 | 7.25 | 300 | 0.6[38] | 3 | −0.14 |
| | | | 306 | 0.19[d] | | |
| 1,2-Dimethylbenzene | 30.55 | 4.28 | 292 | 0.018[d] | | |
| 1,3-Dimethylbenzene | 118 | 4.4 | 292 | 0.057[d] | | |
| 1,4-Dimethylbenzene | | | 293 | 0.062[d] | | |
| 1,2,3-Trimethylbenzene | 102 | 4.88 | 293 | 0.022[d] | | |
| 1,2,4-Trimethylbenzene | 101 | 4.6 | 292 | 0.035[d] | | |
| 1,3,5-Trimethylbenzene | 22.2 | 2.87 | 293 | 0.16[d] | | |
| 1,2,3,4-Tetramethylbenzene | 617 | 6.13 | 323 | 0.04[d] | | |
| $^3$He | | | 2.25 | 0.0406 | | 0.9[3] |
| $^4$He | | | 4.2 | 0.0216 | | 1.05[3] |
| Ne | | | 22 | 0.0016 | | 0.5[22] |
| Ar | | | 82 | 475 | | −0.33[25] |
| | | | 82 | 440 | | −0.45[32] |
| Kr | | | 117 | 1800 | | −0.78[32] |
| Xe | | | 163 | 2200 | | −0.63[32] |
| H$_2$, D$_2$ | | | 21 | 0.02, 0.01[18] | | −1.0[19] |

[a] W. F. Schmidt and G. Bakale, Chem. Phys. Lett., 17, 617 (1972).
[b] A. O. Allen and R. A. Holroyd, J. Phys. Chem., 78, 796 (1974).
[c] J.-P. Dodelet, K. Shinsaka, and G. R. Freeman, J. Chem. Phys., 59, 1293 (1973).
[d] K. Shinsaka and G. R. Freeman, J. Chem. Phys., to be published.
[e] G. Freeman, private communication.
[f] W. Tauchert and W. F. Schmidt, Z. Naturforsch. 29A, 1526 (1974); S. Noda and L. Kevan, J. Chem. Phys., 61, 2467 (1974).

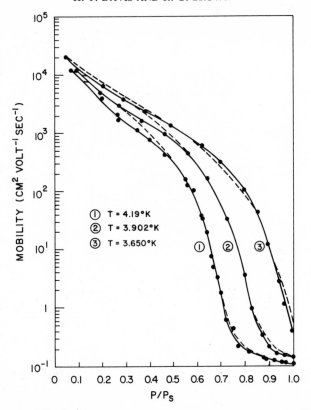

**Fig. 4** Electron mobility in ⁴He gas versus pressure at constant temperature (Ref. 10). The solid curves have no theoretical significance.

Although the high- and low-mobility regions could be understood in terms of quasi-free and bubble states, the nature of the mobility transition and the behavior of the mobility in the transition region was not given a quantitative description for several years. However, as we discuss in later sections, the transition has been described in terms of a percolation model[12,13] relying on density fluctuations for electron entrapment (i.e., Anderson localization) and, alternatively, by a model[14] in which the transition is related to a sharp increase in the fraction of self-trapped electron states with density.

Harrison and Springett[15] recently found a high- to low-mobility transition of electrons in hydrogen gas at 30 and 31.7°K. Their data are shown in Fig. 6. An interesting feature of their data is that the high- and low-mobility branches were observed simultaneously over part of the density range covered. In a

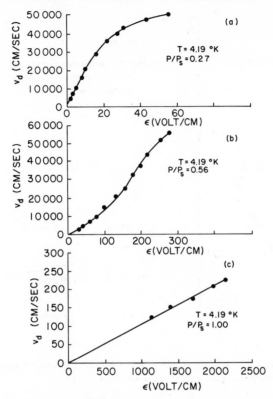

**Fig. 5** Drift velocity of electrons in ⁴He gas versus field strength for the temperatures and pressures indicated (Ref. 10). The pressures are given in units of the saturated vapor pressure $p_s$. Curve $a$ is typical of the low-density region, curve $b$ of the intermediate-density region, and curve $c$ of the high-density region.

similar study, Jahnke and Silver[16] observed a high- to low-mobility transition in helium gas. Their helium data, plotted in Fig. 7, also have a density region in which high- and low-mobility species are simultaneously observed. Harrison and Springett and Jahnke and Silver found that the relative intensities of the currents of the high- and low-mobility branches are density dependent, the current going continuously from a purely high-mobility signal to a purely low-mobility signal over the coexistence density region. An important observation of Jahnke and Silver was that the range of the co-existence density region depended on the drift field and, therefore, on the time the charged species spent traversing the system: the longer the drift time for a given gas density, the greater was the intensity of the low-mobility current relative to that of the high-mobility current. Thus the Jahnke–Silver

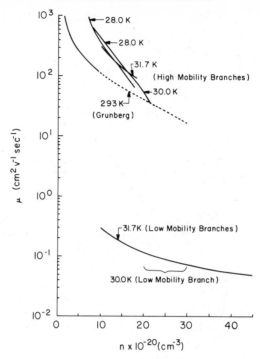

**Fig. 6** Mobility of negative particles in gaseous hydrogen versus number density at fixed temperatures (marked on curves in degrees Kelvin). (Ref. 15).

experiment implies that the two branches are not in equilibrium. The high-mobility species (quasi-free electrons) undergo in the coexistence region a transition to the low-mobility state with a relaxation time comparable to the drift time of the high-mobility species.

Harrison and Springett and Jahnke and Silver concluded that it was consistent with their data to interpret the low-mobility species as a bubble-trapped electron. However, by careful control of the oxygen impurity level Bartels (*Phys. Lett.* **45A**, 491 (1973)) provided rather convincing evidence that the relaxation process observed by these investigators was the formation of oxygen ions by electron attachment to oxygen impurities. Bartels photo-injected a short pulse of electrons (with a 30 nsec ultraviolet light pulse) between a pair of electrodes. If the high-mobility electrons undergo attachment with the attachment frequency $v_a$, then the number $N$ of high-mobility electrons left after a time $t$ is

$$N = N_0 e^{-v_a t}$$

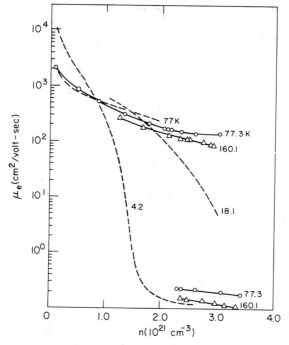

**Fig. 7** Zero-field electron mobilities as a function of number density in dense helium gas (Ref. 16). The dashed lines represent the earlier data of Harrison and Springett for helium [*Phys. Lett.*, **35A**, 73 (1971)]. Note the existence of high- and low-mobility branches in both cases.

where $N_0$ is the number of initially injected electrons. The ratio of the integrated high-mobility and low-mobility current intensities will be

$$\frac{1 - e^{-v_a t_-}}{v_a t_-},$$

where $t_-$ is the drift time of the high mobility species and the currents are integrated over a time long enough to collect all of the initially injected charge.

The values of $v_a$ determined by Bartels in helium at 77.6°K are plotted in Fig. 8. Comparing the dependence of $v_a$ on helium density for helium gas with less than 5 ppm oxygen impurity and with about 50 ppm oxygen impurity, Bartels found that the ratio of $v_a$ for the two impurity levels was about 1/5 and was a constant, independent of the helium density. This result implies an attachment reaction with oxygen. As further evidence that the attachment to oxygen is involved, Bartels pointed out that a peak in $v_a$ occurs

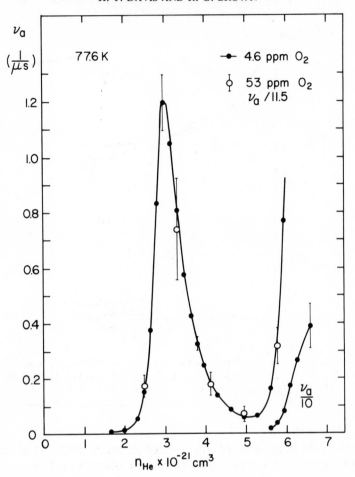

**Fig. 8** The attachment frequency $v_a$ for the formation of $O_2^-$ as a function of the helium density. The oxygen concentration in the helium was 4.6 and 53 ppm. From A. Bartels, *Phys. Lett.* **45A**, 491 (1973).

at a helium density of $3 \times 10^{21}$ atoms/cm$^3$, for which the work function or lowest energy state $V_0$ of a quasi-free electron lies between the energies 0.081 and 0.101 eV of the spin-orbit split vibrational ($v = 4$) doublet state of $O_2^-$. (The value of $V_0$, estimated from the optical potential approximation $V_0 = 2\pi\hbar^2 a n_{He}/m$, with $a$ the electron-helium atom scattering length that is 0.62 Å at low energies, is 0.090 eV at $3 \times 10^{21}$ atoms/cm$^3$.) Thus a peak in $v_a$ occurs when the helium density is adjusted so that the ground state energy,

$V_0$, quasi-free electron coincides with an acceptor state of oxygen. In addition to the peak at a density of $3 \times 10^{21}$ cm$^{-3}$, another peak, corresponding to the $v = 5'$ doublet state of $O_2^-$, is expected to occur at a density of about $7 \times 10^{21}$ cm$^{-3}$. The onset of such a peak is indicated in Fig. 8 by Bartels' data.

Further verification of Bartels' interpretation of the two branch behavior has recently been provided by Jahnke, Silver, and Hernandez (*Phys. Rev.*, submitted for publication). Examining the relative currents of the high- and low-mobility branches in helium gas for wide ranges of temperature and density, these investigators concluded that the assumption that electron-oxygen attachment resonances occur at about $3 \times 10^{21}$ and $7 \times 10^{21}$ cm$^{-3}$ is consistent with the data they took at 52.8, 77.3 and 160°K. At 20.3°K these investigators found a mobility curve much like the low temperature data of Sanders and Levine, Fig. 4, that is, no two-mobility coexistence region was observed. Also, the helium density at which the $v = 4$ attachment peak occurs was found to decrease with temperature, ranging from about $2.8 \times 10^{21}$ cm$^{-3}$ at 52.8°K to about $3.1 \times 10^{21}$ cm$^{-3}$ at 160°K. Jahnke, Silver, and Hernandez interpret the temperature dependence of the density at which $v_a$ peaks, and the disappearance of the two-branch behavior at low temperature, as implying that it is primarily self-trapped electrons (i.e., ones trapped in cavities or bubbles) which undergo attachment to oxygen. With this interpretation, $v_a$ will be proportional to the number of self-trapped electrons as well as to the oxygen concentration. Then, since for a given density the number of self-trapped electrons relative to number of quasi-free ones decreases with decreasing temperature, the disappearance of the two branch region with decreasing temperature may be explained as an increase in $v_a$ due to an increase in the number of self-trapped electrons.

That the presence of oxygen is responsible for the observed coexisting high- and low-mobility species in gaseous helium and hydrogen—Bartels found the $v = 5$ peak in hydrogen at 77.6°K (Thesis, Hamburg, 1973)—appears solidly established at this point. However it seems to the authors that even though Jahnke, Silver, and Hernandez's interpretation is quite suggestive, the role of the self-trapped states in the attachment reaction itself has not been established. Mobility experiments under conditions of extremely low and known oxygen concentrations would be invaluable at this point. In hydrogen, even the existence of cavity or bubble states is questionable in view of the fact that the work function of liquid hydrogen has been observed to be $-1$ eV. The low temperature data ($T \leq 4.2°$K) of Sanders and Levine, Fig. 4, is probably not subject to oxygen impurity complications since the vapor pressure of oxygen at these low temperatures is exceedingly small. Thus, in this case, the only low-mobility states available will be the self-trapped states.

As illustrated by the helium and hydrogen studies mobility and current studies in gases, whose work functions are known functions of temperature and density, may be quite fruitfully exploited to determine electron attachment cross-sections of known ion states and to determine the energy spectrum for unknown ion states.

Like liquid helium, liquid hydrogen and neon are composed of molecules of low polarizability, so that the electron work function could be expected to be positive and the electron-bubble state to be stable. The measured[17,18] electron mobility in liquid hydrogen is $\sim 0.02$ cm$^2$/(V) (sec), a small enough value to indicate a bubble or ion state. However, the work function for liquid hydrogen has been deduced by Raz and Jortner[20] from spectroscopic data to be about $-1$ eV a result implying the instability of the bubble state relative to the quasi-free state. Thus the nature of the localized state in liquid hydrogen seems to differ from that in liquid helium. A great deal more experimental work ought to be done on liquid hydrogen including separate verification of the value of the work function and careful investigation into possible impurity problems in the mobility measurements.

In liquid neon, however, the spectroscopically deduced[20] work function is about 0.5 eV. And the electron mobility behaves in a fashion consistent with the bubble model.[21,22] In Fig. 9a and b are shown mobility data obtained by Loveland, LeComber, and Spear[22] for positive and negative carriers in liquid neon. As was the case in liquid helium, the mobility of the negative species is lower than that of the positive species in liquid neon. As seen in Fig. 9b, the positive species obeys Walden's rule, $\mu_+ \eta \approx$ constant, based on the approximate version of Stokes' law $\mu_+ \approx e/6\pi\eta R_+$, where $R_+$ is an effective radius of the positive ion. However, the product $\mu_- \eta$ for the negative species decreases with increasing temperature (and decreasing density), consistent with the interpretation that $\mu_- \approx e/6\pi\eta R$, where $R$ is the bubble radius which increases with increasing temperature or decreasing density.

The negative carrier mobility data of Bruschi, Mazzi, and Santini[20] was also of the magnitude expected for localized electrons. Their data, however, seem to be almost a factor of 2 lower than those of Loveland, LeComber, and Spear. The reason for the disagreement is not apparent.

The behavior of electrons in liquid argon, krypton, and xenon offers quite a contrast to that in liquid helium. Malkin and Schultz,[23] Williams,[1] and Swan,[24] measuring mobilities at high electric fields (greater than 1 kV/cm) found negative carriers in liquid argon that had high, field-dependent mobilities. Their findings are illustrated in Figs. 10 and 11, taken from Swan's paper.[24] Mobilities computed from the drift velocity data in Fig. 10 are of the order of 25 cm$^2$/(V) (sec), a very high value compared to the value $\sim 2 \times 10^{-3}$ cm$^2$/(V) (sec) observed[1] for positive ions in argon in the field range 24 to 187 kV/cm. Such high mobilities argued strongly for the

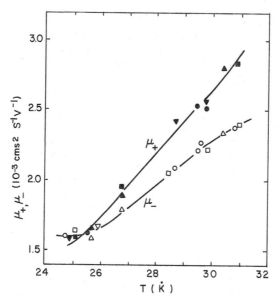

**Fig. 9a** Drift mobilities of positive and negative charge carriers as a function of temperature for different species in liquid neon (Ref. 22).

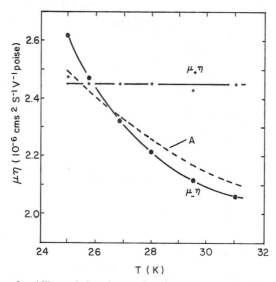

**Fig. 9b** Product of mobility and viscosity as a function of temperature for positive and negative species in liquid neon (Ref. 22). Curve $A$, $\mu_- \eta$ according to formula $\mu_- \eta \approx e/3.2\pi R$, with cavity radius $R$ computed theoretically by T. Miyakawa and D. L. Dexter [*Phys. Rev.*, **184**, 166 (1969)].

TABLE II

Comparison of Observed Electron Mobilities
in Solid and Liquid Inert Gases
(Refs. 22 and 32)

| | $T$ (°K) | Solid, $\mu_S$ [cm²/(V)(sec)] | Liquid, $\mu_L$ [cm²/(V)(sec)] |
|---|---|---|---|
| He | 4.2 | — | 0.02 |
| Ne | 25 | 600 | 0.0016 |
| Ar | 85 | 1000 | 475 |
| Kr | 115 | 3600 | 1800 |
| Xe | 161 | 4000 | 1900 |

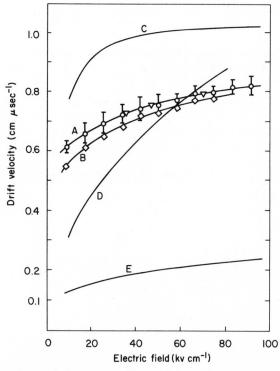

**Fig. 10**  Electron drift velocity in liquid argon. *A*, Commercial-grade argon: ○, $d = 0.1203$ cm; △, $d = 0.172$ cm. *B*, Commercial-grade argon after passage through a cold trap: $d = 0.1203$ cm. *C*, Williams, 1957. *D*, Malkin and Schultz (Ref. 23). *E*, Pure gaseous argon. (Ref. 24).

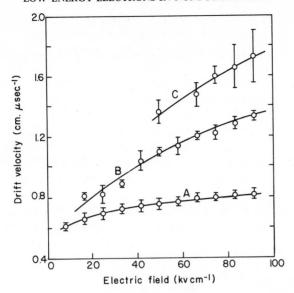

**Fig. 11**  Electron drift velocity in argon–nitrogen mixtures. *A*, Argon; *B*, argon plus 0.07% nitrogen; *C*, argon plus 0.4% nitrogen (Ref. 24).

interpretation that at least at high fields the charge carriers observed were quasi-free electrons. The disagreement among the mobilities measured by the different investigators and the inability of these investigators to obtain the high-mobility negative species at low fields indicated that impurities were involved in the observed data. Swan tested this possibility by purifying liquid argon the best he could and then adding controlled amounts of selected impurities (hydrogen, nitrogen, and oxygen). His results for nitrogen impurities in argon, shown in Fig. 11, show a strong impurity dependence. In fact, the mobility increases with increasing impurity concentration. Inelastic scattering with molecular impurities reduces the average energy and speed of the electrons and can cause an increase in the mobility of quasi-free electrons in two ways. First, the effective electron-atom cross section may increase with increasing electron energy. Swan speculated that this was happening, since a Ramsauer effect is observed for electrons in gaseous argon. Second, since the quasi-free mobility is roughly inversely proportional to the electron speed, lowering the average electron speed increases the mobility. We shall see later that these considerations are qualitatively correct, although the quantitative picture is considerably more complicated and coherent scattering effects are not mentioned in the early discussions of the data. Swan estimated that oxygen impurities of 1 part in $10^8$ would prevent low-field observations of quasi-free electrons because of ion formation.

The first low-field, impurity-independent mobility data collected from liquid argon and krypton were obtained by Schnyders, Rice, and Meyer[25,26] (SRM). In addition to the standard procedure of passing chemically pure argon and krypton over activated charcoal, SRM also purified the liquids by electrolysis for about 2 hours at a voltage of 100 V/cm (using a $^{210}$Po source to produce ions and electrons in the drift space). Below 200 V/cm, SRM observed field-independent high mobilities for the negative species. Some of their low-field data are illustrated in Fig. 12. In the temperature range shown in the figure, the mobility is of the order of 400 cm$^2$/(V) (sec) and decreases with temperature.

The theory of quasi-free electronic conduction of Cohen and Lekner[27,28]— which is discussed in detail in a later section—predicts a low-field mobility of the form

$$\mu = \frac{2}{3n} \left( \frac{2}{\pi m k T} \right)^{1/2} \frac{e}{4\pi a^2 S(0)} \qquad (2.4)$$

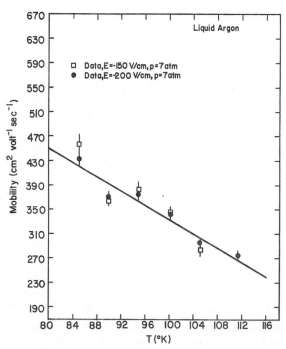

Fig. 12　Electron mobilities in liquid argon as a function of temperature. (Ref. 26).

where $4\pi a^2$ is the thermal energy electron–atom scattering cross section, and $S(0)$ $(= nkT\kappa_T$, $\kappa_T$ = isothermal compressibility) is the low-momentum $(K \rightarrow 0)$ limit of the liquid structure function

$$S(K) = 1 + 4\pi \int [g(r) - 1] \frac{\sin Kr}{Kr} r^2 \, dr \qquad (2.5)$$

where $\hbar K$ is the electron momentum change in an electron–atom collision, and $g(r)$ is the pair correlation function of the fluid. Equation (2.4) is based on the assumption that the electrons behave as plane waves or wave packets coherently scattered [thus the origin of $S(K)$] by the atoms of the fluid. Equation (2.4) is essentially the result given by Bardeen and Shockley[29] for electrons in nonpolar crystals. Also, the basic form of (2.4) was derived and applied to inert gas liquids by SRM[25,26] before Cohen and Lekner's work. The high-field formula of SRM,[26] however, contained the square of a momentum mean free path, whereas the correct form has been shown by Cohen and Lekner's work to be the product of a momentum and an energy mean free path, the latter quantity differing by a structure factor from the former.[30]

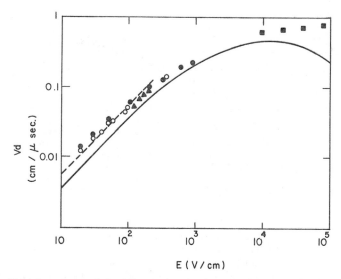

**Fig. 13** Field dependence of the drift velocity of electrons in liquid argon. Squares, Data of D. W. Swan [*Proc. Phys. Soc.*, **83**, 659 (1964)]; triangles, data from Refs. 25 and 26, taken at 85°K; circles, present work, obtained at 85°K with two cells. The line is the single-scatterer theory taken from Ref. 28, and the dashed line is the same corrected for multiple scattering in the Ohmic region. (Ref. 31).

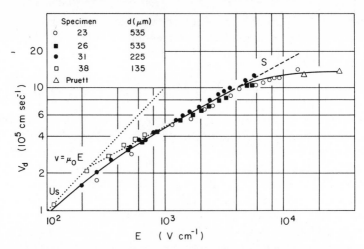

**Fig. 14a** The field dependence of the electron drift velocity in solid argon at 82°K. The dotted lines show the extrapolated linear and $E^{1/2}$ regions intersecting at 1.5 $u_s$. Curve $S$ was calculated from the Shockley theory [W. Shockley, *Bell Syst. Tech. J.*, **30**, 990 (1951)] and fits the results up to $E = 5$ kV cm$^{-1}$. The sound velocity in the solid is denoted by $u_r$. (Ref. 32).

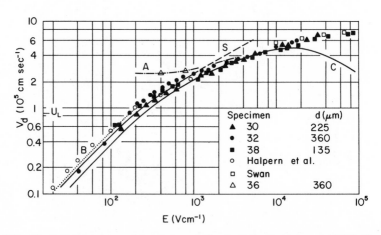

**Fig. 14b** The field dependence of the electron drift velocity in liquid argon at 85°K. Curve $S$, Calculated from Shockley theory; curve $C$, theory of Cohen and Lekner (Refs. 27 and 28); curve $B$ is corrected for multiple scattering and also indicates the linear field dependence; curve $A$, typical results from a specimen with short electron lifetime (200 nsec). The sound velocity in the liquid is denoted by $u_L$. (Ref. 32).

With a scattering length $a \sim 0.78$ Å, (2.4) accounts for the data shown in Fig. 12. The observed scattering length has been correctly predicted by Lekner[28] with a model that takes into account many-body polarization effects on the electron–atom scattering potential.

The Cohen–Lekner theory predicts the onset of field-dependent mobility behavior for fields $E \gtrsim E_c$, where the characteristic fields $E_c$ obey the relation

$$\tfrac{1}{3}(eE_c\Lambda_1)^2 S(0) \approx \frac{2m}{M} (kT)^2 \qquad (2.6)$$

where the momentum mean free path $\Lambda_1$ is $1/n4\pi a^2 S(0)$ to the approximations that $a$ is constant and $S(K) \approx S(0)$. In careful numerical calculations of course, neither of these approximations need be made, although the latter is quite accurate for thermal electrons. For $a \approx 0.78$ Å, $E_c \sim 200$ V/cm. Thus one expects the field dependence of the mobility to commence around 200 V/cm. Halpern et al.[31] demonstrated the correctness of Cohen and Lekner's predictions for electrons in liquid argon at 85°K, as shown in Fig. 13.

In an extensive investigation of electron mobilities in liquid and solid argon, krypton, and xenon near their melting points, Miller, Howe, and Spear[32] demonstrated that the quasi-free electron theory of Cohen and Lekner accounts for the major features of the observed drift velocities in the inert gas solids as well as in the liquids. Their results are summarized in Figs. 14–16 and Table II. If one assumes that the electron–atom scattering

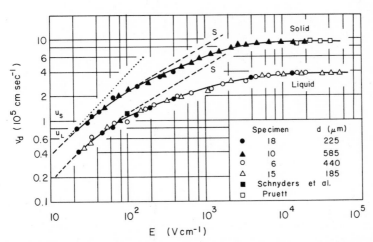

**Fig. 15** The field dependence of the electron drift velocity in solid krypton (113°K) and liquid krypton (117°K) (Ref. 32). Dotted line, Linear field dependence; curves $S$, calculated from Shockley theory.

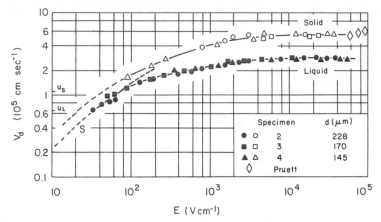

**Fig. 16** The field dependence of the electron drift velocity in solid xenon (157°K) and liquid xenon (163°K) (Ref. 32). Curve $S$, Calculated from Shockley theory.

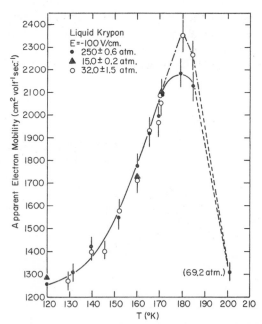

**Fig. 17** Apparent electron mobility in liquid krypton as a function of temperature. All points are measured at a constant electric field strength of $-100$ V/cm. Note that the point at 200.9°K was measured at a pressure of 69.2 atm. This point is displayed to indicate a possible maximum in the temperature dependence of the apparent electron mobility in liquid krypton. This is valid, since the liquid krypton data is insensitive to pressure. (Ref. 26).

350

lengths do not differ significantly in the liquid and solid phases at nearly the same temperatures, then (2.4) predicts that

$$\frac{\mu_s}{\mu_L} \simeq \frac{n_L S_L(0)}{n_s S_s(0)} \tag{2.7}$$

where $\mu_s/\mu_L$ is the ratio of the electron mobilities in the liquid and solid phases, respectively. Experimentally, (2.7) is roughly correct for argon (lhs $\simeq 2.2$, rhs $\simeq 2$). Although data on $S(0)$ are not available for krypton and xenon, one expects from the law of corresponding states that the rhs of (2.7) is the same for argon, krypton, and xenon near the melting point. Thus, since $\mu_s/\mu_L \simeq 2$ for krypton and xenon, it appears that (2.7) is obeyed for all three cases.

For liquid krypton and argon above 114°K, SRM[26,30] found a striking increase in the low-field mobility. The krypton data of SRM are shown in Fig. 17. Some of the results of Jahnke, Meyer, and Rice's extensive study[30] of the temperature and density dependence of the low-field electron mobility in liquid argon are summarized in Figs. 18 and 19. Lekner's[33,34] mean potential electron–atom scattering model explains the behavior shown in Figs. 17 to 19 in terms of a zero in the mean field scattering amplitude. In the gas phase the electron–atom scattering length is negative, because of the contribution of the electron–atom polarization interaction. At low temperatures in the liquid phase, Lekner's model predicts a positive scattering length, the polarization terms largely canceling out because of superposition of the field of neighboring molecules. Thus Lekner reasoned that somewhere between the low-temperature dense liquid state and the higher-temperature dilute gas state the mean field electron–atom scattering length must go through a zero to undergo a sign change. In the region where the mean field scattering length is zero, molecule fluctuations will give a nonzero mean square scattering length and therefore a nonzero cross section, since the cross section varies as the square of the scattering length. Jahnke, Meyer, and Rice have concluded that Lekner's theory accounts quantitatively for the low-field mobilities in liquid argon at densities higher than 1 g/cm$^3$ and qualitatively for some aspects of the low-field mobilities below 1 g/cm$^3$. We discuss Lekner's model in detail in Section VII.B.

The mobility data alone provide conclusive evidence that the quasi-free electron state is more stable than the bubble state in argon, krypton, and xenon. This conclusion is also consistent with a stability criterion developed by Springett, Cohen, and Jortner,[35] which states that the quasi-free state is favored over the bubble state in nonpolar fluids if the work function $V_0$ is negative. The observed work functions for argon, krypton, and xenon, given in Table I, are indeed negative. However, as we shall see when we discuss nonpolar hydrocarbon fluids, the condition that $V_0$ be less than zero is not a

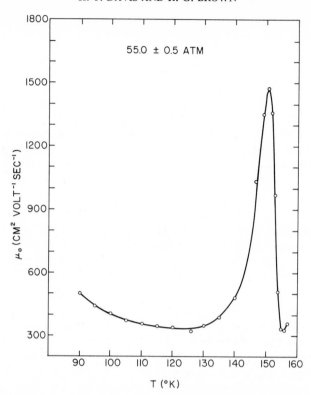

**Fig. 18a** A representative curve of electron zero-field mobility in fluid argon as a function of temperature. Note that the minimum on the low-temperature side of the maximum has a small depth compared to the height of the maximum. Also note the minimum on the high temperature side of the maximum. (Ref. 30).

sufficient condition for the stability of the quasi-free state relative to a localized state in a molecular fluid. In this sense the division of fluids into nonpolar (composed of molecules with little or no permanent dipole moment) and polar (composed of molecules with permanent dipole moment) classes is too crude. The magnitude of the electron–atom polarizability and the details of the local potential fluctuations (associated with intra- and intermolecular coordinates) are probably important in determining the density of states of electrons in molecular fluids.

Chemically, it was expected that unattached electrons could exist in nonpolar hydrocarbons as well as in inert gas liquids. However, owing to the higher melting points of most of the hydrocarbons, impurity attachment prevented observation of nonionic electrons for several years after they were

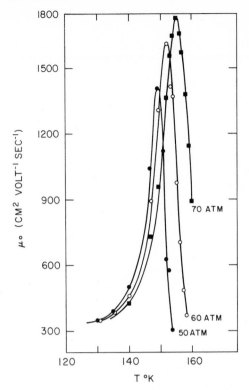

**Fig. 18b**   Fluid argon zero-field mobility maxima as a function of temperature on three isobars. (Ref. 30).

observed in inert gas liquids. The first evidence, to our knowledge, of free-electron motion was provided by an "anomalous" rapidly decaying current "spike" reported by Tewari and Freeman[36] in x-ray pulse studies in liquid neopentane. About a year later, Minday, Schmidt, and Davis[37,38] and Schmidt and Allen[39,40] independently reported time-of-flight measurements of fast negative species in several nonpolar hydrocarbon liquids. The former investigators used a double-shutter apparatus[38] to measure the time of flight of photoinjected electrons across a given drift space subjected to an electric field, whereas the latter investigators used a parallel-plate diode cell to measure the time of flight of electrons created by pulsed x-radiation. A little later, Fuochi and Freeman[41] and Dodelet and Freeman[43] determined electron mobilities in several nonpolar hydrocarbons by x-radiolysis free-ion yield studies, and Beck and Thomas[44] by reaction-rate studies. Many other mobility studies followed these early ones, and in fact are still underway—

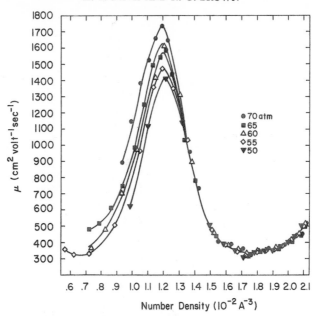

**Fig. 19** Zero-field mobilities as function of number density in fluid argon. This figure represents a summary of zero-field data obtained in Ref. 30.

such is the variety of hydrocarbons. A summary of the mobility data up to the present time is presented in Table I.

In Fig. 20a and b drift velocities versus electric field are displayed for electrons in several hydrocarbons. Also plotted are the mobilities of ions in n-hexane and electrons in argon. In Fig. 21 low-field mobility is plotted versus temperature for several fluids. In Figs. 22 and 23 the composition and temperature dependence of the low-field mobility of electrons in n-hexane–neopentane and methane–ethane mixtures are shown. From Figs. 20 to 23 and the tabulated mobility and work function data of Table I, the following general features emerge.

1. The mobilities are extremely sensitive to the structure of the hydrocarbon molecule; for example, 0.07 cm²/(V) (sec) in n-pentane compared to 70 cm²/(V) (sec) in neopentane, and 0.029 in *trans*-butene compared to 2.2 in *cis*-butene. The trend is that, as the constituent molecule of the fluid becomes more compact geometrically and more symmetric, the mobility increases.

2. In most hydrocarbon liquids studied, the electron mobilities obey an Arrhenius plot with respect to temperature $T$, that is,

$$\mu = \mu_0 e^{-E_A/RT} \tag{2.8}$$

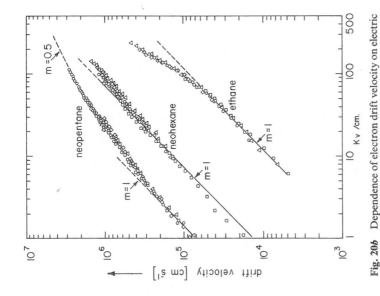

**Fig. 20b** Dependence of electron drift velocity on electric field strength in liquid neopentane ($T = 295°K$), neohexane ($T = 295°K$), and ethane ($T = 200°K$). Different symbols represent different experiments and cells; each point is the average of 5 to 20 determinations. $m$, Slope of the dashed line. (Ref. 49).

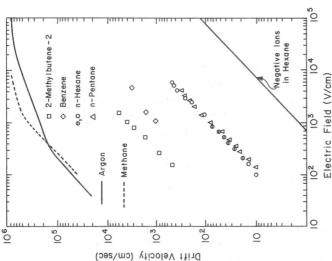

**Fig. 20a** Electron drift velocity versus field in several hydrocarbon liquids at saturation pressure at 300°K. Also shown for comparison are curves for electron in liquid argon (87°K) and ions in $n$-hexane (300°K). Drift velocity in hydrocarbons was measured in single-shutter (open data points) and double-shutter (closed data points) apparatuses. (Refs. 47 and 45).

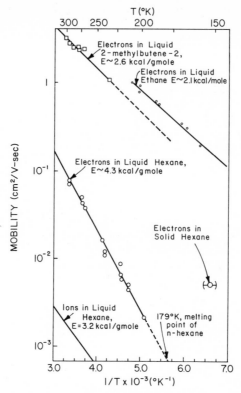

**Fig. 21**  Plots of the logarithm of mobility versus $1/T$ for several systems. (Ref. 38).

where $\mu_0$ is of the order of 100 cm²/(V) (sec) and the activation energy $E_A$ ranges from essentially zero [for methane,[45] neopentane,[46,47] and probably tetramethylsilane (TMS)] to about 0.2 eV. Such temperature dependence implies some sort of localization of the excess electrons. The magnitude of $\mu_0$ suggests that quasi-free electronic states also contribute to excess electron transport.

3. The electron mobilities in binary hydrocarbon mixtures obey an Arrhenius plot of the form

$$\mu_{\text{mix}} = \mu_0^{\text{mix}} e^{-(x_1 E_{A1} + x_2 E_{A2})/RT} \tag{2.9}$$

where $\mu_0^{\text{mix}}$ is a weak function of composition and of the order of 100 cm²/(V) (sec), $x_i$ is the mole fraction of component $i$, and $E_{Ai}$ is the activation energy of the electron mobility in pure liquid $i$. This result, obtained for $n$-hexane–neopentane,[46] $n$-hexane–isooctane,[44] and methane–ethane mixtures,[48] respectively, points toward two important conclusions. One is that the trap

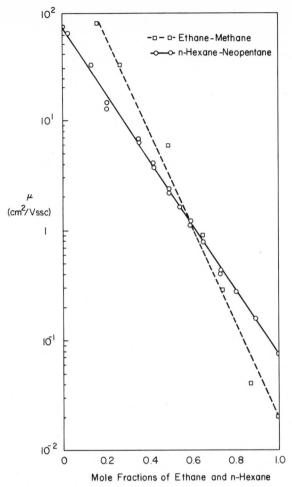

**Fig. 22** Plot of the logarithm of electron mobility versus: mole fraction of $n$-hexane in neo-pentane on hexane mixture at 300°K (Ref. 46); mole fraction of ethane on methane–ethane mixture at 111°K (Ref. 48).

providing the mobility activation energy is a collective trap (if a single-ion state were involved, the activation energy in the $n$-hexane–neopentane mixture, for example, would be roughly equal to $E_{A(\text{hex})}$, since $E_{A(\text{neop})} \approx 0$, and not $x_{\text{hex}} E_{A(\text{hex})}$ as observed). The other conclusion is that the collective trap behaves as an ideal solution with respect to the activation energy. The hydrocarbon mixtures discussed here are known to form ideal solutions, so that the fact that $E_{A(\text{mix})}$ also behaves ideally implies that the trapped electron does not substantially change the structure of the medium around it.

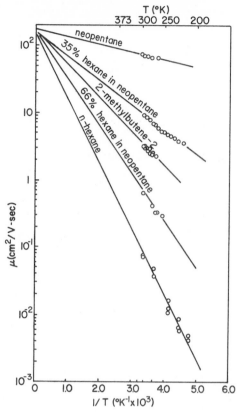

**Fig. 23** Logarithm of electron mobility versus $1/T$ for several compositions of neopentane and $n$-hexane mixture. (Ref. 46).

4. The field dependence of electron mobility in methane[45,48] and neopentane[49] is very similar to that in inert gas liquids, except that the onset of field-dependent mobilities occurs at fields high enough to indicate important inelastic energy losses in the hydrocarbon liquids.

5. The electric fields at which the mobility becomes field-dependent is extremely high for those fluids in which the mobility definitely obeys an Arrhenius expression. For example, the onset of field dependence occurs at 80 kV/cm in ethane,[49] 140 kV/cm in neohexane,[49] and above 100 kV/cm in $n$-pentane.[40]

6. A small permanent dipole moment ($\sim 0.3$ D) does not seem to affect the mobility, judging from the comparison of benzene and toluene results.[46]

7. The mobilities and work functions are correlated, increasing $\mu$ being accompanied by decreasing $V_0$,[50–53] methane being a notable exception.

The mechanism proposed by Minday, Schmidt, and Davis[38] to explain the Arrhenius temperature dependence of the electron mobilities in many hydrocarbons is one in which the electron is periodically localized for a short time in a shallow trap and moves (by thermal promotion) from trap to trap in the quasi-free state. With this two-state model they suggested that the form of the electron mobility is

$$\mu = \mu_f P + \mu_l(1 - P) \qquad (2.10)$$

where $P$ is the probability that the electron is in the quasi-free state, and $\mu_f$ and $\mu_l$ are the mobilities characteristic of quasi-free and localized motions, respectively. Assuming the electrons to be distributed between the high- and low-mobility states according to an equilibrium distribution, one may write

$$P = \frac{N_f}{N_f + N_l} = \frac{1}{1 + N_l/N_f} = \left(1 + \exp\frac{W_l - V_0}{kT}\right)^{-1} \qquad (2.11)$$

where $N_l$ and $N_f$ are the ensemble average numbers of electrons in the localized and free states, respectively (for a system with $N = N_l + N_f$ electrons), $W_l$ is the reversible work required to put a localized electron (or a mole of electrons if a mole base is used) into the fluid, and $V_0$ (the work function) is the reversible work required to put a quasi-free electron into the fluid.

In fluids in which appreciable localization occurs, $P \ll 1$ [see (2.11)] may be written in the form

$$P \approx e^{-E_A/RT} \qquad (2.12)$$

so that

$$\mu \simeq (\mu_f - \mu_l)e^{-E_A/RT} - \mu_l$$
$$\simeq \mu_f e^{-E_A/RT} \qquad (2.13)$$

where

$$E_A \equiv V_0 - W_l \qquad (2.14)$$

and where the second equality of (2.13) is obtained by assuming that $\mu_f$ and $\mu_f e^{-E_A/RT}$ are large compared to $\mu_l$. If $\mu_l \sim$ ionic mobilities, then the observed $\mu$ is large compared to $\mu_l$ and the final form of (2.13) follows. Thus, assuming that the localized electron mobility is of the order of ionic mobilities, Minday, Schmidt, and Davis' (MSD) model yields a form in agreement with experiment if $\mu_f$ is estimated from the quasi-free theoretical formula (2.4). And within the context of their model the activation energy is just the difference between the chemical potentials $W_l$ and $V_0$. A two-state model similar to that of MSD was proposed by Schiller, Vass, and Mandics[53] (SVM), although they proposed a different method of computing $P$. In still another two-state model approach, Kestner and Jortner[54] (KJ) pointed out that when two conductance channels (one of high mobility and one of low mobility) exist, percolation effects must be included in estimating the relative roles of the high- and low-mobility channels. Their model is similar to that proposed by Eggarter and Cohen[12,13] for electrons in helium gas.

In Section IX, we shall discuss the models of MSD, SVM, and KJ as well as the more general approach which assumes a continuum of localized and quasi-free states rather than the two-state approximation. Unfortunately, at this time the details of the localized states are not known, and all the models proposed involve some parametric adjustment.

From the order of magnitude and the qualitative features of the field dependence of the electron mobilities in methane and neopentane, it appears that the electron motion is predominantly quasi-free in these fluids. Davis, Schmidt, and Minday[55] derived the polyatomic version of Cohen and Lekner's Boltzmann equation, and Davis, Schmidt, and Brown[47] derived the approximate formula

$$\mu_f = \tfrac{2}{3}e\left(\frac{2}{\pi mkT}\right)^{1/2}\Lambda_1 \tag{2.15}$$

for the low-field mobility and the expression

$$\tfrac{1}{3}(eE_c\Lambda_1)^2 S(0) \simeq \lambda(kT)^2 \tag{2.16}$$

to estimate the characteristic field $E_c$ above which field-dependent mobilities are observed. The quantity $\lambda$ represents the average fractional energy loss per electron–molecule collision. For monatomic particles,

$$\lambda \simeq \frac{2m}{M} \tag{2.17}$$

where $M$ is the mass of the atom, in which case (2.16) reduces to (2.6). Combining (2.15) and (2.16), one may write the equation for $E_c$ in the form

$$E_c{}^2 \approx \frac{8\lambda}{3\pi mn\kappa_T\mu_f{}^2} \tag{2.18}$$

Estimating from experimental data (Fig. 20) that $E_c \approx 1500$ and $7000$ V/cm and $\mu_f \simeq \mu = 70$ and $400$ cm$^2$/(V) (sec) for methane and neopentane, respectively, one can obtain from (2.18) the estimates $\lambda = 8.4 \times 10^{-4}$ and $1.68 \times 10^{-4}$ for methane and neopentane, respectively. The elastic values of $\lambda$ ($\equiv 2m/M$) in these two cases are $6.82 \times 10^{-5}$ and $1.52 \times 10^{-5}$. Thus the contribution of inelastic energy losses to $\lambda$ is about 90% of the total contribution, if the assumption that electrons are quasi-free in methane and neo-pentane is correct.

Most of the work function data reported in Table I for nonpolar liquids were determined by the photoinjection technique used by Woolf and Rayfield[56] for studying liquid helium. By this technique one measures the current $i$ produced in a phototube as a function of the frequency $v$ of the radiation striking the electron-emitting electrode. The current $i$ obeys the equation

$$i = \alpha A T^2 F\left(\frac{hv - \phi}{kT}\right) \qquad (2.19)$$

where $\alpha$ is the fraction of electrons that absorb a photon, $A$ is a universal constant, $h$ is Planck's constant, $\phi$ is the reversible work of transferring an electron from within the emitting electrode into the phototube, and $F$ is an exponential function known as Fowler's function. The work function $V_0$ of a liquid is defined as

$$V_0 = \phi_{\text{liq}} - \phi_{\text{vac}} \qquad (2.20)$$

where $\phi_{\text{liq}}$ and $\phi_{\text{vac}}$ are the values of $\phi$ determined with a liquid and a vacuum, respectively, in the phototube. Typical data are shown in Fig. 24. The quantity $\ln(i/T^2) - \ln \alpha A^2 = \ln F[(hv - \phi)kT]$ is a universal function

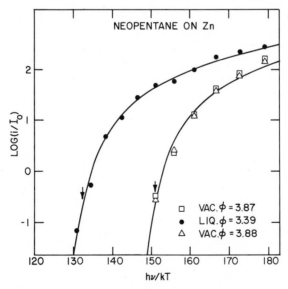

**Fig. 24** Response of neopentane-filled cell, zinc electrode, $\phi = 3.39$, $E = 10^4$; vacuum response: before $\phi = 3.87$, after $\phi = 3.88$. (Ref. 50).

of $(hv - \phi)kT$. Thus, if at a given temperature the vacuum curve for $\ln i$ versus $hv/kT$ is translated along the $\ln i$ axis and the $hv/kT$ axis so as to be superimposed on the liquid curve for $\ln i$, then the value of $\Delta(hv)$ corresponding to the translation along the $hv/kT$ axis is equal to $V_0$.

Sommer[8] deduced the value of $V_0$ for liquid helium from a measurement of the fraction of electrons transmitted from the vapor to the liquid phase as a function of electric field. Raz and Jortner[20] deduced values of $V_0$ for hydrogen, argon, krypton, and xenon from spectroscopic data. Their method is discussed in Section IV.A.

## III. ELECTRICAL PROPERTIES OF IRRADIATED NONPOLAR LIQUIDS

The role played by electrons in electrical phenomena of irradiated liquids has received much attention over the last few years. One of the quantities of interest is $G_{fi}^0$, the zero (applied) field free-ion yield of an irradiated liquid. Most of the ions formed in the irradiation of a liquid never become free, but disappear by a rapid "initial" recombination.[57,58] $G_{fi}^0$ is the number (per 100 eV of radiation energy absorbed) of ions escaping initial recombination with no applied field. The quantity $G_{fi}^0$ can be determined by extrapolation to zero field[50] of the measured high-field values of the field-dependent free-ion yield $G_{fi}^E$, although care must be exercised in such extrapolations.

A method used quite widely[60-64,36] in the early 1960s was to determine $G_{fi}^0$ from measurements of ion mobilities and of the steady-state conductivity under irradiation at essentially zero applied field. Typical of this approach is the procedure used by Allen and Hummel.[60-62] With an applied voltage $V$ between a pair of electrodes of separation $L$, each of area A, one observes under steady-state irradiation a current $i_c$. If one assumes the presence of only two different charged species, the conductivity $\kappa$ of the cell obeys the equation

$$\kappa = \frac{i_c L}{VA} = \mu_{\pm} c \tag{3.1}$$

where $\mu_{\pm} = \mu_+ + \mu_-$ is the sum of the mobilities of the positive and negative species, and $c$ is the concentration of either species. At steady state, with a voltage so small that the number of ions drawn to the electrode is small compared to the number disappearing by recombination, the following balance can be made:

$$\frac{IG_{fi}^0}{100 \text{ eV}} = \imath c^2 \tag{3.2}$$

where $I$ is the rate of absorption of energy per unit volume from the radiation beam, $G_{fi}^0$ is the number of ion pairs produced per 100 eV of radiation absorbed at the field $E = V/L \simeq 0$, and $\imath$ is the ion recombination rate coefficient. By combining (3.1) and (3.2), the free-ion yield may be expressed as

$$\frac{G_{fi}^0}{100\text{ eV}} = \frac{\imath\kappa^2}{I\mu_\pm{}^2} \tag{3.3}$$

Thus the free-ion yield may be determined from measured values of $I$, $\imath$, $\kappa$ and $\mu_\pm$.

Allen and Hummel[60–62] determined $\imath/\mu_\pm$ in a transient-current experiment, in which the radiation beam is suddenly turned off. Then the ions disappear according to the equation

$$-\frac{dc}{dt} = \imath c^2 \tag{3.4}$$

Solving this expression for $c$, one finds from (3.1) that the transient current observed after the irradiation is turned off obeys the relation

$$\frac{1}{i_c} = \frac{1}{i_c(t=0)} + \frac{\imath L}{\mu_\pm AV}t \tag{3.5}$$

which yields the quantity $\imath/\mu_\pm$.

To determine the ion mobilities, Allen and Hummel irradiated for a short time a region near one electrode (all the ions formed were within about 2 mm of the electrode). Under a given polarity across a pair of electrodes, ions of one sign are quickly drawn to the irradiated electrode. Ions of the opposite sign then provide a constant current until they drift to the other side and are collected. Thus the drift time $t_d$ and, consequently, the mobility $\mu = v_d/(V/L) = L^2/t_d V$ can be determined for each species. In the early work,[60–64] only ionic mobilities were observed. This was presumably because the levels of electron-scavenging impurities were so high in the liquids studied that all electrons were converted to ions before crossing the electrode drift space. In 1968, however, Tewari and Freeman[36] observed in neopentane a rapidly decaying current spike later understood to be the part of the transient current due to quasi-free electrons instead of ions. Tewari and Freeman showed that the addition of trace amounts of electron-scavenging impurities eliminate a rapidly varying current spike, thus demonstrating that the low-mobility negative species observed up to that time in neopentane were impurity ions. The later electron mobility studies of Minday, Schmidt, and Davis[37,38] and Schmidt and Allen[39,40] demonstrated that the low-mobility negative species observed in all nonpolar hydrocarbons were impurity ions.

The coexistence of electrons and negative impurity ions causes a difficulty in the above-described method of determining $G_{fi}^0$. It was assumed that only one negative species exists in (3.1) and (3.4). If the electrons are either scavenged by impurities at a rate large compared with the electron–ion recombination rate or are not scavenged at all, then $G_{fi}^0$ may be determined as described above. In fact, in these two extreme cases a simple theory[65] predicts that

$$\frac{\imath}{\mu_\pm} = \frac{4\pi e}{D} \tag{3.6}$$

where $e$ is the electronic charge and $D$ is the dielectric constant. To obtain (3.6), one assumes that recombination occurs rapidly whenever the negative and positive species are within the distance $r_c$ at which the Coulomb energy of attraction equals the thermal agitation energy $kT$, that is, within

$$r_c = \frac{e^2}{DkT} \tag{3.7}$$

Then, assuming a diffusion rate–limited recombination reaction $\imath = 4\pi r_c$ $(D_+ + D_-)$ and combining this with Einstein's relation $D_i/\mu_i = kT/e$, $i = +, -$, between the diffusion coefficient of the $i$th species and its mobility, one obtains (3.6). The agreement between experiment and (3.6) is an indication that either no electrons are scavenged or all are scavenged rapidly compared to electron–ion recombination rates. For nonpolar hydrocarbons $D \simeq 2$ and $4\pi e/D \approx 0.96 \times 10^{-6}$ V cm a value close to the observed values of $\imath/\mu_\pm$ in many of the early experiments in which the electrons were effectively scavenged by impurities (e.g., $(\imath/\mu_\pm)_{exp} = 1.4 \times 10^{-6}$ V cm in $n$-hexane in Ref. 60 and $=0.96 - 1.04 \times 10^{-6}$ in $n$-hexane in Ref. 36).

The clearing-field method developed by Schmidt and Allen[59,66] provides a direct measurement of $G_{fi}^0$ without the uncertainties described above. As stated in Ref. 59, "The method consists of ionizing the liquid by a short pulse of radiation, and then immediately applying a clearing field of such magnitude that all free ions are drawn to the electrodes before any appreciable amount of homogeneous recombination has had time to occur." During the pulse of radiation, the buildup of ions obeys the equation

$$\frac{dc}{dt} = \frac{IG_{fi}^0}{100 \text{ eV}} - \imath c^2 \tag{3.8}$$

If the total number of ions produced is kept small enough so that $IG_{fi}^0/100 \text{ eV} \gg \imath c^2$, then at the termination time of a pulse of duration $\delta t$, the concentration of charge will be

$$c = \frac{G_{fi}^0}{100 \text{ eV}} \int_0^{\delta t} I \, dt$$

If all the charge produced is collected by turning on the clearing field, then a plot of collected charge versus absorbed radiation dose $v \int_0^{\delta_t} I \, dt$ yields a straight line whose slope yields $G_{fi}^0$ from the expression[59]

$$G_{fi}^0 = \frac{10^{10}}{\rho v} \times \text{slope} \qquad \text{in C/mrad} \qquad (3.9)$$

where $\rho$ is the mass density of the fluid, and $v$ is the irradiated volume of the cell.

Table III contains free-ion yield data obtained for many nonpolar hydrocarbon fluids by the clearing-field method.[59,66–70] Values of $G_{fi}^0$ for the hydrocarbons lie roughly between 0.1 and 1.0. The dilute gas–phase value of $G_{fi}^0$ is about 4 for many of the alkanes[70,71] in Table III. Thus the liquid surrounding an ionized pair greatly aids the rapid "initial" recombination of ion pairs formed in the irradiation process.

The free-ion yields correlate with the electron mobilities (low-field mobilities), higher yields occurring in fluids having higher electron mobilities and in a given fluid increasing with temperature. In Table IV ion yields and mobilities are compared for several compounds. For the higher-mobility compounds neopentane and 2,2,3,3-tetramethylpentane, $G_{fi}^0$ is almost a linear function of $\mu$, whereas for lower-mobility compounds the relationship is nonlinear (the nonlinearity increasing with decreasing mobility). The correlation between $\mu$ and $G_{fi}^0$ suggests that electron–medium interactions and scattering parameters that determine the transport of thermalized electrons are similar to those that determine the energy losses and escape probability of the epithermal electrons of the free-ion pairs. The mobility correlation with free-ion yield does not carry over from hydrocarbons to inert gas fluids as witnessed by the fact that the free-ion yield in argon is over twice that in methane, whereas the electron mobilities are about the same for these liquids. Undoubtedly, epithermal electrons lose energy more efficiently in hydrocarbons than in argon, because of inelastic losses to the internal molecular vibrational and rotational modes present in the former and absent in the latter. Thus the probability that a hot electron will escape its parent ion is expected to be smaller in hydrocarbons than in argon, in agreement with the fact that $G_{fi}^0$ is smaller for hydrocarbons than for argon.

The free-ion yield data have usually been translated into a range parameter $b$ through an escape model proposed by Schmidt and Allen.[59] They assumed that the probability that an ionized electron will have been thermalized when it reaches a distance between $r$ and $r + dr$ from the parent ion is given by the Gaussian distribution

$$p(r) = \frac{4\pi r^2}{(\pi b^2)^{3/2}} e^{-r^2/b^2} \qquad (3.10)$$

TABLE III

Free-Ion Yield Data and Range Parameters $b$ [Determined from Eq. (3.12)] for Several Nonpolar Liquids[a]

| Liquid | $T$ (°K) | Free-ion yield, $G_{fi}$ | Dielectric constant, $D$ | Critical radius from Eq. (3.7), $r_c$ (Å) | Range parameter, $b$ (Å) | Arrhenius fit of temperature dependence of free-ion yield, $G_{fi} = G_{fi,0} \exp(-E_A/RT)$ | | |
|---|---|---|---|---|---|---|---|---|
| | | | | | | $G_{fi,0}$ | $E_A$ (kcal/mole) | $F$-level |
| Ar | 87 | 2 | 1.52 | 1267 | 1300[41] | | | |
| Ar + 1.5 mole % $O_2$ | 87 | 0.7 | 1.52 | 1267 | 570[41] | | | |
| Pure $O_2$ | 87 | 0.013 | 1.49 | 1292 | 158[41] | | | |
| $CH_4$ | 120 | 0.8 | 1.67 | 836 | 500[70] | | | |
| $C_2H_6$ | 120 | 0.06 | | | 129[70] | 0.53 | 0.5293 | 23.4 |
| | 148 | 0.08 | | | 118[70] | | | |
| | 183 | 0.13 | 1.80 | 508 | 118[70] | | | |
| $C_3H_8$ | 120 | 0.026 | 2.05 | 680 | 97[41,70] | 0.63 | 0.7698 | 388.4 |
| $C_4H_{10}$ | 296 | 0.193 | 1.76 | 319 | 83.9[66] | | | |
| $n$-Pentane | 296 | 0.145 | 1.842 | 306 | 71.5[66] | | | |
| $n$-Hexane | 296 | 0.131 | 1.885 | 299 | 67.4[66] | | | |
| $n$-Heptane | 296 | 0.131 | 1.926 | 293 | 66.0[66] | | | |
| $n$-Octane | 296 | 0.124 | 1.944 | 290 | 64.2[66] | | | |
| Butene-1 | 293 | 0.093 | 1.88 | 304 | 54[69] | 0.4774 | 0.9623 | 259.3 |
| $trans$-Butene-2 | 293 | 0.08 | 1.84 | 311 | 53[69] | 0.9211 | 1.445 | 755.2 |
| $cis$-Butene-2 | 293 | 0.23 | 2.00 | 286 | 74[69] | 1.886 | 1.261 | 142.3 |
| Isobutene | 293 | 0.25 | 2.08 | 275 | 74[69] | 1.417 | 1.087 | 77.3 |
| Butadiene-1,3 | 269 | 0.038 | 2.05 | 304 | 39[69] | 0.1786 | 0.8437 | 170.7 |
| Butyne-2 | 293 | 0.32 | 1.94 | 295 | 97[69] | | | |
| TMS | 296 | 0.74 | 1.84 | 313 | 158.9[66] | | | |
| Neopentane | 296 | 0.857 | 1.777 | 318 | 178.4[66] | | | |
| | 273 | 1.02 | 1.84 | 333 | 214[43] | | | |
| | 294 | 1.09 | 1.80 | 316 | 215[43] | | | |
| | 302 | 1.13 | 1.78 | 312 | 219[43] | | | |
| | 312 | 1.18 | 1.76 | 305 | 222[43] | | | |
| | 322 | 1.27 | 1.74 | 299 | 229[43] | | | |
| | 333 | 1.35 | 1.71 | 294 | 239[43] | | | |
| | 294 | 1.0 | 1.82 | 313 | 198[41] | | | |
| Isopentane | 296 | 0.17 | 1.838 | 307 | 76[66] | | | |
| Neohexane | 293 | 0.40 | 1.87[36] | | | | | |
| 2,2-Dimethylbutane | 296 | 0.304 | 1.926 | 293 | 92[66] | | | |
| 3-Methylpentane | 296 | 0.146 | 1.901 | 297 | 69.6[66] | | | |
| 2,2,3,3-Tetramethylbutane | 379 | 0.8 | 1.84 | 240 | 130[43] | | | |
| 2,2,4,4-Tetramethylpentane | 295 | 0.83 | 1.98 | 287 | 158[43] | 2.810 | 0.6973 | 68.9 |
| | 385 | 1.19 | 1.80 | 242 | 174[43] | | | |
| 2,2,5,5-Tetramethylhexane | 293 | 0.67 | 1.97 | 290 | 138[43] | 5.232 | 1.196 | 1859 |
| Cyclopentane | 296 | 0.155 | 1.96 | 288 | 68.9[66] | | | |
| Cyclohexane | 296 | 0.148 | 2.022 | 279 | 66.1[66] | 5.776 | 2.122 | 67.2 |
| Cyclohexene | 293 | 0.20 | 2.23 | 256 | 62[69] | 1.435 | 1.157 | 70 |
| Benzene | 296 | 0.053 | 2.278 | 248 | 42.1[66] | | | |

[a] The temperature dependence of the free-ion yield is summarized, where sufficient data exist, by an Arrhenius fit whose statistical $F$ level is also included in the table.

TABLE IV

Comparison of Observed Temperature Trends of Electron Mobilities,
Free-Ion Yields, and Ranges $b$

| | $T$ (°K) | Electron mobility, $\mu$ [cm$^2$/(V)(sec)] | Free-ion yield, $G_{fi}$ | Most probable electron separation distance, $b$ (Å) |
|---|---|---|---|---|
| C$_2$H$_6$ | 120 | 0.0258[48] | 0.06[70] | 129[70] |
| | 148 | 0.131 | 0.08 | 118 |
| | 183 | 0.53 | 0.13 | 118 |
| C$_3$H$_8$ | 120 | 0.00395[47] | 0.026[70] | 97[70] |
| | 148 | 0.0304 | 0.043 | 94 |
| | 183 | 0.113 | 0.076 | 93 |
| | 230 | 0.724 | 0.12 | 90 |
| Neopentane | 273 | 63[37] | 1.03[70] | 214[70] |
| | 294 | 67 | 1.10 | 215 |
| | 302 | 69 | 1.14 | 219 |
| | 312 | 71 | 1.19 | 222 |
| | 322 | 74 | 1.28 | 229 |
| 2,2,3,3-Tetramethylpentane | 295 | 5.2[43] | 0.44[43] | 102[43] |
| | 318 | 5.9 | 0.48 | 101 |
| | 346 | 7.5 | 0.54 | 100 |
| | 375 | 9 | 0.62 | 103 |
| | 431 | 11 | 0.83 | 114 |
| Cyclohexane | 294 | 0.45[43] | 0.25[43] | 67[43] |
| | 315 | 0.62 | 0.29 | 71 |
| | 340 | 0.89 | 0.35 | 74 |
| | 380 | 1.4 | 0.44 | 78 |
| | 395 | 1.7 | 0.48 | 79 |

where $3b^2/2$ is the characteristic mean square displacement of the escaping electron undergoing thermalization. Once thermalized, the probability that the electron will escape initial recombination with the parent ion[73] is $e^{-r_c/r}$, where the critical escape distance parameter $r_c$ is defined by (3.7). Thus in terms of (3.10) the probability $P$ of an ion pair becoming a free-ion pair is

$$P = \int_0^\infty e^{-r_c/r} p(r)\, dr = \frac{4}{\pi^{1/2}} \int_0^\infty e^{-(x^2 + r_c/bx)} x^2\, dx \qquad (3.11)$$

This probability is obtained experimentally from the ratio

$$P = \frac{G_{fi}^0}{G_0} \qquad (3.12)$$

where $G_0$ is the total number of ion pairs formed per 100 eV energy input. It is generally assumed[66] that $G_0$ may be equated to the number of ion pairs determined in vapor-phase experiments.

In attempting to justify the form of (3.10) (i.e., to justify the assumption that a single dispersion parameter independent of the initial electron energy characterizes the distribution of thermalized electrons), Schmidt and Allen[59] appealed to Mozumder and Magee's theoretical computations[73] in which it had been concluded that most of the initial energy of an electron is lost in a distance short compared to $r_c$, so that escaping electrons have little memory of their initial energy. At this point in time, however, the Gaussian model must be considered an ad hoc ansatz, rigorous justification of the model being unavailable.

Values of $b$ necessary to fit (3.11) to the experimental ratio $G_{fi}^0/G_0$ are given in Table III. The $b$ values tend to be 20 to 30% of $r_c$ and correlate roughly with electron mobilities—$b$ being larger in fluids with larger electron mobilities. However, as illustrated in Table IV, although the electron mobility in a given hydrocarbon usually increases with increasing temperature, rapidly for low mobility fluids are less rapidly the higher the mobility, the range $b$ is relatively insensitive to the temperature. The small changes in $b$ that are observed increase with $T$ in some cases and decrease in other cases. Schmidt and Allen[66] noted that the product of the range and the mass density is about the same for normal alkanes from $C_4$ to $C_{14}$. However, the trend is broken by propane and does not extend to branch-chain molecules.

The problem of giving meaning to $b$, or the more basic problem or relating $G_{fi}^0$ to the basic electron relaxation processes in irradiated liquids, has not been solved completely at this time, although the work of Mozumber and Magee,[73] to be discussed later, is a promising start. A related and similarly unsolved problem is that of describing the high-field behavior of the drift velocities of excess electrons in hydrocarbon liquids. Understanding the nature of inelastic energy losses is the key to solving both these problems.

In closing this section, it is perhaps worthwhile to mention two successful techniques for measuring electron mobilities in irradiated fluids. One technique is that used by Schmidt and Allen.[40,45] The drift space between a pair of diodes is uniformly irradiated, and a current $i(0)$ is drawn. At time $t = 0$ the radiation is turned off and the current decays as a superposition of the components

$$i_e = i_e(0)(1 - t/t_{d,e}), \quad i_n = i_n(0)(1 - t/t_{d,n}), \text{ and } i_p = i_p(0)(1 - t/t_{d,p}),$$

where $i_e$, $i_n$, and $i_p$ are the currents due to electrons, negative ions, and positive ions, respectively. The negative ions arise from impurity scavenging. Since the drift time $t_{d,e}$ of electrons is so much smaller than the ionic drift times $t_{d,n}$ and $t_{d,p}$, $t_{d,e}$ can be determined from a current-versus-time trace. Another technique has been used extensively by Freeman and co-workers. A liquid between electrodes subject to the field $E$ is given a pulse ($\sim 10^{-6}$ sec long) of x rays. The initial peak current is due essentially to electrons, so that the electron mobility can be determined from the expression[74,67,68]

$$\frac{6.24 \times 10^{20} i_0}{\text{Dose } AE} = G_{fi}^E \mu \qquad (3.13)$$

where $i_0$ is the peak observed current, the pulse dose is expressed in electron-volts per cubic centimeter, $A$ is the area of the collecting electron, and $G_{fi}^E$ is the free-ion yield at the field $E$.

## IV. SOME SPECTROSCOPIC STUDIES

### A. Optical Absorption

Optical absorption studies have been carried out on a few irradiated hydrocarbon liquids.[75-77] A broad absorption band has been observed in the wavelength range from about 600 to 2200 nm. Gillis et al.[75] examined the optical absorption spectrum of pulse-irradiated liquid propane at $-185$ and $-165°C$. Their data, illustrated in Fig. 25, indicate an absorption peak

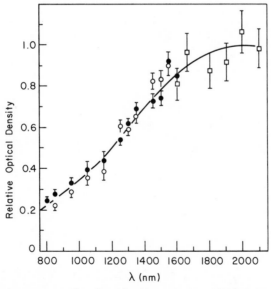

**Fig. 25** Absorption spectrum of irradiated liquid propane at $-185°C$. Points ○ were obtained with the germanium photodiode 160 nsec after the start of a 45-nsec pulse, optical path was 2 cm; points ● were obtained with the InAs detector 2 μsec after the start of a 1-μsec pulse, optical path was 2 cm; points □ were obtained with the InAs detector 2μsec after the start of a 1-μsec pulse, optical path was 5 mm. Each point is an average of usually three measurements, and the error bars represent the uncertainty in the average. (Ref. 75).

at wavelengths $\gtrsim 2000$ nm. The shape of the absorption band was found to be the same for a sample observed 160 nsec after a 45-nsec pulse, and one observed 2 $\mu$sec after a 1-$\mu$sec pulse. Since the electron concentrations should be quite different for the samples studied, their results indicate that the spectrum does not change with decay.

Gillis et al.[75] also found that the time rate of decay of the band is independent of the initial electron concentration. The results of their studies of the rate of decay of the absorption intensity at 1525 nm are shown in Fig. 26. The decay curves for three substantially different doses are the same. This suggests that most electrons are predestined to combine with particular geminate ions.

The band half-life $t_{1/2}$ estimated from the decay curves of Gillis et al. is $\lesssim 95$ nsec for propane at $-185°$C and $\lesssim 35$ nsec at $-165°$C. The viscosity $\eta$ of propane[78] decreases from 8.8 cp at $-185°$C to 2.5 at $-165°$C,[78], the similarity of the ratios $t_{1/2}(-185°C)/t_{1/2}(-165°C)$ and $\eta(-185°C)/\eta(-165°C)$ suggesting that perhaps viscosity-limited processes (such as diffusion of cations) may be involved in the decay of the absorption band.

The absorption band observed in liquid propane is remarkably similar in

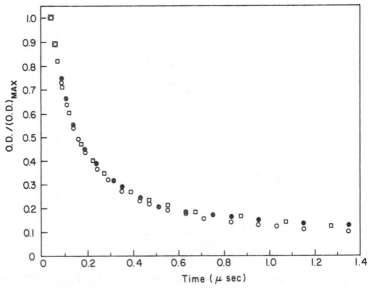

**Fig. 26** Kinetics of decay of absorption at 1525 nm in irradiated liquid propane at $-185°$C. Pulse width was 70 nsec, and times were measured from midpulse. Optical path length was 2 cm. For points $\bigcirc$, dose $= 7.0 \times 10^{16}$ eV/cc, (O.D.)$_{max}$ = 0.0344; for points $\square$, dose $= 3.00 \times 10^{16}$ eV/cc, (O.D.)$_{max}$ = 0.0152; for points $\bullet$, dose $= 1.92 \times 10^{16}$ eV/cc, (O.D.)$_{max}$ = 0.0096. (Ref. 75).

shape and location to that observed in hydrocarbon glasses. For example, in Fig. 27 is shown the absorption spectrum observed by Klassen, Gillis, and Teather[79] in pulse-irradiated 3-methylhexane (3MH) glass [they observed similar results in 3-methylpentane (3MP) glass]. The absorption curves observed 35 to 3000 nsec after a 40-nsec pulse are essentially the same in 3MH and look very much like the absorption curves observed in liquid propane. The maximum absorption wavelength in the early time period is at about 2000 nm, in agreement with the propane result. After times of the order of several microseconds, the spectrum begins to undergo a blue shift, as shown in Fig. 27. Such a blue shift is of course not observed in liquid propane, since the half-life of the absorption band in liquid propane is small compared to several microseconds.

A photobleaching study by Miller and Willard[80] provides some insight into the nature of the electron absorption band in irradiated hydrocarbon glasses. These investigators observed, with repetitive scanning, the esr signal height from gamma-irradiated samples of 3MP glass at 77°K. Their results, Fig. 28, indicate a photobleaching threshold at a wavelength of about 2200 nm, that is, there is a zero photobleaching yield for illumination wavelengths greater than 2200 nm. As Miller and Willard pointed out, their data

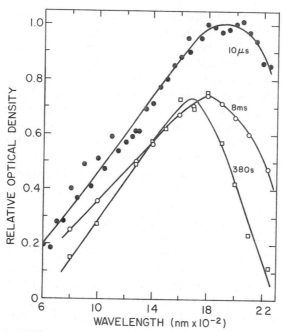

**Fig. 27** Spectral shift in the optical absorption spectrum of $e_t$ in 3MP glass at 76°K. The data points were measured at the times indicated after 1-$\mu$sec or 3-$\mu$sec pulses. (Ref. 79).

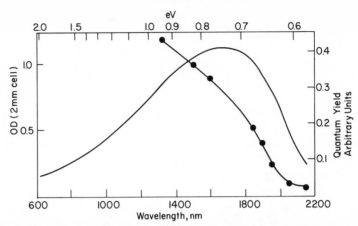

**Fig. 28** Relative photobleaching quantum yields in 3MP glass, and the absorption spectrum for trapped electrons in 3MP glass; gamma dose $2 \times 10^{19}$ eV/g. The relative quantum yield scale is set at 0.4 at 1300 nm to be in approximate correspondence with the reported yield at this wavelength (Ref. 80). Curve with solid circles denotes photobleaching quantum yields.

may be interpreted as indicating the photodetachment continuum for electrons bound in traps lying about 0.53 eV below the ground state of the conduction band (the continuum band being composed of quasi-free states plus the localized tail). This interpretation is consistent with the fact that the absorption band shape (after the initial blue shift) does not change appreciably during thermal decay[81,79] or photobleaching.

The electrons observed in the optical absorption experiments seem therefore to be in much deeper traps than those observed in drift time and fast current response mobility experiments. The activation energies for the observed electron mobilities were seen in such experiments to lie in the range from 0 to about 0.15 eV. An electron in a trap with $\frac{1}{2}$ eV of trapping energy would presumably have to move as a massive, low-mobility entity, much as the electron in a helium bubble does, since thermal promotion to the conduction band would be extremely unlikely at temperatures usually encountered in mobility experiments.

Thus a conjectured picture of the behavior of electrons in hydrocarbon liquids can be put forth at this point. Electrons can occupy high-mobility quasi-free states [mobilities of the order of 100 cm$^2$/(V)(sec)], the intermediate mobility of the localized tail on the conduction band [mobilities between 0.01 and 1 cm$^2$/(V)(sec)], or the very-low-mobility deep traps [mobilities of the order $10^{-4}$ to $10^{-3}$ cm$^2$/(V)(sec)] observed in optical absorption experiments. As discussed in Section IX.A, the localized tail states result from disorder in the fluid arising from thermal fluctuations.

The mechanism of electronic motion of electrons in these states is thermal promotion among the localized states (phonon-assisted hopping) and/or to the quasi-free conduction band. An electron undergoing either phonon-assisted hopping or quasi-free motion is not localized for sufficiently long times to perturb the structure of the fluid. However, in deep traps the electron is probably localized for sufficiently long times to reorder the solvent molecules around itself, that is, the electron may help "dig" its own cavity much as the electron in liquid helium "digs" its own bubble. In this picture the blue shift of the hydrocarbon glass absorption band can be understood as a deepening of the electron trap due to reordering of the fluid molecules around the electron. This shift is difficult to observe in the liquid state, because of the rapid recombination decay of the band.

If one assumes that deep traps require molecular reordering or the presence through thermal fluctuation of relatively unusual molecular configurations, then the fact that high-mobility species are observed in photoinjection time of flight (where no cations are present) and transient-current experiments may be understood to mean that the relaxation time from quasi-free and localized tail states to deep trap states is long compared to time of flight times occurring in the usual mobility experiments. Slow negative species are in fact always observed along with fast species. The former have generally been supposed to be anions of electron-scavenging impurities and not given much attention. Perhaps high-pressure experiments would show slow negative species behaving differently from ions. Deep traps could be long-lived ion states of scavengers. It is also possible that the deep traps are caused by the presence of geminate ions in irradiated materials. Then the deep traps would not be relevant to the behavior of electrons in the absence of cations.

The exciton spectrum of solids and liquids provides another source of information concerning quasi-free electron states. Raz and Jortner[20] have recently reviewed the spectroscopic data on impurity Wannier[82] exciton states in several simple liquids and have deduced electron work functions from the data. It is convenient to discuss the exciton spectrum of an impurity molecule in terms of the diagram in Fig. 29 for an impurity (guest) in a solid host. If the conduction band is assumed to be parabolic, the highly excited impurity levels $E_n$ are given by the Wannier series[82]

$$E_n = E_G - \frac{B}{n^2} \tag{4.1}$$

where $n = 1, 2, \ldots$. The impurity energy gap $E_G (= \lim_{n \to \infty} E_n)$ may be written in the form

$$E_G = I_g + P_+ + V_0 \tag{4.2}$$

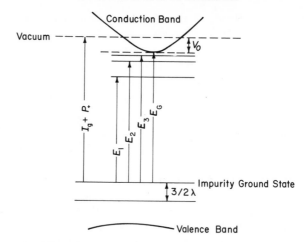

**Fig. 29**  Energy-level diagram for an impurity in an insulating solid where the conduction band is parabolic.

where $I_g$ is the gas-phase ionization potential, $P_+$ is the electrostatic polarization energy of the medium, and $V_0$ is the work function or the energy at the bottom of the conduction band relative to the vacuum state. The excitation binding energy $B$ is:

$$B = 13.6 \frac{m^*}{mD^2} \, \text{eV} \tag{4.3}$$

where $m^*$ is the effective mass of an electron near the bottom of the conduction band of the medium, and $D$ is the dielectric constant of the medium. If the excited electrons are from the valence band rather than from an impurity band, the excitation energy obeys (4.1), except that $m^*$ is the reduced mass of the effective masses of a hole in the valence band and a conduction electron and $E_G$ is:[83]

$$E_G = I_g + P_+ + V_0 + E_v \tag{4.4}$$

where $E_v$ is the half-width of the valence band.

Assuming (4.1) is valid for liquids as well as solids, one can deduce $V_0$ and $m^*$ from exciton spectra. The quantity $I_g$ can be measured in the gas phase, $P_+$ estimated theoretically,[84] and $E_v$ determined by comparing pure system and impurity spectra. Typical absorption spectra for xenon impurity in solid and liquid argon are shown in Fig. 30. The energy shift of the peaks in going from the liquid to the solid phase follows the shift in the bottom of the conduction band.

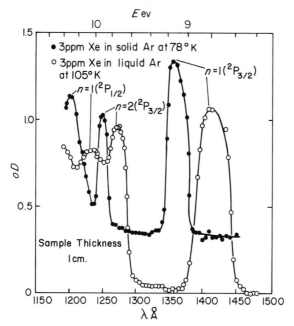

**Fig. 30** Absorption spectrum of xenon impurity in solid and in liquid argon. [B. Raz and J. Jortner, *Proc. Roy. Soc.*, **A317**, 113 (1970)].

In Tables V and VI, Raz and Jortner's[20] experimental spectroscopic results are presented for xenon and $CH_3I$ impurities in solid neon, argon, and krypton. The electron effective mass is about one-half the electron mass in all three solids and is independent of the impurity molecule. The work functions are ordered similarly to liquid-state ordering and are only slightly different in magnitude ($\sim 0.1$ eV) from the corresponding liquid-state values. Liquid-state data for several solids and for liquid argon and krypton are presented in Table VII (See, however, note added in proof to Table VI.) A summary of Raz and Jortner's spectroscopic estimates of $V_0$ for liquid hydrogen and all the inert gas liquids is included in Table I. The spectroscopically determined $V_0$ for liquid $^4$He agrees with the photoinjection determination. For liquid argon and solid xenon, the spectroscopic values seem to be lower than the photoinjection values, although the quoted experimental errors are as large as the differences between the two sets of values.

It is interesting that the valence bandwidth estimates given in Table VI are almost zero for neon and argon, a result implying little overlap of the valence orbitals of the neon and argon atoms. Consistent with this result

TABLE V

Energetic Data for Large-Impurity States of $CH_3I$ in Rare Gas Solids
(Ref. 20)[a]

|  | Ne matrix | Ar matrix | Kr matrix | Guest molecule |
|---|---|---|---|---|
| $E_G$ | 12.15 | 10.54 | 10.40 |  |
| $B$ | 4.4 | 2.4 | 1.72 | Xe |
| $m^*/m$ | 0.5 | 0.5 | 0.6 | $I_g = 12.13$ eV |
| $P_+ + V_0$ | 0.02 | $-1.59$ | $-1.73$ |  |
| $P_+{}^b$ | $-0.58$ | $-1.29$ | $-1.11$ |  |
| $V_0{}^c$ | 0.6 | $-0.3$ | $-0.62$ |  |
| $E_G$ | 9.51 | 8.58 | 8.13 |  |
| $B$ | 4.4 | 2.2 | $-2.10$ | $CH_3I$ |
| $m^*/m$ | 0.5 | 0.45 | 0.50 | $I_g = 9.49$ eV |
| $P_+ + V_0$ | 0.02 | $-0.91$ | $-1.36$ |  |
| $V_0{}^d$ | 0.60 | $-0.3$ | $-0.62$ |  |
| $P_+{}^e$ | $-0.62$ | $-0.6$ | $-0.74$ |  |

[a] All energies in electronvolts.
[b] W. Fowler, *Phys. Rev.*, **151**, 657 (1966).
[c] Calculated from the $P_+$ data.
[d] Adopted from Xe impurity data.
[e] Estimated using Xe impurity data (footnote d).

is the fact that the hole mobility in solid argon is small compared to that in xenon.

Another interesting feature of the exciton spectra[84] is that the relatively large radius ($n > 3$) Wannier states are observable in the condensed phase. It is possible that such large-radius states play a role in the absorption band observed in irradiated hydrocarbon liquids.

## B. Photoionization

Similar to the exciton spectral studies are the photoionization studies of Jarnagin and co-workers[85–87] and more recently of Holroyd.[88] These investigators measured photoionization currents of tetramethylparaphenyl-enediamine (TMPD) impurity in nonpolar hydrocarbon liquids. Jarnagin and co-workers demonstrated that photoionization in these systems can occur by a monophotonic process:

$$(a) \quad T + h\nu \rightarrow T^+ + e^- \tag{4.5}$$

or by a biphotonic mechanism described by the following processes.

| | Process | Rate |
|---|---|---|
| (b) | $T + h\nu \rightarrow T^1$ | $\varepsilon_1[T]I$ |
| (c) | $T^1 \rightarrow T^3$ | $(\phi_T/\tau_S)[T^1]$ |
| (d) | $T^3 + h\nu \rightarrow T^+ + e^-$ | $\varepsilon'_2[T^3]I$ |
| (e) | $e^- + T^+ \rightarrow T$ | $\imath[T^+][e^-]$ |
| (f) | $e^- + X \rightarrow X^-$ | $\tau_F^{-1}[e^-]$ |
| (g) | $T^+ + X^- \rightarrow T + X$ | $\imath_I[T^+][X^-]$ |

$T$, $T^1$, $T^3$, and $T^+$ represent TMPD in the ground, singlet, triplet, and singly ionized state, $I$ is the radiation intensity, X scavenging impurities, and $\varepsilon_1$, $\phi_T/\tau_S$, $\eta\varepsilon_2 \equiv \varepsilon'_2$, $\tau_F^{-1}$, and $\imath_I$ are the characteristic rate constants for the processes e to g. In the monophoton processes b, c, and d are replaced by a, but the remaining events still occur. For either process the current measured with electrodes of area $A$ and a separation $L$, and subject to the voltage drop $V$, is

$$i = (e/L)AV[\mu_E[e^-] + \mu_{T^+}[T^+] + \mu_{X^-}[X^-]] \qquad (4.6)$$

TABLE VI

Energy of the Quasi-Free Electron State in Solid Rare Gases Derived from Spectroscopic Data (Ref. 20)[a]. Note added in proof: By photoemission, N. Schwentner, F. J. Himpsel, V. Saile, M. Skibowski, W. Steinmann and E. E. Koch, *Phys. Rev. Lett.*, **34**, 528 (1975), determined the values of $V_0$ to be $+1.4$, $+0.4$, $-0.3$ and $-0.5$ eV for neon, argon kyrpton and xenon, respectively.

| Source of data | Medium | $E_G$ | $I_g$ | $P_+$[b] | $E_v$ | $V_0$ |
|---|---|---|---|---|---|---|
| Solid neon[c,d] | Ne (solid) | 21.42 | 21.56 | $-0.68$ | $0 \pm 0.1$ | $+0.54 \pm 0.1$ |
| Xe impurity in neon[e] | Ne (solid) | 12.15[f] | 12.13 | $-0.68$ | — | $+0.66 \pm 0.1$ |
| Solid argon[f] | Ar (solid) | 14.17 | 15.76 | $-1.10$ | $0 \pm 0.1$ | $-0.49 \pm 0.1$ |
| Xe impurity in Ar[e] | Ar (solid) | 10.54 | 12.13 | $-1.10$ | — | $-0.49 \pm 0.1$ |
| Solid krypton[f] | Kr (solid) | 10.67 | 14.00 | $-1.10$ | $-0.57$ | $-0.61 \pm 0.2$[h] |
| Xe impurity in Kr[e] | Kr (solid) | 10.40 | 12.13 | $-1.10$ | — | $-0.63 \pm 0.2$ |
| Solid Xe[f] | Xe (solid) | 9.28 | 12.13 | $-1.3$ | $-0.95$ | $-0.6 \pm 0.2$[h] |

[a] All energies in electronvolts.

[b] W. Beall Fowler, *Phys. Rev.*, **151**, 657 (1966).

[c] E. Boursey, J. Y. Roncin, and H. Damany, *Phys. Rev. Lett.*, **25**, 1279 (1970).

[d] R. Hensel, G. Keitel, E. E. Koch, N. Kosuch, and M. Skibowsky, *Phys. Rev. Lett.*, **25**, 1281 (1970).

[e] G. Baldini, *Phys. Rev.*, **137**, A508 (1965).

[f] G. Baldini, *Phys. Rev.*, **128**, 1562 (1962).

[g] Computed according to the assignment of the 10.95-eV transition to $n = 2$ state.

[h] Derived: B. Raz and J. Jortner, *Chem. Phys. Lett.*, **4**, 155 (1969); from x-ray exciton spectrum in solid krypton and xenon: R. Hensel, G. Keitel, E. E. Koch, M. Skibowsky, and P. Schreiber, *Phys. Rev. Lett.*, **23**, 1160 (1969).

TABLE VII

Energy of the Quasi-Free Electron State in Simple Solids and Liquids as derived from Spectroscopic Data. (Ref. 20)[a]

| Medium | Source of data for calculating $E_G$ | $E_G$ | $I_g$ | $P_+ + V_0$ Eq. (4.2) | $P_+$[g] | $V_0$, spectroscopic | $V_0$, other sources |
|---|---|---|---|---|---|---|---|
| Ar(s), 80°K | $n = 2$, Wannier state in Xe/Ar(s)[b,c] | $10.65^b$ | 12.13 | $-1.45$ | $-1.10$ | $-0.35 \pm 0.1$ | |
| Ar(l), 105°K | $n = 2$, Wannier state in Xe/Ar(l)[c] | $10.51^b$ | 12.13 | $-1.55$ | $-1.10$ | $-0.45 \pm 0.2$ | $-0.33 + 0.1^h$ $-0.45^i$ |
| Kr(s), 70°K | $n = 2$, Wannier state in Xe/Kr(s)[b,c] | $10.40^e$ | 12.13 | $-1.73$ | $-1.10$ | $-0.63 \pm 0.1$ | |
| Kr(s), 20°K | X-ray excitons in Kr(s)[d] | $92.2^f$ | 93.9 | $-1.70$ | $-1.10$ | $-0.60 \pm 0.1$ | |
| Kr(l), 135°K | $n = 2$, Wannier state in Xe/Kr(l)[c] | $10.25^e$ | 12.13 | $-1.88$ | $-1.10$ | $-0.78 \pm 0.2$ | |
| Xe(s), 20°K | X-ray excitons in Xe(s)[d] | $65.6^f$ | 67.55 | $-1.95$ | $-1.32$ | $-0.63 \pm 0.2$ | $-0.39^j$ |
| H₂(s), 6°K | Wannier states in Xe/H₂(s) | $10.11^k$ | 12.13 | $-2$ | $-1$ | $-1$ | |
| D₂(s), 6°K | Wannier states in Xe/D₂(s) | $10.20^k$ | 12.13 | $-2$ | $-1$ | $-1$ | |

[a] All energy values in electronvolts.

[b] G. Baldini, *Phys. Rev.*, **A508**, 137 (1965).

[c] B. Raz and J. Jortner, *Proc. Roy. Soc.*, **A137**, 113 (1970).

[d] R. Hensel, G. Keitel, E. E. Koch, M. Skibowsky, and P. Schreiber, *Optics Commun.*, **2**, 59 (1970).

[e] Estimated from Eq. 4.1 (see text).

[f] *d*-shell ionization potentials: J. A. Bearden and A. F. Burr, *Rev. Mod. Phys.*, **39**, 125 (1967).

[g] W. B. Fowler, *Phys. Rev.*, **151**, 657 (1966) and footnote *k*.

[h] Experimental value from electron injection experiments: B. Halpern, J. Lekner, S. A. Rice, and R. Gomer, *Phys. Rev.*, **156**, 351 (1967).

[i] Theoretical value: J. Lekner, *Phys. Rev.*, **158**, 130 (1967).

[j] Experimental value photoemission data: J. F. O'Brien and K. J. Teegarden, *Phys. Rev. Lett.*, **17**, 919 (1966).

[k] A. Gedanken, B. Raz, and J. Jortner, *Chem. Phys. Lett.*, **14**, 326 (1972).

where $[e^-]$, $[T^+]$, $[X^-]$, and so on, represent the concentrations of electrons and species $T^+$, $X^-$, and so on.

In a pulsed experiment the peak transient current will be proportional to the electron mobility $\mu_e$ if $\mu_e$ is large compared to the ionic mobilities $\mu_{T^+}$ and $\mu_{X^+}$. Thus the ratios of peak currents in photon-pulsed TMPD-containing fluids are equal to the ratios of the electron mobilities if the intensities $I$, field $V/L$, and lifetimes $\tau_S$ and $\tau_F^{-1}$ are the same. By assuming this to be the case and setting the electron mobility in TMS to be the value 93 cm$^2$/(V)(sec) measured by Schmidt and Allen,[40] Tekeda, Houser, and Jarnagin deduced from their current peaks the electron mobilities given in Table VIII for hexamethyldisilane (HMDS), tetraethylsilane (TES), and $n$-hexane. The

TABLE VIII

Constants and Parameters Determined from Photoionization Studies (Ref. 86)

|  | TMS | HMDS | TES | HX |
|---|---|---|---|---|
| $\mu_p$ (cm$^2$/(V)(sec) | 93 | 20 | 0.5 | 0.018 |
| $\mu_T$ (cm$^2$(V)(sec) $\times$ 10$^4$ | 12.0 | 4.7 | 6.5 | 7.6 |
| $\mu_x$ cm$^2$/(V)(sec) $\times$ 10$^4$ | 14.5 | 5.6 | 7.5 | 10.1 |
| $\tau_s$ (nsec)$^a$ | 4.3 | 4.3 | 4.3 | 4.3 |
| $\tau_{T3}$ ($\mu$sec)$^b$ | 5–40 | 2–30 | 10–15 | 1–10 |
| $\tau_F$ (nsec)$^b$ | 30–80 | 30–60 | 30–150 | 500–3000 |
| $\varepsilon_1$ (347) $\times$ 10$^{18}$ (cm$^2$) | 2.07 | 2.07 | 2.07 | 2.07 |
| $\langle \varepsilon_1 \rangle \times$ 10$^{13}$ (cm$^2$)$^c$ | 1.16 | 1.16 | 1.16 | 1.16 |
| $\varepsilon_1'$ (260)$^d$ | 1.4 $\times$ 10$^{-21}$ | 1.6 $\times$ 10$^{-22}$ | $\approx$10$^{-24e}$ | $\approx$10$^{-24e}$ |
| $\varepsilon_2'$ (347) $\times$ 10$^{18}$ (cm$^2$)$^d$ | 38 | 11 |  | 2.0 |
| $\langle \varepsilon_2' \rangle / \langle \varepsilon_2' \rangle$ TMS$^f$ | 1.00 | 0.45 | 0.12 | 0.070 |
| $D$ | 1.92$^i$ | 2.02$^h$ | 2.09$^i$ | 1.89 |
| $\phi_T{}^g$ | 0.25 | 0.25 | 0.25 | 0.25 |

$^a$ Supplied by Dr. I. Berlman of Argonne National Laboratory.

$^b$ Specimen-dependent; ranges encountered are given.

$^c$ Intensity weighted average over Xe flash wavelength distribution.

$^d$ Obtained from steady-state measurements.

$^e$ The current at 260 nm in these fluids was contributed by both one and two photon mechanism.

$^f$ The average $\langle \varepsilon_2' \rangle$ obtained in the various fluids are more accurately given relative to that found in TMS under similar experimental conditions. To obtain reasonable absolute values, the ratios should be scaled to 2.0 $\times$ 10$^{-17}$ cm$^2$. This value was obtained as the weighted average of $\varepsilon_2'(\lambda)$ as determined from steady-state techniques (see Ref. 6) and using the filtered xenon flash distribution as weighting function.

$^g$ Measured in 2-methylpentane at 77°K: K. D. Kadogan and A. C. Albrecht, J. Chem. Phys., **72**, 929 (1968).

$^h$ Obtained from the refractive index: O. Wichterle, *Organic Silicon Compounds*, Academic, New York, 1965.

$^i$ A. P. Altstuller and L. Rosenblum, J. Am. Chem. Soc., **77**, 272 (1955).

trends are very similar to those observed in similar hydrocarbons, and the value for hexane is within a factor of 4 of the value determined by time-of-flight measurements.[38,40] Tekeda, Houser, and Jarnagin also determined $\mu_{T+}$ and $\mu_{X+}$ by time-of-flight measurements; they found from the long time behavior of the transient current that the ion recombination rate was diffusion-limited, that is, given by

$$\imath_I = (\mu_{T+} + \mu_{X-}) \frac{4\pi e^2}{DkT} \tag{4.7}$$

They obtained $\tau_S$, $\tau_{T^3}$ (the lifetime of triplet TMPD), and $\varepsilon_1$ from optical data, assumed the electron-ion recombination rate $\imath$ to be

$$\imath = (\mu_e + \mu_{T+}) \frac{4\pi e^2}{DkT}$$

determined $\varepsilon_2'$ from steady-state current measurements, and obtained the scavenging lifetime $\tau_F$ by fitting the transient current data to solutions of the rate equations corresponding to processes b to g. Their summary of these data is given in Table VIII.

Tekeda, Houser, and Jarnagin also observed photoionization thresholds for the monophotonic ionization of TMPD which depended on the host fluid. The threshold in TMS was observed to be about $\frac{1}{2}$ eV lower than in $n$-hexane. The threshold can be understood in terms of the impurity excitation spectrum shown in Fig. 29. The monophotonic photoionization threshold energy corresponds to the impurity gap energy given by (4.2), that is, the threshold energy is

$$E_G = I_g + P_+ + V_0 \tag{4.8}$$

where $I_g$ is the gas-phase ionization potential of TMPD, $P_+$ is the polarization energy of the TMPD positive ion, and $V_0$ is the work function or the energy of a quasi-free electron at the bottom of the quasi-free conduction band. Thus the difference between the ionization threshold energies of TMPD in two nonpolar liquids whose polarization energies $P_+$ are roughly equal (i.e., whose number densities and dielectric constants are roughly the same) is equal to the difference between the electron work function of the two liquids. The measured work function difference for TMS and $n$-hexane is about 0.6 eV, in fair agreement with the photoionization threshold difference of 0.5 eV measured by Tekeda Houser, and Jarnagin. The observed trend for the photoionization rate constant $\varepsilon_2'$ of the triplet state of TMPD, given in Table VIII for 347-nm light, can also be understood in terms of an increase in the ionization energy gap (this time starting from the triplet level) with increasing work function. Also, since the work function of $n$-hexane is larger than TMS, the electron photoionization yield is probably lower in n-hexane than in TMS. This would account for Takeda, Houser, and Jarnagin's underestimate of the electron mobility in n-hexane, since they assumed the yields were the same in converting their peak current data into mobilities.

The work of Jarnagin and co-workers suggested the possibility of using the threshold energies of the monophotonic photoionization of TMPD to determine the work functions of various hydrocarbon liquids. Holroyd[88] carried out such a study. Using known values of $I_g$ and $V_0$, he determined from threshold measurements of $E_G$ for TMPD in n-pentane and TMS that $P_+ = -1.6$ eV for both these liquids. He then estimated the values of $P_+$ for any other hydrocarbon liquid under the assumption that $P_+$ is proportional to the function $(D - 1)/D$ of the dielectric constant of the liquid. The photoionization determinations of $V_0$ are seen in Table IX to agree quite

TABLE IX

Work Functions determined from Photoionization
Threshold Shifts (Ref. 88)[a]

| | $V_0^b$ exptl | Ref. 50 | Ref. 51 |
|---|---|---|---|
| n-Hexane | +0.03 +0.09 | +0.04 | 0.0 |
| n-Pentane | 0 +0.02 | −0.01 | +0.04 |
| 3-Methylpentane | −0.06 +0.01 | | 0.0 |
| 2,2,4-Trimethylpentane | −0.26 −0.15 | −0.18 | −0.14 |
| Cyclopentane | −0.31 −0.18 | −0.28 | −0.17 |
| Neopentane | −0.38 −0.43 | −0.43 | −0.35 |
| TMS | −0.63 −0.61 | −0.62 | −0.55 |
| 2,2-Dimethylbutane | −0.29 −0.24 | | |
| Methylcyclohexane | −0.06 +0.14 | | |
| n-Decane | +0.06 +0.22 | | |

[a] All energy values are in electronvolts.
[b] Values of $V_0$ determined from $E_G - I_g - P_f$.

well with those determined previously by photoinjection experiments. Peterson and coworkers (S. H. Peterson, M. Yaffe, J. A. Schultz and R. C. Jarnagin, J. Chem. Phys., submitted for publication) have recently found that the photoionization thresholds are solute dependent, a finding that brings into question the accuracy of using photoionization to determine work functions.

### C. Photoexcitation of Electrons Trapped in Bubbles in Liquid Helium

In order to test directly the bubble model for excess electrons in liquid helium, Northby and Sanders[89,90] and Zipfel and Sanders[91,92] investigated the response to electromagnetic radiation of a dc current of excess electrons in liquid HeII. In their experiment a current of electrons is generated by applying a dc electric field $E$ between a $^{210}$Po alpha source and a collector grid. On a pair of grids between the collector grid and the alpha source, a square-wave field is generated so that during half of the period the field between the grids is $E$ and during the other half of the period the field is reversed. If the square-wave frequency is adjusted so that for an electron in the lowest energy state the transit time between the pair of grids is slightly greater than one half-period, then no current will reach the collector. (The grids act as doublt shutters which allow charges to pass only if the drift time between the grids is an integral multiple of the half-period of the square wave. The double-shutter principle was used for the mobility measurements reported in Refs. 2, 3, 5, 25, 26, 37, 38, and 46.) If the drift space is illuminated, then electrons, if trapped in bubbles, may be optically excited to higher mobility states (to the quasi-free continuum, for example) with the accompanying appearance of a collector current.

In the experiments performed by Sanders and co-workers, monochromatic light was mechanically modulated, and the component of the collected current synchronous with the chopped light was detected using a phase-sensitive detector. The output signal power $S$ of the detector in such an experiment is given by

$$S = \sigma(\lambda)I(\lambda)\alpha(\lambda) \tag{4.9}$$

where $\sigma(\lambda)$ is the cross section for a radiation-induced electron transition from the normal bubble state to a more mobile state, $I(\lambda)$ is the intensity (power/unit area) of light of wavelength $\lambda$, and $\alpha(\lambda)$ is a weighting function related to the collection frequency. Sanders and co-workers adopted the ratio $S/I$ as the normalized signal.

An example of their results is given in Fig. 31. The main features are the following: (a) the peaks undergo a blue shift with increasing pressure; (b) three peaks are resolved in the photon energy range available to the investigators (0.5 to 3 eV), a large peak at the low-energy end ($\sim 0.6$ to 0.8 eV), a small peak on the high-energy edge of the large peak, and a small peak in the energy range 1.2 to 2 eV; (c) at higher pressures the small peak at the edge of the large peak and the large peak are no longer resolved; and (d) at very low pressures[90] (not shown in Fig. 31), the high-energy peak is much larger than the low-energy peak.

Zipfel and Sanders[92] explained their results in terms of a model in which the low-energy large peak in Fig. 31 represents a transition of the electron

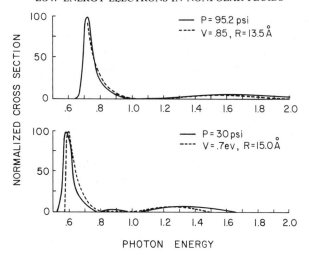

**Fig. 31** Normalized photoejection current $S/I$ of electrons in liquid helium versus photon energy of illuminating radiation (Ref. 92). The dashed curves represent the cross-section prediction of Zipfel and Sanders (Ref. 92) explained in the text.

from a ground-state bubble trap (considered to a square well of depth $V_d$ and radius $R$) to the continuum. Given that assignment, the values of $V_d$ and $R$ can be determined from the low- and high-energy peaks (the intermediate peaks have to be considered an anomaly in this assignment) of the spectrum. The low-pressure value of $V_d$ was found to be about 0.6 eV. In the context of the bubble model for the electron, the square well depth should equal the barrier of the fluid to injection of an electron from the vacuum (polarization effects are very small in helium) to the liquid, that is, $V_d$ should equal the work function. Since the measured low-pressure value of the work function has been determined to be 1 to 1.3 eV, the spectral assignment of Zipfel and Sanders poses a problem.

An alternative interpretation of the observed photoejection spectrum has been suggested by Miyakawa and Dexter.[93] In their model, to be discussed in more detail in Section VIII, the low-energy peak is assigned to the $1s$-to-$2p$ transition of an electron in a square-well trap. The small peak on the high-energy edge of the low-energy peak is interpreted as the $1s$ to the lowest-energy continuum band (the smallness of the peak arising from a small cross section for this transition). The high-energy peak is interpreted as the $1s$ to the next-highest-energy continuum band. The low-pressure well depth and bubble radius implied by Miyakawa and Dexter's spectral assignments are 0.95 eV and 17 Å. Thus the value of their depth is consistent with measured work functions. Why the $1s$-to-$2p$ transition would be accompanied by a current flow in the photoejection experiment is not fully understood. It was suggested by Miyakawa and Dexter that the thermal energy released in the process (estimated[94] to be $\sim 0.05$ eV during bubble relaxation after excitation) increases the local temperature and thermally dissociates the bubble from the entrained quantum vortex lines. This interpretation is consistent with the experimental observations that (a) the photoejection current decreases[92] rapidly above 1.3°K, and (b) from analysis[95] of the lifetimes of electronic bubbles trapped on vortex lines in rotating liquid helium, it appears that the bubbles thermally dissociate from the vortex lines above 1.34°K. Incidentally, the rotating liquid helium experiments yield an estimate of the bubble radius (14.5 to 16 Å at 1.7°K) that is in reasonable agreement with the values deduced from mobility[3] (12.7 Å at 3°K, 14 Å at 4.2°K) and photoejection studies.

### D. Reflection and Emission Spectra of Liquids

The main subject of this article is the behavior of excess low-energy electrons. To enter into a full discussion of the excited-state chemistry, the photochemistry as it were, of liquids lies outside the scope and the page limitations of this article, as well as outside our areas of competence. However, two interesting experiments are mentioned since, as in the case of exciton absorption spectra discussed in Section IV.A, they bring out strong similarities between liquid- and solid-state behavior. The implications of this similarity for theory are obvious.

The reflection spectrum of solid and liquid xenon was investigated by Beaglehole,[96] and his results are illustrated in Fig. 32. Much of the exciton structure observed in the solid is also found in the liquid. The total width of the exciton manifold is about the same in the liquid and in the solid, but resolution of the level structure in the liquid (or in the higher-temperature solids included in Fig. 32 for that matter) is not possible.

The spectral line widths $\Delta v$ have been estimated by Rice and Jortner[97] to be:

$$\Delta v \simeq \frac{nh^2 \tilde{a}^2 k}{m\pi} S(\mathbf{k}) \qquad (4.10)$$

when $n$ is the atomic density of the system, $h$ is Planck's constant, $\tilde{a}$ is the electron-atom scattering length, $k = |\mathbf{k}| = m|\mathbf{v}|/h$ with $|\mathbf{v}|$ the bound excited electron orbital velocity, $m$ is the mass of electron, and $S(\mathbf{k})$ is the structure factor of the medium. Equation (4.10), applicable whenever the electron mean free path in the medium is large compared to the orbital radius, represents an estimate of the broadening of spectral lines arising from weak scattering of the orbiting electron by the surrounding medium. For liquid xenon one estimates $\Delta v \sim 0.1$ eV, a value that may account for the

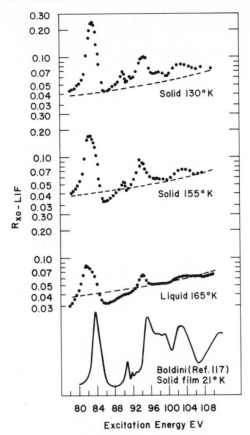

**Fig. 32** Reflection spectrum of liquid and solid xenon. (Ref. 96).

lack of resolution of the level structure (with spacing $\sim 0.2$ eV) in the liquid and in high-temperature solids.

Given the optical absorption results on simple liquids and solids discussed in Section IV.A and the reflection spectroscopic results just mentioned, it seems reasonable to conclude that exciton states exist in liquids, that is, that molecular scattering events do not so shorten the lifetime of collective excitations that they cease to be meaningful in the description of the liquid. Thus it is to be expected that energy can be transferred over long distances.

Excitation energy transfer must play a role in the rapid recombination processes and ion-free yields of irradiated substances. Very little is currently known (to us at any rate) concerning excitation energy transfer in liquids. One interesting experimental investigation was carried out by Meyer et al.[98]

They investigated the emission spectra of liquid helium, neon, argon, krypton, and xenon with alpha particles. The emission spectra of all the liquids were red-shifted (from 2 to 6 eV) and, by comparison with known gas spectra, the emitting species were identified as excimers $He_2^*$, $Ne_2^*$, and so on, of the liquid under investigation. If the lifetime of the molecular excited state is long relative to the time required for molecular displacement, and if an excimer can be formed, it seems likely that excitation energy can be self-trapped with high efficiency. Clearly, the situation in which self-trapping is generated by exciton-fluid coupling is analogous to the electron self-trapping in liquid helium. Just as the excess electron states of helium and argon are fundamentally different, corresponding to trapped and free electrons, one must expect to find liquids in which energy transfer does occur with ease. Each case must be examined separately.

Of course, if the excimer species is long-lived, it can serve as a carrier of energy. Indeed, a phenomenon attributable to energy transfer via the intermediacy of $He_2^*$ was discovered by Meyer, et al.[99] When liquid helium is doped with nitrogen and oxygen (present as small solid particles), the emission spectrum of the alpha-particle-irradiated liquid arises from the transitions $A^3\Sigma_u^+ \to X^1\Sigma_g^+$ and $C^3\Sigma_u^+ \to X^3\Sigma_g^-$ of nitrogen and oxygen, respectively. These are the transitions that would be excited by triplet $He_2^*$ and, since the lowest triplet state of helium has a very long lifetime (many seconds), it is reasonable to suppose that diffusion of $He_2^*$ can serve to transfer electronic excitation energy over long distances in liquid helium. It should be noted that, although unlikely, one cannot at present rule out atom interchange–energy exchange:

$$(He_a He_b)^* + He_c \to (He_a He_c)^* + He_b$$

or other mechanisms of energy transfer.

## V. SOME ELECTRON-MOLECULE REACTION RATE STUDIES

The reaction rates of electrons in fluids with scavengers provide another probe for examining the electronic states of the fluids. The literature on recombination and scavenging reactions is vast but will not be dealt with in great detail in this paper. In this section recent measurements of reaction rates of electrons with selected solutes in a variety of fluids are discussed. The trends observed bring out interesting qualitative differences between high-mobility fluids such as neopentane and low-mobility fluids such as n-hexane, as well as qualitative differences among the rates for various solutes in the same fluid.

Beck and Thomas[44] studied reaction rates of electrons with oxygen $CCl_4$, and biphenyl in several hydrocarbon liquids using two-photon laser photolysis of small amounts of anthracene and pyrene in the liquids. They photo-ionized either anthracene or pyrene with a 20-nsec laser pulse in a drift space subject to an electric field. The current observed in such an experiment is proportional to

$$A = \sum_j n_j |z_j| \mu_j \qquad (5.1)$$

where $n_j$ is the concentration of the $j$th charged species, $z_j$ is its valence, and $\mu_j$ is its mobility. Since the mobilities of electrons in nonpolar hydrocarbons are much larger than ionic mobilities, the quantity $A$ is well approximated by

$$A \simeq n_e \mu_e \qquad (5.2)$$

until electron scavenging reduces the electron concentration $n_e$ well below the ionic concentrations. Thus, as in the transient experiments mentioned earlier,[67,68,74] the peak current in Beck and Thomas' pulse experiment is proportional to the electron yield and the electron mobility. Using known mobilities, Beck and Thomas found that the photoionization yields were within a factor of 2 of one another in nonpolar hydrocarbons with electron mobilities ranging from about 0.07 ($n$-hexane) to about 13 cm$^2$/(V)(sec) (dimethylbutane). The higher yields were observed in fluids with lower work functions (and higher mobilities). The mobility data given in Table I for an isooctane–$n$-hexane mixture were obtained by Beck and Thomas from peak current measurements under the assumption that the electron yields were the same at all compositions. With this assumption the ratio of the peak currents in different fluids gives the electron mobility ratio for these fluids. The absolute values given in Table I for isooctane–$n$-hexane mobilities were determined by scaling the data to the value for $n$-hexane measured in time-of-flight experiments. (The fact that the assumption of constant electron yields is not strictly true represents a basic drawback in Beck and Thomas' mobility determinations.)

Beck and Thomas determined the reaction-rate constants of added scavengers from the dependence of the decay time of the current on scavenger concentration. They established that the scavenging reaction is the first-order process

$$e^- + \text{scavenger} \rightarrow \text{scavenger ion}$$

Table X includes a summary of their measured rate constants $\imath$, where

$$\imath \equiv \frac{\mathscr{K}}{n_s} \qquad (5.3)$$

$\mathscr{K}$ being the decay rate constant and $n_s$ the scavenger concentration (expressed in units of molarity for the entries in Table X. The rate observed in

TABLE X

Rate Constants $\imath$ of Electron-Molecule Reactions at $T = 23 \pm 1°C$
(Refs. 44 and 100)[a]

| Reactant | Solvent | $\mu$ [cm$^2$/(V)(sec)] | $\imath \times 10^{-12}$ (mole$^{-1}$ sec$^{-1}$) |
|---|---|---|---|
| Oxygen | n-Hexane | 0.078 | 0.11 |
| Oxygen | Cyclohexane | 0.24–0.33 | 0.11 |
| Oxygen | Methylcyclohexane | 0.068 | 0.17 |
| Oxygen | Neopentane | 70 | 1.2 |
| Biphenyl | Ethanol | $10^{-4}$–$10^{-3}$ | 0.0043 |
| Biphenyl | n-Hexane | 0.078 | $0.77 \pm 0.1$ |
| Biphenyl | Cyclohexane | 0.24–0.33 | $2.6 \pm 0.4$ |
| Biphenyl | Benzene | 0.6, 0.068[b] | 0.8 |
| Biphenyl | p-Xylene | 0.062 | 0.6 |
| Biphenyl | Cyclopentane | 1.1 | $3.5 \pm 0.5$ |
| Biphenyl | 2,2,4-Trimethylpentane | 7 | $11 \pm 5$ |
| Biphenyl | 2,2-Dimethylbutane | 10.9 | $13 \pm 4$ |
| CCl$_4$ | n-Hexane | 0.078 | 0.85 |
| CCl$_4$ | n-Pentane | 0.14 | 1.8 |
| CCl$_4$ | Cyclohexane | 0.24 | 1.8 |
| CCl$_4$ | Cyclopentane | 1.1 | $3 \pm 1$ |
| CCl$_4$ | Neopentane | 70 | 28.6 |
| CCl$_4$ | TMS | 90 | 54 |
| CH$_3$I | Cyclohexane | 0.24–0.33 | 2.0 |
| C$_2$H$_5$Br | n-Hexane | 0.078 | 1.47 |
| C$_2$H$_5$Br | n-Pentane | 0.14 | 1.6 |
| C$_2$H$_5$Br | 2,2,4-Trimethylpentane | 7 | 6.3 |
| C$_2$H$_5$Br | Neopentane | 70 | 0.32 |
| C$_2$H$_5$Br | TMS | 90 | 0.042 |

[a] Scavenger (reactant) concentrations ranged from about $10^{-7}$ to $10^{-4}$ M. The rate constants are given in units of mole$^{-1}$ sec$^{-1}$, that is, $\imath \equiv \mathcal{K}/n_s$, where $n_s$ denotes the molarity of the reactant. Error estimates for $\imath$ are entered wherever available.

[b] The origin of the disagreement between the mobility of 0.6 reported in Ref. 46 and 0.068 reported in the reference in footnote $d$ of Table I is not known. The lower number is supported by the observed value of $\imath$ for benzene, a point made earlier in Ref. 44.

ethanol is given for comparison. Electrons in ethanol are trapped in cavities and have mobilities in the ionic range. As might be expected, the rate constants are larger in fluids having larger electron mobilities. However, the process is not diffusion-controlled, since the rate constant does not increase linearly with electron mobility. In fact, Beck and Thomas suggested that the rate constant for the electron-biphenyl reaction in the different hydrocarbons they studied increases as the square root of the mobility.

Allen and Holroyd[100] measured reaction rates using a modification of the drift velocity technique used earlier by Schmidt and Allen.[40] With an applied electric field $E$ between parallel electrodes of spacing $d$, a steady beam of x rays is suddenly turned on uniformly between the electrodes. The transient-current subsequently observed if a first-order reaction is occurring (rate = $-\mathcal{K}n_e$, where $n_e$ is the local electron concentration) is given by

$$i = \frac{eav_d}{\mathcal{K}d}\left[\left(d - \frac{v_d}{\mathcal{K}}\right)(1 - e^{-\mathcal{K}t}) + v_d t\, e^{-\mathcal{K}t}\right] \tag{5.4}$$

where $a$ is the electron volumetric generation rate in electrons per second per cubic centimeter. The current reaches a steady-state value at $t = d/v_d = d/\mu E$. Thus analysis of transient-current data yields the electron mobility $\mu$ and the electron reaction rate $\mathcal{K}$.

Allen and Holroyd studied electron reaction rates with $CCl_4$, $CH_3I$, oxygen, and $C_2H_5Br$ as a function of temperature in a variety of liquids. Their results are summarized in Table X. As observed by Beck and Thomas, the reactions are not diffusion-controlled. With the data shown in Fig. 33a, Allen and Holroyd also suggested that the rate is proportional to the square root of the mobility in the case of $CCl_4$, $CH_3I$, biphenyl and, with a lower magnitude, almost for oxygen. However, the electron reaction rate with $C_2H_5Br$ actually decreases with increasing mobility, indicating a behavior totally different from that of the other compounds studied. The temperature dependence of the reaction rate with $C_2H_5Br$ is also unusual, $\mathcal{K}$ increasing with $T$ in $n$-hexane and decreasing with $T$ in neopentane and TMS, whereas for the other reactants $\mathcal{K}$ increased with $T$ in all cases. These trends are shown in Fig. 34.

The assumption that the electron scavenger reaction rate is proportional to the square root of the electron mobility for all the scavengers except $C_2H_5Br$ discussed does not stand up under careful scrutiny of Beck and Thomas' and Allen and Holroyd's data. An important implication of the square-root dependence is that the activation energy $E_R$ of the reaction-rate constant, defined by the expression $\imath = A \exp(-E_R/RT)$, equals about one-half of the mobility activation energy $E_A$. A summary of observed values of $E_R$ is presented in Table XI. In $n$-hexane and cyclopentane, with mobility activation energies of 4.06 and 3.5 kcal/mole, the reaction-rate activation energy should be 2.06 and 1.75 kcal/mole, respectively. This is in strong disagreement with the experimental values $E_R = 5.96$ and 4.6 kcal/mole for the electron–biphenyl reaction. In neopentane and TMS, $E_A \simeq 0$, whereas $E_R$ for the electron–$CCl_4$ reaction is found experimentally to be 3.54 and 2.21 kcal/mole in these solvents, again in strong disagreement with the square-root model.

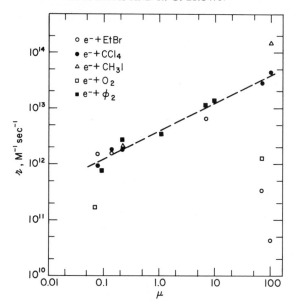

**Fig. 33a** Electron-molecule rate constants $\imath = \kappa/n_s$, where $n_s$ is the concentration of added scavenger, for several scavengers in different hydrocarbons plotted versus electron mobilities in the hydrocarbons at 23°C. (Refs. 100 and 44).

Based on observations to follow, we suggest that the observed trends of reaction rate versus mobility are a result of the transition from a diffusion-controlled rate in low-mobility solvents, in which case $\imath = \imath_d \propto \mu$, to an activation energy–controlled rate in high-mobility solvents, in which case $\imath = \imath_a$, where $\imath_a$ is the reaction-rate constant in a solvent in which the rate of diffusion of the electron to within some interaction distance $\delta$ (e.g., the van der Waals radius for a hard-sphere cross section model may be a rough measure of $\delta$) of the scavenger molecule is fast compared to the rate at which the electron–molecule interacting pair undergo the transition to the ground-state negative ion. $\imath_a$ is a function of local electron scavenger interactions. For the low concentrations of electrons and scavengers studied, the overall rate constant for an intermediate case is:

$$\imath^{-1} = \imath_d^{-1} + \imath_a^{-1} \tag{5.5}$$

for the model we propose here. [Beck and Thomas[44] suggested the possibility of explaining their data in terms of (5.5), but did not follow through on the suggestion.]

Although the nature of the solvent may affect $\imath_a$, the effect is probably small compared to the variation in $\imath_d$ with solvent, since $\imath_d$ is proportional

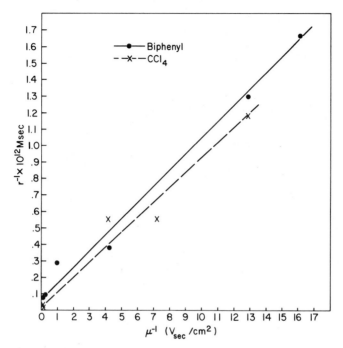

**Fig. 33*b*** Inverse of reaction rate $\imath$ versus inverse of electron mobility for electrons scavenged by biphenyl and $CCl_4$ in various hydrocarbon solvents at 23°C.

TABLE XI

Activation Energies and Preexponential Factors Determined by Fitting Data for Electron-Molecule Rate Constants $\imath$ to the Arrhenius Expression $\imath = \imath_0 \exp(-E_R/RT)^a$

| Reactant | Solvent | $\imath_0$ (mole$^{-1}$ sec$^{-1}$) | $E_R$ (kcal/mole) | $E_A$ (kcal/mole) |
|---|---|---|---|---|
| Biphenyl | *n*-Hexane | $1.64 \times 10^{16}$ | 5.96 | 4.06 |
| Biphenyl | Cyclopentane | $10^{16}$ | 4.6 | 3.5 |
| Biphenyl | 2,2,4-Trimethylpentane | $3 \times 10^{13}$ | 0.8 | |
| $C_2H_5Br$ | *n*-Hexane | $5.33 \times 10^{14}$ | 3.47 | 4.06 |
| $C_2H_5Br$ | Neopentane | $1.30 \times 10^9$ | $-3.22$ | 0 to 0.5 |
| $C_2H_5Br$ | TMS | $7.92 \times 10^9$ | $-0.95$ | $\sim 0$ |
| $CH_3I$ | TMS | $1.71 \times 10^{14}$ | 0.052 | $\sim 0$ |
| $CCl_4$ | Neopentane | $1.15 \times 10^{16}$ | 3.54 | 0–0.5 |
| $CCl_4$ | TMS | $2.20 \times 10^{15}$ | 2.21 | $\sim 0$ |

$^a$ Data taken from Refs. 44 and 89. Electron mobility activation energies $E_A$ are given for comparison in the last column.

391

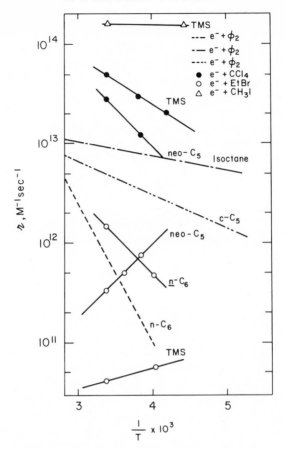

**Fig. 34** Electron-molecule rate constants $\imath = \kappa/n_s$ as a function of $T$ for several scavengers in $n$-hexane, neopentane, and TMS. (Refs. 100 and 44).

to the solvent-sensitive electron mobility. Thus one expects from (5.5) a linear plot of $\imath^{-1}$ versus $\mu^{-1}$. In Fig. 33$b$, $\imath^{-1}$ is plotted versus $\mu^{-1}$ for the reactants biphenyl and $CCl_4$. Although there is some scatter, due perhaps to a solvent dependency of $\imath_a$, the linear relationship predicted by (5.5) is roughly obeyed. (Errors reported for $\imath$ range as high as $\pm 45\%$.)

In their rate-versus-temperature studies on the reactant biphenyl in $n$-hexane, Beck and Thomas found that, within their experimental scatter which implies a scatter of $E_R$ values of about 1 kcal, $\imath$ varies linearly with $\mu$. This result implies that in $n$-hexane the reaction is diffusion-controlled, that is, $\imath \approx \imath_d$ or $\imath_a \gg \imath_d$. If the reaction sphere radius $\delta$ is defined by $\imath_d = (6.02 \times 10^{20}$ molecules/(cm$^3$)(mole) $4\pi\delta D_e$, where the electron-diffusion coefficient may be determined by $D_e = kT\mu/e$, then setting $\imath_d$ equal to the experimental rate constant for the electron–biphenyl reaction in $n$-hexane yields the value $\delta = 5.04$ Å at $T = 296°$K. This value for $\delta$ is not unreasonable, since it is of the order of the mean spacing between solvent and reactant molecules. The fact that the electron scavenger reaction-rate constants are approximately the same [roughly $10^{12}/$(mole)(sec)] in $n$-hexane for all the reactants included in Table X except oxygen indicates that the scavenging reaction is diffusion-controlled in $n$-hexane for all scavengers except oxygen if it is diffusion-controlled for the biphenyl case. The ratio of the rate constant in cyclohexane to that in $n$-hexane is for all scavengers studied except oxygen of the order of 2 or 3, in rough agreement with the mobility ratio in these two solvents. Thus it seems reasonable to conclude that the reaction is diffusion-controlled in cyclohexane, except for the scavenger oxygen. The linearity between $\imath$ and $\mu$ does not hold for cyclopentane which has a mobility of the order 1 cm$^2$/(V) (sec). It appears that somewhere between $\mu = \frac{1}{4}$ to 1 cm$^2$/(V) (sec) the transition from a diffusion-controlled to an activation energy–controlled reaction begins for the reactants biphenyl, $CCl_4$, $CH_3I$, and $C_2H_5Br$.

Oxygen and $C_2H_5Br$ present special cases. Since the observed $\imath$ for oxygen is less than for biphenyl, it is consistent with the present interpretation to assume that $\imath_a \ll \imath_d$ for the electron–oxygen reaction. This assumption agrees with the result that $\imath$ is about the same for oxygen in $n$-hexane, methylcyclohexane, and cyclohexane, even though a three- or fourfold mobility change is involved in this sequence of solvents. However, the rate constant for oxygen in neopentane is 10 times higher than in $n$-hexane, a solvent sensitivity that is surprising but that perhaps could be understood in terms of a resonance attachment process that occurs readily if the bottom of the conduction band ($V_0$) lines up energetically with an excited state of the oxygen ion.

Similarly, for the electron–$C_2H_5Br$ reaction, the values of $\imath$ in $n$-hexane and $n$-pentane are consistent with the diffusion-controlled rate, whereas for the higher-mobility solvents presented in Table X the reaction-rate constant actually decreases with increasing mobility. In terms of the present model, this trend implies that $\imath_a \ll \imath_d$ for the higher-mobility solvents and that $\imath_a$ decreases with increasing $\mu$. Such an effect could arise from solvent modifications of the structure of the scavenger, or because the state of the electrons in the low-mobility solvents is different from that in the high-mobility solvents. The latter possibility seems more likely, since the solvent–solute interactions do not differ very much for the sequence of solvents studied. If one supposes that a localized electron reacts more readily with $C_2H_5Br$ than does a quasi-free or extended-state electron, then the trend of decreasing $\imath$ with increasing $\mu$ can be understood as the result of an increasing probability that the electron is quasi-free in the sequence 2,2,4-trimethylpentane, neo-pentane, and TMS. The negative activation energies for the electron–$C_2H_5Br$ reaction in neopentane and TMS can be understood in terms of the same idea, that is, as $T$ increases, the probability that the electron is localized decreases so that $\imath$ decreases. The implication of this idea is exciting. From the magnitudes, and field and temperature dependences of electron mobilities in neopentane and TMS, it appears that localized states play only a minor role in electron transport, that is, the electron behaves almost as a quasi-free electron in its drift or diffusive motions. However, the reaction rate of electrons with $C_2H_5Br$ in the high-mobility fluids may be determined primarily by the fraction of time (or the probability) that the electron is in the localized state. Thus reaction rates may turn out to be extremely sensitive probes for determining aspects of the electronic states overlooked in work function and mobility studies.

A few experimental results and much speculation have been presented in this section. More extensive experimental scavenging-rate studies and quantitative investigations into the local chemistry of electron scavenger interactions provide challenging and probably very fruitful possibilities for future work. Gas-phase data on electron scavenging reactions will also be quite useful in putting together a quantitative picture of the liquid-phase scavenging processes.

## VI. THEORY OF ENERGY STATES OF QUASI-FREE ELECTRONS

### A. A Wigner–Seitz Pseudopotential Model

A completely rigorous determination of the spectrum of quasi-free states (better referred to as extended or delocalized states) in a dense disordered medium has not yet been attained. Springett, Jortner, and Cohen (SJC),[35] however, have developed a model theory for monatomic fluids which is at least qualitatively correct and has the appealing features of simplicity and a certain intuitive reasonableness. These investigators combine an average lattice representation of a fluid with a pseudopotential model in which the electron potential is a sum of the mean field-polarization potential and a hard-core potential representing the repulsive Hartree–Fock pseudo-potential. Their theory is outlined here, and its usefulness for polyatomic fluids is considered.

The basic problem in determining quasi-free energy levels is to solve the one-electron Schrödinger equation

$$\left[ -\frac{\hbar^2}{2m} \nabla^2 + V(r) \right] \psi = E\psi \qquad (6.1)$$

for the extended states (i.e., for unnormalized wave functions). $V(r)$ denotes the one-electron potential exerted by the unperturbed medium. If the electron were trapped in long-lived states, its energy determination would have to involve a self-consistent calculation of the reordering of the molecules of the medium caused by electron–medium interactions. Presumably, molecular rearrangements respond too slowly to follow the motion of quasi-free electrons.

In a crystal the potential $V(\mathbf{r})$ has the translational symmetry of the lattice, and the wave functions $\psi_{\mathbf{k}}$ corresponding to the energy state $E_{\mathbf{k}}$ obey the Floquet–Bloch theorem:

$$\psi_k = u_k(\mathbf{r})e^{i\mathbf{k}\cdot\mathbf{r}} \tag{6.2}$$

where $u_k(\mathbf{r})$ has the same translational symmetry as $V(\mathbf{r})$. Insertion of $\psi_{\mathbf{k}}$ into the Schrödinger equation yields for the ground state $(\mathbf{k} = 0)$ the equation

$$-\frac{\hbar^2}{2m}\nabla^2 u_0 + Vu_0 = E_0 u_0 \tag{6.3}$$

and for excited states $(\mathbf{k} \neq 0)$ the equation

$$-\frac{\hbar^2}{2m}[\nabla^2 u_{\mathbf{k}} + 2i\mathbf{k}\cdot\nabla u_{\mathbf{k}}] + Vu_{\mathbf{k}} = \left(E_k - \frac{\hbar^2 k^2}{2m}\right)u_{\mathbf{k}} \tag{6.4}$$

In the theory[101] of conduction electrons in metals, it is assumed that $\mathbf{k}$ is small so that the term $i\mathbf{k}\cdot\nabla u_{\mathbf{k}}$ can be treated as a perturbation. Expanding $u_{\mathbf{k}}$ in the perturbation series:

$$u_{\mathbf{k}} = u_0 + u_1 + u_2 + \cdots \tag{6.5}$$

one obtains the first-order equation

$$-\frac{\hbar^2}{2m}\nabla^2 u_1 + (V - E_0)u_1 = -i\frac{\hbar^2}{m}\mathbf{k}\cdot\nabla u_0 \tag{6.6}$$

where $E_0$ and $u_0$ are to be determined from (6.3). The energy to the order given by $u_0 + u_1$ is of the form[101]

$$E_k = E_0 + \frac{\hbar^2 k^2}{2m^*} \tag{6.7}$$

where the effective mass $m^*$ is determined from $u_0$ and $u_1$. For quasi-free electrons one expects $m^*$ to be of the order of the electronic mass.

Boundary conditions must be specified for (6.3) and (6.6). If one imagines each atom to be the center of identical and space-filling polyhedra (the translational replication of any one of which would yield the crystal lattice), then it suffices to determine the wave function for an electron in one of these polyhedra. Symmetry conditions for the wave function on the surface of a polyhedron serve as boundary conditions.[102] For example, if it is assumed that the potential has the symmetry

$$V(\mathbf{r}) = V(-\mathbf{r}) \tag{6.8}$$

in the polyhedron then, because of the translational symmetry of the lattice, the derivative of the ground-state wave function $u_0$ must vanish on the surface of the polyhedron.

In a theory of metals, Wigner and Seitz[103] simplify the boundary conditions by replacing the polyhedron by a sphere of equal volume and satisfy the boundary conditions on the surface of this sphere. The radius $r_s$ of the sphere is given by the relation

$$\tfrac{4}{3}\pi r_s^{3} n = 1 \tag{6.9}$$

where $n$ is the number density of the medium. Assuming the potential to be spherically symmetric, that is, $V = V(r)$, the Wigner–Seitz boundary condition on $u_0$ becomes

$$\left. \frac{\partial u_0}{\partial r} \right|_{r=r_s} = 0 \tag{6.10}$$

Then, since the rhs of (6.6) is an odd function, $u_1$ must be odd, a condition yielding the boundary condition

$$u_1 |_{r=r_s} = 0 \tag{6.11}$$

when combined with the translational condition of the Floquet–Bloch theorem

$$u_1(\mathbf{r}_s) = u_1(-\mathbf{r}_s) \tag{6.12}$$

where $|\mathbf{r}_s| = r_s$.

For the case $V = V(r)$, the Wigner–Seitz solution for $\psi_{\mathbf{k}}$ to first order in $\mathbf{k} \cdot \nabla u_{\mathbf{k}}$ can be summarized in terms of Bardeen's results:[101]

$$\psi_{\mathbf{k}} = [u_0 + i\mathbf{k} \cdot \mathbf{r}(v_1 - u_0)]e^{i\mathbf{k} \cdot \mathbf{r}} \tag{6.13}$$

where $u_0$ satisfies (6.3), with the boundary condition given by (6.10) and $v_1 (\equiv u_1/i\mathbf{k} \cdot \mathbf{r} + u_0)$.

$$\frac{d^2}{dr^2}(r^2 v_1) - 2v_1 + \frac{2m}{\hbar^2}(E_0 - V)r^2 v_1 = 0 \tag{6.14}$$

with the boundary condition

$$v_1(r) = u_0(r) \qquad \text{at} \quad r = r_s \tag{6.15}$$

In terms of this solution, Bardeen's formula for the effective mass in (6.7) is

$$\frac{m}{m^*} = \frac{4\pi}{3} r_s^3 [u_0(r_s)]^2 \left[ \frac{1}{rv_1} \frac{d(r^2 v_1)}{dr} - 1 \right]_{r=r_s} \tag{6.16}$$

$\psi_k$ is normalized in the Wigner–Seitz cell, that is,

$$\int_{r \leq r_s} |\psi_k|^2 \, d^3r = 1 \tag{6.17}$$

One of the assumptions of SJC is that, in computing the quasi-free electron states for a fluid, the fluid structure can be approximated by a lattice structure for which the Wigner–Seitz solution of the Schrödinger equation is valid. Such an assumption will make sense if either the phenomena under investigation is not sensitive to the long-range nature of a lattice structure, or the short-range order (nearest- and next-nearest-neighbor order, for example) present in dense fluids as well as in solids determines the structure-dependent part of the phenomena. The assumption cannot be given a priori justification at this time, so its validity must be tested by comparison of its results with experiment.

To estimate the electron-medium potential, SJC made use of the conclusions of pseudopotential theory. The short-range potential between an electron and an atom should in principle be taken to be the self-consistent field Hartree–Fock potential. In practice the Hartree–Fock potential is difficult and time-consuming to calculate. However, pseudopotential theory[104–106] uses the orthogonality restrictions, imposed between the core and excess electron states by the exclusion principle, to simplify the potential calculation greatly. The Hartree–Fock potential is replaced by a fictitious potential (pseudopotential) which projects the core states. The effect of the core-state projections, which tend to keep the wave function out of the core region, is (a) to add onto the large attractive Hartree–Fock potential a large repulsive term, and (b) to replace the excess electron wave function by a pseudo-wave function which is smooth in the core region and equal to the actual wave function outside the core region. In the theory of metals,[107,108] the pseudopotential has been adequately described as a constant negative potential in the ion core plus the Coulomb potential outside the core. This pseudopotential is of course not adequate for excess electrons in fluids of neutral atoms. From electron–atom scattering experiments and theory, it has been found that the core pseudopotential must be positive for inert gas atoms.[109,35] In fact, O'Malley[109] showed that observed low-energy electron–atom scattering cross sections for inert gas atoms can be accounted for by adding to the negative scattering length, arising from the polarization potential, a positive scattering length arising from the core pseudopotential, a result implying that the core pseudopotential is positive for inert gas atoms. Also, numerical calculations of the scattering of electrons from the Hartree–Fock fields of neon and argon (excluding polarization effects) yield positive scattering lengths[110] and an energy-independent scattering cross section $\hat{\sigma}$ where s-wave scattering is dominant.[111] SJC noted that the scattering results are consistent with scattering from a hard-core potential of the form

$$
\begin{aligned}
v_a &= +\infty && r < \tilde{a} \\
&= 0 && r > \tilde{a}
\end{aligned}
\tag{6.18}
$$

where the hard-core radius is given by

$$
\tilde{a} = \left( \frac{\tilde{\sigma}}{4\pi} \right)^{1/2}
\tag{6.19}
$$

That is, the hard-core radius is the Hartree–Fock scattering length. Thus they proposed (6.18) for the Hartree–Fock core pseudopotential.

Outside the atom core electron–atom interactions give rise to the polarization potential which can be approximated as

$$
\begin{aligned}
v_p(\mathbf{r} - \mathbf{r}_i) &= -\frac{\alpha e^2}{2|\mathbf{r} - \mathbf{r}_i|^4} && |\mathbf{r} - \mathbf{r}_i| > d \\
&= 0 && |\mathbf{r} - \mathbf{r}_i| < d
\end{aligned}
\tag{6.20}
$$

where $\alpha$ is the atomic polarizability, $\mathbf{r}$ is the position of the electron, and $\mathbf{r}_i$ is the position of the $i$th atom. $d$ is a cutoff parameter [of $\mathcal{O}\,(10^{-8}\ \text{cm})$] which can be chosen to fit the polarization potential in the core.[112,113] The polarization potential of an electron in the Wigner–Seitz sphere centered on an atom at $\mathbf{r}_0$ is:

$$
U_p = v_p(\mathbf{r} - \mathbf{r}_0) + \left\langle \sum_{i \neq 0} v_p(\mathbf{r} - \mathbf{r}_i) \right\rangle
\tag{6.21}
$$

where $\left\langle \sum_{i \neq 0} v_p(\mathbf{r} - \mathbf{r}_i) \right\rangle$ denotes the ensemble average of polarization potential between the electron and the atoms lying outside the reference Wigner–Seitz sphere. The ensemble average can be written in the form[110]

$$
\left\langle \sum_{i \neq 0} v_p(\mathbf{r} - \mathbf{r}_i) \right\rangle = n \int v^p(\mathbf{r} - \mathbf{r}')\delta(\mathbf{r} - \mathbf{r}')g(\mathbf{r} - \mathbf{r}')\, d^3r'
\tag{6.22}
$$

where $g$ is the pair correlation function of the fluid, and $\delta$ is the screening function which accounts for the fact that the dipole induced in a given atom by the electron is affected by the dipoles the electron induces in the surrounding atoms. In a theoretical work to be discussed in Section VI.B, Lekner[110] showed that $\vartheta$ in argon is close to the Lorentz local-field form

$$\vartheta(r) = 1 \qquad r < r_s$$
$$= \frac{1}{1 + (8\pi/3)n\alpha} \qquad r > r_s \tag{6.23}$$

SJC obtained

$$\left\langle \sum_{i \neq 0} v_p(\mathbf{r} - \mathbf{r}_i) \right\rangle = -\frac{3\alpha e^2}{2r_s^4}\left(1 + \frac{8\pi}{3}n\alpha\right)^{-1} \tag{6.24}$$

Finally, assuming the electron charge to be equally distributed in the range $\frac{1}{2}r_s < |\mathbf{r} - \mathbf{r}_0| < r_s$, SJC obtained the mean field estimate

$$v_p(\mathbf{r} - \mathbf{r}_0) \approx -\frac{\frac{1}{2}\alpha e^2 \int_{r_s/2}^{r_s}(d^3r/r^4)}{\int_{r_s/2}^{r_s} d^3r}$$

$$\approx -\frac{12\alpha e^2}{7r_s^4} \tag{6.25}$$

which when added to (6.24) yields their estimate of the electron–polarization interactions:

$$U_p = -3\frac{\alpha e^2}{r_s^4}\left[\tfrac{8}{7} + (1 + \tfrac{8}{3}\pi n\alpha)^{-1}\right] \tag{6.26}$$

Thus the final form of the SJC pseudopotential in the Wigner–Seitz cell centered at $\mathbf{r} = 0$ is

$$V(r) = v_a(r) + U_p \tag{6.27}$$

where $v_a(r)$ is given by (6.18) and $U_p$ by (6.26). The pseudo-wave function is obtained by solving (6.3) and (6.14) with this pseudopotential. The results are:

$$u_0 = \beta_0 \frac{\sin k_0(r - \tilde{a})}{r} \tag{6.28}$$

with $k_0$ determined by

$$\tan k_0(r_s - \tilde{a}) = k_0 r_s \tag{6.29}$$

$$v_1 = \frac{\beta_1}{r}[n_1(k_0\tilde{a})j_1(k_0 r) - j_1(k_0\tilde{a})n_1(k_0 r)] \tag{6.30}$$

where $j_1$ and $n_1$ are spherical Bessel and Neumann functions. $\beta_0$ and $\beta_1$ are normalization constants connected by the relation

$$\beta_0 \sin k_0(r_s - \tilde{a}) = \beta_1[n_1(k_0\tilde{a})j_1(k_0r_s) - j_1(k_0\tilde{a})n_1(k_0r_s)] \qquad (6.31)$$

Equation (6.31) and the normalization condition

$$\int_{r<r_s} |\psi_{\mathbf{k}}(\mathbf{r})|^2 \, d^3r = 1 \qquad (6.32)$$

where $\psi_{\mathbf{k}}$ is given by (6.13), determine $\beta_0$ and $\beta_1$. The ground-state energy is of the form

$$E_0 = \frac{\hbar^2 k_0{}^2}{2m} + U_p \equiv T_0 + U_p \qquad (6.33)$$

and the effective mass $m^*$ is determined from the formula

$$\frac{m}{m^*} = \frac{4\pi}{3} r_s{}^3 (k_0 r_s)[u_0(r_s)]^2 \left[ \frac{n_1(k_0\tilde{a})j_1'(k_0r_s) - j_1(k_0\tilde{a})n_1'(k_0r_s)}{n_1(k_0\tilde{a})j_1(k_0r_s) - j_1(k_0\tilde{a})n_1(k_0r_s)} \right] \qquad (6.34)$$

According to (6.33), the ground-state energy of the electron is just the sum of the mean field-polarization energy $U_p$ and the zero-point energy $T_0$ arising from the excluded volume restriction of the atom hard core in the Wigner–Seitz cell. Since injection of quasi-free electrons into a liquid should be accompanied by little perturbation to the fluid (as has also been assumed in the present model), the measured work function $V_0$ should be equal to the ground-state electron energy $E_0$. Estimating the hard-core scattering lengths for helium, neon, and argon from Hartree–Fock calculations,[111] SJC predicted the values of $E_0$ compared in Table XII with experimental values of $V_0$. The agreement between experiment and theory is remarkable in view of the sensitivity of $T_0$ to the estimated value of $\tilde{a}$ and considering the number of simplifying assumptions underlying the model.

Since Hartree–Fock scattering lengths are not available for the hydrocarbon molecules, predictions of $E_0$ cannot be made for these systems. However, partial tests of the theory can be made. Assume first that the Hartree–Fock pseudopotential of the nonspherical hydrocarbons can be represented by an angle-averaged spherical potential of the form of (6.18), where now $\frac{4}{3}\pi\tilde{a}^3$ represents the average volume excluded to the electron by the molecule in the Wigner–Seitz sphere. Next, assume that the electron–molecule polarization potential can be angle-averaged so that the $\alpha$ in $U_p$ is taken to be the isotropic polarizability of the molecule. Then by equating the experimental $V_0$ to the theoretical expression for $E_0$, one can determine the cutoff length $\tilde{a}$. Cutoff lengths have been determined for n-hexane, neopentane, and TMS. The value of 1.93 Å obtained for neopentane is very

TABLE XII

The Ground-State Energy of the Quasi-Free Electron State in Liquid Rare Gases and Hydrogen
(Ref. 35)

| | $r_s$ (Å) | $\alpha$ (Å$^3$) | $\tilde{a}$ (Å) | $k_0$ (Å$^{-1}$) | $T$ (eV) | $U_p$ (eV) | $V_0$, theory (eV) | $V_0$, experiment (eV) |
|---|---|---|---|---|---|---|---|---|
| $^4$He | 2.22 | 0.20 | 0.751 | 0.66 | 1.66 | $-0.341$ | $+1.32$ | $+1.05 \pm 0.05^a$ |
| | | | | | | | | $+1.3 \pm 0.02^b$ |
| Ne | 1.86 | 0.39 | 0.556 | 0.71 | 1.92 | $-1.17$ | $+0.75$ | $0.5^c$ — |
| Ar | 2.22 | 1.65 | 0.794 | 0.70 | 1.87 | $-2.50$ | $-0.63$ | $-0.33 \pm 0.05^d$ |
| $H_2$ | 2.2 | 0.790 | 1.056 | 0.72 | 2.00 | $-1.10$ | 0.90 | $-1.0^c$ |

$^a$ From Ref. 56.
$^b$ From Ref. 8.
$^c$ From Ref. 20.
$^d$ From Ref. 31.

close to the 2.22-Å scattering length required to fit the quasi-free mobility prediction (2.4) to an experiment for electrons in neopentane. This is consistent with Lekner's finding[28] that the scattering length in liquid argon is positive and is determined primarily by the Hartree–Fock core potential, polarization scattering largely canceling out with polarization simply lowering the bottom of the conduction band (by the amount $U_p$ in the present theory). His theoretically determined scattering length for liquid argon is 0.751 Å, while the Hartree–Fock value for $\tilde{a}$ is 0.794 Å. Estimating the structure factor $S(0) = nkT\kappa_T$ for TMS to be about the same as for neopentane (which is 0.06 at 296°K)[114] and using the experimental mobility of 90 cm$^2$/(V) (sec) in the quasi-free mobility formula (2.4), one estimates a scattering length of 2.2 Å for TMS. This result compares favorably with the value of 2 Å determined from the work function data.

Another indirect test of the theory of the ground-state energy is to compare values of $E_0$ predicted for mixtures with measured work functions. Brown and Davis[115] applied the theory to binary mixtures by representing a mixture by a pseudo-one-component fluid whose parameters are determined by Amagat's mixing law.[116] The density of a mixture according to Amagat's law (which is a fairly good approximation for dense fluids of similar molecules) is given by the expression

$$\frac{1}{n} = \frac{x_1}{n_1} + \frac{x_2}{n_2} \tag{6.35}$$

where $n_i$ and $x_i$ are the density of pure $i$ and the mole fraction of component $i$ of the mixture. According to (6.35), the molar volume of the mixture is the mole fraction weighted sum of the molar volumes of pure liquids 1 and 2. Analogously, we define the electron–molecule hard-core excluded volume of the pseudo-one-component fluid by the relation

$$\tfrac{4}{3}\pi a^3 = x_1 \tfrac{4}{3}\pi \tilde{a}_1{}^3 + x_2 \tfrac{4}{3}\pi \tilde{a}_2{}^3 \tag{6.36}$$

where $\tilde{a}_i$ denotes the electron–molecule hard-core cutoff length of the $i$th component. And the effective fluid polarizability (the polarizability being, roughly speaking, a measure of molecular volume) is defined by

$$\bar{\alpha} = x_1 \bar{\alpha}_1 + x_2 \bar{\alpha}_2 \tag{6.37}$$

Using the pure fluid data given in Table XIII to determine $n$, $a$, and $\bar{\alpha}$ from (6.35) to (6.37), and $r_s$ from (6.9), Brown and Davis predicted the work function

TABLE XIII

Data Used for Estimating Work Functions From Eq. (6.33) for Pseudo-One-Component Fluid Defined by Eqs. (6.35)–(6.37)

| Compound | $n$ ($10^{-3}$ Å$^{-3}$) | $\alpha$ (Å$^3$) | $V_0$ (eV) | $r_s$ (Å) | $\tilde{a}$ (Å) |
|---|---|---|---|---|---|
| $n$-Hexane | 4.53[a] | 11.8[b] | 0.10[e] | 3.73 | 2.17 |
| Neopentane | 5.11[a] | 10.0[b] | −0.35[e] | 3.60 | 1.93 |
| TMS | 4.38[c] | 12.8[d] | −0.60[e] | 3.79 | 2.04 |

[a] *Handbook of Chemistry and Physics*, R. C. Meast, ed., 53rd ed., Chemical Publishing, Chicago, 1972.

[b] J. O. Hirschfelder, C. F. Curtiss, and R. B. Bird, *Molecular Theory of Gases and Liquids*, Wiley, New York, 1954.

[c] Reference 86.

[d] A. P. Altshuller and L. Roseblum, *J. Am. Chem. Soc.*, **77**, 272 (1955).

[e] R. A. Holroyd and W. Tauchert, *J. Chem. Phys.*, **60**, 3715 (1974).

versus mole fraction for neopentane–$n$-hexane and TMS–$n$-hexane binary mixtures. As seen in Fig. 35, the model yields quite good predictions of the composition dependence of $V_0$ for the two mixtures considered. Of course, the fact that the parameters $\tilde{a}_1$ and $\tilde{a}_2$ were fitted to experiment in the pure fluid limits ensures that the theory fits at each end of the concentration range. Nevertheless, since the value of $V_0$ is a small difference of two large numbers, it is significant that the concentration dependence of the model agrees so well with observed behavior.

An interesting question is whether the average electron–molecule hard-core radius $\tilde{a}$ determined from $V_0$ is related to the average molecule–molecule

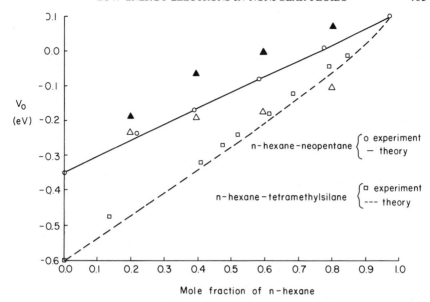

**Fig. 35** Comparison of predicted [from (34a) for pseudo-one-component fluid defined by (75) to (77)] and experimental work functions versus composition. Experimental values taken from R. A. Holroyd and W. Tauchert [*J. Chem. Phys.*, (to appear)]. For significance of open and filled triangles see Section IX.C.

hard-core radius. If they are related, then, by dimensional analysis, one would expect the ratio of $\tilde{a}$ to the cube root of the critical molar volume $V_c$ to be a universal constant for the hydrocarbon fluids. The ratios of "observed" $\tilde{a}$ [determined by equating measured $V_0$'s to $E_0$ in (6.33)] to $V_c^{1/3}$ are given in Table XIV. The ratios $\tilde{a}/V_c^{1/3}$ have a mean value of 0.305 Å/$(cm^3/mole)^{1/3}$ and are not widely dispersed about the mean. The last column in Table XIV contains the values of $\tilde{a}$ determined from the mean value of $\gamma$. They agree quite well with the "observed" $\tilde{a}$ values. For comparison, the molecule–molecule, hard-core radius can be estimated from the van der Waals model, according to which $V_c = 3N_0 b = 12N_0[\frac{4}{3}\pi a_{vdW}^3]$. Thus $a_{vdW}/V_c^{1/3} = 0.318$ Å/$(cm^3/mole)^{1/3}$, a value remarkably close to the average value of $\tilde{a}/V_c^{1/3}$.

Effective masses of the electrons can be computed from (6.34). Values predicted for the ratio of the effective to the free-electron mass are given in Table XV. The values of $\tilde{a}$ used for $^4$He and one of the entries for argon were those obtained theoretically from the Hartree–Fock field.[35,111] The methane $\tilde{a}$ was estimated by fitting measured mobility to (2.4), For the hydrocarbons

TABLE XIV

Correlation of Work Function–Determined Electron-molecule Hard-Core radii $a$, with the Critical Volumes $V_c$[a]

| Compound | Observed work functions, $V_0$ (eV) | Observed radii, $a$ (Å) | $a/V_c^{1/3}$ [Å/(cm$^3$/mole)$^{1/3}$] | $\tilde{a} = 0.305\, V_c^{1/3}$ (Å) |
|---|---|---|---|---|
| $n$-Pentane | $-0.01$[b] | 2.03 | 0.301 | 2.05 |
| Neopentane | $-0.43$[b] | 1.92 | 0.287 | 2.04 |
| Cyclopentane | $-0.28$[b] | 1.85 | 0.292 | 1.94 |
| $n$-Hexane | $0.04$[b] | 2.17 | 0.304 | 2.17 |
| 2,2-Dimethylbutane | $-0.15$[c] | 2.13 | 0.300 | 2.16 |
| 2-Methylpentane | $0.01$[d] | 2.16 | 0.303 | 2.17 |
| Benzene | $-0.14$[c] | 1.97 | 0.308 | 1.94 |
| Toluene | $-0.22$[c] | 2.14 | 0.313 | 2.08 |
| Methylcyclohexane | $0.14$[d] | 2.23 | 0.326 | 2.08 |
| 2,2,4-Trimethylpentane | $-0.18$[b] | 2.36 | 0.305 | 2.39 |
| $n$-Decane | $0.22$[d] | 2.68 | 0.317 | 2.57 |

[a] The work function data were collected at room temperature ($\sim 300°$K) and at the saturated vapor pressure of the fluid being studied.

[b] From Ref. 5.

[c] From Ref. 53.

[d] From Ref. 88.

the second entry for argon, krypton, and xenon the values for $\tilde{a}$ were determined by fitting theoretical $E_0$ values to measured work functions. The effective electron masses predicted for inert gas liquids and methane are nearly the free-electron mass (only 15 to 30% higher). However, in neopentane and $n$-hexane the effective masses are predicted to be substantially higher (three- to fourfold) than the free-electron value. Exciton spectral studies[20,117] indicate experimental values of $m^*/m \approx \frac{1}{2}$ in liquids argon, krypton, and xenon. Thus the SJC theory seems to overestimate the effective mass of quasi-free electrons systematically.

A final interesting feature of the SJC theory is that the values of $\tilde{a}$ to fit $E_0$ to experimental $V_0$ values for liquid argon, krypton, and xenon are 0.88, 1.00, and 1.19, respectively, and for solids are the similar values 0.895, 1.03, and 1.20 Å, where the liquids values used for $V_0$ are given in Table I and the solid values used were $-0.49$, $-0.61$, and 0.6 eV. In considerable disagreement with these solid values for $V_0$ (derived from spectroscopic data and presented in Ref. 20), Schwentner and coworkers (N. Schwentner, F. J. Himpsel, V. Saile, M. Skibowski, W. Steinmann and E. E. Koch, *Phys. Rev. Lett.* **34**, 528 (1975)) have determined $V_0$ directly, from photoemission

# LOW-ENERGY ELECTRONS IN NONPOLAR FLUIDS 405

TABLE XV

Quasi-Free Electron Effective Masses $m^*$ and Normalization Constants $\beta_0, \beta_1$ Determined from the Springett–Jortner–Cohen theory (Ref. 35)

| Liquid | $r_s$ (Å) | $\tilde{a}$ (Å) | $k_0$ (Å$^{-1}$) | $\beta_0$ (Å$^{-1/2}$) | $\beta_1$ (Å$^{-1/2}$) | $\dfrac{m^*}{m}$ |
|---|---|---|---|---|---|---|
| $^4$He | 2.22 | 0.751 | 0.66 | 0.425 | −0.211 | 1.229 |
| Ne | 1.86 | 0.556 | 0.71 | 0.483 | −0.159 | 1.163 |
| Ar | 2.22 | 0.794 | 0.70 | 0.414 | −0.253 | 1.291 |
| Ar$^b$ | 2.22 | 0.895 | 0.802 | 0.385 | −0.373 | 1.503 |
| Kr$^b$ | 2.39 | 1.00 | 0.775 | 0.350 | −0.527 | 1.812 |
| Xe$^b$ | 2.54 | 1.19 | 0.836 | 0.323 | −0.559 | 1.915 |
| CH$_4$$^c$ | 2.51 | 0.85 | 0.58 | 0.402 | −0.198 | 1.216 |
| Neopentane$^b$ | 3.60 | 1.92 | 0.714 | 0.236 | −0.723 | 2.750 |
| $n$-Hexane$^b$ | 3.73 | 2.17 | 0.803 | 0.211 | −0.991 | 3.688 |

$^a$ The quantities are given for electron energies near the bottom of the conduction band (i.e., $kr_s \ll 1$).

$^b$ Value of $\tilde{a}$ chosen to fit ground-state energy $E_0$ to experimental work functions. $V_{0(exp)} = -0.33, -0.78, -0.63, -0.40$ and $0.10$ eV for Ar, Kr, Xe, neopentane, and $n$-hexane, respectively.

$^c$ Value of $\tilde{a}$ chosen to fit mobility formula [see Eq. (2.4)], to $\mu_{exp} = 400$ cm$^2$/(V)(sec).

studies, to be $+0.4$, $-0.3$ and $-0.5$ eV, for argon, krypton and xenon, respectively.

## B. Lekner's Effective-Potential Model

The electron–medium potential model used by Lekner[110] to compute the ground-state energy, the effective mass, and the scattering cross sections of quasi-free electrons in liquid argon is substantially more rigorous than the SJC model. He uses the known Hartree potential[118] to account for the short-range electron–atom interactions, accounts for exchange and correlation effects through the polarization potential

$$v_p(r) = -\frac{\alpha e^2}{2(r^2 + r_\alpha^2)^2} \tag{6.38}$$

and introduces a self-consistent method for calculating the many-body screening of the polarization potential. The quantity $r_\alpha$ is adjusted so that the predicted zero-energy electron–argon atom scattering length $a$ agrees with the gas-phase experimental value $-1.5a_0$, where $a_0$ ($=0.528$ Å) is the Bohr radius. The value required of $r_\alpha$ for argon is:

$$r_\alpha = 1.22a_0 = 0.65 \text{ Å} \tag{6.39}$$

An electron at a large distance $r$ from an isolated atom of polarizability $\alpha$ induces on the atom a dipole of strength $\alpha e/r^2$. This dipole in turn interacts with the electron with a force of magnitude $2\alpha e^2/r^5$, so that the long-range polarization potential of the electron-isolated atom pair is $-\alpha e^2/2r^4$. At short distances the effects of exchange dominate the correlation effects accounted for by $-\alpha e^2/2r^4$. This fact is adjusted for by the parameter $r_\alpha$ in the modified polarization potential of (6.38).

Consider now the problem of computing many-body screening of the polarization potential in a condensed medium. Lekner noted that, since

$$v_A \ll v_e \ll v_{Ae} \qquad (6.40)$$

where $v_A$, $v_e$, and $v_{Ae}$ are velocities of atoms, quasi-free electrons, and atomic (bound) electrons, the motion of the atoms can be ignored in calculating the mutual screening effect of neighboring atoms, and the motion of the free electrons can be ignored in calculating the induced polarizations of the atomic electrons.

In the absence of neighboring atoms, atom 1 at a distance $r_1$ from a point charge $-e$ experiences an electric field $e/r_1{}^2$ directed along $\mathbf{r}_1$. In the presence of other atoms, the average local field experienced by the subject atom is

$$\frac{e}{r_1{}^2}\, \mathcal{A}(r_1)\hat{\mathbf{r}}_1 \qquad (6.41)$$

where $\mathcal{A}(r_1)$ is a screening function defined to account for the effect of the field of the induced dipoles of the other atoms. The averaged local field on atom 1 is along $\hat{\mathbf{r}}_1(\equiv\mathbf{r}_1/r_1)$, and $\mathcal{A}(r_1)$ is a function only of the magnitude of $\mathbf{r}_1$ since the fluid is isotropic. The field given by (6.41) is equal to the direct field $e/r_1{}^2$ plus the contribution to the field at $\mathbf{r}_1$ arising from the induced dipoles of the surrounding atoms. The average field on a neighboring atom at $\mathbf{r}_2$ is $(e/r_2{}^2)\mathcal{A}(r_2)\hat{\mathbf{r}}_2$ and induces a dipole of average strength $\alpha(e/r_2{}^2)\mathcal{A}(r_2)$. This dipole contributes an electric field at position $\mathbf{r}_1$. Multiplying this field (computed under the assumption that the dipole behaves as a point dipole) by $ng(r_{21})\, d^3r_2$, the probable number of atoms lying between $\mathbf{r}_{21}$ and $\mathbf{r}_{21} + d\mathbf{r}_2$ from atom 1 and integrating $d^3r_2$ over all possible positions (i.e., over the volume of the system), one obtains the induced part of the field to add to $e^2/r^2$ to obtain the total field equal to (6.41). The resulting equation for $\mathcal{A}(r)$ is

$$\mathcal{A}(r) = 1 - \pi n\alpha \int_0^\infty \frac{ds}{s^2}\, g(s) \int_{r-s}^{r+s} \frac{dt}{t^2}\, \mathcal{A}(t)\chi(r, s, t) \qquad (6.42)$$

where

$$\chi(r, s, t) = \frac{3}{2s^2}(s^2 + t^2 - r^2)(s^2 + r^2 - t^2) + (r^2 + t^2 - s^2) \quad (6.43)$$

The quantity $g(r_{21})$ is the pair correlation function of the fluid, and $n$ is the atomic density of the fluid.

For a structureless fluid the solution to (6.43) is the Lorentz local field result

$$\mathscr{J}_L = [1 + \tfrac{8}{3}\pi n\alpha]^{-1} \quad (6.44)$$

To compute $\mathscr{J}(r)$ for a real fluid one needs the pair correlation function $g(r)$. Lekner used the analytical result obtained from the Percus–Yevick theory[119] for hard-sphere systems. He showed that, by suitable adjustment of the hard-sphere diameter $d$ entering the theory, the x-ray scattering structure function $S(K)$ of (2.5) was predicted accurately over the whole range of $K$ for liquid argon. (The value chosen for $d$ was 3.44 Å for argon at 84°K.) Several perturbation schemes[120,121] of computing $g$ have been developed since Lekner's original work, which support the validity of using the diameter-adjusted Percus–Yevick hard-sphere theory for $g(r)$, for $r_{in} < r < \infty$ (where $r_{in}$ is the position of the minimum in the pair potential of the atoms). The error made in using the Percus–Yevick result in the region $0 < r < r_{in}$ in the integrand of (6.42) is not very large, because $g(r)$ goes to zero rapidly for $r$ only 15 to 20% less than $r_{in}$.

Lekner's result for $\mathscr{J}(r)$ for liquid argon at 84°K is shown in Fig. 36. Interestingly, $\mathscr{J}(r)$ is approximated fairly well by the Lorentz value $\mathscr{J}_L$ for distances $r$ greater than the nearest-neighbor separation of the fluid atoms.

Taking into account screening, one can express the polarization potential between an electron and an atom in the fluid in the form

$$\frac{1}{2}\left[\frac{\alpha e^2}{r^2}\mathscr{J}(r)\right]\left[-\frac{e}{r^2}\right] = \frac{\alpha e^2}{2r^4}\mathscr{J}(r) \quad (6.45)$$

if it is assumed that the polarizability is independent of $r$ [as has been assumed in deriving (6.42) for $\mathscr{J}$]. This assumption would break down sufficiently close to the nucleus (for $r \lesssim 2r_\alpha$ according to Lekner's conclusion), but there the Coulomb potential dominates the exchange and correlation interactions. Thus Lekner assumed $\mathscr{J}$ to be adequately given by (6.42) and took the electron-atom screened potential to be of the form

$$v_P{}^{\mathscr{J}}(r) = \frac{\alpha e^2 \mathscr{J}(r)}{2(r^2 + r_\alpha{}^2)^2} \quad (6.46)$$

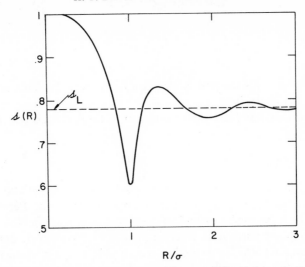

**Fig. 36** Local-field function $\mathscr{A}(R)$ due to point charge calculated from the Percus–Yevick hard-core pair correlation function corresponding to liquid argon density at the triple point and a hard-core diameter $\delta = 3.44$ Å. (Ref. 28).

where the exchange parameter is determined, as mentioned above, from the gas-phase zero-energy electron–atom scattering length. The total potential between an electron and an atom in the fluid is, according to Lekner's model,

$$v_1(r) = v_H(r) + v_P{}^{\circ}(r) \tag{6.47}$$

where $v_H$ is the Hartree potential.

The total potential energy of an electron at position $\mathbf{r}'$ near the $i$th atom at $\mathbf{r}_i$ is

$$V(r) = v_1(\mathbf{r} - \mathbf{r}_i) + \sum_{j \neq i}^{N} v_1(\mathbf{r} - \mathbf{r}_i) \tag{6.48}$$

For an electron in the Wigner–Seitz sphere centered on atom $i$, Lekner approximates the quantity $\sum_{j \neq i}^{N} v(|\mathbf{r} - \mathbf{r}_i|)$ by the ensemble average value

$$v_2(\mathbf{r} - \mathbf{r}_i) = \left\langle \sum_{j \neq i}^{N} v(\mathbf{r} - \mathbf{r}_j) \right\rangle' \tag{6.49}$$

where the prime indicates that the average is taken over all atoms except $i$. Thus the potential of an electron near atom $i$ (i.e., roughly speaking within the Wigner–Seitz sphere on $i$) is approximated by $\tilde{V}$, where

$$\tilde{V}(\mathbf{r} - \mathbf{r}_i) \equiv v_1(\mathbf{r} - \mathbf{r}_i) + v_2(\mathbf{r} - \mathbf{r}_i) \tag{6.50}$$

The exact domain of validity of (6.50) is not yet specified. In a Wigner–Seitz average lattice representation of the fluid, the potential is fully specified if it is assumed that

$$V = \tilde{V}(r) \qquad 0 < r \leq r_s \tag{6.51}$$

where $r$ is the distance of the electron from the center of any atom. The potential is then periodically extended over the average lattice. Alternatively, the potential about any atom can be assumed to be

$$V = \tilde{V}(r) \qquad 0 < r < r_m$$
$$= \tilde{V}(r_m) \equiv \tilde{V}_m \qquad r_m < r < r_s \tag{6.52}$$

where $r_m$ is obtained from the condition

$$\frac{\partial \tilde{V}}{\partial r}(r) = 0 \qquad \text{at } r = r_m \tag{6.53}$$

Equation (6.52) is also periodically extended. This second choice, which is the one used by Lekner, has the advantage that the electron-atom forces are continuous as the electron leaves the force field on one atom and enters that of another. This is more acceptable physically than the discontinuous-force field introduced by (6.51).

It turns out, however, that $r_m$ and $r_s$ are almost equal, as shown in Fig. 37 in which is shown Lekner's computations of $v_1(r)$, $v_2(r)$, and $\tilde{V}(r)$. The level line $\tilde{V}_m$ shown in the plot represents the maximum value of $\tilde{V}(r)$, that is, $\tilde{V}_m \equiv \tilde{V}(r_m)$. The quantity $v_2(r)$ was computed from the formula

$$v_2(r) = \frac{2\pi n}{r} \int_0^\infty ds\, sg(s) \int_{|r-s|}^{r+s} dt\, tv_1(t) \tag{6.54}$$

which is obtained by carrying out the integrations in (6.49) with the aid of the bipolar coordinate system. The quantities shown in Fig. 37 were obtained using the theoretical Hartree potential[122] for $v_H$ for argon [see (6.46)] for $v_p{}^\delta$, and the Percus–Yevick hard-sphere $g(r)$ with the hard-sphere diameter 3.44 Å. For $\alpha$ the argon value 1.65 Å$^3$ was used. The density used was that of saturated liquid argon at 84°K, namely, 0.02123 Å$^{-3}$.

Bardeen's Wigner–Seitz solution of the one-electron Schrödinger equation was outlined in Section VI.A. For the potential given by (6.52), the ground-state $u_0$ and first excited wave functions are obtained from (6.3) and (6.14) with the boundary conditions (6.10) and (6.15). The effective mass is determined by (6.16). For saturated liquid argon at 84°K, Lekner solved the Wigner–Seitz equations for the potential $\tilde{V}$ given by (6.52). He found

$$E_0 = -0.46 \text{ eV} \tag{6.55}$$

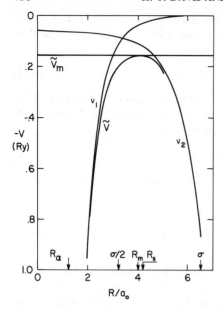

**Fig. 37** Electron–atom potentials in liquid argon. $v_1$, $v_2$, $\tilde{V}$ and $\tilde{V}_m$ are defined by (6.47), (6.49), (6.50), and (6.52), respectively.

and

$$m^* = 0.43 \text{ m} \qquad (6.56)$$

This value for $E_0$ is in good agreement with the values $-0.33 \pm 0.1$ and $-0.45 \pm 0.2$ eV determined for the work function by photoinjection and spectroscopic methods, respectively (see Table VII). As mentioned in Section IV.A, effective masses in liquids and solid argon have been estimated from spectral studies to be about 0.5 m. Thus, unlike the SJC model, Lekner's potential model yields the correct effective mass for argon.

### C. Construction of a Pseudopotential

In the simple model of Section VI.A a hard-sphere cutoff pseudopotential was introduced to account for the electron–molecule interactions. In a rigorous pseudopotential treatment of electron–molecule interactions, the pseudopotential must be well defined and must be constructed in a consistent fashion from the actual electron–molecule interaction potential. The one-electron self-consistent field wave function of an excess electron–molecule system is of the form

$$(T + V)\psi = E\psi \qquad (6.57)$$

where $T = -(\hbar^2/2m)\nabla_r^2$ is the kinetic energy operator of the electron, and $V$ is the electron–molecule self-consistent field interaction.

In an atom or molecule the large negative potential energy $V$ of an electron in the core region of an atom and the large kinetic energy characteristic of the oscillations of $\psi$ in that region largely cancel out. The oscillations of $\psi$ arise from the restrictions of the Pauli principle and the requirement that the excess electron wave function be orthogonal to the core electron wave functions. In solid-state physics it has proved useful[104–106,123] to introduce the concept of a pseudopotential and a pseudo-wave function $\phi$ in which the core oscillations have been removed. The SJC model treated in Section IV.A exploits the ideas inherent to the pseudopotential approach. In this section is described the more rigorous approach of Kestner, Jortner, Cohen and Rice (KJCR),[124] which must provide the basis of such simplified models as that of SJC.

In the pseudopotential method one solves instead of (6.57), the equation

$$(T + V + V_R)\phi = E\phi \tag{6.58}$$

where $V_R$ is a nonlocal potential introduced to project the oscillations of $\psi$ in the atom cores, and $\phi$ is a pseudo-wave function which is equal to $\psi$ at large distances from the core, but inside the core has the oscillations of $\psi$ removed. The sum $V + V_R$ defines the pseudopotential

$$V_{ps} = V + V_R \tag{6.59}$$

There is a degree of arbitrariness in the choice of the pseudo-wave function and/or pseudopotential.[125,126] Austin, Heine, and Sham[127] showed that if $V_R$ is of the form

$$V_R\phi = \sum_c \chi_c \langle \chi_c | F | \phi \rangle \tag{6.60}$$

where $\chi_c$ are the core orbitals and $F$ is an arbitrary operator, then (6.58) has the same eigenvalues as (6.57) and for a given eigenvalue the eigenfunctions $\psi$ and $\phi$ may be related by the expression

$$\psi = \phi - \sum_c b_c \chi_c \tag{6.61}$$

where $b_c$ obeys, owing to the orthogonality condition $(\psi, \chi_c) = 0$, the relation

$$b_c = (\chi_c, \phi) \tag{6.62}$$

Thus, choosing a convenient operator $V_R$, one determines $\phi$ from (6.58) and constructs $\psi$ from (6.61) and (6.62).

Cohen and Heine[125] showed that the smoothest pseudo-wave functions (in the sense of having the smallest kinetic energy) are obtained if $F$ is chosen such that

$$V_R\phi = -\sum_c \chi_c \langle \chi_c | V - \bar{V} | \phi \rangle \tag{6.63}$$

where

$$\overline{V} \equiv \frac{\langle \phi | V + V_R | \phi \rangle}{\langle \phi | \phi \rangle} \tag{6.64}$$

The similar Austin, Heine, and Sham[127] choice of $V_R$, denoted herein as $V_R{}^A$, and defined by the expression

$$V_R{}^A \phi = - \sum_c \chi_c \langle \chi_c | V | \phi \rangle \tag{6.65}$$

enjoys the advantage of simplicity over the Cohen–Heine choice. And, since $\overline{V}$ is small as a result cancelation of $V$ and $V_R$, the pseudo-wave function determined by $V_R{}^A$ is also smooth compared to $\psi$.

The object of introducing the pseudopotential problem is to obtain certain desired information by solving a simpler problem than that posed by (6.57). For example, since $\psi$ and $\phi$ are identical far from the atom (molecule), $\phi$ suffices to describe electron–atom (molecule) scattering amplitudes. Similarly, since the eigenvalues of the pseudopotential Schrödinger equation are identical to those of the self-consistent-field one-electron Schrödinger equation, the energy levels can be computed from the pseudopotential equation.

In their calculation of a pseudopotential for helium, KJCR used the form $V_R{}^A$ for $V_R$. The potential $V$ was constructed from the sum of the nuclear, Coulomb, and exchange potentials arising from distortion of the atom by the electronic charge. Thus

$$V = V_A + V_p$$

$$= - 2\frac{e^2}{r_3} + 2e^2 \int \frac{\chi_c{}^2(r_1)}{r_{13}} d^3 r_1 - e^2 \int \frac{\chi_c(r_1)\chi_c(r_3)}{r_{13}} d^3 r_1 \mathscr{P}_{13} + V_p(r_3) \tag{6.66}$$

where $r_3$ is the distance of the excess electron from the nucleus, $r_{13}$ is the distance between the excess and a core electron, $V_p(r_3)$ is the polarization potential, and $\mathscr{P}_{13}$ is an operator which permutes electrons 1 and 3 in functions to its right. The core wave function used was the Bagus–Gilbert[128] nominal basis set

$$\chi_c = 0.18159 \varphi_{1S}(2.906) + 0.84289 \varphi_{1S}(1.453) \tag{6.67}$$

where $\varphi_{1S}(z)$ is a normalized Slater $1S$ function with orbital exponent $z$. For the polarization potential, KJCR used the adiabatic approximation and the assumption that each core electron polarizes independently as if it were in the field of an effective nuclear charge $z_{\text{eff}}$. Accordingly, their potential was taken to be of the form

$$V_p = V_{p,0} + c_1 V_{p,1} + V_{p,2} \tag{6.68}$$

where the $l = 0$ and $l = 1$ components, $v_{p,0}$ and $v_{p,1}$, computed by Reeh[129] and Bethe,[130] respectively, are complete in the sense that core penetration of excess electrons is included in their determination. The component $v_{p,2}$ used is the nonpenetration form given by LeBahn and Callaway.[131] The explicit forms of the components of $v_p$ are

$$V_{p,1} = -\frac{4.5}{x^4}\left[1 - \frac{e^{-2x}}{3}\left(1 + 2x + 6x^2 + \frac{20}{3}x^3 + \frac{4}{3}x^4\right) - \tfrac{2}{3}e^{-4x}(1 + x)^4\right]$$

(6.69)

$$V_{p,0} = -\frac{15}{x^6}\left[1 - e^{-2x}\left(1 + 2x + 2x^2 \right.\right.$$
$$\left.\left. + \frac{4}{3}x^3 + \frac{2}{3}x^4 + \frac{4}{15}x^5 + \frac{4}{45}x^6 + \frac{4}{225}x^7\right)\right]$$

(6.70)

$$V_{p,2} = 2e^{-2x}\left[-\frac{2}{x^2} + \frac{1}{2x} + \frac{5}{2} + x + 2(\ln 2 + c)\left(\frac{1}{x^2} + \frac{1}{x}\right)\right.$$
$$+ Ei(-2x)\left(1 - \frac{1}{x^2}\right) + \ln x\left(-2x + 1 + \frac{2}{x} + \frac{1}{x^2}\right)$$
$$\left. - \alpha\left(\frac{2}{x} + 4 + 4x\right) + \beta(2 + 4x)\right] + \frac{4\alpha}{x} + 2\left(1 - \frac{1}{x}\right)Ei(-2x)$$
$$+ 2e^{-4x}\left[\frac{2}{x^2} + \frac{7}{2x} + \frac{5}{2} + x - \overline{Ei}(2x)\left(\frac{1}{x^2} + \frac{3}{x} + 4 + 2x\right)\right.$$
$$\left. + \left(\frac{1}{x^2} + \frac{2}{x} + 1\right)\ln x\right]$$

(6.71)

where

$$\alpha = -(1 + 1/x)e^{-2x}(\ln 2 + c + \tfrac{1}{2}\ln x) + \tfrac{1}{2}(1/x - 1)Ei(-2x)$$
$$+ e^{-2x}(1/4x + \tfrac{1}{2} - \tfrac{1}{2}x),$$
$$\beta = (1 + 1/x)[e^{-2x}(\tfrac{1}{2}Ei(2x) - \tfrac{1}{2}\ln x - \ln 2 - c) - \tfrac{1}{4}]$$
$$+ \tfrac{1}{2}(1/x - 1)[(Ei(-2x) - \ln x) + e^{-2x}(1/4x + \tfrac{1}{2} - \tfrac{1}{2}x)],$$

and

$$c = 0.577215665 \cdots = \text{Euler's constant},$$

$$Ei(-x) = -\int_x^\infty \frac{e^{-t}}{t}\,dt, \qquad \overline{Ei}(x) = \mathscr{P}\int_{-\infty}^x \frac{e^t}{t}\,dt$$

$x = 1.6875r_3/a_0$ and the $v_{p,l}$ are given atomic units. The quantity $c_1$ was set equal to 1.1219 so that (6.68) has the correct value $\alpha e^2/2r_3^4$ at large distances.

For low-energy electrons it is expected that the pseudo-wave function will vary slowly over the atomic core region, so that $V_R^A$ may be approximated by a local potential $V_R^L$, that is,

$$V_R^A \phi \simeq - \sum_c \chi_c \langle \chi_c | V \rangle \phi \equiv V_R^L \phi \qquad (6.72)$$

Similarly, the exchange term may be approximated by a local potential

$$\int \frac{\chi(r_1)\chi(r_3)}{r_{13}} d^3r_1 \, \mathscr{P}_{13} \phi(r_3) \simeq \left[ \int \frac{\chi(r_1)\chi(r_3)}{r_{13}} d^3r_1 \right] \phi(r_3) \qquad (6.73)$$

With the approximations of (6.72) and (6.73), KJCR obtained a local pseudo-potential of the form

$$V_{ps}^L(r) = V^L(r) + V_R^L(r) \qquad (6.74)$$

$$V^L(r) = V_A^L(r) + V_p(r) \qquad (6.75)$$

$$V_A^L = -6.116632 \frac{e^{-\Lambda r}}{r} + 0.535625 \frac{e^{-2\Lambda r}}{r} + 3.3831894 \frac{e^{-3\Lambda r}}{r}$$

$$+ 0.1978492 \frac{e^{-2\Lambda r}}{r} + 2.2064609 e^{-2\Lambda r} + 2.026883 \frac{e^{-3\Lambda r}}{r}$$

$$+ 0.191650 e^{-4\Lambda r} \qquad (6.76)$$

$$V_R^L = 5.6346 e^{-\Lambda r} + 3.434869 e^{-2\Lambda r} + 1.107442 e^{-1.6875r} \qquad (6.77)$$

where $\Lambda = 1.453$. $V_A^L$, $V_R^L$, and $r$ are expressed in atomic units (6.76) and (6.77).

The potential $V_R^L$ is strongly repulsive, so that the local pseudopotential is a strongly repulsive potential at distances greater than the Bohr radius. In Fig. 38 the local pseudopotential is compared with the local potential [i.e., $V$ with the approximation of (6.73)]. Low-energy electrons are excluded from the core region by the strongly repulsive part of $V_{ps}$, thus justifying to some extent the SJC model in which the pseudopotential is represented by a hard-core cutoff.

The zero-energy gas-phase scattering length predicted by $V_{ps}^L$ is $1.193a_0$, compared to the experimental value[109] $1.19a_0$. In the energy region from 0 to about 0.2 eV, KJCR found that $V_{ps}^L$ predicts a sudden increase in the scattering cross section in the region 0 to 0.05 eV followed by a region 0.05 to 0.2 eV (the upper limit of their investigations) of constant cross section. This behavior seems to be qualitatively correct, although the magnitude of the predicted cross section in the region of constancy is about 30% higher than the experimental one.[132]

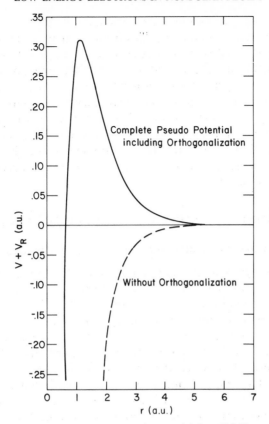

**Fig. 38** The local electron–helium atom pseudopotential [see (6.74)] versus separation $r$. The dashed curve represents the local electron–atom potential [(6.6) plus the approximation of (6.73)]. (Ref. 124).

As a refinement of the SJC calculation of the ground-state energy of quasi-free electrons in liquid helium, one could solve the Wigner–Seitz equation (6.3) with the pseudopotential given by (6.74) (the polarization effect on $V_{ps}^L$ in helium is so small that the many-body correction to the polarization part need not be considered). The Wigner–Seitz equation has not been solved for $V_{ps}^L$, although as a crude estimate, the Born approximation for the plane-wave electron $\phi(\mathbf{r}) \approx e^{i\mathbf{k}\cdot\mathbf{r}}/\Omega^{1/2}$ at low energies $k \approx 0$ leads to the estimate

$$E_0 \approx \langle \phi | V_{ps} | \phi \rangle \frac{N}{\Omega} \int V_{ps}\, d^3r = \frac{N}{\Omega} \int V_{ps}^L\, d^3r = 1.26 \text{ eV} \qquad (6.78)$$

for saturated liquid helium at 1 atm. The quantity $\Omega$ is the volume of the system, and $N$ is the number of atoms in the system. In the special case of constant $\phi$, used in (6.78), the matrix element $\langle \phi | V_{ps} | \phi \rangle$ of the full nonlocal pseudopotential is the same as that of $V_{ps}^L$. The agreement between the Born approximation to $E_0$ and the experimental value (between 1 and 1.3 eV) is fortuitously good, for the zero-point kinetic energy arising from the multiple scattering of the quasi-free electrons has been neglected by letting $\phi$ be a plane wave. However, the agreement does serve to give one some confidence in the applicability of the KJCR pseudopotential for describing excess electrons in liquid helium, or other liquids if the appropriate pseudopotentials have been constructed.

## VII. KINETIC THEORY OF QUASI-FREE ELECTRONS

### A. Boltzmann Equation for Single-Scatterer Approximation

As has been seen in previous sections, electron mobilities in argon, krypton, and xenon solids, of the order of 1000 cm$^2$/(V)(sec), are more than $10^5$ times ionic mobilities in the same systems. The electron mobilities are 25 to 100 times the simple kinetic theory formula $\mu = e/(mkT)^{1/2}n4\pi a^2$ in which it is assumed that the only differences between diffusing electrons and the ions are their different masses and cross sections. Schnyders, Rice, and Meyer[25,26] correctly inserted the structure factor [see (2.5)] into the denominator of the simple formula to obtain a result similar to (2.4). The structure factor accounts for diffraction (i.e., coherent scattering) effects ignored in the classical formula. A satisfying theoretical description of the magnitude, field dependence, and temperature and density maxima of quasi-free electron mobilities in condensed inert gases has been generated in terms of the Boltzmann equation derived by Cohen and Lekner.[27] In this section, a version of their theory generalized to polyatomic substances by Davis, Schmidt, and Minday[55] is described.

The assumptions of the theory are that (1) a Boltzmann equation is valid, and (2) the quasi-free electron can be treated as a scattered plane wave whose total scattering amplitude at a point in the system is the coherent sum of amplitudes scattered singly from individual scattering centers. Under the single-scatterer approximation, the scattering amplitude $F_{l/i}$ for scattering by a many-molecule system of a plane-wave electron from the momentum state $\hbar k$ through a solid angle $\Omega$ fixed in the element of solid angle $d\Omega = \sin\theta\, d\theta\, d\sigma$ and into the momentum state $\hbar k'$ is of the form

$$F_{l/i} = \sum_{v=1}^{N} F_{l_v/i_v}\, e^{i\mathbf{K}\cdot\mathbf{r}_v} \tag{7.1}$$

where $F_{l_v/i_v}$ is the scattering amplitude of the $v$th scattering center which goes from state $i_v$ to $l_v$ in scattering the plane-wave electron from state $\mathbf{k}$ into state $\mathbf{k}'$. The momentum vector $\mathbf{K}$ is by definition $\mathbf{K} = \mathbf{k} - \mathbf{k}'$. The cross section per scatterer per unit solid angle for the $i \rightarrow l$ transition is

$$\left(\frac{d\sigma}{d\Omega}\right)_{l/i} = \frac{k'}{kN}|F_{l/i}|^2 \tag{7.2}$$

so that the total cross section per scatterer per unit solid angle arising from energy-conserving transitions of a system initially in state $i$ is

$$\left(\frac{d\sigma}{d\Omega}\right)_i = \sum_l \int d(\hbar\omega)\,\delta(E_i - E_l - \hbar\omega)\left(\frac{d\sigma}{d\Omega}\right)_{l/i} \tag{7.3}$$

Here $\hbar\omega(=\hbar^2 k^2/2m - \hbar^2 k'^2/2m)$ is the energy lost by the electron to the system. From (7.3) it follows that the differential cross section corresponding to the state $i$ of the system is

$$\left(\frac{d^2\sigma}{d\Omega\,d\hbar\omega}\right)_i = \sum_l \delta(E_i - E_l - \hbar\omega)\left(\frac{d\sigma}{d\Omega}\right)_{l/i}$$

$$\equiv (2\pi\hbar)^{-1} \sum_l \int_{-\infty}^{\infty} dt\, e^{-i(\hbar\omega - E_i + E_l)t/\hbar}\left(\frac{d\sigma}{d\Omega}\right)_{l/i} \tag{7.4}$$

where the second equality arises from a representation of the Dirac delta function. If $H_0$ denotes the Hamiltonian of the system (excluding the electron) and $|j\rangle$ the eigenfunction of $H_0$ corresponding to the eigenvalue $E_j$, then (7.4) can be rewritten in the form

$$\left(\frac{d^2\sigma}{d\Omega\,d\hbar\omega}\right)_i = \frac{k'}{2\pi\hbar kN}\sum_{v=1}^{N}\sum_{v'=1}^{N}\int_{-\infty}^{\infty} dt\, e^{-i\omega t}\cdot\langle i|\hat{F}_v(t)e^{i\mathbf{K}\cdot\mathbf{r}_v(t)}\hat{F}_{v'}^*(0)e^{-i\mathbf{K}\cdot\mathbf{r}_{v'}(0)}|i\rangle \tag{7.5}$$

where $\hat{F}_v$ is a scattering amplitude operator whose matrix elements are defined by the relation

$$\langle l|\hat{F}_v|i\rangle = F_{l_v/i_v}\prod_{\substack{\alpha \neq v}}^{N}\delta_{l_\alpha, i_\alpha} \tag{7.6}$$

$\hat{F}^*$ is the Hermitian conjugate of $\hat{F}$. In the spirit of the single-scatterer approximation, (7.6) supposes one can describe the state of the system in terms of individual scatterer quantum numbers. The Heisenberg operator notation has been used in (7.5), that is, time-dependent operators are defined as:

$$\hat{F}_v(t)e^{i\mathbf{K}\cdot\mathbf{r}_v(t)} \equiv e^{iH_0 t/\hbar}\hat{F}_v e^{i\mathbf{K}\cdot\mathbf{r}_v}e^{-iH_0 t/\hbar} \tag{7.7}$$

Finally, in a system at thermal equilibrium the appropriate differential cross section is the following canonical ensemble average of (7.5):

$$\frac{d^2\sigma}{d\Omega\,dh\omega} = \frac{k'}{2\pi\hbar kN} \sum_{\nu=1}^{N} \sum_{\nu'=1}^{N} \int_{-\infty}^{\infty} dt\, e^{-i\omega t} \cdot \langle \hat{F}_{\nu}(t) e^{i\mathbf{K}\cdot\mathbf{r}_{\nu}(t)} \hat{F}_{\nu'}^{*}(0) e^{-i\mathbf{K}\cdot\mathbf{r}_{\nu'}(0)} \rangle_T$$

(7.8)

where

$$\langle \alpha \rangle_T \equiv \sum_i e^{-E_i/kT}\, \frac{\langle i\alpha i \rangle}{\sum_i e^{-E_i/kT}} \equiv \frac{\mathrm{Tr}\,(e^{-H_0/kT}\alpha)}{\mathrm{Tr}\,(e^{-H_0/kT})}$$

(7.9)

Although the single-scatterer approximation has been used to obtain (7.8), the result is relatively general. The scattered waves are not restricted to electron waves; the expression is valid for neutrons and x rays, for example. The scattering centers may be the atoms of a monatomic substance, the individual atoms of a polyatomic substance, the individual molecules of a polyatomic system, many-molecule clusters, and so on. The quantity $\mathbf{r}_\nu$ locates the $\nu$th scattering entity; $\hat{F}_\nu$, representing the scattering amplitude of the $\nu$th entity, may be further decomposable in terms of the substructure and scattering amplitudes of constituents of $\nu$. Unfortunately, there is some uncertainty in specifying the appropriate scattering entity for a given situation. The choice should of course be such as to enhance the likelihood that the single-scatterer approximation is accurate. A necessary condition for the validity of the single-scatterer approximation is that the average distance (mean free path $\Lambda$) between scattering events be large compared with the de Broglie wavelength $\Lambda_{dB}$ of the electron. This condition is also sufficient for the validity of the Boltzmann equation. In actual computations for liquid argon, Lekner[28] found good agreement between theory and experiment even where $\Lambda \sim \Lambda_{dB}$, indicating cancellation of errors arising from the single-scattering approximation.

Consider a dilute solution of electrons (dilute enough that electron–electron interactions may be neglected) in a system of $N$ identical neutral molecules. Assume that a constant external electric field $\mathbf{E}$ is applied to the system. One can derive a balance equation for the momentum distribution function $f(\mathbf{p})$ of the electrons at steady state. At steady state the action of the field $\mathbf{E}$ on the electrons causes the following rate of change in $f(\mathbf{p})\,d^3p$, the probable number density of electrons with momenta between $\mathbf{p}$ and $\mathbf{p} + d\mathbf{p}$ (i.e., in momentum state $\mathbf{p}$):

$$e\mathbf{E} \cdot \frac{\partial f}{\partial \mathbf{p}}(\mathbf{p})\,d^3p$$

(7.10)

The balance equation for $f(\mathbf{p})$ is obtained by equating (7.10) to the rate of change of $f(\mathbf{p})\,d^3p$ arising from collisions between electrons and molecules.

Consider first the rate that electrons are lost from the state $\mathbf{p} = \hbar\mathbf{k}$ by the scattering process in which the electrons are scattered into the solid angle $d\Omega$, attain the momentum $\mathbf{p}' = \hbar\mathbf{k}'$, and incur an energy change between $\hbar\omega$ and $\hbar\omega + d\hbar\omega$. This rate is

$$\left(\frac{d^2\sigma\,(\mathbf{p} \to \mathbf{p}';\,\omega)}{d\Omega\,d(\hbar\omega)}\,d\Omega\,d(\hbar\omega)\right)\frac{p}{m}\,f(\mathbf{p})\,d^3pn \qquad (7.11)$$

Equation (7.11) is built up by multiplying $[d^2\sigma/d\Omega\,d(\hbar\omega)]\,d\Omega\,d(\hbar\omega)$, the scattering probability per scatterer, by $(p/m)f(\mathbf{p})\,d^3p$, the flux of electrons in state $\mathbf{p}$, to obtain the rate of loss of electrons from state $\mathbf{p}$ per scattering molecule. Multiplication by $n$, the number density of scatterers, yields (7.11). This equation is then integrated over all $\Omega$ and $\hbar\omega$ to obtain the following total rate of change in $f(\mathbf{p})$ due to "loss" collisions:

$$d^3pn \iint d\Omega\,d(\hbar\omega)\,\frac{p}{m}\,\frac{d^2\sigma(\mathbf{p} \to \mathbf{p}';\,\omega)}{d\Omega\,d(\hbar\omega)}\,f(\mathbf{p}) \qquad (7.12)$$

Next consider the rate that electrons are gained in the state $\mathbf{p}$ by the scattering process in which electrons initially in the momentum state $\mathbf{p}'$ are scattered through the solid angle $d\Omega'$ and attain the momentum state $\mathbf{p}$. This process is accompanied by the electron energy change $-\hbar\omega$. Integrating the expression analogous to (7.11) but for the transition $\mathbf{p}' \to \mathbf{p}$, we obtain the following total rate of change in $f(\mathbf{p})$ due to "gain" collisions:

$$d^3p'n \iint d\Omega'\,d(\hbar\omega)\,\frac{p'}{m}\,\frac{d^2\sigma\,(\mathbf{p}' \to \mathbf{p};\,-\omega)}{d\Omega\,d(\hbar\omega)}\,f(\mathbf{p}') \qquad (7.13)$$

The conservation of momentum and energy imply that

$$p\,d\Omega'\,d^3p' = p'\,d\Omega\,d^3p \qquad (7.14)$$

and time reversal and isotropy imply the condition of detailed balancing in the form

$$\frac{p}{p'}\,\frac{d^2\sigma\,(\mathbf{p} \to \mathbf{p}';\,\omega)}{d\Omega\,d(\hbar\omega)} = e^{\beta\hbar\omega}\,\frac{p'}{p}\,\frac{d^2\sigma\,(\mathbf{p}' \to \mathbf{p};\,-\omega)}{d\Omega\,d(\hbar\omega)} \qquad (7.15)$$

Combining (7.13) to (7.15) forming the difference between the resulting form of (7.13) and (7.15), and equating this difference to (7.10), one obtains the Boltzmann equation:

$$e\mathbf{E}\cdot\frac{\partial f}{\partial\mathbf{p}} = n \iint d\Omega\,d(\hbar\omega)\,\frac{p'}{m}\,\theta(\mathbf{K},\,\omega)[e^{-\beta\hbar\omega}f(\mathbf{p}') - f(\mathbf{p})] \qquad (7.16)$$

with the notation

$$\theta(\mathbf{K}, \omega) = \frac{p}{p'} \frac{d^2\sigma\,(\mathbf{p} \to \mathbf{p}'; \omega)}{d\Omega\,d(\hbar\omega)} \tag{7.17}$$

Equation (7.16) can be solved formally by expanding $f(\mathbf{p})$ in Legendre polynomials:

$$f(\mathbf{p}) = \sum_{l=0}^{\infty} f_l(\varepsilon) P_l\,(\cos\zeta) \tag{7.18}$$

where $\zeta$ is the angle between the momentum $\mathbf{p}$ and the field $\mathbf{E}$ and $\varepsilon = \hbar^2 k^2/2m = p^2/2m$. Inserting (7.18) into the rhs and lhs of (7.16), we obtain

$$eE\frac{p}{m}\left[P_0\frac{1}{3}\left(f'_1 + \frac{f_1}{\varepsilon}\right) + P_1\,(\cos\zeta)\left(f'_0 + \frac{2}{5}f'_2 + \frac{3}{5}\frac{f_2}{\varepsilon}\right)\right.$$

$$\left. + P_2\,(\cos\zeta)\frac{1}{3}\left(2f'_1 - \frac{f_1}{\varepsilon}\right) + \cdots\right]$$

$$= n\int d\Omega\int d(\hbar\omega)\frac{p'}{m}\theta(\mathbf{K}, \omega)\sum_{l=0}^{\infty}[e^{-\beta\,\omega}f_l(\varepsilon')P_l\,(\cos\zeta') - f_l(\varepsilon)P_l\,(\cos\zeta)] \tag{7.19}$$

The prime of $f'_l$ denotes the derivative with respect to $\varepsilon$. Since the quantities in the integrand of (7.19) do not depend on the azimuthal angle $\varphi$ of $\Omega$, the addition theorem leads—after integration over $\varphi$—to the following form for the rhs of (7.19):

$$2\pi\int d\theta\sin\theta\int d(\hbar\omega)\frac{p'}{m}\theta(\mathbf{K}, \omega)\sum_{l=0}^{\infty}P_l(\cos\zeta)[e^{-\beta\hbar\omega}f_l(\varepsilon')P_l(\cos\theta) - f_l(\varepsilon)] \tag{7.20}$$

Inserting this result in (7.19) and equating factors of $P_l(\cos\zeta)$ on each side of the equation, we find the first three members of the set of equations for $f_l(\varepsilon)$:

$$\frac{eEp}{3m}\left(f'_1 + \frac{f_1}{\varepsilon}\right) = n\int d\Omega\int d(\hbar\omega)\frac{p'}{m}\theta(\mathbf{K}, \omega)[e^{-\beta\hbar\omega}f_0(\varepsilon') - f_0(\varepsilon)] \tag{7.21}$$

$$\frac{eEp}{m}\left(f'_0 + \frac{2}{5}f'_2 + \frac{3}{5}\frac{f_2}{\varepsilon}\right) = n\int d\Omega\int d(\hbar\omega)\frac{p'}{m}\theta(\mathbf{K}, \omega)$$

$$\times\,[e^{-\beta\hbar\omega}f_1(\varepsilon')P_1(\cos\theta) - f_1(\varepsilon)] \tag{7.22}$$

$$\frac{eEp}{3m}\left(2f'_1 - \frac{f_1}{\varepsilon}\right) = n\int d\Omega\int d(\hbar\omega)\frac{p'}{m}\theta(\mathbf{K}, \omega)[e^{-\beta\hbar\omega}f_2(\varepsilon')P_2(\cos\theta) - f_2(\varepsilon)] \tag{7.23}$$

Following arguments first advanced by Shockley,[133] Cohen and Lekner claim that terms of order $l \geq 2$ in (7.18) may be neglected for both low and high fields. Although Shockley's conclusion is seen below to remain true for the low-field limit, his arguments in the high-field limit depend on the assumption that, for electron energies much less than the electronic excitation energies of the molecules, an electron colliding with a molecule loses an average energy of about $(2m/M)\langle\varepsilon\rangle$ per collision. $M$ is the mass of the molecule, and $\langle\varepsilon\rangle$ is the average energy of the electron. Since an electron scattered by polyatomic molecules can involve energy transfer substantially larger than Shockley's estimate, his conclusion that terms of order $f_l$ for $l \geq 2$ can be neglected in the high-field limit must be treated with some reservation. However, for subionizing, inelastic scattering, the fractional energy lost by the electron per collision $\lambda$, although larger than the monatomic value $\lambda \sim 2m/M$, may still be sufficiently small to justify Shockley's approximation. Certainly, the approximation should be accurate for fields up to the onset of field dependence of mobility, since the electrons have only slightly more than thermal energies and therefore should excite only weak inelastic modes. From experimental results in methane and neopentane, the fractional energy loss $\lambda$ was estimated in Section II to be about $8.4 \times 10^{-4}$ and $1.6 \times 10^{-4}$, respectively, only about 10 times greater than $2m/M$ for these systems.

With the neglect of $f_l, l \geq 2$, and under the condition $\lambda \ll 1$, (7.21) and (7.22) can be solved explicitly. The procedure is to expand $p'$ and $\Theta(K, \omega)$ in terms of the elastic limit of $K$, namely, $K_0$ given by

$$K_0{}^2 = \frac{2p^2}{\hbar^2} (1 - \cos\theta) \tag{7.24}$$

From definition,

$$\hbar^2 K^2 = p^2 + p'^2 - 2pp' \cos\theta \quad \text{and} \quad p' = \sqrt{p^2 - \hbar\omega(2m)}$$

so that

$$p' = p\left[1 - \frac{\hbar\omega}{2\varepsilon} + \mathcal{O}\left(\frac{\hbar\omega}{\varepsilon}\right)^2\right] \tag{7.25}$$

$$K = K_0\left[1 - \frac{\hbar\omega}{4\varepsilon} + \mathcal{O}\left(\frac{\hbar\omega}{\varepsilon}\right)^2\right] \tag{7.26}$$

and

$$\Theta(K, \omega) = \Theta(K_0, \omega) + (K - K_0)\frac{\partial\Theta}{\partial K_0}(K_0, \omega) + \cdots = \Theta(K_0, \omega)$$

$$- \frac{\hbar\omega}{4\varepsilon} K_0 \frac{\partial\Theta}{\partial K_0}(K_0, \omega) + \mathcal{O}\left(\frac{\hbar\omega}{\varepsilon}\right)^2 \tag{7.27}$$

Neglect of the terms of order $(\hbar\omega/\varepsilon)^2$ in using (7.25) and (7.27) in the integrands of (7.21) and (7.22) introduces errors of order $\lambda$ compared to the terms kept.

Using (7.25) and (7.27) and the approximations

$$e^{-\beta\hbar\omega} \simeq 1 - \beta\hbar\omega + \frac{(\beta\hbar\omega)^2}{2} \tag{7.28}$$

and

$$f_l(\varepsilon') \approx f_l(\varepsilon) - \hbar\omega f_l'(\varepsilon) + \frac{(\hbar\omega)^2}{2} f_l'' \tag{7.29}$$

(7.21) and (7.22) reduce to the forms

$$\frac{1}{3}\frac{eE}{\varepsilon}\frac{d}{d\varepsilon}(\varepsilon f_1) = n\int d\Omega \left[\left(-\beta\alpha_1 + \frac{\beta\alpha_2}{2\varepsilon} + \frac{\beta^2\alpha_2}{2} + \frac{\beta}{4\varepsilon}K_0\frac{d\alpha_2}{dK_0}\right)\right.$$

$$\left. \times (f_0 + kTf_0') + \frac{\beta\alpha_2}{2}(f_0' + kTf_0'') \right] \tag{7.30}$$

and

$$eEf_0' = -n\int d\Omega\, f_1\alpha_0[1 - P_1(\cos\theta)] \tag{7.31}$$

where

$$\alpha_n \equiv \int d(\hbar\omega)(\hbar\omega)^n\Theta(K_0,\omega) \qquad n = 0, 1, 2 \tag{7.32}$$

As will be clearer later, the quantities $\alpha_n$ are related to correlation functions involving the structure of the medium and the scattering amplitudes of the molecules. $\alpha_1$ and $\alpha_2$ are essentially the mean and mean-square energy loss to the medium by electrons scattered into the solid angle $\Omega$ with a momentum change to $K_0$. It is useful at this point to introduce the approximate relationship[134]

$$\alpha_2 \approx 2kT\alpha_1 \tag{7.33}$$

which involves neglect of terms of order $\lambda\varepsilon/kT$. With the aid of (7.33), and the relationship

$$K_0\frac{d\alpha_2}{dK_0} = 2\varepsilon\frac{d\alpha_2}{d\varepsilon} \tag{7.34}$$

(7.30) can be integrated once with respect to $\varepsilon$ and the result combined with (7.31) to yield an equation whose solution is

$$f_0 = c \exp\left[ -\int_0^x \frac{y}{y + b(y)}\, dy \right] \tag{7.35}$$

where $c$ is a normalization constant,

$$x \equiv \frac{\varepsilon}{kT} \tag{7.36}$$

and

$$b(x) = \frac{1}{3} \frac{(eE\Lambda_0)(eE\Lambda_1)}{\lambda(kT)^2} \tag{7.37}$$

The quantities $\Lambda_0$ and $\Lambda_1$ are mean free paths defined by

$$\Lambda_0^{-1} = \frac{1}{N} \sum_v n \int d\Omega \, (1 - \cos\theta) \langle |\hat{F}_v(0)|^2 \rangle_T \tag{7.38}$$

and

$$\Lambda_1^{-1} = n \int d\Omega \, (1 - \cos\theta)\alpha_0$$

$$= \frac{1}{N} \sum_{v, v'}^N n \int d\Omega \, (1 - \cos\theta) \langle \hat{F}_v(0)e^{i\mathbf{K}_0 \cdot \mathbf{r}_v(0)} \hat{F}_{v'}^*(0)e^{-i\mathbf{K}_0 \cdot \mathbf{r}_{v'}(0)} \rangle_T \tag{7.39}$$

The second expression for $\Lambda_1^{-1}$ is obtained by noting that the integral over $d\hbar\omega$ in (7.32) introduces the delta function $\delta(t)$ so that the subsequent time integration yields the correlation function shown in (7.39).

$\Lambda_0$ represents the mean free path for completely incoherent scattering, whereas $\Lambda_1$ is the momentum mean free path which contains coherent as well as incoherent scattering contributions.

The quantity $\lambda$ is defined by

$$\lambda = \frac{n \int d\Omega \, (\beta/2)\alpha_2}{e\Lambda_0^{-1}} \tag{7.40}$$

and to the approximation given by (7.33) is

$$\lambda = \frac{n \int d\Omega \, \alpha_1}{e\Lambda_0^{-1}} \tag{7.41}$$

in which form the interpretation of $\lambda$ as the fractional energy loss is appropriate. The integral over $d\hbar\omega$ in the expression for $\alpha_1$ yields the first derivation of the delta function $\delta(t)$. The time integration then yields the result

$$\lambda = -\frac{n}{\varepsilon \Lambda_0^{-1}} \frac{1}{N} \sum_{v,v'}^{N} \int d\Omega \, \langle [H_0, \hat{F}_v(0)e^{i\mathbf{K}_0 \cdot \mathbf{r}_v(0)}]\hat{F}_{v'}^*(0)e^{-i\mathbf{K}_0 \cdot \mathbf{r}_{v'}(0)} \rangle_T \qquad (7.42)$$

where $[H_0, A]$ denotes the commutator of $H_0$ and $A$.

The drift velocity is defined by the expression

$$\mathbf{v}_d = \int f(\mathbf{p}) \frac{\mathbf{p}}{m} \, d^3 p \qquad (7.43)$$

which in terms of the expansion given by (7.18) for $f$ yields

$$\mathbf{v}_d = \int f_1 \frac{\mathbf{E} \cdot \mathbf{pp}}{Epm} \, d^3 p = \frac{1}{3} \int f_1 \frac{\mathbf{p}}{m} \, d^3 p \, \frac{\mathbf{E}}{E} \qquad (7.44)$$

With the aid of (7.31), written in the form

$$f_1 = -e\Lambda_1 E f_0' \qquad (7.45)$$

the drift velocity becomes

$$\mathbf{v}_d = \frac{1}{3} \int e\Lambda_1 \frac{\mathbf{p}}{m} f_0' \, d^3 p \, \mathbf{E} \qquad (7.46)$$

a form that demonstrates the validity of interpreting $\Lambda_1$ as the momentum mean free path. In terms of the reduced energy $x = \varepsilon/kT$, the formulas for the magnitude of the drift velocity and average electronic energy can be written in forms

$$v_d = -\frac{Ee}{3} \left(\frac{2kT}{m}\right)^{1/2} \frac{\int_0^\infty x\Lambda_1 f_0'(x) \, dx}{\int_0^\infty x^{1/2} f_0(x) \, dx} \qquad (7.47)$$

and

$$\langle \varepsilon \rangle = kT \frac{\int_0^\infty x^{3/2} f_0(x) \, dx}{\int_0^\infty x^{1/2} f_0(x) \, dx} \qquad (7.48)$$

In the limit that the parameter $b$ given by (7.37) is small, $f_0$ is Maxwellian, that is, $f_0 = ce^{-x}$, the mobility $\mu \equiv v_d/E$ will be field-independent and $\langle \varepsilon \rangle = \frac{3}{2}kT$. For large $b$ the distribution $f_0$ becomes field-dependent, with an accompanying increase in average electron energy and a field-dependent mobility, the trend of the mobility depending on the behavior of $\Lambda_1$ and $b$ with electron energy. The criterion for the onset of field-dependent mobilities is clearly that $b$ be of order unity, that is,

$$\frac{1}{3} \frac{(eE\Lambda_0)(eE\Lambda_1)}{\lambda(kT)^2} = \mathcal{O}(1) \qquad (7.49)$$

To appreciate the structure of the quantities entering $\Lambda_1$ and $\lambda$, consider a scattering amplitude of the form

$$\hat{F}_v = \bar{F}^e + \Delta F^e(I_v) + \hat{F}^{\text{in}}(I_v)$$

where $a$ is the mean elastic scattering amplitude of a molecule, defined by $a = \langle \hat{F}_v \rangle_T$; $\Delta a(I_v)$ is the angular dependent elastic scattering length which is a function of the internal molecular coordinates $I_v$ (such as the Euler angles of orientation of the molecule) which do not change appreciably during an electron–molecule collision; and $F^{\text{in}}$ represents the inelastic electron–molecule scattering amplitude. Assume that the center-of-mass coordinates are independent of the internal molecular degrees, so that the Hamiltonian $H_0$ may be written in the form $H_0^{CM} + H_0{}^I$, where $H_0^{CM}$ is a Hamiltonian containing the center-of-mass translational kinetic energy and the interactions depending only on center-of-mass separations, and $H_0{}^I$ is a Hamiltonian depending only on internal degrees of freedom of the molecules. For this model (7.39) yields

$$\Lambda_1^{-1} = n \int d\Omega \, (1 - \cos\theta)\tilde{\sigma}_{\text{inc}} + n \int d\Omega \, (1 - \cos\theta)\tilde{\sigma}_{\text{coh}} S^{CM}(K_0) \quad (7.50)$$

where

$$\tilde{\sigma}_{\text{inc}} \equiv \langle |\hat{F}_1|^2 \rangle_T - \langle |\hat{F}_1 \hat{F}_2| \rangle_T \quad (7.51)$$

$$\tilde{\sigma}_{\text{coh}} \equiv \langle |\hat{F}_1 \hat{F}_2| \rangle_T \quad (7.52)$$

and

$$S^{CM}(K_0) = 1 + N\langle e^{i\mathbf{K}_0 \cdot \mathbf{r}_{12}} \rangle_T$$

$$= 1 + n \int [g^{CM}(\mathbf{r}_1, \mathbf{r}_2) - 1] e^{i\mathbf{K}_0 \cdot \mathbf{r}_{12}} \, d^3r + (2\pi)^3 n \, \delta(\mathbf{K}_0) \quad (7.53)$$

$\tilde{\sigma}_{\text{inc}}$ is an incoherent scattering cross section resulting from orientational parts of the scattering amplitude. For a monatomic system in which $\hat{F}_v = \bar{F}^e$, $\tilde{\sigma}_{\text{inc}}$ vanishes. The quantity $\tilde{\sigma}_{\text{coh}}$ represents the cross section for the coherent part of the scattered electron wave and as such is accompanied by the structure factor $S^{CM}(K_0)$. [The delta function in (7.53) is of no consequence here, since the forward-scattering contribution for $K_0 = 0$ vanishes in the kinetic expressions.] If $\hat{F}_v = \bar{F}^e$, then $\tilde{\sigma}_{\text{coh}} = |\bar{F}^e|^2$ and the expression for $\Lambda_1^{-1}$ becomes

$$\Lambda_1^{-1} = n \int d\Omega \, (1 - \cos\theta) |\bar{F}^e|^2 S^{CM}(K_0) \quad (7.54)$$

the form used by Lekner in his theory of quasi-free electrons in monatomic systems.

For the model being considered the fractional energy loss parameter becomes

$$\lambda = \frac{2m}{M} + \lambda^i$$

where $2m/M$ is the elastic part of $\lambda$ obtained from (7.42) with the aid of the result

$$\langle [H_0{}^T, e^{i\mathbf{K}_0 \cdot \mathbf{r}_\nu(0)}] e^{-i\mathbf{K}_0 \cdot \mathbf{r}_{\nu'}(0)} \rangle_T = \frac{\hbar^2 K_0{}^2}{2M} \delta_{\nu\mu} \tag{7.55}$$

where $M$ is the molecular mass, and $\lambda^i$ is the elastic part:

$$\lambda^i = -\frac{n}{\varepsilon \Lambda_0^{-1}} \frac{1}{N} \sum_{\nu, \nu'}^N \int d\Omega \, \langle \{H_0{}^I, [\Delta F^e(I_\nu) + F^{\text{in}}(I_\nu)]\}$$

$$\times [\Delta F^e(I_{\nu'}) + F^{\text{in}}(I_\nu)] \rangle_T \langle e^{i\mathbf{K}_0 \cdot \mathbf{r}_{\nu\nu'}} \rangle_T \tag{7.56}$$

For low-energy electron scattering from multipole (excluding dipole) interactions and polarization interactions, one can show that $\tilde\sigma_{\text{coh}}$ is often large compared to $\tilde\sigma_{\text{inc}}$, whereas $\lambda^i$ is large compared to $2m/M$. Thus the inelastic scattering contribution may be small in the mean free paths $\Lambda_0$ and $\Lambda_1$ and yet quite large in the fractional energy loss parameter. For example, in nitrogen gas,[135] electron-quadrupole scattering gives $\lambda \approx 50$ $(2m/M)$, whereas the quadrapolar value for $\tilde\sigma_{\text{inc}}$ is only about 2% of the elastic value 4.5 Å$^2$ for $\tilde\sigma_{\text{coh}}$. Of course, since $S(K_0) \sim$  in the condensed phase for small $K_0$ (low electron energies, the contribution of $\tilde\sigma_{\text{coh}}$ is lowered relative to $\tilde\sigma_{\text{inc}}$ for liquids and solids so that the two terms could make comparable contributions to $\Lambda_1$.

## B. Monatomic Systems

In a monatomic system the electron–atom scattering amplitude $F_\nu$ in $\Lambda_0$, $\Lambda_1$, and $\lambda$ depends only on the electron energy $\varepsilon$ and the scattering angle $\theta$. In this case the expressions for these quantities reduce to the Cohen–Lekner results:

$$\Lambda_0^{-1} = n \int d\Omega \, (1 - \cos\theta) \tilde\sigma_{\text{coh}}(\varepsilon, \theta) \equiv n\sigma_0 \tag{7.57}$$

$$\Lambda_1^{-1} = n \int d\Omega \, (1 - \cos\theta) \tilde\sigma_{\text{coh}}(\varepsilon, \theta) S(K_0) \equiv n\sigma_1 \tag{7.58}$$

$$\lambda = \frac{2m}{M} \tag{7.59}$$

where

$$\tilde\sigma_{\text{coh}} = |F^e|^2 \tag{7.60}$$

$F^e$ is the electron–atom scattering amplitude. Thus the problem of computing the distribution $f_0$ from (7.35) and the drift velocity and average electron energy from (7.47) and (7.48) is reduced to the problem of determing the electron–atom differential cross sections $\tilde{\sigma}_{coh}$.

The cross section to be determined is that of a plane-wave (free) electron scattered by an atom of the medium. Lekner argued[110] that to calculate the scattering amplitude (or phase shift $\delta_l$) by the usual partial-wave method it is necessary to subtract the constant $\tilde{V}_m$ in the electron–atom potential of a dense medium. The quantity represents a uniform lowering of the potential of the electron due to many-body polarization effects and should not be included in the scattering computation, because the electron "sees" only differences in the potential. Thus Lekner introduced the effective electron–atom scattering potential

$$V_{eff}(r) = \tilde{V}(r) - \tilde{V}_m \qquad 0 < r < r_m$$
$$= 0 \qquad r > r_m \qquad (7.61)$$

where $\tilde{V}$, $\tilde{V}_m$, and $r_m$ are defined in Section VI.B. This choice of $V_{eff}$ is the so-called muffin tin model.

Using the determination of $\tilde{V}$ outlined in Section VI.B for $V_{eff}$, Lekner computed the differential cross sections required in (7.57) and (7.58) by the method of partial waves for liquid argon at 84°K. His results for the cross section $\sigma_0$, defined by (7.57), are shown in Fig. 39. There are several interesting features. First, the zero-energy scattering length $a \equiv \lim_{\epsilon \to 0} F$, determined by $V_{eff}$, is positive and equal to 0.76 Å, a value close to the Hartree–Fock value 0.794 Å used as the hard-core cutoff in the SJC pseudo-potential model. As a consequence of the positive scattering length, the Ramsaur minimum, observed in the dilute gas, is suppressed liquid phase (the zero-energy scattering length in a dilute gas is $-0.782$ Å for argon). The change in sign of the scattering length in going from dilute gas to dense liquid is caused by the many body screening of the polarization potential (thus converting it from a fairly long-ranged potential to the short-ranged $V_{eff}$). The cross section $\sigma_0$ is almost constant for electron energies up to about 4 eV, a behavior similar to that of the corresponding low-energy, electron–hard core cross section, indicating perhaps why the SJC pseudo-potential model is moderately successful.

Lekner's results for the drift velocity, the average electron energy, the de Broglie wavelength, and $\Lambda_0$ and $\Lambda_1$ are presented in Table XVI. A comparison between experimental and theoretical drift velocities for liquid argon is given in Figs. 13 and 14b. The predictions are quite good up to fields of about $10^4$ V/cm. The observed onset of the field-dependent mobility at about 200 V/cm is well matched by the predicted curve.

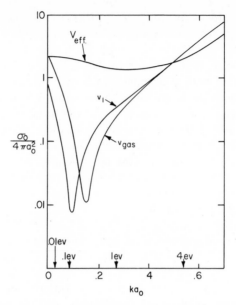

**Fig. 39** Electron scattering cross sections for Lekner's effective electron–fluid potential $V_{\text{eff}}$, [see (7.61)], the potential $v_1$ defined by (6.47), and the gas-phase electron–atom potential $v_{\text{gas}}$ obtained by setting $a \equiv 1$ in (6.47). (Ref. 28).

TABLE XVI

Theoretical Drift Velocities in Argon at 84°K
(Ref. 28)[a]

| $E$ (V/cm) | $v_d$ (cm/$\mu$sec) | $\langle \varepsilon \rangle$ (eV) | $\lambda$ (Å) | $\Lambda_0$ (Å) | $\Lambda_1$ (Å) |
|---|---|---|---|---|---|
| 10 | 0.0039 | 0.011 | 116 | 6.6 | 136 |
| $10^2$ | 0.037 | 0.011 | 116 | 6.6 | 136 |
| $10^3$ | 0.21 | 0.028 | 73 | 6.7 | 140 |
| $10^4$ | 0.47 | 0.51 | 17 | 8.7 | 140 |
| $10^5$ | (0.25) | (2.4) | (8) | (9.0) | (60) |

[a] $E$ = field; $v_d$ = drift velocity; $\langle \varepsilon \rangle$ = mean electron energy; $\lambda$ = de Broglie wavelength; $\Lambda_0$ = energy transfer mean free path; $\Lambda_1$ = momentum transfer mean free path.

428

In discussing the divergence between theory and experiment for fields greater than $10^4$ V/cm, Lekner claimed (with reference at that time to Swan's[24] work in Fig. 13) that the experimental data were uncertain in that region, and he stated that his computed cross sections were uncertain for electron energies above 2 eV. The experimental values in Fig. 13 are apparently firm, since Miller, Howe, and Spear[32] reproduced Swan's results as seen in Fig. 14$b$.

In the low-field limit $\hbar K_0 = \mathcal{O}(\sqrt{mkT})$ and $S(K_0) \simeq S(0)$, so that $\sigma_1 \simeq \sigma_0 S(0)$. Thus, if $\sigma_0$ is approximately constant, as in the case of argon at 84°K, then the low-field mobility determined by (7.47) is of the form given earlier by (2.4) if the notation $\sigma_0 = 4\pi a^2$ is used. As pointed out in Section II, the predicted ratios of solid-phase to liquid-phase electron mobility agree with the measured ratios for argon, krypton, and xenon near their melting points. Thus the kinetic theory of Cohen and Lekner seems to give a substantially correct description of quasi-free conduction in monatomic systems.

An important point should be raised, however. Aside from cross-section errors, another error inherent in the single-scatterer theory is the neglect of multiple-scattering effects. In fact, it is inconsistent to compute the ground-state energy $E_0$ from the intrinsically multiple-scattering Wigner–Seitz model while determining the mobility from the single-scatterer model. To estimate multiple-scattering effects, Lekner computed the cross sections $\sigma_0$ and $\sigma_1$ assuming the electrons have the Wigner–Seitz momentum $\hbar k_0 \equiv \sqrt{2mE_0}$ $(= 6.63 \times 10^7 \hbar$ cm$^{-1})$ and the observed effective mass $m^* = 0.5$ m. Lekner found a single-scatterer mobility for argon at 84°K of 370 cm$^2$/(V)(sec), whereas his estimated corrections for multiple scattering lead to the value $- 540$ cm$^2$/(V)(sec), both answers in similar agreement with the experimental 450 cm$^2$/(V)(sec). Thus multiple-scattering effects seem to be self-canceling for argon.

At temperatures much higher than melting temperatures, the low-field mobilities of krypton and argon become strongly temperature- or density-dependent, the magnitudes changing by factors of 2 to 6 and going through a peak in the 10 to 20°K temperature range. This behavior is summarized in Figs. 17 to 19.

An explanation of this behavior was provided by Lekner in a variation of his effective-scattering potential model. Near the melting point of argon, Lekner found the scattering length to be positive ($a = 0.76$ Å, $n = 2.1 \times 10^{-2}$ Å$^{-3}$). However, in argon gas at a density of about $5 \times 10^{-4}$ Å$^{-3}$, the scattering length is known[136] to be $- 0.87$ Å. Thus Lekner[33] reasoned that, somewhere in the density range of fluid argon, the scattering length passes through the value zero. If this does happen, the electron mobility will then increase indefinitely at some density, its magnitude limited only by fluctuations in the scattering length arising from fluctuations in the scattering potential.

Expressing the scattering amplitude in the form

$$\hat{F}_v = -\langle a \rangle + (\Delta a)_v \tag{7.62}$$

where $\langle a \rangle$ is the ensemble average scattering length, and $(\Delta a)_v$ is the scattering length of the $v$th atom due to a fluctuation in the local potential, one obtains from (7.51) and (7.52):

$$
\begin{aligned}
\tilde{\sigma}_{\text{inc}} &= \langle (\langle a \rangle + \Delta a)^2 \rangle - [\langle (\langle a \rangle + \Delta a) \rangle]^2 \\
&= \langle (\Delta a)^2 \rangle
\end{aligned} \tag{7.63}
$$

and

$$\tilde{\sigma}_{\text{coh}} = (\langle a \rangle)^2$$

assuming that the fluctuations are random so that

$$\langle (\Delta a)_1 (\Delta a)_2 \rangle = \langle \Delta a_1 \rangle \langle \Delta a_2 \rangle = 0 \tag{7.64}$$

In the low-field limit it is reasonable to assume that $S(K_0) \simeq S(0)$ and that $\tilde{\sigma}_{\text{inc}}$ and $\tilde{\sigma}_{\text{coh}}$ are not strong functions of $\varepsilon$ and $\theta$. Thus the low-field mobility may be approximated by the simple formula

$$\mu = \frac{2}{3}\left(\frac{2}{\pi m k T}\right)^{1/2} \frac{e}{n\sigma_1} \tag{7.65}$$

where[33]

$$\sigma_i = 4\pi[\langle (\Delta a)^2 \rangle + (\langle a \rangle)^2 S(0)] \tag{7.66}$$

For a given fluctuation $\Delta V$ in the local potential of an atom, the Born approximation ($K_0$ small) gives the corresponding $\Delta a$:

$$\Delta a = \frac{2m}{\hbar^2} \int_0^\infty dr \, r^2 \Delta V \tag{7.67}$$

$\Delta V$ being a fluctuation, it should be sufficiently small to justify the Born approximation. Lekner assumed that the fluctuation $\Delta V$ could be approximated as a constant in the Wigner–Seitz sphere of the atom, and as zero outside the sphere. Thus

$$\Delta a = \frac{2m}{\hbar^2}\frac{r_s^3}{3}\Delta V = \frac{m}{2\pi n \hbar^2}\Delta V \tag{7.68}$$

and

$$\langle (\Delta a)^2 \rangle = \left(\frac{m}{2\pi n \hbar^2}\right)^2 \langle (\Delta V)^2 \rangle \tag{7.69}$$

The fluctuations in the potential are due to fluctuations in the positions of the molecules in the liquid. Thus, if $\Delta\mathbf{R}$ is the fluctuation in the separation of a pair of atoms, then $\Delta V = \Delta\mathbf{R} \cdot \nabla_R V + \mathcal{O}(\Delta R)^2$ and

$$\langle(\Delta V)^2\rangle = \tfrac{1}{3}\langle(\Delta R)^2\rangle 4\pi n \int_0^\infty dR\, R^2 g(R) \left(\frac{dV}{dR}\right)^2 + \mathcal{O}(\Delta R)^4 \qquad (7.70)$$

if it is assumed that the fluctuation $\Delta\mathbf{R}$ is independent of the atomic separation $R$.

Lekner made the further rather drastic assumptions that (a) the mean-square displacement can be estimated from the Einstein model:

$$\langle(\Delta R)^2\rangle = \tfrac{1}{3}r_s{}^2 nkT\kappa_T = \tfrac{1}{3}r_s{}^2 S(0) \qquad (7.71)$$

(b) $g(R)$ may be approximated by the step function $g = 0$, $R < \delta$, and $g = 1$, $R > \delta$, for hard spheres of diameter $\delta$, and (c) $dV/dR \simeq (2\alpha e^2/R^5)\mathfrak{d}_L$ for $R > \delta$, where $\mathfrak{d}_L$ is the Lorentz local field screening function given by (6.44). With these assumptions, Lekner obtained

$$\langle(\Delta a)^2\rangle = \frac{4}{63} S(0) \frac{1}{\pi n\delta^3} \left(\frac{r_s}{\delta}\right)^2 \left(\frac{\alpha\mathfrak{d}_L}{a_0\delta}\right)^2 \qquad (7.72)$$

a result implying the mobility maximum, for the density at

$$\mu^{\text{max}} \approx \frac{21M}{2} e\delta^7 \left(\frac{2\pi}{m}\right)^{1/2} \left(\frac{a_0}{r_s\alpha\mathfrak{d}_L}\right)^2 \frac{c^2}{(kT)^{3/2}} \qquad (7.73)$$

where $c$ is the velocity of sound, related to $S(0)$ by $Mc^2/kT = S(0)$ and $a_0$ ($=0.528$ Å) is the Bohr radius.

Lekner observed that, if one assumes the maximum to appear at an approximately constant density $n_m$, that is, that $\langle a\rangle = 0$ at $n_m$ independently of $T$, then the mobility maximum will be proportional to $c^2 T^{-3/2}$. The argon data of Jahnke, Meyer, and Rice[30] shown in Fig. 19 demonstrate that the mobility maximum occurs at almost the same density for the range of pressures (and the corresponding range of temperatures). The experimental value of $n_m$ obtained from Fig. 19 is 0.012 Å$^{-3}$. With the hard-core radius $\delta = 3.44$ Å used by Lekner[28] in successfully predicting the structure function $S(K)$ of argon from the Percus–Yevick hard-core theoretical formula, Jahnke, Meyer, and Rice (JMR) predicted from (7.73) the maximum mobility versus $c^2 T^{-3/2}$. The predicted slope of $\mu_{\text{max}}$ with $c^2 T^{-3/2}$ is in excellent agreement with experiment, although the predicted magnitude is only 64% of the observed value of $\mu_{\text{max}}$. In view of the approximations made to evaluate $\Delta a$, the agreement is convincing of the validity of the model.

In a semiempirical extension of his model, Lekner[34] expanded $\langle a \rangle$ in a Taylor's series about $n_m$. Keeping the first-order term, he obtained

$$\langle a \rangle = \gamma(n - n_m) \tag{7.74}$$

giving for the momentum cross section of (7.66)

$$\sigma_1 = 4\pi[\langle(\Delta a)^2\rangle + \gamma^2 S(0)(n - n_m)^2] \tag{7.75}$$

JMR tested this relationship by determining $\sigma_1$ from the theoretical equation for $\mu$, (7.65), with their experimental mobility data in Fig. 19 and plotting the quantity $A \equiv \sqrt{\sigma_1/4\pi S(0)}$ versus density. Their results are shown in Fig. 40. As implied by (7.75), for densities sufficiently greater than $n_m$ for $(\langle a \rangle)^2$ to dominate $\langle(\Delta a)^2\rangle$, the quantity $A$ is a linear function of density.

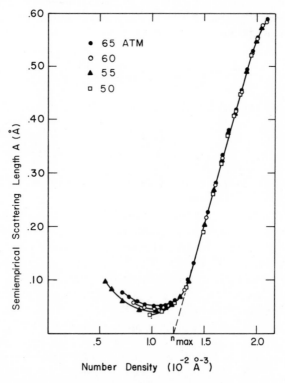

**Fig. 40** Effective scattering length $A \equiv \sqrt{\sigma_1/4\pi S(0)}$, where $S(0) = nkT\kappa_T$ and $\sigma_1$ is determined from the theoretical expression (7.65) combined with experimental values of the mobility. (Ref. 30).

The slope of the plot in this density region gives a value $\gamma = 69.2$ Å$^4$, which is independent of temperature according to the data shown in Fig. 38. An alternative way to obtain $\gamma$ is to compute $\langle(\Delta a)^2\rangle$ according to the theoretical expression (7.72), obtain $\sigma_1$ empirically from the measured mobility and (7.65), and use these values in (7.75) to determine $\gamma$. With this method, JMR found an average value $\gamma = 63.4$ Å$^4$ using their data in the density region $n > n_m$. For $n \lesssim n_m$, the fact that the theoretical value of $4\pi\langle(\Delta a)^2\rangle$ is an overestimate of $\sigma_1$ leads to the impossible conclusion that $\gamma^2 < 0$.

TABLE XVII

Theoretical Calculation of $\mu$, with $\gamma = 65.0$, from (7.65) Compared with low-field experimental Mobilities for Liquid Argon [Ref. 30]

| $T$ (°K) | $n$ (Å$^{-3}$) | $\langle\Delta a^2\rangle$ (Å$^2$) | $nkT\chi_T\gamma^2(n - n_m)^2$ (Å$^2$) | $\mu^{th}$ [cm$^2$/(V)(sec)] | $\mu^{exp}$ [cm$^2$/(V)(sec)] |
|---|---|---|---|---|---|
| 90.1 | 0.02101 | 0.0001 | 0.0186 | 507.7 | 500 |
| 95.0 | 0.02056 | 0.0002 | 0.0197 | 477.3 | 440 |
| 100.0 | 0.02009 | 0.0002 | 0.0208 | 451.1 | 406 |
| 105.0 | 0.01960 | 0.0002 | 0.0218 | 429.4 | 375 |
| 110.0 | 0.01909 | 0.0003 | 0.0228 | 412.1 | 358 |
| 115.1 | 0.01854 | 0.0003 | 0.0236 | 399.1 | 348 |
| 120.1 | 0.01797 | 0.0004 | 0.0242 | 391.6 | 343 |
| 126.0 | 0.01724 | 0.0006 | 0.0245 | 391.1 | 321 |
| 130.0 | 0.01671 | 0.0007 | 0.0243 | 398.5 | 347 |
| 135.0 | 0.01596 | 0.0010 | 0.0232 | 423.4 | 385 |
| 140.0 | 0.01510 | 0.0014 | 0.0205 | 486.5 | 476 |
| 146.6 | 0.01355 | 0.0029 | 0.0102 | 880.1 | 1032 |
| 149.1 | 0.01270 | 0.0043 | 0.0027 | 1745.0 | 1350 |
| 150.6 | 0.01202 | 0.0062 | 0.0001 | 2031.7 | 1470 |
| 151.7 | 0.01141 | 0.0082 | 0.0089 | 788.3 | 1355 |
| 152.8 | 0.01048 | 0.0132 | 0.0897 | 142.6 | 966 |
| 153.8 | 0.00884 | 0.0294 | 0.8420 | 19.9 | 510 |
| 154.8 | 0.00719 | 0.0394 | 1.733 | 12.0 | 332 |
| 155.9 | 0.00618 | 0.0485 | 1.607 | 14.9 | 326 |
| 156.9 | 0.00562 | 0.0554 | 1.415 | 18.4 | 358 |

Using a weighted-average value $\gamma = 65.0$ Å$^4$, JMR computed the mobilities of electrons in argon from (7.65), (7.72), and (7.75). Some of their results are shown in Table XVII. The agreement is good quantitatively ($\lesssim 15\%$ error) in the density range 0.021 down to 0.014 Å$^{-3}$, and is not too bad ($\lesssim 50\%$ error) down to 0.0114 Å$^{-3}$. However, for densities less than 0.01 the predicted mobilities are low by a factor of the order of 20. Similarly, above 0.014 Å$^{-3}$ the predicted field dependence of the drift velocity agrees with JMR's experimental results, and below $0.011$ $\text{Å}^{-3}$ the predicted dependence disagrees. Within the context of Lekner's model, the drift velocity can be predicted from (7.47) using

$$\Lambda_0^{-1} = 4\pi n[\langle(\Delta a)^2\rangle + (\langle a\rangle)^2] \tag{7.76}$$

$$\Lambda_1^{-1} = 4\pi n[\langle(\Delta a)^2\rangle + S(K_0)(\langle a\rangle)^2] \tag{7.77}$$

and $\lambda = 2m/M$ to compute the field-dependent parameter $b$ entering (7.35) for $f_0$.

The results described in the preceding paragraphs of this section allow the following conclusions to be made concerning the single-scatterer kinetic theory of electrons in dense inert gas systems:

*a.* Lekner's mean potential model accounts for the magnitudes and field dependence of electron mobilities in liquids and solids near their triple points.

*b.* Including the contribution of fluctuations through Lekner's formula [see (7.72)] and using the semiempirical extension (7.74) of the average scattering length, the observed mobilities in argon can be accounted for quantitatively for densities ranging from the triple-point density to $0.014$ $\text{Å}^{-3}$.

*c.* The maximum in the mobility is explained by Lekner's fluctuation model which postulates a zero of the average scattering length at a certain density $n_m$. This density was deduced to be $0.012$ $\text{Å}^{-3}$ by JMR.

*d.* Below densities of $0.01$ $\text{Å}^{-3}$ the theory predicts low-field mobilities in argon that are low by a factor of about 20. Correspondingly, the predicted field dependence is in poor agreement in this density region.

There are three possible explanations for the theoretical discrepancies:

1. The assumptions made in the estimations of $\langle\Delta a^2\rangle$ are too severe to warrant the validity of (7.22) at low densities.

2. Fluctuations in the liquid affect the mobility in some manner not accounted for by Lekner's zero-energy scattering length model.

3. Incipient electron localization influences the mobility on the high-density side of the maximum.

Better approximations to $\langle\Delta a^2\rangle$ may be made by avoiding the limiting assumptions in the functions $\mathscr{A}(R)$ and $g(s)$, but it is noted that these assumptions become more valid at lower densities. The estimates for the mean-square displacement used in (7.72) do, however, become invalid at these lower densities, and it is probably in this direction that improvements in $\langle\Delta a^2\rangle$ could be made. Interestingly, if it is assumed that $\langle a\rangle \equiv 0$ at low densities and $\sigma_1 = 4\pi\langle(\Delta a)^2\rangle$, mobilities predicted by (7.65)—1100, 590, 523, 510, and 470 $\text{cm}^2/(\text{V})(\text{sec})$ for the densities 0.01048, 0.00884, 0.00719, 0.00618, and 0.00562 $Å^{-3}$, respectively—are in fair agreement with the experimental data given in Table XVII. This result, if it is more than coincidence, implies that the mobility maximum represents a point of transition from coherent scattering [where $(\langle a \rangle)^2 S(0)$ dominates $\sigma_1$] to incoherent scattering [where $\langle (\Delta a)^2 \rangle$ dominates $\sigma_1$].

An important item in the preceding theoretical analysis of the mobility maximum was Lekner's hypothesis that the average scattering length $\langle a \rangle$ goes through zero as a function of density. In an attempt to understand the mechanism responsible for the behavior of the average scattering length, Jahnke, Holzwarth, and Rice[137] (JHR) investigated the implications of Lekner's effective-potential model for fluid argon for densities ranging from the triple-point value to about half the critical-point value. They solved the Wigner–Seitz equations given in Section VI.A using the potential $\tilde{V}$ given by (6.52) to obtain the ground-state energy $E_0 = V_m + \hbar^2 k_0^2 / 2m$ and the effective mass $m^*$ of the quasi-free electrons. For the corresponding muffin-tin scattering potential (7.61), they computed the scattering length $\langle a \rangle$.

The results of JHR are summarized in Table XVIII. There are no experimental values to compare with the work functions and effective masses given in the table for any of the densities given except the highest, which almost coincides with the value Lekner used.

The predicted scattering length $\langle a \rangle$ varies very slowly with density, going from 1.458 $a_0$ at 0.021 $Å^{-3}$ to 1.136 $a_0$ at 0.0042 $Å^{-3}$. Thus Lekner's effective-potential model fails to yield the zero of $\langle a \rangle$, which was successfully used to explain the mobility peak in terms of the single-scatterer kinetic theory. JHR found that attempts to correct for multiple-scattering effects by evaluating $\langle a \rangle$ at the Wigner–Seitz zero-point energy $\hbar^2 k_0^2 / 2m$ lowered the values of $\langle a \rangle$, but by no more than 30%. Thus, if multiple scattering effects cause the zero of $\langle a \rangle$, they cannot be handled in the manner just described. (The last column in Table XVIII contains $k_0$ computed from the optical approximation to the zero-point energy, namely,

$$\frac{\hbar^2 k_0^2}{2m} \equiv \frac{4\pi n \hbar^2 \langle a \rangle}{2m}$$

The optical approximation for $k_0$ and the Wigner–Seitz approximation agree to within 30% in Table XVIII.)

The slow variation in the scattering length with respect to the fluid density is inherent in a muffin-tin potential, since it is short-ranged. Thus, as JHR point out, it is probably in this aspect of Lekner's model that the problem lies. Near the triple point the fluid is dense and highly ordered, so that the long-ranged tail on the polarization interaction largely cancels out because of dielectric screening. This would account for the success of Lekner's

## TABLE XVIII

Scattering Lengths, Zero-Point Energies, and Effective Masses Computed for Fluid Argon with Lekner's Effective Potential Model (Ref. 137)[a]

| Fluid density | | Muffin-tin radius [Eq. (6.53)], $R_m$ (bohrs) | Wigner-Seitz radius, $R_s$ (bohrs) | Lorentz screening factor, $\partial_L$, [Eq. (6.44)] | Maximum average potential | Calculated scattering length, $\langle a \rangle$ (bohrs) | Effective mass, $m^*/m$ | Wigner-Seitz, $k_0$ (bohr$^{-1}$) | Optical potential, $k_0$ (bohr$^{-1}$) |
|---|---|---|---|---|---|---|---|---|---|
| $n$(Å$^{-3}$) | $n$ (bohr$^{-3}$) | | | | | | | | |
| 0.02113 | 0.003131 | 4.00 | 4.24 | 0.774 | 0.157 | 1.458 | 0.75 | 0.330 | 0.239 |
| 0.01903 | 0.002820 | 4.16 | 4.39 | 0.794 | 0.121 | 1.360 | 0.78 | 0.295 | 0.219 |
| 0.01682 | 0.002492 | 4.25 | 4.58 | 0.813 | 0.109 | 1.334 | 0.81 | 0.270 | 0.204 |
| 0.01482 | 0.002196 | 4.29 | 4.77 | 0.832 | 0.099 | 1.309 | 0.84 | 0.248 | 0.190 |
| 0.01176 | 0.001743 | 4.38 | 5.16 | 0.862 | 0.087 | 1.279 | 0.88 | 0.213 | 0.167 |
| 0.00808 | 0.001197 | 4.54 | 5.84 | 0.901 | 0.070 | 1.232 | 0.92 | 0.167 | 0.136 |
| 0.00422 | 0.000625 | 4.82 | 7.25 | 0.945 | 0.047 | 1.136 | 0.97 | 0.110 | 0.094 |
| (Gas) | ($\lesssim 4 \times 10^{-6}$) | ($\infty$) | ($\infty$) | (1.000) | (0.000) | $-1.647^b$ | | | |

[a] For the Wigner–Seitz model $k_0 = \sqrt{2m(E_0 - \bar{V}_m)/\hbar^2}$, and for the optical approximation $k_0 = \sqrt{4\pi n \langle a \rangle}$. 1 ry = 13.6 eV and 1 bohr = $a_0$ = 0.528 Å.

[b] See Ref. 115.

model in computing $\langle a \rangle$ near the triple point. Away from the triple point, there is more disorder in the fluid, with a corresponding loss of radial symmetry of the instantaneous scattering potential "seen" by the electron and a decrease in the degree of cancelation of the long-ranged tail of the polarization potential, which tail is known to give a negative scattering length in the gas.

The calculations of JHR point up the need for work on the determination of a more accurate electron-medium potential than that proposed by Lekner, and perhaps on a more accurate treatment of multiple scattering than has been accomplished at this point in time. On a purely empirical basis, we have shown that the behavior of $\langle a \rangle$ can be accounted for by a potential of the form of Eq. (6.46) times an exponential screening factor $e^{-\kappa r}$, where $\kappa$ is an adjustable function of density. The resulting potential is much longer-ranged than Lekner's muffin-tin potential in the moderate and low density region.

## VIII. THEORY OF ELECTRONIC BUBBLES

From the thermodynamics the condition that one electron state be stable relative to another is that the free energy of the former be lower than that of the latter. Thus, for an electron in the bubble state to be stable relative to the quasi-free state, the condition is

$$E_t < V_0 \qquad (8.1)$$

where $V_0$ is the work function of a quasi-free electron, and $E_t$ is the reversible work of putting an electron from the vacuum into a bubble. $E_t$ can be expressed in the form

$$E_t = E_e + E_s + E_v + E_p \qquad (8.2)$$

where $E_e$ is the ground-state energy of the localized electron, $E_s$ and $E_v$ are the surface and volume work necessary to create the bubble, and $E_p$ is the energy of polarization of the medium surrounding the electronic bubble.

For a cavity of radius $R$, thermodynamical estimates of $E_s$ and $E_v$ are

$$E_s = 4\pi R^2 \gamma \qquad (8.3)$$

and

$$E_v = \tfrac{4}{3}\pi R^3 P \qquad (8.4)$$

where $\gamma$ is the surface tension (which may be curvature-dependent[138,139] for small cavities) and $P$ is the pressure of the fluid. If the density surrounding the cavity differs from bulk, $E_v$ must include the free energy corresponding to the density variation. A rigorous evaluation of $E_p$ is very difficult, requiring performance of the integral

$$\int V(\mathbf{r}, \mathbf{r}_1, \ldots, \mathbf{r}_N) |\psi(r)|^2 \rho_N(\mathbf{r}_1, \ldots, \mathbf{r}_N) \, d^3 r \, d^3 r_1 \cdots d^3 r_N \qquad (8.5)$$

where $V$ is the electron–fluid interaction potential, $\psi(r)$ is the localized electron wave function, and $\rho_N$ is the distribution function of the fluid around the cavity. $\psi$ and $\rho_N$ must be determined self-consistently, since the electron affects the molecular distribution around the cavity and the molecules around the cavity affect the electron. If negligible molecular rearrangement takes place outside the cavity (i.e., if the density outside the cavity is the same as the bulk density far from the cavity) and the electron distribution $|\psi(r)|^2$ is zero outside the cavity, the continuum approximation to $E_p$ is

$$E_p = -\frac{D-1}{2D}\frac{e^2}{R} \qquad (8.6)$$

An important feature of $E_p$ is that it can be negative so that, even if the work function is negative, a bubble may be stable for sufficiently large molecular polarizabilities [or large $D$ if a continuum model such as (8.6) is used].

The forms of $E_s$ and $E_p$ given by (8.3) and (8.4) are used in the theory of stability of electronic bubbles to be discussed in the remainder of this section. In the formal analysis the only assumption made about $E_p$ is that it is a function of $R$. The model used in this section is that presented by Springett, Jortner, and Cohen,[35] except that they neglected $E_p$ (a legitimate omission for helium and perhaps hydrogen, but probably not for any other fluid) and is quite similar to the original model of Kuper.[7b]

The density distribution around the localization cavity is taken to be

$$\begin{aligned} n &= 0 & r < R \\ &= n_0 & r > R \end{aligned} \qquad (8.7)$$

where $n_0$ is the bulk fluid density. The validity of assumption has been verified for liquid helium by a more rigorous cavity model to be discussed below (Eq. (8.22)). In gases, the density around the cavity differs from the bulk value. If an electron were promoted from the cavity to the vacuum (without changing the structure of the cavity and the surrounding fluid), the potential-energy change of the electron would be $-E_p$. The kinetic energy necessary to eject the electron from the vacuum back into the bottom conduction band is $V_0$. Thus in effect an electron is localized in the cavity by a potential barrier of height $V_0 - E_p$. Thus the localized electron wave function may be determined from the equation

$$-\frac{\hbar^2}{2m}\nabla^2\psi = E_e\psi \qquad r < R$$

$$-\frac{\hbar^2}{2m}\nabla^2\psi + V_0'\psi = E_e\psi \qquad r > R$$

(8.8)

where

$$V_0' = V_0 - E_p \qquad (8.9)$$

For $E_p = 0$, this is the model SJC applied to hydrogen and helium; for $E_p$ given by (8.6), it corresponds to a model presented by Miyakawa and Dexter.[140]

The solution to (8.8) is

$$\psi = \frac{\sin kr}{r} \qquad r < R$$

$$= \frac{A}{r}e^{-k'r} \qquad r > R$$

(8.10)

where

$$k = \left(\frac{2mE_e}{\hbar^2}\right)^{1/2}$$

$$k' = \left[\frac{2m(V_0' - E_e)}{\hbar^2}\right]^{1/2}$$

(8.11)

Note that if $V_0' < 0$, (8.8) has no localized solution. The boundary conditions (that $\psi$ and $d\psi/dr$ are continuous at $r = R$) yield the relation

$$\cot kR = -\frac{k}{k'} \qquad (8.12)$$

which determines $k$ for a given $R$.

The equilibrium bubble radius is obtained from the condition

$$\frac{\partial E_t}{\partial R} = 0 \qquad \text{at } R_m \qquad (8.13)$$

Thus, if $V_0$ is known and $E_s$, $E_v$, and $E_p$ are known as functions of $R$, the equilibrium radius and lowest-energy bubble state may be determined from

(8.13). In the special case that $E_p$ is independent of $R$ and that $E_s$ and $E_v$ are given by (8.3) and (8.4) with $\gamma$ independent of $R$, condition (8.13) yields

$$4\pi\beta \frac{\gamma}{(V_0')^2} = \frac{X^2(1 - X^2)^{1/2}}{\kappa R_m[\kappa R_m(1 - X^2)^{1/2} + 1]} - 2\pi\beta^{3/2} \frac{P}{(V_0')^{5/2}} \kappa R_m \quad (8.14)$$

where $\kappa R$ and $X$ are related through the equation

$$\cot g \, \kappa R X = -\left[\frac{(1 - X^2)^{1/2}}{X}\right] \quad (8.15)$$

and the following notations have been introduced:

$$\kappa = \sqrt{2mV_0'/\hbar^2}$$

$$k = X\kappa \qquad k' = (1 - X^2)^{1/2}\kappa$$

$$\beta = \frac{\hbar^2}{2m} \quad (8.16)$$

Equation (8.15) is (8.12) written in the notation of (8.16). In the same notation $E_t$ becomes

$$E_t = V_0'\left[X^2 + 4\pi\beta \frac{\gamma}{V_0'^2}(\kappa R)^2 + \frac{4}{3}\pi\beta \frac{P}{(V_0')^{5/2}}(\kappa R)^3\right] + E_p \quad (8.17)$$

If $E_p$ is a function of $R$, then the actual value of $E_t(R_m)$ will be lower than the value obtained from the $R_m$ given by (8.14).

For low-pressure liquids the terms arising from $E_v$ are negligible in (8.14) and (8.17). In this case the stability condition $E_t - V_0 < 0$ becomes

$$\frac{4\pi\beta\gamma}{V_0'^2} < \frac{1 - X^2}{\kappa R} \quad (8.18)$$

SJC found numerically that the maximum value of the rhs of (8.18) for $X$ and $\kappa R$ obeying (8.15) is obtained for $X = 0.78$ and $\kappa R = 2.9$. Thus the criterion for stability of the bubble state of the electron, when $V_0' > 0$ and $E_v$ is small, may be expressed as

$$\frac{2\hbar^2\gamma}{m(V_0')^2} < 0.047 \quad (8.19)$$

corresponding to an upper limit of 0.78 on $X$ and a lower limit of 2.9 on $\kappa R$.

Using experimental values for $V_0$ and computing $E_p$ iteratively from (8.6), one can determine the electronic bubble energy $E_t$ and the radius $R_m$ of the equilibrium cavity from (8.14), (8.15), and (8.17). The values predicted for

TABLE XIX

Bubble Radius and Energy Computed from (8.14), (8.15), and (8.17)[a]

| Liquid | $T$ (°K) | $n \times 10^2$ ($\text{Å}^{-3}$) | $\gamma$ (erg/cm²) | $V_0$ (eV) | $-E_p$ (eV) | $\dfrac{2\hbar^2\gamma}{m(V_0 - E_p)^2}$ | $R_m$ (Å) | $E_t$ (eV) |
|---|---|---|---|---|---|---|---|---|
| ⁴He | 3 | 2.13 | 0.55 | 1.3 | 0 | 0.00961 | 16.8 | 0.237 |
| Ne | 25 | 3.72 | 5.57 | 0.5 | 0.21 | 0.0314 | 7.54 | 0.383 |
|  |  |  |  |  | 0.12 | 0.0433 | 6.93 | ∼0.01 |
| H₂ | 19 | 2.17 | 2.0 | 0.90 | 0.13 | 0.0072 | 10.7 | 0.283 |

[a] The densities, surface tensions, and work functions are experimental values. The values of $E_p$ are equal to the values obtained from (8.6) for the given $R_m$. For hydrogen, the theoretical value of Table XII is used for $V_0$.

liquid ⁴He and neon are given in Table XIX. The values 16.8 and 9.0 Å predicted for the cavity radii of ⁴He and neon, respectively, are in reasonable agreement with the ranges 10 to 16 Å for ⁴He and 5 to 8 Å for neon deduced from drift mobility and quantum vortex studies.

Hydrogen poses a serious problem to the current theoretical understanding of electronic bubbles in simple liquids. If one uses the SJC theoretical value of $V_0$ (Table XII), then one predicts a stable electronic bubble with an equilibrium radius of 10.7 Å. From Stokes' law the mobility of a bubble is $\mu = e/\xi\eta R$, where $\xi$ is between $4\pi$ and $6\pi$ depending on whether slip or no-slip boundary conditions hold. Using $\xi = 6\pi$, $\eta$ (21°K) = $1.5 \times 10^{-4}$ P, and $R = 10.7$ Å, one predicts a mobility of about $5.4 \times 10^{-3}$ cm²/(V)(sec) compared to the observed value of $5 \times 10^{-3}$ cm²/(V)(sec). This result seems to confirm the bubble model for electrons in liquid hydrogen. Also, the high-to-low-mobility transformation observed in hydrogen vapor (Fig. 6) is very similar to that observed in helium vapor where bubbles seem to be confirmed, the similarity hinting that bubbles exist in hydrogen. The problem, however, is that the work function deduced from spectroscopic data is $-1$ eV in liquid hydrogen, in disagreement with the theoretical value 0.9 eV used in Table XIX. With $V_0 = -1$ eV, the electronic bubble will not be stable according to the present theory. Possible explanations of the discrepancy are that (1) the measured work function is in error by 1 or 2 eV (unlikely since photoinjection and spectroscopic work functions agree within about $\pm 0.1$ eV where comparisons have been made), (2) the negative species in hydrogen is not an electronic bubble but is some other low-mobility species, for example, an ion or a polaron, or (3) the ionization potential of hydrogen in the liquid is shifted in some way not accounted for by (4.2) used to estimate $V_0$ from spectroscopic data. Of these three possibilities, which are not exhaustive, the first seems the least likely. Photoinjection experiments would be useful at this point to perhaps assess the third possibility. Also, studies of the pressure and temperature dependence of electron mobilities in liquid hydrogen would be helpful in clarifying the situation.

Since $E_p$ is dependent on $R$, it is not consistent to use (8.14) in determining the minimum value of $E_t$, since this equation is derived under the assumption of constant $E_p$. For $E_p$ of the form given by (8.6), Miyakawa and Dexter[119] (MD) solved numerically (8.15) and (8.17) for $R_m$ and $E_t(R_m)$ for hydrogen and neon. In their model, however, $V_0$ was taken to be the zero-point Wigner–Seitz energy $\hbar^2 k_0{}^2/2m$, where $k_0$ is determined by (6.29). In their analysis they account for the polarization contribution to $V_0$ by judicious choice of the cutoff parameter $\tilde{a}$ used in computing $k_0$. Of the values they examined for neon, the choice of $\tilde{a} = 0.206$ Å gives the best agreement between theoretical (0.45 eV, $T = 25°K$) and experimental (0.5 eV, $T = 25°K$) work functions. Using this choice of $\tilde{a}$ and the interpolation formula $\gamma = 15.20[1 - (T/44.38°K)]^{1.216}$ ergs/cm$^2$, MD obtain results such as: $V_0 = 0.46$, 0.37, and 0.33 eV; $R_m = 6.68$, 8.45, and 9.13 Å; $E_t = 0.44$, 0.295, and 0.24 eV for liquid neon at $T = 25$, 35, and 39°K, respectively. The radii are in agreement with the value 7.54 Å given in Table XIX, which was determined without varying $E_p$ in the minimization procedure. The difference between $E_t(25°K) = 0.383$ eV given in Table XIX and the MD value of 0.44 eV is due partly to the fact that, for arriving at the value in Table XIX, $V_0$ was taken to be 0.5 eV, larger by 0.04 eV than the value of MD. These investigators perhaps also used a slightly different dielectric constant in computing $E_p$ ($D = 1.28$ at 25°K was used for computing $E_p$ in Table XIX).

Predicting $\mu\eta$ versus $T$ from the Stokes' formula $\mu = e/\xi\eta R_m(T)$ with $\xi = 3.2\pi$, and the MD values of $R_m(T)$, Loveland, Le Comber, and Spear[22] obtained fairly good agreement with experimental results for electrons in liquid neon. The comparison is shown in Fig. 9b.

In disagreement with experimentation similar to the SJC model, MD predicted for liquid hydrogen values of $V_0$ ranging from 0.5 to 2 eV, depending on their choice of $\tilde{a}$.

In the gas phase the surface term $E_s$ is negligible, so that the stability condition $E_t - V_0 < 0$ for the formation of a bubble becomes

$$4\pi\left(\frac{\hbar^2}{2m}\right)^{3/2}\frac{P}{3(V_0')^{5/2}} \leq \frac{1 - X^2}{(\kappa R)^3} \qquad (8.20)$$

From this relation and values of $V_0'$ determined by the SJC theory in Section VII.A, the densities at which quasi-free electrons in 4°K $^4$He gas and 20°K hydrogen gas become unstable relative to the bubble state are about $10^{21}$ and $2 \times 10^{21}$ molecules/cm$^3$, respectively. In $^4$He the high-to-low-mobility transition is observed (Fig. 7) in the region $0.6 \times 10^{21}$ to $1.2 \times 10^{21}$ molecules/cm$^3$, and in hydrogen it is observed (Fig. 6) in the region $10^{21}$ to $3 \times 10^{21}$ molecules/cm$^3$.

Up to this point in this section, only the ground-state behavior of electronic bubbles has been treated. Consider now the excited state. As alluded to in Section IV.C, the photoejection spectra of Sanders et al.[90-92] have been explained by Miyakawa and Dexter[93] with the aid of the cavity model being considered in this section. With the form of $E_p$, $E_s$, and $E_v$ given by (8.6), (8.3), and (8.4), the total energy for an electron in the $j$th state is

$$E_{t,j} = E_{e,j} + \tfrac{4}{3}\pi R^3 P + 4\pi R^2 \gamma - \frac{D-1}{2D}\frac{e^2}{R} \qquad (8.21)$$

where $E_{e,j}$ is the $j$th energy level of an electron in a square well of radius $R$ and depth $V_d(P)$ (equated variously to $E_0$, $V_0$, or $V_0 + [(D-1)/2D](e^2/R)$ in the above calculations), which is a function of pressure through the density. In the MD theory of the photoejection spectrum, $V_d$ was equated to the Wigner–Seitz energy $E_0 = \hbar^2 k_0{}^2/2m$, with $k_0$ determined by (6.29). By assuming a fixed value of the cutoff length $\tilde{a}$, the pressure dependence of $V_d$ then arises from the density dependence of $E_0$. For a given value of $V_d$, MD computed the equilibrium cavity radius $R_m$ from the minimum-energy condition $\partial E_{t,j}/\partial R = 0$ for $E_{e,j}$ corresponding to the $1s$ bound state of the cavity. With the assignment that the low-energy peak in Fig. 31 corresponds to the $1s$-to-$2p$ transition, MD found that the scattering length $\tilde{a}$ corresponding to the zero-pressure value $V_d(0) = 0.948$ eV gave the best prediction of the position of the assigned peak versus pressure. A few of their predicted values for the equilibrium cavity radius are $R_m = 17.1$, $14.2$, $12.7$, and $10.9$ Å for $P = 0$, $5$, $10$, and $25$ atm, respectively, for liquid $^4$He at $1.25°$K. These radii can be compared with those determined[142] from trapping lifetimes of electronic bubbles by vortex lines at $1.7°$K: $R_m = 16.0$, $11.8$, and $10.2$ Å at $0$, $10$, and $20$ atm, respectively.

Using the Wigner–Seitz value of $\tilde{a}$ corresponding to $V_d(0) = 0.948$ eV, MD calculated the cross sections for the transitions $1s$ to $1p$, $1s$ to $2p$, $1s$ to lowest-energy continuum band, and $1s$ to next-to-lowest-energy continuum band as a function of pressure. In Fig. 41 the energies corresponding to the maxima of these cross sections are plotted versus pressure. Also plotted are the peak energies of the low- and high-energy peaks observed in the photoejection experiment.[89] The observed peak positions of the low-energy peak agree very well with the predictions for the $1s$-to-$2p$ transition, and those of the high-energy peak agree quite well with curve $w$ which corresponds to the $1s$ to next-to-lowest-energy continuum band. Curve $b$ corresponds to the $1s$ to lowest-energy-continuum band transition and, as observed experimentally, merges with the peak for the $1s$-to-$2p$ transition as the pressure

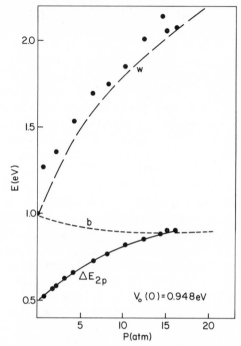

**Fig. 41** Calculation of peak positions of the absorption cross section for the $1s$-to-$2p$ transition and the $1s$-to-continuum transitions $b$ and $w$ described in the text (Ref. 93). The experimental points were taken from Ref. 92.

increases from zero. The relative magnitudes of the predicted cross sections (Fig. 4 of Ref. 93) for the transitions represented in Fig. 41 appear visually to agree with experiment, although MD do not present qualitative comparisons.

An interesting prediction of MD is that the cross section for the $1s$-to-$1p$ transition is at least 50 times greater than the $1s$-to-$2p$ cross section. The transition occurs, according to their calculations, in the energy range 0.1 to 0.2 eV. Thus it would be interesting to carry out photoejection experiments with light in the 10 to 25 $\mu$ region to reinforce the present understanding of the behavior of electronic bubbles.

The nature of the electronic bubble has been described in terms of the simplest of cavity models. There are many more refined versions of the simple model that are available in the literature. Corrections for the curvature dependence of the surface tension have been made.[138] The density distribution around the cavity has been allowed a more realistic form than that of

(8.7). For example, the form

$$n = 0 \qquad r < R_0$$
$$= n_0\{1 - [1 + \alpha(r - R_0)] \exp [-\alpha(r - R_0)]\} \qquad r > R_0 \quad (8.22)$$

where $\alpha$ and $R_0$ are variational parameters, has been used.[138] The most sophisticated calculations available are those of Hiroike, Kestner, Rice, and Jortner (HKRJ),[141] who use the mathematical isomorphism between a pair product form of the wave function and the pair distribution function of a classical liquid in an external field. The interfacial energy arises naturally in the formalism without introducing a surface tension. They use a "soft" bubble of the form of (8.22) and the Austin, Heine, Sham[127] pseudopotential constructed from the interaction potential given by $V_A$ of (6.66). The refined calculations lead to the conclusion that the qualitative features of the trapped state are independent of whether or not the bubble boundary is sharp [see (8.7)] or soft [see (8.22)], and of the gross magnitude of the surface energy. The sharpness of the pseudopotential is perhaps responsible for the fact that the simple square-well model actually gives reasonably good quantitative results for the behavior of electronic bubbles.

## IX. THE ROLE OF LOCALIZED STATES IN ELECTRON TRANSPORT IN DISORDERED MEDIA

### A. Some General Features of Conduction Bands in Disordered Systems

In the theoretical models of the previous sections it was tacitly assumed that the motion of electrons in liquids involves either extended (quasi-free) states (as in argon, krypton, and xenon) or localized (trapped) states (as in helium, hydrogen, and neon). However, what has emerged in recent years as an apparently universal feature of disordered media is that there exist tails of localized states on the conduction bands of these media.[143] The situation is shown schematically in Fig. 42. The shaded regions below and above $E_c$ and $E_{c'}$ represent the localized states. It is expected[144] that the electronic mobility $\mu(E)$ will be much smaller for energies $E$ in the region of localized states than in the region of extended states. This behavior is also illustrated in Fig. 42.

An order-of-magnitude estimate of the difference between the mobilities in the localized and extended states is obtained if one assumes that the electron moves in the localized state by hopping (with a frequency $v$) between neighboring atoms (with a mean jump distance $l$). The diffusion coefficient for this process is $D = vl^2/6$ so that, from the Nernst–Einstein relation $\mu = eD/kT$, the mobility is

$$\mu = \frac{evl^2}{6kT} \qquad (9.1)$$

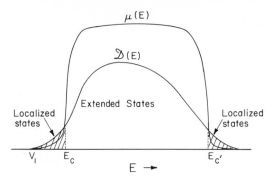

**Fig. 42**   Illustration of the density of states $\mathscr{D}(E)$ and the mobility $\mu(E)$ of a conduction band of a disordered medium.

If $l$ is taken to be of the order of atomic spacing, say 3 Å, and if $v$ is assumed to be less than $10^{15}$/sec, which is roughly the transit time of a thermal electron over the distance $l$, then (9.1) yields the estimate $\mu < 6\ \mathrm{cm}^2/(\mathrm{V})(\mathrm{sec})$ for $T = 300°\mathrm{K}$. This is to be compared to the quasi-free mobilities of the order of $10^2$ to $10^3\ \mathrm{cm}^2/(\mathrm{V})(\mathrm{sec})$ reported in previous sections for dense fluids.

The role of disorder in creating a localized tail on a conduction band is illustrated nicely by the semiclassical approximation to the density of states.[143b] Consider a potential $V(\mathbf{r})$ which is bounded below by $V_1$ and is randomly varying in space (e.g., $V$ may be an effective potential constructed from the actual potential by subtracting a part evaluated for an average lattice model of the system, or $V$ may be the local optical or Wigner–Seitz potentials whose variations arise from local density fluctuations). To the semiclassical approximation the density of states corresponding to a given profile for $V(\mathbf{r})$ is

$$\mathscr{D}_V(E) = \frac{1}{(2\pi)^3\hbar^3} \int_\Omega d^3r\, d^3p\, \delta\!\left[ E - \frac{p^2}{2m} - V(\mathbf{r}) \right]$$

$$= \frac{(2m)^{3/2}}{(2\pi)^2\hbar^3} \int_\Omega d^3r\, \eta(E - V)(E - V)^{1/2} \qquad (9.2)$$

where $\eta(x) = 0,\ 1$ for $x < 0$ and $> 0$, respectively. If the probability that $V$ lies between $V$ and $V + dV$ is $p(V)\, dV$, then the ensemble average value of the density of states $\mathscr{D}(E)$ is

$$\mathscr{D}(E) = \int p(V)\mathscr{D}_V(E)\, dV = \frac{(2m)^{3/2}\Omega}{(2\pi)^2\hbar^3} \int_{V_1}^{E} p(V)(E - V)^{1/2}\, dV \qquad (9.3)$$

where the second is obtained by permuting the integrations over $d^3r$ and $dV$ and noting that the ensemble average value of a function of $E - V$ is independent of $\mathbf{r}$. Thus the integral over $d^3r$ simply yields the volume $\Omega$ of the system.

Consider a completely ordered system described by $p(V) = \delta(V - \overline{V})$, where $\overline{V}$ is some average potential. Then $\mathcal{D}(E)$ has the well-known quasi-free electron from $(E - \overline{V})^{1/2}\eta(E - V)$ with a band edge at $\overline{V}$ (i.e., $d\mathcal{D}/dE \to \infty$ as $E \to \overline{V}$ from above). However, if $p(V)$ is a piecewise continuous distribution, $d\mathcal{D}(E)/dE$ has no singularity and therefore no sharp band edge. The low-energy tail to the band will be qualitatively of the form shown in Fig. 42 for any piecewise continuous distribution $p(V)$.

Aside from the density of states in a system with a randomly varying potential, the question of the nature of the electronic motion arises. Ziman[145] has pointed out that one is dealing here with a percolation problem. To see this, consider an electron energy $E$. For a given potential profile $V(\mathbf{r})$, the electron is allowed access to the point $\mathbf{r}$ if $E > V(\mathbf{r})$ and it is prohibited access to the point of $E < V(\mathbf{r})$ (quantum mechanical tunneling is ignored in this discussion). The average volume fraction $C(E)$ of the system allowed to an electron of energy $E$ is given by

$$C(E) = \int_{V_1}^{E} p(V) \, dV \qquad (9.4)$$

If the allowed regions and prohibited regions are distributed randomly throughout the system then, according to percolation theory,[146] there exists a nonzero critical volume fraction $C^*$, the percolation threshold fraction, such that if

$$C(E) < C^* \qquad (9.5)$$

there is *zero* probability that a continuous channel of allowed region will span the entire system. If $C(E) < C^*$, an electron cannot traverse the system and will therefore have a zero mobility. Thus the localization edge $E_c$ in Fig. 42 is determined from the condition

$$C(E_c) = C^*$$

in this treatment.

If the mobility in the allowed region is denoted by $\mu_f(E)$, then the observed mobility is expected to be of the form

$$\mu(E) = \mu_f(E)h(E) \qquad (9.6)$$

where $0 < h < 1$ and

$$\begin{array}{ll} h = 0 & C < C^* \\ \neq 0 & C > C^* \end{array} \qquad (9.7)$$

For $C$ only slightly larger than $C^*$, $h$ will be very small, since most of the allowed region will be surrounded by the prohibited region and only a small tortuous portion of the allowed region will provide an open channel for the electron. In a dynamical picture of the motion, the electron moves back and forth across an allowed region until a fluctuation converts a prohibited region to an allowed region. The newly converted allowed region allows the electron to move further from its starting point. Continuing this process the electron diffuses through the system with a diffusion coefficient that is larger the greater the average distance an allowed region projects along the drift direction. As $C - C^*$ increases, the average distance an allowed region spans in the drift direction increases. Thus one might expect $h \propto (C - C^*)^\gamma$, where $\gamma > 0$.

In computer studies of randomly connected conducting networks, Kirkpatrick[146] has found that $C^*$ ranges from about 0.1 to 0.3, depending on the connectivity of the network (i.e., where the conductors form a square, cubic, face-centered cubic, etc., lattice), and for $C$ near $C^*$ he found

$$h \propto (C - C^*)^\gamma \tag{9.8}$$

where $\gamma \sim 1.5$ to 1.6. The quantity $C$ in the network represents the fraction of connective bonds having unit conductance, and $1 - C$ is the fraction having zero conductance. For a face-centered cubic structure, Kirkpatrick found $C^* \simeq 0.18$, a value representing perhaps a reasonable guess for a continuous medium.

The model as described thus far provides a good qualitative picture of the mobility drop as $E$ approaches $E_c$ from above. However, in a more general picture of electronic conduction, $C(E)$ represents the volume fraction of the system in which the electron moves with a high mobility $\mu_f(E)$, and $1 - C(E)$ is the volume fraction of the system in which the electron moves with low mobility $\mu_l(E)$. The system mobility $\mu(E)$ may then be some function of $C(E)$, $\mu_f(E)$, and $\mu_l(E)$. Neither the existence nor the nature of such a functional relationship has been demonstrated rigorously although, in the works discussed in Sections IX.B and IX.C two versions of such a function are introduced.

The localized states just described arise from intrinsic fluctuations of the potential field, that is, they are a function only of the medium. The equilibrium bubble states, by contrast, may be described as self-trapping states and are the result of self-consistent interaction between the electron and the medium. Of course, an electron localized by intrinsic fluctuations may relax into a self-trapped state if the self-trapping relaxation rate is great enough. Presumably, the localized tails on the conduction bands in liquid argon, krypton, and xenon are so narrow and represent such a small fraction of the band that they do not contribute measurably to the electron transport process.

## B. Mobility of Electrons in Helium Gas

The ideas presented in Section IX.A were used by Eggarter and Cohen[12,13] in describing the mobility transition observed in helium gases. As a fluctuating potential $V(\mathbf{r})$, they take the Wigner–Seitz potential $V_{WS}(n(\mathbf{r})) \equiv E_0 = \hbar^2 k_0^2/2m$, with $k_0$ determined by (6.29) for the local density $n(\mathbf{r})$. Since polarization effects are small in helium, $V_{WS}(n(\mathbf{r}))$ represents the local potential for injecting an electron from the vacuum to a system of density $n(\mathbf{r})$.

Fluctuations in the local density are accompanied by fluctuations in the local potential $V_{WS}$. To compute the fluctuations, however, one must introduce a characteristic distance $L$ and a corresponding volume $L^3$ in which the fluctuations of the density are computed. On the basis of the uncertainty principle, they chose $L$ to be approximately the de Broglie wavelength, that is, they set

$$L = 2\pi c\left(\frac{3\hbar^2}{2mE}\right)^{1/2} \tag{9.9}$$

where $c$ is an adjustable parameter of order 1. From grand canonical ensemble theory, the mean-square fluctuation $\sigma_n^2$ of the number density in the box of volume $L^3$ is known to be

$$\sigma_n^2 = nL^3 kT \left(\frac{\partial n}{\partial P}\right)_T \tag{9.10}$$

which yields the corresponding mean-square fluctuation in the potential in the box

$$\sigma_V^2 = \left(\frac{\partial V_{WS}}{\partial n}\right)^2 \sigma_n^2 \tag{9.11}$$

Since there are many atoms in a box of volume $L^3$, the distributions of $n$ and, correspondingly, $V$ can be represented by the Gaussian distribution, which for $V$ has the form

$$p(V) = \frac{1}{\sqrt{2\pi\sigma_V^2}} e^{-(V-\bar{V})^2/2\sigma_V^2} \tag{9.12}$$

where $\bar{V} = V_{WS}$ for the average number density of the fluid and $\sigma_V^2$ is given by (9.11). Putting this result into the semiclassical approximation (9.3), for $\mathscr{D}(E)$, one obtains

$$\mathscr{D}(E) = \frac{(2m)^{3/2}\Omega}{4\pi^2} \sigma_V^{1/2} \mathscr{F}\left(\frac{E-\bar{V}}{\sigma_V}\right) \tag{9.13}$$

where

$$\mathscr{F}(x) = (2\pi)^{-1/2} \int_0^\infty z^{1/2} e^{-(x-z)^2/2} \, dz \qquad (9.14)$$

Similarly, the average fraction of the system allowed to the electron, defined by (9.4), is found to be

$$C(E) = (2\pi)^{-1/2} \int_{-\infty}^{(E-\bar{V})/\sigma_V} e^{-z^2/2} \, dz \qquad (9.15)$$

Eggarter and Cohen estimate the mobility $\mu_f(E)$ of electrons in an allowed region from the formula

$$\mu_f(E) = \frac{4}{3} \frac{e}{m} \tau(E) \qquad (9.16)$$

where the collision time is given by

$$\tau(E) = \lambda(E) v_a^{-1} \qquad (9.17)$$

$\lambda(E)$ is an effective mean free path of the form

$$\lambda(E) = \frac{\lambda_c \lambda_p}{\lambda_c + \lambda_p} \qquad (9.18)$$

where $\lambda_c = (n\sigma)^{-1}$ is the ordinary electron–atom mean free path in a gas, and $\lambda_p \simeq LC(E)/[1 - C(E)]$ is the mean free path of electron scattering off the prohibited regions. The quantity $v_a^{-1}(E)$ is the average of the inverse of the electron velocity $v(E) \equiv \sqrt{\hbar^2(E - V)/2m}$ in the allowed region, that is, $v_a^{-1} \equiv \int_{V>E} v^{-1}(E) p(V) \, dV$.

In addition to the quasi-free motion in an allowed region, a low-energy electron localized in a low-density region, a "pseudobubble," by the higher-density fluid surrounding the region can move with the entire low-density region in much the same way as an electronic bubble moves in liquid helium. Eggarter and Cohen compute the mobility $\mu_l$ of this pseudobubble from the semihydrodynamical formula (2.3) suggested by Levine and Sanders[9,10] to explain the low-mobility portion of their gas-phase mobility data.

To estimate the overall mobility $\mu(E)$ of electrons of energy $E$ in helium gas, Eggarter and Cohen took into account percolation effects by expressing $\mu(E)$ in the form

$$\mu(E) = \mathscr{P}(E)\mu_f(E) + [1 - \mathscr{P}(E)]\mu_l(E) \qquad (9.19)$$

where $\mathscr{P}(E)$ is the percolation probability. For $\mathscr{P}(E)$ they used the form $\mathscr{P}(E) = 0$ if $C(E) \leq C^* = 0.3$, and $\mathscr{P}(E) = 1 - e^{-25[C(E) - 0.3]}$ if $C(E) > 0.3$, as determined by Frisch, Hammersley, and Welsh[147] from computer experiments on a simple cubic conductance network. In view of Kirkpatrick's[146] computer studies, $\mathcal{P}(E)$ should probably be replaced by a function $h(E)$ which also goes to zero as $C \to C^*$ from above but, as indicated in (9.8), more slowly than $\mathcal{P}(E)$.

The radius $R_{ps}$ of the pseudobubble, needed to estimate $\mu_l$ from (2.3), was estimated to be $R_{ps} = c'(3\Delta N/4\pi n)^{1/3}$, where $\Delta N \approx L^3[n - n_{max}(E)]$, where $n_{max}(E)$ is the greatest density for which $V_{WS}(n) < E$. Thus $\Delta N$ is an estimate of the number of atoms displaced to keep $V_{WS} < E$ in the pseudobubble.

With the formulas outlined above, the general theoretical expression

$$\mu = \frac{\int \mathcal{D}(E)\mu(E)e^{-\beta E}\,dE}{\int \mathcal{D}(E)e^{-\beta E}\,dE} \tag{9.20}$$

the thermodynamical values of $\sigma_n{}^2$ and $n$, the value $4\pi(0.65 \text{ Å})^2$ for the cross section $\sigma$, and adjusted values of $c$ and $c'$, the theory gives excellent agreement with experiment for the density dependence of the mobility of electrons in helium gas for temperatures ranging from 3 to 18.1°K. The parameters $c$ and $c'$ are slowly varying and of the order of 1 (e.g., $c = 1.4$ and $c' = 0.4$ at 4°K, and $c = 1.02$ at 18.1°K). The dashed curves in Fig. 4 are typical of the theoretical predictions.

Recently, Hernandez[14] developed a theory of electrons in gases similar to the theory of Eggarter and Cohen, but with the following important differences: (1) The characteristic length $L$ is taken to be the momentum mean free path $(n\sigma)^{-1}$ instead of the de Broglie wavelength of (9.9). (2) The density of states includes self-trapping states by adding to the quasi-free electron contribution to the density of states of the $i$th cell (of volume $L^3$ and having a potential $V_i$)

$$2L^3 \frac{(2m)^{3/2}}{4\pi^2\hbar^3}(E - V_i)^{1/2}\eta(E - V_i) \tag{9.21}$$

and self-trapping contribution

$$2L^3 \frac{(2m^*)^{3/2}}{4\pi^2\hbar^3}E_K^{1/2}\delta(V_i - E - F_R - \alpha_R) \tag{9.22}$$

which is important when $E \ll \bar{V}$. $m^*$ is the effective mass of the self-trapped electron, $E_K$ is the kinetic energy of the center of mass of the entity moving with the trapped electron, $F_R$ is the extra energy of the medium due to local distortion caused by the electron, and $\alpha_R$ is the electronic binding energy.

Hernandez obtains three contributions to the electronic mobility, two contributions coming from (9.21) and giving, according to him, the contributions of the quasi-free states and the Anderson localized states (which are the ones considered by Eggarter and Cohen) due to intrinsic potential fluctuations. The third contribution, coming from (9.22), represents self-trapping effects and must be computed self-consistently just as the electronic bubble computations were. Hernandez credits the effect of the Anderson localized states for the slight difference between quasi-free predictions and measured mobilities on the low-density side of the mobility transition, but argues that a precipitous increase in the effective mass of the self-trapped electron is responsible for the large drop in the mobility in the transition density range. Since effective-mass calculations seem difficult if not impossible at this point it is difficult to test Hernandez's mobility theory in the transition region. However, it does seem to differ from Eggarter and Cohen's version, although the difference is difficult to quantify since both theories imply that $\mu$ is approximately the quasi-free value (2.2) for densities under the transition density range and approximately the hydrodynamic value (2.3) for densities above this range. Perhaps the nonequilibrium overshoot mobility results[15,16] (Figs. 6 to 8) provide some test of the theories, although we do not see how at this point.

### C. Electron Mobility in Hydrocarbon Liquids

Kirkpatrick[146] found the percolation phenomena observed in his computer studies can be explained quantitatively for values of $C$ ranging from 1 down to at least 0.3, and qualitatively over the entire range 0 to 1 (in his studies $C$ is the fraction of connective bonds having unit conductance, and $1 - C$ is the fraction having zero conductance) by the effective-medium theory developed by Bruggeman[148] and Landauer[149] for solid mixtures of conducting substances (e.g., disordered alloys). Assuming a medium to be composed of a random mixture of two substances, one substance having an electron mobility $\mu_0$ and the other a mobility $\mu_1$, one obtains for the effective-medium theory the following expression for the mobility of the mixture:

$$\frac{\mu}{\mu_0} = \mathscr{A} + \left[ \mathscr{A}^2 + \frac{1}{2} X \right]^{1/2} \tag{9.23}$$

where

$$X = \frac{\mu_1}{\mu_0} \tag{9.24}$$

and

$$\mathscr{A} = \tfrac{1}{2}[(\tfrac{3}{2}C - \tfrac{1}{2})(1 - X) + \tfrac{1}{2}X] \tag{9.25}$$

where $C$ represents the volume fraction of the substance with mobility $\mu_0$. Equation (9.23) can be obtained from Landauer's expression for conductance by noting the correspondence between Ohm's law and Fick's law. Thus the conductance ratios obtained by Landauer can be replaced by corresponding ratios of the self-diffusion coefficients, which by the Nernst–Einstein law are equal to mobility ratios. Thus (9.23) is obtained.

Effective-medium theory has been successful[150] in describing the metal–nonmetal transition in one-component (e.g., mercury vapor) and two-component (e.g., sodium–ammonia solutions) systems with the interpretation that $\mu_0$ and $\mu_1$ represent, respectively, electron mobility in the high- and low-mobility regions of the fluid. Kestner and Jortner have suggested that (9.23) is applicable to electron mobility in a hydrocarbon liquid, with the interpretation that $C$ is the average volume fraction of the system in which electrons have a high mobility $\mu_0$ and $1 - C$ is the average volume fraction in which the electrons are in localized states of low mobility $\mu_1$. The low-mobility regions probably arise from potential wells caused by orientational fluctuations of the polyatomic molecules rather than density fluctuations. The fact that electrons are much more mobile in fluids composed of symmetric molecules (e.g., methane, neopentane) than of unsymmetric molecules (e.g., ethane, n-hexane) supports the hypothesis that orientational fluctuations cause localization in liquid hydrocarbons.

To apply (9.23) to mobility in hydrocarbons, Kestner and Jortner made the following added assumptions:

1. The potential energy distribution $p(V)$ is Gaussian, so that

$$C(E) = \frac{1}{(2\pi\sigma_V^2)^{1/2}} \int_{-\infty}^{E} e^{-(V-\bar{V})^2/2\sigma_V^2}\, dV \qquad (9.26)$$

2. The energy distribution of the electrons is strongly peaked about the average $\langle E \rangle$, and therefore $C(E) \simeq C(\langle E \rangle) \equiv C$.

3. The work function $V_0$ is given by

$$V_0 = T_0 + \bar{V}_p + \bar{V}_d \qquad (9.27)$$

where $T_0$ is the zero-point kinetic energy of a quasi-free electron, and $\bar{V}_p$ and $\bar{V}_d$ are the values of the polarization and dipolar potentials of the electron–medium interaction averaged over the sampling length $L$ associated with the distribution $p(V)$. (Whether $L$ is the de Broglie wavelength, quasi-free mean free path, and so on, is a problem not resolved at this point, since adjustable parameters absorb the $L$ dependence.) Equation (9.27) can be combined with the expression $\bar{V} = \bar{V}_a + \bar{V}_p + \bar{V}_d$, where $\bar{V}_a$ is the average of the repulsive part of the electron–atom potential, to obtain

$$\bar{V} = V_0 + \alpha \qquad (9.28)$$

where $\alpha = \bar{V}_a - T_0$. Thus $C$ can be written in the form

$$C = (2\pi)^{-1/2} \int_{-\infty}^{(E_t - V_0)/\sigma_V} e^{-z^2/2} \, dz \qquad (9.29)$$

where $E_t = \langle E \rangle - \alpha$.

4. The parameters $\mu_0$, $\mu_1$, $E_t$ and $\sigma_V$ are the same for all the hydrocarbon liquids.

With assumptions 1 to 4, a suitable adjustment of the four parameters of the model, and experimental values of the work functions, Kestner and Jortner[54] predicted with (9.23) electron mobilities for several hydrocarbons. The results are compared with experiment in Fig. 43. The values of the parameters used are given in the legend for the figure. The agreement with experiment is encouraging and, in view of the model, implies that the drop from

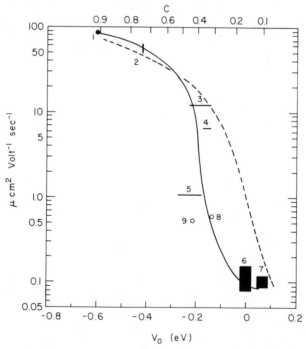

**Fig. 43** Electron mobility versus work function. Smooth curve is effective medium theory [see (9.20)], with the parameters $\mu_0 = 100$ cm$^2$/(V) (sec), $\mu_1 = 0.06$ cm$^2$/V, $E_t = -0.27$ eV, $\sigma_V = 0.26$ eV. Dashed curve is Schmiller, Vass, Mandics theory [see (2.10) and (9.27)] with $\sigma_c = 0.12$ eV, $E_t = -0.26$ eV, $\mu_f = 80$ cm$^2$/(V) (sec) and $\mu_l = 0$. Experimental points are (1) TMS, (2) neopentane, (3) 2,2,-dimethylbutane, (4) 2,2,4-trimethylpentane, (5) cyclopentane, (6) n-pentane, (7) n-hexane, (8) benzene, and (9) toluene.

high to low mobility in going from liquids of higher to lower molecular symmetry is a percolation effect.

In spite of the agreement between theory and experiment illustrated in Fig. 43, the use of four adjustable parameters in explaining eight data points gives cause for reservation in accepting the effective-medium model without further tests. One piece of evidence supporting the model is that quasi-free mobilities predicted for neopentane and $n$-hexane from (2.4) with the "work function" hard-core radii given in Table XIV are 94 and 170 cm$^2$/(V)(sec), in reasonable agreement with the value $\mu_0 = 100$ cm$^2$/(V)(sec) used for the model predictions. A test of Kestner and Jortner's assumption that the parameters $\mu_0$, $\mu_1$, $E_t$, and $\sigma_V$ are the same for all hydrocarbons can also be tested. By comparing the experimental mobilities of $n$-hexane–neopentane mixtures[46] at various compositions with (9.23) for the assumed values of the parameters, one can construct the work function, as a function of composition, required to fit the theory to experiment. The open triangles in Fig. 35 represent the work function required to fit theory and experiment. The agreement with the measured work function is not very good, suggesting either a failure of the effective-medium model or Kestner and Jortner's assumption 4. Further experimental tests of the model and basic theoretical work on the model parameters are in order at this time. High-pressure, dense-gas experiments look promising, allowing the possibility for observing mobility behavior in the same fluid under widely differing temperature and density conditions.

In another two-state model of electron mobility in hydrocarbons, Schiller, Vass, and Mandics[53] (SVM) assume the form given by (2.10), in which the probability $P$ that the electron is in the quasi-free state is assumed to be of the Gaussian form

$$P = \frac{1}{(2\pi)^{1/2}} \int_{(V_0 - E_t)/2\sigma_c}^{\infty} e^{-z^2/2}\, dz \qquad (9.30)$$

where $\sigma_c^2 = kT^2 C_V$, $C_V$ being the constant-volume specific heat, $V_0$ the work function, and $E_t$ the energy of a trapped electron. In this model it is assumed that the system may be subdivided into $N$ cells which in the absence of an electron have the energy distribution $p(e) = (2\pi\sigma_c^2)^{-1/2} \exp[-(e - \bar{e})/2\sigma_c^2]$, where $e$ is the energy of a cell, and $\bar{e}$ is the mean energy of a cell. The free energy of the system with a quasi-free electron is $F_f = N\bar{e} + V_0$, whereas with a localized electron it is $F_l = (N - 1)\bar{e} + e + E_t$. Thus an electron will be quasi-free in the system if $F_f < F_l$, or if $e - \bar{e} < V_0 - E_t$. Accordingly, (9.30) gives the probability that an electron in the system will be quasi-free.

Assuming $\mu_f \gg \mu_l$, estimating from experimental data that $\sigma_c = 0.12$ eV at 300°K for all the hydrocarbons considered, and assuming the common values

$E_t = -0.26$ eV and $\mu_f = 80$ cm$^2$/(V)(sec) for these hydrocarbons, SVM predicted the dashed curve shown in Fig. 43. The effective-medium theory seems to agree somewhat better with experiment than the SVM theory, although the fact that SVM adjusted two parameters fewer than Kestner and Jortner should be borne in mind. The SVM model predicts mobility activation energies of 0.29, 0.04, and $\sim 0$ eV for $n$-hexane, 2,2-dimethyl-butane, and TMS. Experimentally, the values are about 0.18, 0.052, and $\sim 0$ eV, in sufficiently good agreement with experiment to warrant further testing of the SVM model for hydrocarbons. Values of $V_0$ needed to adjust the SVM theoretical mobilities to the experimental mobilities for electrons in $n$-hexane–neopentane mixtures are shown (solid triangles) in Fig. 35. The agreement with experimental $V_0$ values is not very good, although it is slightly better than that found by a similar comparison for the Kestner–Jortner model.

It should be noted that methane data ($\mu \simeq 400$ cm$^2$/V sec, $V_0 \simeq 0$) do not correlate with either the Kestner–Jortner or the SVM model. This indicates either a failure of the models themselves or the simplifying assumptions about the nature of the parameters entering the models.

Finally, in the process originally suggested by Minday, Schmidt, and Davis,[47] the mobility is given by (2.10), $\mu_f$ estimated from the quasi-free formula, and the activation energy given by (2.14), in which a model for the localized state must be specified to calculate the reversible work $W_l$ of localization. If the cavity model described in the theory of electronic bubbles is assumed, $W_l$ will equal $E_t$ given by (8.2), but with local dipolar interactions and the anisotropic polarization interactions leading to a larger polarization term than that given by (8.6). By assuming $E_p = \omega e^2/R$ and choosing the parameter $\omega$ so that quantity $V_0 - E_t = 0.18$ eV, the observed activation energy for the electron mobility in $n$-hexane, values of $\omega$ and the equilibrium cavity radius $R_m$ are determined from the bubble model of Section VIII to be $\omega = 0.35$ and $R_m = 3.3$ Å. With an anisotropic polarizability of about 2 Å$^3$, one estimates the anisotropic polarization interaction of about six molecules surrounding the cavity could provide the stabilization energy necessary for cavity entrapment of an electron in $n$-hexane; this conclusion is reached by setting

$$\frac{\omega e^2}{R_m} = \frac{D-1}{2D}\frac{e^2}{R_m} + 6\,(2\,\text{Å}^3)\frac{e^2}{R_m{}^4}$$

Thus the possibility that cavity formation is involved in electronic conduction in hydrocarbons cannot be entirely ruled out. Again high-pressure gas experiments may be useful to test such possibilities. Of course, the model of Minday, Schmidt and Davis depends only upon the existence of high- and low-mobility states. The low-mobility states need not be bubbles.

## X. REMARKS ON THE THEORY OF RADIATION FREE-ION YIELDS

As discussed in Section III, many experimental data have been collected on free-ion yields $G^0_{fi}$ in liquids. These data have been discussed mostly in terms of the "thermalization" range parameter $b$ introduced empirically through (3.10). No adequate a priori theory of $b$ exists presently. However, several theoretical articles of Magee, Mozumder, and co-workers have shed light on the ion yield process.

An a priori calculation of the free-ion yield of a given radiation pulse is extremely complicated. For example, consider a primary 1-MeV electron as the radiation source. In the course of its penetration through a molecular medium, the primary electron creates ion pairs by knocking out electrons from the molecules. By the time the primary electron has thermalized, it will have created many secondary electrons with initial energies ranging from zero to many kiloelectronvolts. These secondary electrons can in turn ionize more molecules, creating tertiary electrons with an energy distribution that depends on the energy of the secondary electrons. And so the process continues until no more ionization can occur. In Monte Carlo simulations of high-energy electron tracks, Mozumder and Magee[150,73] made the important discovery that most of the energy of a high-energy (1-MeV in the case studied) primary electron is deposited in producing secondary electrons with energies too low (less than about 100 eV) to produce further ionization with appreciable efficiency. For example, from Table II of Ref. 73, the total ion pair yield of a 1-MeV primary electron in $n$-hexane, computed by the Monte Carlo scheme of Mozumder and Magee, is 4.46 ion pairs per 100 eV of energy lost by the primary electron, and of this yield 98.4 % of the electrons produced have energies less than 100 eV. Thus, in analysis and theory of free-ion yields, one can probably ignore higher energy ($\geq 100$ eV) ion pair formation without serious error. This has indeed been done in correlating free-ion yield data with (3.11) and (3.12).

The relaxation or thermalization of an electron in a fluid involves three sequential processes of energy loss: (1) Electronic processes in which the electron loses energy by ionization and electronic excitation of the fluid molecules. (2) Intramolecular vibrational excitation occurring from energies below the electronic excitation threshold ($\sim 6$ eV) down to the vibrational threshold ($\sim 0.4$ eV). (3) Subvibrational processes, which include elastic collisions, excitation of molecular rotations, and excitation of intermolecular vibrations. Final thermalization of the electron is the result of the third process. If it is assumed that the spatial displacement of an electron is governed by a Gaussian distribution in each of these processes, then the overall process is described by the Gaussian distribution

$$p_E(\mathbf{r})\, d^3r = \frac{1}{(\pi b^2)^{3/2}}\, e^{-r^2/b^2}\, d^3r \tag{10.1}$$

where $p_E(\mathbf{r})\, d^3r$ represents the probability that an electron beginning at $\mathbf{r} = 0$ (or between $\mathbf{r} = 0$ and $d\mathbf{r}$) with energy $E$ will be at $\mathbf{r}$ (or between $\mathbf{r}$ and $\mathbf{r} + d\mathbf{r}$) when it has been thermalized. Under the assumption that the three processes are Gaussian, it follows that

$$b^2 = \sigma_e{}^2 + \sigma_v{}^2 + \sigma_{sv}^2 \tag{10.2}$$

where $\frac{3}{2}\sigma_e{}^2$, $\frac{3}{2}\sigma_v{}^2$, and $\frac{3}{2}\sigma_{sv}^2$ are the mean-square displacements of an electron involved in processes 1, 2, and 3, respectively.

Mozumder, Chatterjee, and Magee[151,152] have concluded that process 1 does obey a Gaussian and have computed $\sigma_e$ versus energy $E$ of the electron. At low energies $\sigma_e$ is rather small, for examples, for $n$-hexane the following estimates are obtained:[152] $\sigma_e = 4.5, 6.4$, and $8.7$ Å for $E = 50, 75$, and $100$ eV, respectively. Assuming for process 2 a cross section of $\sim 40$ Å$^2$, an average vibrational quantum of $\sim 0.2$ eV, a number density equal to $n$-hexane at $293°$K, and that $E$ falls from 6 to 0.4 eV, Mozumder and Magee obtain the rough estimate $\sigma_v \sim 17$ Å for electrons in $n$-hexane. Since $\sigma_e$ depends primarily on the density of carbon and hydrogen atoms in the fluid and $\sigma_v$ on the number of C—H bonds per molecule, one would not expect the $\sigma_e$ and $\sigma_v$ values of neopentane to differ markedly from those of $n$-hexane. It will be convenient below to assume they are the same for the two fluids.

Since the estimated value of $\sqrt{\sigma_e{}^2 + \sigma_v{}^2}$ is about 18 Å, and since the measured values of $b$ are about 67 and 178 Å, respectively, for $n$-hexane and neopentane, the Gaussian model represented by (10.1) and (10.2) implies that $\sigma_{sv} \sim 64.5$ Å for $n$-hexane and $\sim 177$ Å for neopentane. These estimates pose a serious problem for the Gaussian approximation to the dispersion of subvibrational electrons (energies from $\sim 0.4$ eV to $kT$), that is, for process 3. The relaxing electrons are quasi-free in $n$-hexane down to trapping or localization energies (estimated to be $\sim 0.15$ eV from mobility data) and in neopentane down to thermal energies ($kT \sim 0.025$ eV at $296°$K). Thus since the momentum mean free path $\Lambda_1$ is of order 100 Å for quasi-free electrons, the displacement of electrons in process 3 will obey a Gaussian distribution only if $\sigma_{sv} \gg \Lambda_1$. If the number $v$ of momentum scattering events in thermalization during process 3 is large compared to unity, then process 3 will obey a Gaussian distribution with $\sigma_{sv} \simeq \sqrt{\frac{2}{3}v}\, \Lambda_1$. Otherwise, the process will not be Gaussian. In fact, if the thermalization range of an electron in process 3 is of the order of $\Lambda_1$ or less—as is actually the case since $b$, a measure of the root-mean-square thermalization range, is $\sim 100$ Å—then it will be a better approximation to assume a linear trajectory for the relaxing electron than to assume a Gaussian distribution.

If it is assumed that processes 1 and 2 are Gaussian and that the electron loses its excess energy in process 3 without deflection, that is, in a linear trajectory, then the formula predicting the free-ion yield is of the form

$$\int \frac{e^{-R^2/(\sigma_e{}^2 + \sigma_v{}^2)}}{[\pi(\sigma_e{}^2 + \sigma_v{}^2)]^{3/2}} e^{-r_c/(R+l)} d^3R \tag{10.3}$$

rather than of the form given by (3.11). $R$ is the distance the electron travels during processes 1 and 2, and $l$ is the distance the electron travels during the final thermalization step, process 3. Although (10.3) cannot be tested quantitatively without theoretical values of $\sigma_e$, $\sigma_v$, and $l$, certain empirical observations can be made. In view of the items discussed in the proceeding paragraphs, it is reasonable to approximate $\sqrt{\sigma_e{}^2 + \sigma_v{}^2}$ as 20 Å for the electrons of the majority ion pairs produced in a radiation pulse experiment. Equating (10.3) to the experimental free-ion ratio $G_{fi}^0/G_0 = 0.131/4$ and $0.857/4$ given in Table III for $n$-hexane and neopentane at 296°K and assuming $\sqrt{\sigma_e{}^2 + \sigma_v{}^2} = 20$ Å, one concludes that $l_h = 64$ Å and $l_{np} = 180$ Å for $n$-hexane and neopentane, respectively, for the model to reproduce the experimental values. The values used for $r_c$ are given in Table III, and $G_0$ was set equal to 4. The difference between $l$ for the fluids can be rationalized in terms of the ideas set forth earlier in explaining mobility trends. In $n$-hexane the linear motion of the relaxing electron is expected to maintain until the electron energy has dropped to some characteristic value $E_h{}^t$ that is low enough for trapping to occur. At the onset of trapping the momentum mean free path of the electron will begin to decrease drastically so that the linear trajectory approximation will be invalid. A Gaussian distribution with a root-mean-square dispersion of the order of $\sigma \sim 5$ Å would be a better description of the thermalization process for electrons in the trapping energy region. Since $l_h \gg 5$ Å, the range of the subexcitation electron in $n$-hexane is determined primarily by the energy loss in the quasi-free region. Moreover, in the context of the present model, the difference $l_{np} - l_h = 114$ Å arises from the fact that the electron remains quasi-free even at thermal energies in neopentane. Thus the distance $l_{np} - l_h$ represents the distance an electron in neopentane travels in relaxing from the energy $E_h{}^t$ to thermal energy. Assuming the mean free path $\Lambda_0$ for inelastic energy losses is about 5 Å, then the number $N$ of energy-loss collisions (not direction-changing collisions since the momentum mean free path is $\sim 100$ Å for subvibrational electrons in neopentane) that the electron in neopentane undergoes in thermalizing from the energy $E_h{}^t$ may be estimated by the relation $N\Lambda_0 = l_{np} - l_h$, yielding $N \sim 23$ collisions. Estimating $E_h{}^t$ as the mobility activation energy of electrons in $n$-hexane, that is, $E_h^t \sim 0.176$ eV and defining $\langle \varepsilon \rangle$, the average electron energy lost per inelastic collision in neopentane, by the expression $N\langle \varepsilon \rangle = E_h^t - \frac{3}{2}kT$, $T = 296°$, one obtains the estimate $\langle \varepsilon \rangle \simeq 0.00654$ eV. This estimate of average inelastic energy losses of subvibrational electrons (energies $< 0.4$ eV) is consistent with Raman spectral studies[153] that indicate the existence in liquid hydrocarbons of vibrations having characteristic energies in the range 0.004 to 0.01 eV.

The preceding paragraph represents no more than a plausibility exercise in support of the model implied by (10.3). Quantitative computations of $\sigma_e$, $\sigma_v$, and $l$ must be carried out before (10.3) can be adequately tested. However, we believe the model is the most promising one available currently and urge future work on it.

As is obvious from the preceding paragraphs, a rigorous treatment of the ion pair recombination process is not yet available. Theoretical developments on this problem will probably go hand in hand with those on the problem of describing the energy losses and field dependence observed in electron mobility studies and in spectral absorption studies.

Owing to the lack of time and space, we have not considered the roles of multiple ionizations (as opposed to ionization of isolated ion pairs) and of scavenging reactions on radiative ionization processes. Nor have we described the transient behavior of irradiated fluids. For these topics the interested reader should read the articles and bibliographies published recently by Hummel et al.,[154–155] Schuler et al.,[156–157] and Mozumder.[158]

## References

1. R. L. Williams, *Can. J. Phys.*, **35**, 134 (1957).
2. (a) L. Meyer and F. Reif, *Phys. Rev.*, **110**, 279 (1958); (b) **123**, 727 (1961).
3. L. Meyer, H. T. Davis, S. A. Rice, and R. J. Donnelly, *Phys. Rev.*, **126**, 1927 (1962).
4. P. de Magistris, I. Modena, and F. Scaramuzzi, in J. G. Daunt, D. O. Edwards, F. J. Milford and M. Yaqub, Eds., *Low Temperature Physics LT9, Proceedings of the 9th International Conference on Low Temperature Physics*, Plenum, New York, 1965, pp. 249–252.
5. L. Meyer, in J. G. Daunt, D. O. Edwards, F. J. Milford and M. Yaqub, Eds., *Proceedings of the 9th International Conference on Low Temperature Physics, Columbus, Ohio, Aug. 31–Sept. 4, 1964*, Plenum, New York, 1965, pp. 338–341.
6. R. J. Donnelly, *Phys. Rev. Lett.*, **14**, 39 (1965); P. E. Parks and R. J. Donnelly, *Phys. Rev. Lett.*, **16**, 45 (1966).
7. (a) R. A. Ferrel, *Phys. Rev.*, **108**, 167 (1957); (b) C. G. Kuper, *Phys. Rev.*, **122**, 1007 (1961).
8. W. T. Sommer, *Phys. Rev. Lett.*, **12**, 271 (1964).
9. J. L. Levine and J. M. Sanders, Jr., *Phys. Rev. Lett.*, **8**, 159 (1962).
10. J. L. Levine and T. M. Sanders, Jr., *Phys. Rev.*, **154**, 138 (1967).
11. A. M. Tyndall, *The Mobility of Positive Ions in Gases*, Cambridge University Press, New York, 1938.
12. T. P. Eggarter and M. H. Cohen, *Phys. Rev. Lett.*, **25**, 807 (1970); **27**, 129 (1971).
13. T. P. Eggarter, *Phys. Rev.*, **A5**, 2496 (1972).

14. J. P. Hernandez, *Phys. Rev.*, **A7**, 1755 (1973).
15. H. R. Harrison and B. E. Springett, *Chem. Phys. Lett.*, **10**, 418 (1971).
16. J. A. Jahnke and M. Silver, *Chem. Phys. Lett.*, **19**, 231 (1973).
17. H. R. Harrison and B. E. Springett, *Phys. Lett.*, **35**, 73 (1971).
18. B. Halpern and R. Gomer, *J. Chem. Phys.*, **43**, 1069 (1965).
19. A. Gedanken, B. Raz, and J. Jortner, *Chem. Phys. Lett.*, **14**, 326 (1972).
20. B. Raz and J. Jortner, in J. Jortner and N. R. Kestner, Eds., *Electrons in Fluids—Proceedings of Colloque Weyl III*, Springer-Verlag, New York, 1973, pp. 413–422.
21. L. Bruschi, G. Mazzi, and M. Santini, *Phys. Rev. Lett.*, **28**, 1504 (1972).
22. R. J. Loveland, P. G. Le Comber, and W. E. Spear, *Phys. Lett.*, **39A**, 225 (1972).
23. M. S. Malkin and H. L. Schultz, *Phys. Rev.*, **83**, 1051 (1951).
24. D. W. Swan, *Proc. Phys. Soc.*, **83**, 659 (1964).
25. H. Schnyders, S. A. Rice, and L. Meyer, *Phys. Rev. Lett.*, **15**, 187 (1965).
26. H. Schnyders, S. A. Rice, and L. Meyer, *Phys. Rev.*, **150**, 127 (1966).
27. M. H. Cohen and J. Lekner, *Phys. Rev.*, **158**, 305 (1967).
28. J. Lekner, *Phys. Rev.*, **158**, 130 (1967).
29. J. Bardeen and W. Shockley, *Phys. Rev.*, **80**, 72 (1950).
30. J. A. Jahnke, L. Meyer, and S. A. Rice, *Phys. Rev.*, **A3**, 734 (1971).
31. B. Halpern, J. Lekner, S. A. Rice, and R. Gomer, *Phys. Rev.*, **156**, 351 (1967).
32. L. S. Miller, S. Howe, and W. E. Spear, *Phys. Rev.*, **166**, 871 (1968).
33. J. Lekner, *Phys. Lett.*, **27A**, 341 (1968).
34. J. Lekner, *Phil. Mag.*, **18**, 1281 (1968).
35. B. E. Springett, M. H. Cohen, and J. Jortner, *J. Chem. Phys.*, **48**, 2720 (1968).
36. P. H. Tewari and G. R. Freeman, *J. Chem. Phys.*, **49**, 4394 (1968).
37. R. M. Minday, L. D. Schmidt, and H. T. Davis, *J. Chem. Phys.*, **50**, 1473 (1969).
38. R. M. Minday, L. D. Schmidt, and H. T. Davis, *J. Chem. Phys.*, **54**, 3112 (1971).
39. W. F. Schmidt and A. O. Allen, *J. Chem. Phys.*, **50**, 5037 (1969).
40. W. F. Schmidt and A. O. Allen, *J. Chem. Phys.*, **52**, 4788 (1970).
41. P. G. Fuochi and G. R. Freeman, *J. Chem. Phys.*, **56**, 2333 (1972).
43. J. -P. Dodelet and G. R. Freeman, *Can. J. Chem.*, **50**, 2667 (1972).
44. G. Beck and J. K. Thomas, *Chem. Phys. Lett.*, **13**, 295 (1972); *J. Chem. Phys.*, **57**, 3649 (1972); *J. Chem. Phys.*, **60**, 1705 (1974).
45. G. Bakale and W. F. Schmidt, *Z. Naturforsch.*, **28**, 511 (1973).
46. R. M. Minday, L. D. Schmidt, and H. T. Davis, *J. Phys. Chem.*, **76**, 442 (1972).
47. H. T. Davis, L. D. Schmidt, and R. G. Brown in J. Jortner and N. R. Kestner, Eds., *Electrons in Fluids*, Springer-Verlag, New York, 1973, pp. 393–411.
48. W. F. Schmidt, G. Bakale, and W. Tauchert, *1973 Annual Conference on Electrical Insulation and Dielectric Phenomena, 1973 Annual Report*, National Academy of Science, Washington, D.C. (1974).
49. G. Bakale and W. F. Schmidt, *Chem. Phys. Lett.*, **22**, 164 (1973).
50. R. A. Holroyd and M. Allen, *J. Chem. Phys.*, **54**, 5014 (1971).
51. R. A. Holroyd, B. F. Dietrich, and H. A. Schwartz, *J. Phys. Chem.*, **76**, 3794 (1972).
52. R. A. Holroyd, *J. Chem. Phys.*, **57**, 3007 (1972).
53. R. Schiller, S. Vass, and J. Mandics, *Report* KFKI-72-50, Hungarian Acad. Sci., Budapest.
54. N. R. Kestner and J. Jortner, *J. Chem. Phys.*, **59**, 26 (1973).
55. H. T. Davis, L. D. Schmidt, and R. M. Minday, *Phys. Rev.*, **A3**, 1027 (1971).
56. M. A. Woolf and G. W. Rayfield, *Phys. Rev. Lett.*, **15**, 235 (1965).
57. G. Jaffe, *Ann. Phys.*, **25**, 257 (1908).
58. G. Jaffe, *Le Radium*, **10**, 126 (1913).
59. W. F. Schmidt and A. O. Allen, *J. Phys. Chem.*, **72**, 3730 (1968).
60. A. O. Allen and A. Hummel, *Discuss. Faraday Soc.*, **36**, 95 (1963).

61. A. Hummel and A. O. Allen, *J. Chem. Phys.*, **44**, 3426 (1966).
62. A. Hummel, A. O. Allen, and F. H. Watson, Jr., *J. Chem. Phys.*, **44**, 3431 (1966).
63. G. R. Freeman, *J. Chem. Phys.*, **39**, 988 (1963).
64. G. R. Freeman and J. M. Fayadh, *J. Chem. Phys.*, **43**, 86 (1965).
65. P. Debye, *Trans. Electrochem. Soc.*, **82**, 265 (1942).
66. W. F. Schmidt and A. O. Allen, *J. Chem. Phys.*, **52**, 2345 (1970).
67. P. G. Fuochi and G. R. Freeman, *J. Chem. Phys.*, **56**, 2333 (1972).
68. J. -P. Dodelet and G. R. Freeman, *Can. J. Chem.*, **50**, 2667 (1972).
69. J. -P. Dodelet, K. Shinsaka, U. Kortsch, and G. R. Freeman, *J. Chem. Phys.*, *59*, 2376 (1973).
70. M. G. Robinson, P. G. Fuochi, and G. R. Freeman, *Can. J. Chem.*, **49**, 3657 (1971).
71. G. R. Freeman and E. D. Stover, *Can. J. Chem.*, **46**, 3235 (1968).
72. T. A. Stoneham, D. R. Ethridge, and G. G. Meisels, *J. Chem. Phys.*, **54**, 4054 (1971).
73. A. Mozumder and J. L. Magee, *J. Chem. Phys.*, **47**, 939 (1967).
74. E. E. Conrad and J. Silverman, *J. Chem. Phys.*, **51**, 450 (1969).
75. H. A. Gillis, N. V. Klassen, G. G. Teather, and K. H. Lokan, *Chem. Phys. Lett.*, **10**, 481 (1971).
76. J. T. Richards and J. K. Thomas, *Chem. Phys. Lett.*, **10**, 317 (1971).
77. J. H. Baxendale, C. Bell, and P. Wardman, *Chem. Phys. Lett.*, **12**, 347 (1971).
78. F. D. Rossini, D. D. Wagman, W. H. Evans, S. Levine and I. Jaffe, Eds., *Selected Values of Physical and Thermodynamic Properties of Hydrocarbons and Related Compounds*, Carnegie Press, Pittsburgh, Pa., 1953, p. 287.
79. N. V. Klassen, H. A. Gillis, and G. G. Teather, *J. Phys. Chem.*, **76**, 3847 (1972).
80. J. R. Miller and J. E. Willard, *J. Phys. Chem.*, **76**, 2341 (1972).
81. J. B. Gallivan and W. H. Hamill, *J. Chem. Phys.*, **44**, 1279 (1966).
82. R. S. Knox, *Theory of Excitons*, Academic, New York, 1962.
83. B. Raz and J. Jortner, *Chem. Phys. Lett.*, **9**, 224 (1971).
84. W. B. Fowler, *Phys. Rev.*, **151**, 657 (1966).
85. N. Houser and R. C. Jarnagin, *J. Chem. Phys.*, **52**, 1069 (1970).
86. S. S. Takada, N. E. Houser, and R. C. Jarnagin, *J. Chem. Phys.*, **54**, 3195 (1971).
87. R. C. Jarnagin, *Acc. Chem. Res.*, **4**, 420 (1971).
88. R. A. Holroyd, *J. Chem. Phys.*, **57**, 3007 (1972).
89. J. A. Northby, Ph.D. Thesis, University of Minnesota, Minneapolis, 1966.
90. J. A. Northby and T. M. Sanders, Jr., *Phys. Rev. Lett.*, **18**, 1184 (1967).
91. C. Zipfel and T. M. Sanders, Jr., in J. F. Allen, D. M. Finlayson, and D. M. McCall, Eds., *Proceedings of the 11th International Conference on Low Temperature Physics*, Vol. 1, University of St. Andrews Printing Department, St. Andrews, Scotland, 1969, p. 296.
92. C. Zipfel, Ph.D. Thesis, University of Michigan, Ann Arbor, 1969.
93. T. Miyakawa and D. L. Dexter, *Phys. Rev.*, **A1**, 513 (1970).
94. W. Beall Fowler and D. L. Dexter, *Phys. Rev.*, **176**, 337 (1968).
95. B. E. Springett, *Phys. Rev.*, **155**, 139 (1967).
96. D. Beaglehole, *Phys. Rev. Lett.*, **15**, 551 (1965).
97. S. A. Rice and J. Jortner, *J. Chem. Phys.*, **44**, 4470 (1966).
98. L. Meyer, J. Jortner, S. A. Rice, and E. G. Wilson, *J. Chem. Phys.*, **42**, 4250 (1965).
99. L. Meyer, J. Jortner, S. A. Rice, and E. G. Wilson, *Phys. Rev. Lett.*, **12**, 415 (1964).
100. A. O. Allen and R. A. Holroyd, *J. Phys. Chem.*, **78**, 796 (1974).
101. J. Bardeen, *J. Chem. Phys.*, **6**, 367 (1938).
102. J. M. Ziman, *Electrons and Phonons*, Oxford University Press, London, 1963.
103. E. Wigner and F. Seitz, *Phys. Rev.*, **43**, 804 (1933); **46**, 509 (1934).
104. J. C. Phillips and L. Kleinman, *Phys. Rev.*, **116**, 187 (1959).
105. M. H. Cohen and V. Heine, *Phys. Rev.*, **122**, 1821 (1961).

106. B. J. Austin, V. Heine, and L. J. Shaw, *Phys. Rev.*, **127**, 276 (1962).
107. M. H. Cohen, in J. Friedel and A. Ljuinier, Eds., *Metallic Solid Solutions*, Benjamin, New York, 1963, p. XI. 1.
108. I. V. Aberenkov and V. Heine, *Phil. Mag.*, **12**, 529 (1965).
109. T. F. O'Malley, *Phys. Rev.*, **130**, 1020 (1963).
110. J. Lekner, *Phys. Rev.*, **158**, 130 (1967).
111. N. F. Mott and H. S. W. Massey, *Theory of Atomic Collisions*, 3rd ed., Oxford University Press, New York, 1965.
112. B. Kivel, *Phys. Rev.*, **116**, 926 (1959).
113. M. H. Mittleman and K. M. Watson, *Ann. Phys.* (N.Y.), **10**, 268 (1960).
114. K. D. Luks and J. Kohn, private communication.
115. R. G. Brown and H. T. Davis, *Chem. Phys. Lett.*, **27**, 78 (1974).
116. J. M. Smith and H. C. van Ness, *Introduction to Chemical Engineering Thermodynamics*, McGraw Hill, New York, 1959, pp. 107–112.
117. G. Baldini, *Phys. Rev.*, **128**, 1562 (1962).
118. J. Holtzmark, *Z. Phys.*, **55**, 437 (1929).
119. E. Thiele, *J. Chem. Phys.*, **39**, 474 (1963).
120. J. D. Weeks, D. Chandler, and H. C. Anderson, *J. Chem. Phys.*, **54**, 5237; **55**, 5422 (1971).
121. L. Verlet and J. J. Weis, *Phys. Rev.*, **A5**, 939 (1972).
122. J. Holtzmark, *Z. Phys.*, **55**, 437 (1929).
123. J. Callaway, *Energy Band Theory*, Academic, New York, 1964; P. W. Anderson, *Concepts in Solids*, Benjamin, New York, 1963; W. A. Harrison, *Pseudopotentials in the Theory of Metals*, Benjamin, Reading, Mass., 1966.
124. N. R. Kestner, J. Jortner, M. H. Cohen, and S. A. Rice, *Phys. Rev.*, **A1**, 56 (1965).
125. M. H. Cohen and V. Heine, *Phys. Rev.*, **122**, 1821 (1961).
126. F. Bassani and V. Celli, *J. Phys. Chem. Solids*, **20**, 64 (1961).
127. B. J. Austin, V. Heine and L. J. Sham, *Phys. Rev.*, **127**, 276 (1962).
128. P. Bagus, T. L. Gilbert and C. C. J. Roothaan, *J. Chem. Phys.*, **56**, 5195 (1972).
129. H. Reeh, *Z. Naturforsch.*, **15a**, 377 (1960).
130. H. A. Bethe, *Handbuch der Physik*, Vol. 29, Part 1, Edwards Brothers, Ann Arbor, Mich., 1943, pp. 339 ff.
131. R. W. LaBahn and J. Callaway, *Phys. Rev.*, **135**, A1539 (1964).
132. D. E. Golden and H. W. Bandel, *Phys. Rev.*, **138**, A14 (1965).
133. W. Shockley, *Bell Syst. Tech. J.*, **30**, 990 (1951).
134. I. I. Gurevich and L. V. Tarasox, *Low-Energy Neutron Physics*, North-Holland, Amsterdam, 1968, p. 477.
135. E. Gerjuoy and S. Stein, *Phys. Rev.*, **97**, 1671 (1955).
136. D. E. Golden and H. W. Bandel, *Phys. Rev.*, **149**, 58 (1965).
137. J. A. Jahnke, N. A. W. Holzwarth, and S. A. Rice, *Phys. Rev.*, **A5**, 463 (1971).
138. J. Jortner, N. R. Kestner, S. A. Rice, and M. H. Cohen, *J. Chem. Phys.*, **43**, 2614 (1965).
139. H. Reiss, H. L. Frisch, E. Helfand, and J. L. Lebowitz, *J. Chem. Phys.*, **32**, 119 (1960).
140. T. Miyakawa and D. L. Dexter, *Phys. Rev.*, **184**, 166 (1969).
141. K. Hiroike, N. R. Kestner, S. A. Rice, and J. Jortner, *J. Chem. Phys.*, **43**, 2625 (1965).
142. S. A. Rice, 1, *Accts. Chem. Res.* **1**, 81 (1968).
143. For summaries of the present understanding of the subject, see (a) N. F. Mott and E. A. Davis, *Electronic Processes in Non-Crystalline Materials*, Clarendon, Oxford, 1971; (b) M. H. Cohen, in J. Jortner and N. R. Kestner, Eds., *Electrons in Fluids*, Springer-Verlag, New York, 1973, pp. 257–285.
144. M. H. Cohen, *J. Non-Cryst. Solids*, **4**, 391 (1970).
145. J. M. Ziman, *J. Phys.*, **C1**, 1532 (1968).
146. S. Kirkpatrick, *Rev. Mod. Phys.*, **45**, 574 (1973).

147. H. L. Frisch, J. M. Hammersley, and D. J. A. Welsh, *Phys. Rev.*, **126**, 949 (1962).
148. D. A. G. Bruggeman, *Ann. Phys.* (Leipzig), **24**, 636 (1935).
149. R. Landauer, *J. Appl. Phys.*, **23**, 779 (1952).
150. A. Mozumder and J. L. Magee, *J. Chem. Phys.*, **45**, 3332 (1966).
151. A. Mozumder, A. Chatterjee, and J. L. Magee, unpublished results.
152. A. Mozumder, *J. Chem. Phys.*, **55**, 3020 (1971).
153. *Selected Raman Spectral Data, Ser. No. 255, American Petroleum Institute Research Project 44*, Chemical Thermodynamics Research Center, Texas A & M University, College Station, 1964, (unpublished).
154. A. Hummel, *J. Chem. Phys.*, **49**, 4840 (1968); *J. Chem. Phys.*, **48**, 3268 (1968).
155. S. J. Rzad, R. H. Schuler, and A. Hummel, *J. Chem. Phys.*, **51**, 1369 (1969).
156. S. J. Rzad, P. P. Infelta, J. M. Warman, and R. H. Schuler, *J. Chem. Phys.*, **52**, 3971 (1970).
157. P. P. Infelta and R. H. Schuler, *J. Phys. Chem.*, **76**, 987 (1972).
158. A. Mozumder, *J. Chem. Phys.*, **48**, 1659 (1968); **55**, 3026 (1971).

# SOLUTIONS OF METALS IN MOLTEN SALTS

NORMAN H. NACHTRIEB

*Department of Chemistry, The University of Chicago, Chicago, Illinois*

## TABLE OF CONTENTS

## I. INTRODUCTION

The past 2 decades have witnessed a considerable advance in our understanding of the solutions certain metals form with their corresponding molten salts, particularly halides. There had been occasional interest in such solutions since 1909, when Aten[1] measured the electrical conductance in molten Bi–BiCl$_3$ and Cd–CdCl$_2$ systems. But the modern era began in the late 1950s with the investigations of Bredig[2] and his collaborators on the phase diagrams of alkali metal–alkali halide binaries, and their subsequent measurement of the electrical conductance of several systems as a function of temperature and composition. Their studies established unequivocally that true solutions are formed, whose thermodynamic properties indicate their monodisperse nature, rather than colloidal dispersions as had earlier been conjectured. Moreover, the magnitude of the specific conductance so far exceeds that of the pure molten salts as to make it obvious that charge transport takes place by the superposition on the ionic mechanism of some kind of electronic motion. Excellent reviews by Bredig[3] and Corbett[4] summarize the status of the metal–molten salt field up to about 1964, and the purpose of this article is to attempt a synthesis of that information with newer and older experimental data and concepts.

It has been customary to classify binary metal–molten salts into one or the other of two categories:

1. Those, like $Bi-BiCl_3$ and $Cd-CdCl_2$, whose electrical conductance and its temperature coefficient are typical of ionic solutions.

2. Those, like alkali metal–alkali halide melts, that exhibit mixed ionic and electronic conductance.

In the first class, chemical reactions occur involving electron transfer from the neutral metal to the cation of the salt, leading to cationic species of intermediate valence. Sometimes, as in the $Bi-BiCl_3$ system, compounds of a lower valence state have been isolated and characterized as solids.[5,6] This is not invariably the case, however; evidence points to the existence of the $Cd_2^{2+}$ ion in solutions of cadmium in molten cadmium chloride, but on solidifying the melt exsolves metallic cadmium in a colloidally dispersed form. By contrast, solutions of the second class invariably regenerate the metal and essentially pure salt on freezing.

It appears that the distinction between the two classes of solution lies in the relative depth of the trap for the valence electron of the metallic component. It is evidently deeper for solutions that exhibit no electronic component in their conductivity, and in which charge transfer is virtually complete. The nature of the electron trap has been the subject of much speculation, and a definitive theory accounting for all the known facts (thermodynamic, spectroscopic, magnetic, and transport) is still lacking.

The state of the valence electron of the metallic component is a matter of primary interest. We limit consideration to the second class of systems, in which there is a significant electronic contribution to the conductivity even at low dissolved metal concentration, and in which the electron trap is therefore relatively shallow. Initially, we further confine attention to those alkali metal–alkali halide systems in which miscibility is complete (e.g., cesium in its halides), or to temperatures above the consolute temperature for systems (e.g., sodium or potassium in their halides) for which there is an immiscibility gap in the phase diagram. Such systems exhibit continuous variation in their transport properties, ranging from those characteristic of the pure salt to those of the pure liquid metal. The nonmetal → metal transition occurs somewhere in this interval, and an important question is whether the valence electron is better described in terms of a localized state or by a plane wave.

The electrical conductance provides the first and probably most important clue. Pure liquid metallic potassium has an equivalent conductance[7] of $9.9 \times 10^5$ ohm$^{-1}$ cm$^2$ eq$^{-1}$ at 740°C (i.e., above the consolute temperature, 728°C, of the K–KBr system), while the limiting equivalent conductance*

---

* The ionic contribution to the conductance has been removed by assuming additivity of the electronic and ionic contributions and defining $\Lambda_e = (\Lambda_t - X_s\Lambda_s)/X_m$, where $X_s$ and $X_m$ denote the mole fractions of salt and metal, $\Lambda_t$ is the measured equivalent conductance of the solution, and $\Lambda_s$ is the equivalent conductance of the pure salt.

conductivity changes from positive to negative and the mobility varies inversely as the iodide ion concentration, in accordance with the predicted scattering by impurities of conduction electrons at the top of a degenerate band. In this region the melt is unquestionably metallic. Between 50 and 80% bismuth a transition occurs between thermally activated electron hopping and impurity scattering of free electrons. Magnetic susceptibility studies[14] in this concentration range indicate the presence of neutral bismuth, and it may be that the hopping mechanism involves one-electron transfer from $Bi^0$ to $Bi^+$.

A further extension of the hopping model was made by Emi and Bockris[15] in a calculation of the electron mobility in dilute solutions of sodium in NaCl and of potassium in KCl. They express the mobility by

$$\mu_e = \mu_0 p(U, S) p(W) \tag{2.11}$$

where $\mu_0$ is the usual Einstein relation [see (2.3)], $p(U, S)$ is the probability of a fluctuation in the metal donor–cation acceptor distance and the configuration of their surroundings required to equalize the energy of an electron on the two sites, and $p(W)$ is the electron tunneling probability. For the electron exchange frequency in $\mu_0$, they calculated $v = 2.03 \times 10^{12}/\text{sec}$ for the neutral atom–cation pair $(Na^0-Na^+)$ in Na–NaCl and $v = 1.83 \times 10^{12}/\text{sec}$ for $(K^0-K^+)$ in K–KCl using Mie's approximation.

To satisfy the condition for electron tunneling, the energy of an electron must be the same on the atom and cation sites:

$$E_{\text{atom}} = E_{\text{cation}} \tag{2.12}$$

At equilibrium these energies are not equal, but are determined by the electrostatic potentials (and these in turn are governed by the radial distribution functions). In the absence of such knowledge, Emi and Bockris resort to a Born–Haber cycle to approximate the equilibrium potential of an electron at an atom or cation site:

$$E_{\text{atom}} = E_v - I + P_{\text{atom}} \tag{2.13}$$

and

$$E_{\text{cation}} = E_L - I - P_{\text{cation}} + E'_v \tag{2.14}$$

where $E_v$ is the energy required to transfer an atom from the melt into a vacuum, $I$ is the ionization potential of the metal atom, and $P_{\text{atom}}$ (or $P_{\text{cation}}$) is the work required to return the cation from the vacuum to the metal atom (or cation) site in the melt without a change in the positions of its neighbors. $E_L$ is the Madelung energy for the pure molten salt, and $E'_v$ ($\approx E_v$) is the energy required to transfer a metal atom from a nonequilibrium site [determined by (2.12)] to the vacuum. The difference between the energy of an electron at an equilibrium cation site and its energy at an equilibrium atom site is therefore

$$E_{cation} - E_{atom} = E_L - (P_{atom} + P_{cation}) \qquad (2.15)$$

If $U_0$ is the energy of the $(M^0-M^+)$ pair in their equilibrium state, and $U$ is their energy when electron transfer occurs, then

$$U - U_0 = 0.5(E_{cation} - E_{atom}) \qquad (2.16)$$

where the factor 0.5 arises from the symmetry of the electron transfer barrier. The probability $p(U, S)$ of a configuration fluctuation that satisfies (2.12) is therefore given by

$$p(U, S) = \exp\left(\frac{S - S_0}{k}\right) \exp\left(-\frac{U - U_0}{kT}\right) \qquad (2.17)$$

for which Ichikawa and Shimojii's method[10] for calculating the entropy change for the fluctuation is employed.

To calculate $p(W)$, the electron tunneling probability, Emi and Bockris choose an Eckart potential evaluated for a one-dimensional Coulomb barrier. For the electron mobility in Na–NaCl, they obtain $\mu_e = 0.045$ cm$^2$/(V)(sec) at 1121 K, as compared with 0.034 cm$^2$/(V)(sec) from the experimental studies of Bronstein and Bredig.[16] For the K–KCl system at 1091°K, the theoretical value is 0.032 cm$^2$/(V)(sec), as compared with 0.031 cm$^2$/(V)(sec) from experiment. A small positive temperature dependence is predicted for both systems ($\approx 0.3\%$ per degree), which lies within experimental uncertainty. Emi and Bockris' calculations, like Rice's estimate, refer to infinite dilution of metal in salt, and the agreement with experiment is remarkably good.

## C. F-Center Model

Pitzer[17] proposed a quite different model for the valence electron in alkali metal–alkali halide solutions, based on regular solution theory and the phase diagrams determined by Bredig et. al. He suggested that it may closely resemble an F center, the well-known solid-state defect in which an electron occupies the site of an anion vacancy. Pitzer considered a two-stage process, the first step of which is the conversion of a solid alkali metal into a hypothetical salt having the rock-salt structure, with cations occupying their normal sites and electrons the normal anion sites. The second step is the random mixing of this hypothetical salt with a real alkali halide having the same cation. Kleppa's calorimetric studies[18] showed that the excess enthalpies and excess entropies of mixing of molten alkali halides having a common cation and different anions are small, and one would expect correspondingly small enthalpies and excess entropies of mixing of F centers with anions.

Choosing reasonable values for a spherical F-center cavity radius, Pitzer calculated the excess lattice energies of the hypothetical "electron salt" using the Madelung equation for the electrostatic potential energy and the ground state of a particle in a spherical box for the electron kinetic energy. These calculated lattice energies lie well above the lattice energies of the real alkali metals, the excess enthalpies amounting to 19, 18, 10, 9, and 9 kcal/mole, respectively, for lithium, sodium, potassium, rubidium, and cesium "electronides." He concluded that one should expect quite large positive heats of solution for alkali metals in their corresponding halides. Qualitative considerations of the excess entropies of random mixing of F centers with anions led him to predict that these should also be positive and appreciable in magnitude.

From such considerations, Pitzer undertook an analysis of the phase diagrams of alkali metal–alkali halide binaries, calculating the excess partial molar free energy of each component from its solubility in the other:

$$\Delta \bar{G}_i^E = \Delta \bar{G}_i - RT \ln x_i = RT \ln \gamma_i \qquad (2.18)$$

where the reference state is the pure liquid component. On the salt-rich side of the phase diagram, where the essentially pure salt is in equilibrium with the liquid, the usual freezing-point-lowering equations apply and lead to the activity coefficient and excess partial molar free energy of the dissolved metal. Along the two-liquid-phase equilibrium line where mutual solubilities are low, the activity coefficient of the major component (solvent) may be estimated quite accurately and then calculated for the same component when it is the solute (minor component) in the other phase at the same temperature. As the temperature and the mutual solubilities increase along the two-phase-liquid equilibrium line, this is no longer valid. But if regular solution theory is assumed to apply, it is possible to obtain the solvent activity coefficient by successive approximations using

$$RT \ln \gamma_i = \Delta \bar{G}_i^{\circ}(1 - x_i)^2 \qquad (2.19)$$

where $\Delta \bar{G}_i^{\circ}$ is the excess partial molar free energy of component $i$ at zero concentration or, alternatively, the difference between the chemical potential of the solute and the solvent standard states. Regular solution theory provides another relation for the excess partial molar free energy in terms of the volume fraction, and an alternative way of calculating $\Delta G^E$:

$$RT \ln \gamma_i = \Delta \bar{G}_i^{\circ}(1 - \phi_i)^2 \qquad (2.20)$$

where $\phi_1$ is the volume fraction of component $i$.

By either procedure the excess partial molar free energy of each component may be calculated as a function of the temperature, and of course the negative temperature dependence of this quantity is equal to the excess entropy of solution. For sodium–sodium halides Pitzer found expressions of the form

$$\Delta \bar{G}^{\circ}_{\text{metal}} = \Delta \bar{H}^{\circ}_{\text{metal}} - T \Delta \bar{S}^{\circ}_{\text{metal}} \qquad (2.21)$$

in which $\Delta \bar{H}^{\circ}_{\text{metal}}$ is 18.1, 18.1, and 16.3 kcal/mole for the Na–NaCl, Na–NBr, and Na–NaI systems, respectively, and the excess entropy of solution is about 9 cal/(mole) (deg). These values are in good agreement with predictions based on the formation of the hypothetical sodium electronide "salt" and its random mixing with a sodium halide. Large positive excess free energy, enthalpy, and entropy values are also found for the solution of the salt in metal, and there is approximate agreement with the prediction of regular solution theory:

$$\frac{\Delta \bar{G}_1^{\circ}}{\Delta \bar{G}_2^{\circ}} = \frac{V_1}{V_2} \qquad (2.22)$$

where $V_1$ and $V_2$ are the molar values of the salt and the metal, respectively.

There is no obvious reason why these excess thermodynamic functions should be large and positive when the metal is present in the form of neutral atoms in the molten salt, and the excellent agreement between Pitzer's predictions and the deductions drawn from experimental solubilities constitutes strong support for an F-center-like state for the valence electron.

Further convincing evidence for the F-center model has come from the spectroscopic studies of Gruen, Krumpelt, and Johnson,[19] and from their reinterpretation of the earlier work by Mollwo. Following the pioneering work by Hilsch and Pohl at Göttingen on the absorption spectra and photo-conductivity of color centers in crystalline alkali halides, Mollwo[20] investigated the temperature dependence of the band maxima and their full band-width at half-maximum absorption (FWHM) in nine alkali halides. He made three important observations:

1. $\nu_{\text{max}}$ shifts to lower frequencies with increasing temperature, varying inversely with the square of the cation–anion internuclear distance ($d$ in centimeters)

$$\nu_{\text{max}} = 5.02 \times 10^{-1} \, d^{-2} \, \text{sec}^{-1} \qquad (2.23)$$

for all alkali halides, with only a small trend from fluorides to iodides.

2. At any given temperature from 20 to 600°C, all alkali halides have the same FWHM when the widths are normalized to the frequency of their maximum absorption.

3. The FWHM of the color center in crystalline NaCl and KCl increases with temperature, and is a linear function of the heat content, $\int_0^T Cp \, dT$. The residual bandwidth extrapolated to 0K, appears to depend on crystal perfection.

Subsequently, Mollwo[21] extended his measurements on additively colored NaX and KX (X = Cl, Br, I) just above their melting points, and made three further significant observations:

4. $v_{max}$ is abruptly shifted to lower frequency at the melting point, and the FWHM is greatly broadened.

5. $\tilde{v}_{max}$ in the melt is independent of the anion, having the value 12,600 cm$^{-1}$ (1.57 eV) for sodium salts, and 10,200 cm$^{-1}$ (1.27 eV) for potassium salts at their melting points. (In additively colored crystals $\tilde{v}_{max}$ depends strongly on the anion, varying as I < Cl < Br < F for a given cation).

6. $\tilde{v}_{max}$ in the melts lies lower than the resonance line of the atom by 4400 cm$^{-1}$ for Na–NaX melts, and by 2800 cm$^{-1}$ for K–KX melts. Mollwo nevertheless attributed the absorption in the melts to atoms in the ground state assuming their energy levels to be broadened and shifted by local electric fields (the Stark effect).

Gruen and his co-workers repeated Mollwo's experiment on additively colored NaCl in the crystal and melt, using sapphire cells to avoid the problems Mollwo encountered due to discoloration of glass absorption cells. Their data agreed closely with Mollwo's, nevertheless, and they added another important observation to his:

7. Noting that $d^{-2}$ in Mollwo's relation [equation (2.23)] is proportional to $V_m^{-2/3}$, they reanalyzed his data using accurately measured densities for crystalline and molten alkali halides to calculate their molar volumes as a function of temperature. Such a calculation corrects for the large volume expansion on fusion of the salts, and $v_{max}$ for the melts was found fit the same, essentially linear, relation that applies to crystalline F centers:

$$v_{max} = AV^{-2/3} \qquad (2.24)$$

In effect, correction for the increase in the mean cation–anion separation resulting from thermal expansion of the crystal and its fusion, indicates a common environment of the valence electron in both the crystal and the melt.

This extraordinary observation, coupled with Pitzer's thermodynamic evidence, provides strong support for a model in which the electron is trapped in a cavity defined by a shell of nearest-neighbor alkali metal cations, and is consistent with all of Mollwo's observations. The dependence of $v_{max}$ on $d^{-2}$ or $V_m^{-2/3}$ has the correct form for a particle moving in the central field of a spherical cavity, but this is an obvious oversimplification. It appears to be more reasonable to describe the electron by a molecular orbital, following Kahn and Kittel:[22]

$$\psi = (\tfrac{1}{6})^{1/2} \sum_{i=1}^{6} \psi_i \qquad (2.25)$$

where $\psi_i$ is the atomic orbital for the electron on the $i$th of the nearby alkali cations. Presumably, as in the crystal, there are about six cations with an average octahedral coordination around the electron. These cations (erstwhile atoms) should have a polarized ground state because of the asymmetric field of the vacancy, and it is reasonable to suppose that the ground state wave function $\psi_i$ is an admixture of $s$ and $p$ functions:

$$\psi_i = \psi_s - \frac{\varepsilon}{\sqrt{2}}(\psi_p^1 - \psi_p^{-1}) \tag{2.26}$$

If the electron were localized on a single cation, however, its wave function would be a pure $s$ function, and its $g$ value would be 2.0023. Experimental esr measurements by Sosis[23] in this laboratory lead to $g = 1.9980$ in molten NaCl containing 0.51 mole % sodium at 835°C, and $\Delta g$ is therefore $-0.0043$. For a ground-state wave function having the form of (2.26) and with the spin-orbit interaction given by $\lambda L \cdot S$, Kahn and Kittel calculate the change in the $g$ factor of an F center from the free electron value to be

$$\Delta g = -\frac{4}{3}\frac{\lambda}{\Delta}\left(\frac{\varepsilon^2}{1+\varepsilon^2}\right) \tag{2.27}$$

where $\lambda$ is the spin-orbit splitting (18 cm$^{-1}$ for a gaseous sodium atom), and $\Delta$ is the $3s - 3p$ separation (taken to be the observed $\tilde{v}_{max} = 12,600$ cm$^{-1}$). The value of $\Delta g$ calculated with $\varepsilon = 0.9$ is $\Delta g = -1.0 \times 10^{-3}$. The value calculated by Kahn and Kittel for the F center in solid KCl is lower than Hutchison and Noble's[24] measured value by the same factor of 4. The direction of the shift in the Landé $g$ factor is correct and, within the limits of the rough MO approximation, the agreement is not unreasonable. The choice of of the free-atom value for the spin-orbit splitting is a lower limit, and its correct value is undoubtedly greater.

Further strong support for the F-center model of the electron in melts comes from the broadening of the optical absorption band. As noted by Mollwo (observation 3), the bandwidth at half-maximum is a linear function of the heat content of the F center in the solid. We have observed that addition of the latent heat of fusion to the integrated heat capacity extends this relationship to Na–NaCl, K–KCl, and K–KBr melts, as shown in Fig. 1. This has interesting implications for the vibrational motion of the F center and its molten salt counterpart.

We denote the overlap of the $s$-wave function of the electron in the anion vacancy with $s$, $p$, and other orbitals on a neighboring cation by

$$S_i = \langle \phi_s | \phi_i \rangle \tag{2.28}$$

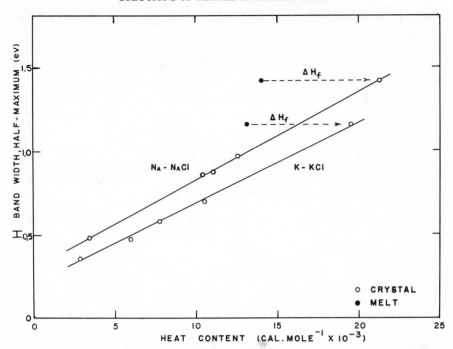

**Fig. 1**  Variation of absorption band width with total heat content for crystalline alkali halide F-centers and for alkali metal–alkali halide melts.

and the sum of the squares of these overlap integrals by

$$\Lambda = \sum_i S_i^2 \tag{2.29}$$

If we postulate that the energies of the ground and lowest exited electronic states of the F center are modulated by variations in $\Lambda$, we anticipate that its logarithmic temperature dependence will be the same as the logarithmic temperature dependence of the bandwidth at half-maximum:

$$\frac{1}{H}\frac{dH}{dT} = \frac{1}{\Lambda}\frac{d\Lambda}{dT} \tag{2.30}$$

In particular, vibrational modes that remove the $O_h$ point group symmetry of the F center and those symmetric modes that change the effective radius of the F-center cavity will modulate both $\Lambda$ and $H$. If $\Lambda_0$ represents the sum over the overlap integrals at some arbitrary reference temperature, we may expand $\Lambda$ in a power series at that temperature, and to first order:

$$\Lambda = \Lambda_0 + \frac{\partial^2 \Lambda}{\partial x^2}\langle u^2 \rangle_{av} \tag{2.31}$$

where $u$ is the amplitude of vibration, and the odd powers of $u$ average individually to zero. Then

$$\frac{1}{H}\frac{\partial H}{\partial T} = \frac{1}{\Lambda}\frac{\partial \Lambda}{\partial T} = \frac{1}{\Lambda}\frac{\partial^2 \Lambda}{\partial x^2}\frac{\partial \langle u^2\rangle_{\text{av}}}{\partial T} \tag{2.32}$$

and it should be possible to calculate the fractional change in the absorption bandwidth with temperature from the product of $(1/\Lambda)(\partial^2 \Lambda/\partial x^2)$ and $\partial \langle u^2\rangle_{\text{av}}/\partial T$. The second factor enters into the Debye–Waller theory for the temperature-broadening of x-ray diffraction lines, and has been used to account for the downfield chemical shift with temperature of nmr in crystalline and molten salts.[25] It may be obtained from Leibfried's approximation[26] for temperatures above the Debye temperature:

$$\langle u^2\rangle_{\text{av}} = \frac{18kT}{M\omega_D{}^2} \tag{2.33}$$

Thus

$$\frac{\partial \langle u^2\rangle_{\text{av}}}{\partial T} = \frac{18k}{M\omega_D{}^2} = \frac{18\hbar^2}{Mk\theta_D{}^2} \tag{2.34}$$

The first factor on the rhs of (2.32) may be obtained from Löwdin's[27] calculation of the overlap integrals as a function of interionic separation for a variety of $s$ and $p$ wave functions for anions and cations in alkali halides. The factor $(1/\Lambda)(\partial^2 \Lambda/\partial x^2)$ has the essentially constant value $12.9 \times 10^{16}/\text{cm}^2$ for alkali halides having the rock salt structure; it is independent of the particular anion and cation orbitals, because the overlap involves only the exponential tails of the radial parts of their wave functions. Table I shows a comparison of the fractional change in bandwidth with temperature calculated from (2.32) and the values experimentally determined by Mollwo[20] and by Markham and Konitzer.[28] The agreement is striking and, in view of

TABLE I

Fractional Increase in Bandwidth with Temperature

| | $\theta_D(°K)$ | $\dfrac{\partial \langle \mu^2\rangle_{\text{av}}}{\partial T}$ $(\text{cm}^2\text{deg}^{-1})$ | $\dfrac{1}{H}\left(\dfrac{\partial H}{\partial T}\right)_{\text{calc}}$ $(\text{deg}^{-1})$ | $\dfrac{1}{H}\left(\dfrac{\partial H}{\partial T}\right)_{\text{exp}}$ $(\text{deg}^{-1})$ | Reference |
|------|------|------|------|------|------|
| NaCl | 281 | $1.89 \times 10^{-20}$ | $2.44 \times 10^{-3}$ | $1.91 \times 10^{-3}$ | 20 |
| | | | | $2.80 \times 10^{-3}$ | 28 |
| KCl | 230 | $2.21 \times 10^{-20}$ | $2.86 \times 10^{-3}$ | $3.18 \times 10^{-3}$ | 20 |
| | | | | $3.94 \times 10^{-3}$ | 28 |
| KBr | 177 | $2.34 \times 10^{-20}$ | $3.02 \times 10^{-3}$ | $3.00 \times 10^{-3}$ | 20 |
| KI | 140 | $2.68 \times 10^{-20}$ | $3.46 \times 10^{-3}$ | $2.56 \times 10^{-3}$ | 20 |
| RbCl | 157 | $2.93 \times 10^{-20}$ | $3.78 \times 10^{-3}$ | $2.51 \times 10^{-3}$ | 20 |

Fig. 1, it is evident that it applies to both the crystalline F center and to the electron in the corresponding melt.

It thus appears to us that the F center is the currently most promising model for the electron at low concentration in alkali halide melts. As in the solid, it may be regarded as a localized state in which the electron is confined to a cavity defined by a nearest-neighbor shell of cations in approximately octahedral coordination. It is probably best considered as a multicenter state, which may be described to first order by an LCAO-MO in which $s$, $p$, and possibly higher orbital states of the cation are combined with the $s$ state of the electron. Such a center, which may be considered an $M_6^{5+}$ species on the average, has a ground state that lies lower than that of the neutral atom in a dielectric medium, and it possesses a bound excited state whose energy lies 1.57 and 1.27 eV above the ground state, respectively, in sodium and potassium halide melts at their melting points. The electrical conductance appears to involve a tunneling mechanism in which, as Rice has proposed,[11] thermal fluctuation prepares a configuration in which an acceptor site has the same potential as the F center.

Promising though the F-center model is, there remain numerous problems to explore. It is remarkable that the energy of the band maximum in the melt is almost completely independent of the anion, suggesting that only nearest-neighbor interactions are important; by contrast, in the solid anions are strongly coupled in the electronic transition. Direct calorimetric determination of the enthalpy of solution of the alkali metals in their molten halides would be valuable as a test of Pitzer's calculations. One would expect a positive partial molar volume of solution for the alkali metal in its halides, and a direct measurement of this quantity would be useful for estimating the volume of the supposed cavity and the extent of relaxation of cations about the electron.

Further theoretical and experimental effort needs to be directed toward an understanding of the electron concentration dependence of both equilibrium and transport properties. For example, the magnetic susceptibility measurements on K–KCl by Bettman,[29] on Na–NaCl by Arendt and Nachtrieb,[30] and on Cs–CsCl by Nachtrieb[31] all show the expected trend toward lower values with increasing metal concentration attributable to electron spin pairing. But more accurate measurements and a more complete theory of the diamagnetic and paramagnetic susceptibility of an electron in a localized multicenter state are needed.

### Acknowledgments

The writing of this article and some of the research described therein were supported by Grant No. AF-AFOSR-71-1962D from the Air Force Office of Scientific Research and by the National Science Foundation through its Contract No. NSF-GH-33636-A1 with The University of Chicago for materials research.

# References

1. A. H. W. Aten, *Z. Phys. Chem.*, **66**, 641 (1909); **73**, 579 (1910).
2. M. A. Bredig, J. W. Johnson, and W. T. Smith, Jr., *J. Am. Chem. Soc.*, **77**, 307 (1955); M. A. Bredig and H. R. Bronstein, *J. Phys. Chem.*, **64**, 64 (1960); J. W. Johnson and M. A. Bredig, *J. Phys. Chem.*, **62**, 604 (1958); M. A. Bredig and J. W. Johnson, *J. Phys. Chem.*, **64**, 1899 (1960); M. A. Bredig, H. R. Bronstein, and W. T. Smith, Jr., *J. Am. Chem. Soc.*, **77**, 1454 (1955); H. R. Bronstein and M. A. Bredig, *J. Am. Chem. Soc.*, **80**, 2077 (1958); H. R. Bronstein and M. A. Bredig, *J. Phys. Chem.*, **65**, 1220 (1961).
3. M. A. Bredig, in M. Blander, Ed., *Molten Salt Chemistry*, Interscience, New York, 1964, p. 367.
4. J. D. Corbett, in B. Sundheim, Ed., *Fused Salts*, McGraw-Hill, New York, 1964, p. 341.
5. J. D. Corbett, *J. Am. Chem. Soc.*, **80**, 4757 (1958).
6. A. Hershaft and J. D. Corbett, *J. Chem. Phys.*, **36**, 551 (1962); A. Hershaft and J. D. Corbett, *Inorg. Chem.*, **2**, 979 (1963).
7. Ref. 3, p. 391.
8. E. G. Wilson, *Phys. Rev. Lett.*, **10**, 432 (1963).
9. M. Shimoji and K. Ichikawa, *Phys. Lett.*, **20**, 480 (1966).
10. K. Ichikawa and M. Shimoji, *Trans. Faraday Soc.*, **62**, 3543 (1966).
11. S. A. Rice, *Discuss. Faraday Soc.*, **32**, 181 (1961).
12. D. O. Raleigh, *J. Chem. Phys.*, **38**, 1677 (1963).
13. L. F. Grantham and S. J. Yosim, *J. Chem. Phys.*, **38**, 1671 (1963).
14. L. E. Topol and L. D. Ransom, *J. Chem. Phys.*, **38**, 1663 (1963).
15. T. Emi and J. O'M. Bockris, *Electrochim. Acta*, **16**, 2081 (1971).
16. H. R. Bronstein and M. A. Bredig, *J. Am. Chem. Soc.*, **80**, 2077 (1958).
17. K. S. Pitzer, *J. Am. Chem. Soc.*, **84**, 2025 (1962).
18. M. E. Melnichak and O. J. Kleppa, *J. Chem. Phys.*, **57**, 5231 (1972).
19. D. M. Gruen, M. Krumpelt, and I. Johnson, in G. Mamontov, Ed., *Molten Salts: Characterization and Analysis*, Dekker, New York, 1969, p. 169.
20. E. Mollwo, *Z. Phys.*, **85**, 56 (1933).
21. E. Mollwo, *Nachr. Ges. Wiss. Göttingen, Math.-Phys. Kl.*, *Fachgruppe II*, **1**, 203 (1935).
22. A. H. Kahn and C. Kittel, *Phys. Rev.*, **89**, 315 (1953).
23. M. Sosis, Ph.D. Dissertation, University of Chicago, Chicago, Illinois, 1974.
24. C. A. Hutchison, Jr., and G. A. Noble, *Phys. Rev.*, **87**, 1125 (1952).
25. S. Hafner and N. H. Nachtrieb, *J. Chem. Phys.*, **40**, 2891 (1964).
26. G. Leibfried, in S. Flügge, Ed., *Handbuch der Physik*, Vol. 7, Part 1, Springer-Verlag, Berlin, 1955, p. 104.
27. P. K. Löwdin, *Advan. Phys.*, **5**, 1 (1956).
28. Markham and Konitzer, *J. Chem. Phys.*, **34**, 1936 (1961).
29. M. Bettman, *J. Chem. Phys.*, **44**, 3254 (1966).
30. R. H. Arendt and N. H. Nachtrieb, *J. Chem. Phys.*, **53**, 3085 (1970).
31. N. H. Nachtrieb, unpublished results.

# AUTHOR INDEX

Numbers in parenthesis are reference numbers and show that an author's work is referred to although his name is not mentioned in the text. Numbers in *italics* indicate the pages on which the full references appear.

# SUBJECT INDEX